P9-DMJ-880

Study Guide
to accompany

Life The Science of Biology
SEVENTH EDITION

Study Guide

to accompany

Life The Science of Biology
SEVENTH EDITION
PURVES • SADAVA • ORIANS • HELLER

Edward M. Dzialowski *University of North Texas*

Betty McGuire *Smith College*

Lindsay Goodloe *Cornell University*

Nancy Guild *University of Colorado at Boulder*

Paula Mabee *University of South Dakota*

 Sinauer Associates, Inc.

 W. H. Freeman and Company

Cover photo © Steve Bloom/stevebloom.com.

Study Guide to accompany *Life: The Science of Biology,* **Seventh Edition**

Copyright © 2004 Sinauer Associates, Inc. All rights reserved.
These pages may not be reproduced in whole or in part without permission.

Address editorial correspondence to:
Sinauer Associates, Inc.
23 Plumtree Road
Sunderland, MA, 01375 U.S.A.
Fax: 413-549-1118
Internet: www.sinauer.com; publish@sinauer.com

Address orders to:
VHPS/W. H. Freeman & Co. Order Department
16365 James Madison Highway,
U.S. Route 15, Gordonsville, VA, 22942 U.S.A.
Internet: www.whfreeman.com

ISBN 0-7167-5811-3
Printed in the U.S.A.

4 3 2 1

To the Student

Biology is an incredibly exciting field of study, but in order to appreciate new discoveries and discussions, it is necessary to have a firm grasp of underlying concepts and ideas. Your textbook is designed to give you a comprehensive overview of important biological phenomena. It will also serve as a resource to you in future studies. Together with your instructor, your textbook will provide you with important information for beginning your study of biology.

This Study Guide is designed to supplement, not to replace, your text and your instructor. It was written for you, the student, in language that you can understand, but it does emphasize proper usage of biological terminology. Important concepts and ideas have been synthesized into short, easy-to-read segments that provide an overview of the biological phenomena discussed in your text. This summary should serve as a preview tool, an understanding check, and a review tool. Each chapter summary is followed by a series of questions. These questions have been grouped into two categories. Knowledge and Synthesis Questions are designed to determine if you have retained information from the chapter, and if you can put together various concepts in order to answer questions. Application Questions ask you to apply the knowledge you gleaned from the chapter to answer questions that are more open-ended. These questions require you to have assimilated several concepts and to think a little beyond what you have just read. Your instructor may ask questions similar to these on your tests, or may use an entirely different approach to questioning your knowledge, but these questions will be a good check of how well you understand the material. Brief answers to the questions are provided at the end of each chapter. These answers are not exhaustive but are instead designed to point to the correct concepts in your book and in the summaries. Because of the nature of many of the Application Questions, your answers should be much more expansive than the short explanations given in the Study Guide.

Strategies for Studying Biology

Each individual has his or her own unique study pattern. However, the authors believe that there are some successful strategies that are universal. We recommend that you first preview a chapter in your text. In this initial preview it is important to note the organization of the chapter and the main points, and to go over the chapter summary at the end. By referring to the Study Guide at this point you will further understand the organization of the material. If you follow these steps, you might have some specific questions in your mind that you will expect your reading to answer.

Reading a text is an active process. We recommend that you always have a pencil and paper available. Jotting notes in margins also serves as an excellent mental trigger when it comes time to review. As you read each section of your text, see if you can summarize it in your own words. Compare your summaries to those provided at the end of the chapter and in your Study Guide. You want to assure yourself that the main points you are noting match those that the author has selected. Refer back to those questions that came up as you previewed the chapter. You should be able to answer your own questions by the time you have finished your reading.

As you are reading, take time to review all the figures and tables. Your text makes use of figures to illustrate points, describe pathways, and give visual representation to complex topics. Often these figures are more helpful than the paragraphs of written explanation. In places where figures might be beneficial to you but are not provided in the text, try to draw them yourself. This is especially important when attempting to understand structures and pathways. You will find that the Study Guide also refers you to specific tutorials and activities on the Student Website (www.thelifewire.com) and CD. Each of these visual tools will aid in your understanding of the material.

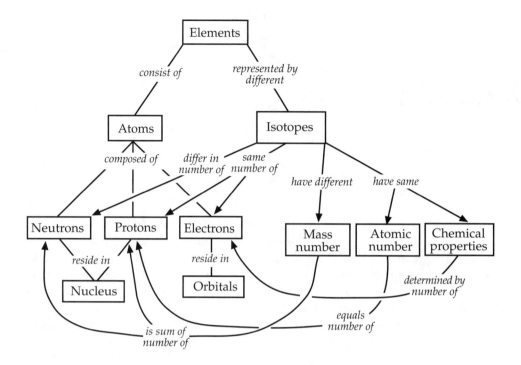

Once you have read the material, summarized it in your own words, and reviewed all of the figures and tables, it is time to do another quick review. Scan the summary in your text again and read over the key concepts in this Study Guide. We recommend further that you construct a "mind map" of the main points of the chapter. This will assure you that you see how concepts are interrelated and that you understand the organization of the material. An example of a "mind map" is shown above. Main concepts are in boxes with supporting concepts listed below them. Arrows are used to connect concepts and show relationships.

Finally, when you are comfortable with chapter content, move to the questions section of this Guide. We recommend working through all of the questions before checking your answers. Any questions that you do not understand or that you answer incorrectly should be flagged as material you need to review. Remember that your goal is to understand the material completely, not just to answer these particular questions.

Experiments have suggested that the average attention span of a reader is approximately 12 minutes. You cannot expect to sit down and master an entire chapter in one sitting. Break up your study into short segments of approximately 30 minutes each. This will give you time to get down to business, maximize your attention, and learn without becoming frustrated or drained. At the end of this time move to some other activity or area of study. When you return to biology you will find your mind is ready to absorb more.

This entire process should take place well in advance of your exams. You cannot learn even one chapter adequately the day before a test. Mastering biology requires daily study and review. If you keep up with learning the concepts as your course proceeds, you will find that reviewing the text and Study Guide summaries and questions is adequate preparation. The key to learning biology well is slow and diligent, daily work.

Doing Well on Biology Exams
Your instructors are your best resource for doing well on exams. They are there to assist you in learning the material that is presented in your textbook. They are experts in their fields and understand how the concepts presented are vital to your biology education. Follow their lead in preparing for tests. Test performance is directly linked to classroom attendance and daily study. Sit front and center in your lecture room and pay close attention to what your instructor writes or provides in presentations. Your instructor will give you guidance about the most important points to study. You should take careful notes and ask for clarification when necessary.

You should always compare your classroom and lecture notes to your text and the Study Guide. Mark in the books those points your instructor emphasizes. Were there specific questions that the instructor brought up in class? Do these correspond to the discussion and application questions at the end of your chapter or in the Study Guide? If so, chances are you will see them again on a test. This practice should be part of your daily study regimen.

Approach a test itself with confidence. Read the directions carefully and read the questions completely. One of the biggest mistakes students make on exams has to do with not following instructions. We recommend that you read carefully through the entire test even before you begin to answer the first question. You may find that early questions are answered partially in later ones, and that an overview of the whole exam helps to trigger your memory of the material.

If your test contains multiple choice questions, treat each of the answer options as if it were a separate true/false question. In other words, mentally fill in the question with each answer and ask yourself if the resulting statement is true or false. Be sure to read each of the options, even if the first one strikes you as the correct one. You may be dealing with a question that has multiple correct answers or asks you to select "all of the above" or "none of the above." If you are unsure about how to answer a question, begin by ruling out answers that you know are incorrect. By eliminating some possibilities, your chance of selecting the right one is greatly improved! If you narrow the choices down to two possible answers and just can't decide between them, go with your gut response. You just may be right. And when you go over your answers, change your response only if you know you answered a question incorrectly. If you aren't sure, it may be best to leave the answer as is. Your first instincts are often correct.

Essay questions require you not only to think through the concepts, but also to decide how you will present them. Organized, concise essay answers are always preferred to rambling answers with little direction. Take a look at the point value of each question. This is often an indication of how many "points" you need to make in your discussion. A 10-point question rarely can be answered with a single sentence. Look carefully at what your instructor is asking. If you are required to "discuss" a concept or a problem, do not present a list or some scattered phrases. Write in complete sentences and paragraphs unless you are asked specifically to list or itemize the material. Be sure that you address all points in the question. Essay questions frequently have multiple parts to test your understanding of the links between various concepts.

Always be aware of the time you have to complete your test. Answer the questions you know swiftly, and allow yourself time to go back and really think about those you are struggling with. Be sure to concentrate on the questions with the highest point value. Those are the questions testing the most critical concepts and are the ones that will have the most influence on your grade.

Once you think you have completed the test, take a few minutes to go back over it. If there are any questions about which you are still unclear, be sure to ask your instructor for clarification.

Learn from Your Mistakes

After your instructor returns your exam or posts the answers, review the mistakes you have made. Take this opportunity to clarify misunderstandings and re-learn material when necessary. You will find this helpful not only for the final exam, but also as you approach upper level courses. Material reviewed, corrected, and re-learned is more likely to be retained in the long run.

Get Help before It Is Too Late

Many students come to college without adequate study habits and/or are ill prepared for the rigor of biology at the college level. But this is a problem that can be solved fairly easily. Many colleges and universities have academic centers that assist with locating tutors, teaching study habits, and acquainting you with a variety of other resources. Make good use of these facilities. They are there for you.

Good luck in your study of introductory biology!

Contents

1 An Evolutionary Framework for Biology

Important Concepts

Living things are organized genetic units capable of metabolism, reproduction, and evolution.

- Metabolism is the coordinated chemical reactions of a living organism. Through metabolic adjustments, organisms maintain a constant internal environment as the external environment changes.

- Life is maintained by reproduction with variation. Errors in genetic material allow for biological evolution. The diversity of environments on Earth has allowed for the large variety of organisms. Organisms are able to live in a variety of environments because of adaptations.

The fact of evolution has changed how biology is approached.

- Buffon was one of the first to claim that descent with modification takes place. This early statement of evolution attempted to explain similarities in mammal limbs that remained, even when the resulting structure was no longer functional.

- Lamarck, a student of Buffon, postulated a theory of acquired characteristics. Though his theory eventually was discredited, it was a shift in thinking to the possibility of change through time.

- Darwin and Wallace independently proposed the theory of evolution through natural selection. This theory is based on three observations and a resulting conclusion:
  Fact: Reproductive rates must be balanced by death rates.
  Fact: Organisms are variable.
  Fact: Offspring inherit features from their parents.
  Inference: Differences among individuals influence how well they survive and reproduce. Those traits that help an individual to survive and reproduce are maintained in later generations. This process is called natural selection.

- Evolutionary change takes place over long periods of time and without any goals.

Certain evolutionary results had to occur before others.

- Life arose approximately 4 billion years ago.

- The aggregation of complex chemicals in the process of chemical evolution led to the existence of life. These initial simple molecules allowed for the production of more complex molecules.

- The first known cells are approximately 3.8 billion years old. A membrane-enclosed cell allows maintenance of an internal environment. Cells can capture and use energy (metabolize) and replicate. Early cells were simple prokaryotes with no internal membrane-enclosed compartments. Over time, increasingly complex cells with diverse patterns of metabolism developed.

- Approximately 2.5 billion years ago, photosynthetic organisms arose. These cells could take light energy from the sun and convert it to chemical energy with oxygen as a by-product. Large numbers of photosynthetic organisms increased atmospheric oxygen levels, allowing for the evolution of aerobic metabolism. Photosynthetic organisms also contributed to the insulating ozone layer that shields Earth from radiation and modifies temperature. This shield eventually made life on land possible.

- Eukaryotic cells arose when larger prokaryotic cells engulfed and assimilated smaller cells, resulting in membrane-enclosed compartments that carry out specific cellular functions.

- Multicellularity evolved about 1 billion years ago. Multicellularity makes it possible for cells to specialize and organism size to increase. The development of resting spores and the ability of cells to clump were the first steps toward multicellularity. Specialized sex cells led to new patterns of cell division and increased possibilities for genetic variation.

- Sexual reproduction brings about genetic recombination and variation. Variation is essential for adaptation to environmental change. Organisms inherit characteristics that increase survival and reproductive success in specific environments.

All organisms are descendants of a single unicellular organism.

- Species are genetically independent groups of organisms.

- Different species come about when two groups of organisms in one species are isolated from each other so that they can no longer mate and produce viable offspring.

- Ancestral organisms are neither "primitive" nor lower than "derived" organisms. All living organisms are successfully adapted to their environment.

- There are as many as 30 million different species on Earth today. This is due to the splitting of populations during speciation events. The evolutionary relationship between all organisms can be described using an evolutionary tree.

Scientists are currently assembling the Tree of Life.

- Scientists are determining the branching pattern of the Tree of Life using the molecular evidence found in organisms. All life is placed into one of three major domains: Archaea, Bacteria, and Eukarya.

- The Eukarya are divided into four groups: Protista, Plantae, Fungi, and Animalia.

- Organisms are referred to by their genus and species names (e.g., humans are *Homo sapiens*).

Scientists ask specific types of questions and reason in a methodical way.

- Scientific inquiry involves the use of tools to help augment the human senses. The hypothesis-prediction approach typically used by scientists begins with particular observations that lead the observer to question why things are as they appear. These questions lead to the formation of a hypothesis. Predictions are made in regard to the hypothesis, and then they are tested.

- A hypothesis that withstands the test of time without being falsified is called a theory.

- Hypotheses are tested by either the comparative method or by conducting controlled experiments. The comparative method involves gathering data to examine the patterns found in nature. Controlled experiments involve isolating the variables of interest, while keeping the other variables that may influence the outcome as steady as possible. Observations are then made to test the hypothesis.

- For scientific results to be accepted, they have to hold up to many rounds of testing. Scientific "fact" is based on many forms of tested evidence.

Biological science has an impact on every human being and is the basis for many public policy decisions.

- The environmental stresses of population growth, the emergence of agricultural and medical technologies, and other modern challenges have increased the need for everyone to understand biological information.

The Big Picture

- Evolution provides us with a context for understanding relationships among organisms. All available evidence points to an origin of life some 4 billion years ago. Heritability of characteristics from generation to generation, death rates that counter reproductive rates, and natural selection have led to the current host of organisms inhabiting the planet. There are specific characteristics of life, and the evolutionary roots of these characteristics can be traced through time.

- All life on Earth is organized in the Tree of Life based on molecular evidence. When studying organisms from an evolutionary perspective, scientists are concerned with evolutionary relatedness and common ancestors.

- Science follows a specific hypothesis-prediction approach. The necessity of a testable and falsifiable hypothesis sets science apart from other methods of inquiry. Control of variables and repeated testing lends credibility to the method.

Common Problem Areas

- Many students have trouble remaining open-minded enough to begin to understand evolutionary biology. Neither your instructors nor this book are asking you to set aside your beliefs. We are asking you to learn science. We encourage you to examine the evidence supporting evolution without prejudice, as you would any other scientific subject.

- Refrain from learning the hypothesis-prediction approach merely as a series of steps. Think about how each step in the process follows from the preceding step. There are two characteristics of science that set it apart from other modes of inquiry. One is the setting of hypotheses. The other is the continual process by which it is carried out. With each conclusion leading to a new set of hypotheses, science has the capacity to alter our overall base of knowledge.

- You most likely have studied a five-kingdom system of classification to this point. The domain structure and six kingdoms may be new to you. Be careful not to assume you know this!

Study Strategies

- Many students find that understanding the hypothesis-prediction approach makes understanding some of the evidence behind evolution easier. If you find you are having trouble, work on that section first.

- Drawing a time line or examining the calendar model of Figure 1.3 will help you understand the chronology of the major evolutionary events discussed in your book.

- Review the following activities on the Student Website/CD:
  Activity 1.1 The Hierarchy of Life
  Activity 1.2 The Major Groups of Organisms

Test Yourself

Knowledge and Synthesis Questions

1. Life arose on Earth approximately how many years ago?
   a. 3.8 billion
   b. 3.8 million
   c. 4000
   d. 1.5 billion

2. _____ are the differences among organisms that allow for life in such a wide variety of environments on Earth.
   a. Prokaryotic cells
   b. Eukaryotic cells
   c. Homeostasis
   d. Adaptations

3. Which of the following statements regarding evolutionary biology is *not* true?
   a. Darwin and Wallace postulated that mortality rates match reproductive rates, that characteristics are heritable, and that those characteristics most suited to the environment are selected for.
   b. Darwin postulated that characteristics could be acquired in a single generation.
   c. Darwin was the first to come up with an evolutionary theory.
   d. Both b and c

4. Photosynthesis was a major evolutionary milestone because
   a. photosynthetic organisms contributed oxygen to the environment, which led to the evolution of aerobic organisms.
   b. photosynthesis led to conditions that allowed life to arise on land.
   c. photosynthesis is the only metabolic process that can convert light energy to chemical energy.
   d. All of the above

5. Sexual reproduction was evolutionarily important because
   a. previously all "daughter" organisms were clones of the parent organism.
   b. a new kind of nuclear division, meiosis, increased variability between generations.
   c. variation is necessary for evolution.
   d. All of the above

6. The evolution of multicellularity led to an increased need for maintaining a constant internal environment. This maintenance of a constant internal environment is referred to as
   a. homeopathy.
   b. homeostasis.
   c. environmental maintenance.
   d. adaptation.

7. Which of the following are necessary for speciation to occur?
   a. Reproductive isolation of two groups of organisms
   b. A large amount of mutation within a population
   c. A reduction in the number of individuals in a population
   d. None of the above

8. Which of the following is *not* a domain in the Tree of Life?
   a. Archaea
   b. Plantae
   c. Eukarya
   d. Bacteria

9. The deepest division of the evolutionary history of life is the
   a. kingdom.
   b. genus.
   c. domain.
   d. species.

10. Organisms are named using first their _____ and then their _____ .
    a. species; genus
    b. genus; species
    c. domain; genus
    d. genus; domain

11. In the frog example in your book, researchers subjected frogs to various levels of UV-B radiation while keeping all other variables constant. This is an example of a _____ experiment.
    a. repeated
    b. controlled
    c. laboratory
    d. None of the above

12. For a hypothesis to be scientifically valid, it must be _____ and it must be possible to _____ it.
    a. testable; prove
    b. testable; reject
    c. controlled; prove
    d. controlled; reject

Application Questions

1. Discuss how the process of scientific inquiry is different from other forms of inquiry. Include in your discussion a description of the hypothesis-prediction approach.

2. Your book states that evolution proceeds without a "goal." With this idea in mind, would you say that organisms that are alive currently are more "advanced" than those that have become extinct?

3. Define adaptation in your own terms. If organisms do not acquire characteristics in a single generation, how do they "adapt" to an environment?

4. Look over a recent newspaper to see how many articles are directly related to biology. Select one article and discuss how the researchers followed or did not follow the hypothesis-prediction approach.

Answers
Knowledge and Synthesis Answers

1. **a.** Available evidence puts the beginning of life at about 3.8 billion years ago.

2. **d.** Adaptations are the differences found in organisms that allow them to live in an environment.

3. **d.** Evolutionary ideas began nearly 100 years before Darwin and Wallace published their ideas on natural selection. For these ideas to emerge, it was necessary to accept that Earth was ancient, because evolution occurs slowly over extended periods of time.

4. **d.** Photosynthesis caused the accumulation of oxygen in the atmosphere and contributed to the formation of an ozone layer. The presence of oxygen made the evolution of aerobic organisms possible, and the ozone layer shielded Earth from harmful radiation. These two phenomena contributed to the evolution of terrestrial life.

5. **d.** Variation among organisms is essential for evolution by natural selection. When an organism divides by mitosis, the only variation that can occur is through random genetic mutation. The processes of meiosis and sexual reproduction, however, involve reorganizing and mixing genes from two organisms, leading to novel genetic combinations in offspring.

6. **b.** Homeostasis is the maintenance of a consistent internal environment.

7. **a.** For speciation to occur, reproductive isolation, such that two groups can no longer reproduce and produce viable offspring, is necessary.

8. **b.** Plantae is a kingdom found in the domain Eukarya.

9. **c.** The knowledge of domains has come primarily from DNA evidence.

10. **b.** An example of this is *Homo sapiens*.

11. **b.** For experiments to be scientifically valid they must be controlled.

12. **b.** Scientific hypotheses are set apart from mere conjecture by being testable and falsifiable.

Application Answers

1. The process of scientific inquiry is different because a hypothesis must be testable and it must be possible to reject it. The hypothesis-prediction approach begins with observations that lead to questions. From the questions, hypotheses are formed that are probable explanations for the phenomena being observed. Predictions are formed from the hypotheses and tested. Conclusions are drawn from the tests. These conclusions may, in turn, lead to additional hypotheses.

2. Extinct organisms were adapted to the environment during their own time. Current organisms are best adapted to the environment at this time. It cannot be said that one is more advanced than the other.

3. An adaptation is a characteristic of an organism that contributes to its ability to survive and reproduce in its environment. Mutation and sexual recombination contribute over time to the variation among organisms in a population. Those organisms that survive and reproduce best leave (on average) the most offspring in the next generation; natural selection removes organisms that are not well adapted to their environment. In this way, populations gradually become adapted to fit their environments.

4. Often an article reporting on science news does not provide enough information for the reader to know whether a particular study followed the hypothesis-prediction approach. It pays to be a smart reader and consumer in this respect.

2 Life and Chemistry: Small Molecules

Important Concepts

All matter is composed of the same chemical elements.

- Atoms combine to form all matter. The nucleus of atoms contains a defined number of positively charged protons and neutral neutrons. Negatively charged electrons move in their outer shells. An element is comprised of only one type of atom. There are more than 100 elements, and they are grouped according to certain characteristics in the periodic table.

- Six elements compose the majority of every living organism: carbon, hydrogen, oxygen, nitrogen, phosphorus, and sulfur.

- The number of protons in an element determines its type and is known as its atomic number. Neutrons are found in every element except hydrogen. The total number of protons and neutrons equals the mass number of the atom.

- Isotopes of an element have the same number of protons but differ in the number of neutrons present. Radioisotopes are unstable isotopes that give off energy. This decay eventually causes the atom to change to a new element. Radioisotopes are important in the medical field.

- Protons and neutrons have mass and contribute to atomic weight. Atomic weights on a periodic chart are averages of all isotopes of an element.

Biologists are interested in chemical reactions and the interactions of electrons.

- Interactions between atoms involve the associations of their electrons. Electrons continuously orbit the nucleus of an atom in a defined space.

- The orbital of an atom is the space in which an electron is found 90 percent of the time. These patterns of orbitals compose a series of electron shells, each with a specific number of electrons. The innermost shell (s orbital) can contain up to two electrons. The second shell can have as many as 4 orbitals, each containing 2 electrons. The number of electrons in the outermost shell determines how the atom interacts with other atoms. Many biologically important atoms, such as carbon and nitrogen, are stable when they have eight electrons in the outermost shell. This phenomenon is referred to as the octet rule.

- Two or more linked atoms compose a molecule resulting in stabilization of the outer electron shell of the atoms.

Chemical bonds hold molecules together.

- Atoms share pairs of electrons to stabilize their outer shells in covalent bonding. This type of bond is very stable and strong and can be broken only with a great deal of energy. All molecules have a three-dimensional shape, and the interactions between a given pair of atoms always have the same length, angle, and direction. This shape of a molecule affects how it behaves.

- Multiple covalent bonds may exist. Although a single covalent bond involves one pair of shared electrons, double bonds involve two pairs of shared electrons. Triple bonds share three pairs, but are rare.

- Covalent bonding between atoms of the same element results in equal sharing of electrons. The attractive force of an atom on electrons is known as electronegativity. Two atoms that have the same electronegativity form nonpolar covalent bonds.

- Bonds between different elements generally result in an unequal sharing of the electrons. Unequal sharing of electrons results in partial (δ) charges in molecules because one nucleus is more electronegative than the other, attracting the electrons more strongly. One atom of a molecule may be partially negative whereas the other is partially positive. This balance of partial charges results in a "polar" molecule with a δ^- pole and δ^+ pole.

- Partial charges allow hydrogen bonding of molecules. This occurs often between molecules of water. The δ^+ of H is attracted to the δ^- of another molecule. This is a weak bond that is easily broken because no electrons are shared, but this type of bond is important in stabilizing the three-dimensional shape of large molecules such as proteins and DNA.

- Ions are formed when atoms lose or gain electrons, resulting in a net positive or negative charge.

  - Cations are positively charged and have fewer electrons than protons.

  - Anions are negatively charged and have more electrons than protons.

- More than one electron can be gained or lost from an atom or molecule. A group of covalently bonded atoms may also gain or lose electrons to form complex ions.

- Ionic bonds form between ions of opposite charge. Ionic bonds are not as strong as covalent bonds. Ions can also attract the partial charges of polar molecules.

- Polar molecules are hydrophilic, or "water loving," because of the partial charges. Nonpolar molecules are hydrophobic, or "water hating." Attractions of nonpolar molecules are enhanced by van der Waals forces. Polar molecules tend to aggregate with other polar molecules, and nonpolar molecules aggregate with other nonpolar molecules.

Chemical reactions occur when atoms combine and form new bonds.

- Reactions involve reactants that are altered to produce products. Matter is neither created nor destroyed, but changed. Some reactions produce energy, and others require energy. Energy is the capacity to cause change and is typically measured in calories or joules.

- Chemical bonds hold potential energy. When these bonds are broken, release of this energy is important to living systems.

Water is an essential molecule composing more than 70 percent of a living organism and providing many biologically important functions.

- Water is a polar molecule and the solvent in which most biological reactions take place. Its ability to form hydrogen bonds and tetrahedral shape give it these properties.

- Ice floats because of the less-dense pattern of hydrogen bonding that results when water molecules are frozen. Ice insulates ponds and lakes in the winter, allowing life to continue beneath the frozen layer.

- Much energy is required to break the hydrogen bonds when molecular ice melts. In contrast, much energy is lost during the freezing of water.

- Large bodies of water help moderate temperature on land and in the atmosphere. The high specific heat of water in the liquid state provides it with a high heat capacity and the ability to moderate temperatures.

- Water has a high heat of vaporization and thus requires the input of energy to break the hydrogen bonds and go from a liquid to a gaseous state. This makes evaporating water an effective coolant.

- The polar nature of water and the formation of hydrogen bonds contribute to the cohesive strength of water and its surface tension, both of which are biologically important.

- Solutions are formed when a substance is dissolved in water or another liquid.

Quantitative terms are important to understand for both biology and chemistry.

- Concentration is the amount of a substance in a given amount of solution.

- A mole is the amount of an ion or compound in grams whose weight is numerically equal to its molecular weight.

- Avogadro's number = 6.02×10^{23} molecules per mole. This number relates the number of molecules of any substance to its weight. A 1 molar solution is made by dissolving 1 mole of a substance into 1 liter of water. Molar concentration of a substance is notated by putting square brackets around the symbol for that substance. For example, [H^+] stands for the molar concentration of hydrogen ions.

Acids and bases are biologically important.

- Acids dissolve in water and release H^+ ions.

- Bases dissolve in water and attach to H^+ ions.

- Acids and bases do not all ionize in the same way. Some ionize easily and completely and are called strong acids or bases. Others only partially or reversibly ionize and are called weak acids or bases.

- When strong acids and bases are ionized, the reaction is irreversible. Ionization of weak acids and bases can be reversible reactions.

- Water is a very weak acid because two molecules are required for ionization. Though ionization of water is uncommon, it is biologically important.

- Solutions are acidic or basic whereas compounds and ions are bases or acids.

- The pH of a solution refers to the concentration of H^+ ions. Specifically, $pH = -\log 10[H^+]$. Lower pH indicates a higher H^+ concentration or a more acidic solution. A higher pH indicates a lower H^+ concentration or a more basic solution.

- Buffers are mixtures of weak acids and their corresponding base that prevent fluctuations in the pH of a solution.

Molecules have specific properties that enhance their biological activity.

- Functional groups are small groups of atoms that have specific properties and are attached to larger molecules. The 20 amino acids found in proteins each have a different functional or R group, giving each amino acid a different property.

- Isomers are molecules that have the same number and kind of atoms but are arranged differently. Structural isomers have different order and binding of the atoms. Optical isomers occur when carbon has four different groups or atoms attached to it.

- Molecules vary in size, three-dimensional shape, and chemical properties, providing a variety of functions.

The Big Picture

- Understanding the chemical building blocks of all matter is essential to understanding the biology of organisms. Atoms, containing protons, neutrons, and electrons, form elements or combine to form molecules. Molecules may be held together with covalent or ionic bonds and stabilized in three-dimensional conformations by hydrogen bonds. Chemical reactions change reactants to products without creating or destroying matter.

- Water has unique properties that make it biologically important. An essential property is its ability to act as a solvent. How a substance ionizes in water determines if it is acidic or basic. Acids have a high concentration of H^+, whereas bases have a low concentration of H^+. The pH scale allows us to compare the concentrations of H^+ in a variety of solutions.

- The properties of molecules determine their function and how they will react with other molecules.

Common Problem Areas

- pH tends to confuse many people. Remember that it is a concentration of H^+ ions. Those that have a higher concentration are acidic, and those that have a lower concentration are basic. Also remember that pH is a logarithmic scale, and that a low pH value is an acid and a high pH value is a base.

- Learning quantitative terminology and how solutions are made is often difficult. Completing practice problems is the best way to learn this material.

- People are sometimes overwhelmed by the periodic table. Consult Figure 2.3 in your book to make sure you understand the components. Do not try to memorize the table.

- Use Table 2.1 to understand the differences between the various types of bonds and interactions.

Study Strategies

- The best way to learn the quantitative terms and pH material is to work on many practice problems. Also take advantage of any laboratory experience making solutions. Every good biologist makes many solutions over time.

- Use the figures in your book to help you visualize what atoms and molecules look like.

- Be careful not to get lost in terminology. Think about what each term means, and focus on understanding the concept rather than memorizing the term.

- Review the following tutorial and activities on the Student Website/CD:
  Tutorial 2.1 Chemical Bonds
  Activity 2.1 Electron Orbitals
  Activity 2.2 Features of Amino Acids
  Activity 2.3 Chemical Functional Groups

Test Yourself

Knowledge and Synthesis Questions

1. The atomic number of an element refers to the number of
   a. protons and neutrons in an atom.
   b. protons in an atom.
   c. electrons in an atom.
   d. neutrons in an atom.

2. Which of the following statements concerning electrons is *not* correct?
   a. Electrons orbit the nucleus of an atom in defined orbitals.
   b. The outer shell of all atoms must contain eight electrons.
   c. An atom may have more than one valence shell.
   d. Electrons are negatively charged particles.

3. The element with which of the following atomic numbers would be most stable?
   a. 1
   b. 3
   c. 12
   d. 18

4. How can you differentiate between an element and a molecule?
   a. Molecules may be composed of different types of atoms, whereas elements are always composed of only one type of atom.
   b. Molecules are composed of only one type of atom, whereas elements are composed of different types of atoms.
   c. Molecules are elements.
   d. Molecules always have larger atomic weights than elements.

5. The strongest chemical bonds occur when
   a. two atoms share electrons in a covalent bond.
   b. two atoms share electrons in an ionic bond.
   c. hydrogen bonds are formed.
   d. van der Waals forces are in effect.

6. You have discovered that a molecule is hydrophilic. What else do you know about this molecule?
   a. It cannot form hydrogen bonds.
   b. It is a polar molecule.
   c. It has a partial positive region and a partial negative region.
   d. Both b and c

7. The stability of the three-dimensional shape of many large molecules is dependent on
   a. covalent bonds.
   b. ionic bonds.
   c. hydrogen bonds.
   d. van der Waals attractions.

8. The molecular weight of glucose is 180. If you added 180 grams of glucose to a 0.5 liter of water, what would be the molarity of the resulting solution? (See Figure 2.3 for a periodic table.)
   a. 18
   b. 1
   c. 9
   d. 2

9. Which of the following properties makes water an excellent solvent?
   a. The pH of water is 7.
   b. Water is a polar molecule.
   c. Water has a high surface tension.
   d. Water has a high heat capacity.

10. Cola has a pH of 3; blood plasma has a pH of 7. The hydrogen ion concentration of cola is _____ than the hydrogen ion concentration of blood plasma.
    a. 4 times greater
    b. 4 times lesser
    c. 400 times greater
    d. 10,000 times greater

11. If solution A has a pH of 2 and solution B has a pH of 8, which of the following statements is true?
    a. A is basic and B is acidic.
    b. A is acidic and B is basic.
    c. A is a base and B is an acid.
    d. A has a greater $[OH^-]$ than B.

12. One mole of glucose ($C_6H_{12}O_6$) weighs
    a. 180 grams.
    b. 42 atomic mass units.
    c. 96 grams.
    d. 342 grams.

13. Which of the following is *not* an organic molecule?
    a. $C_6H_{12}O_6$
    b. CH_3OH
    c. H_2SO_4
    d. CH_3COOH

Application Questions

1. In the following figures, place electrons in the appropriate shells based on the atomic numbers of the elements carbon ($_6$C), nitrogen ($_7$N), sodium ($_{11}$Na), chlorine ($_{17}$Cl), argon ($_{18}$Ar), and oxygen ($_8$O).

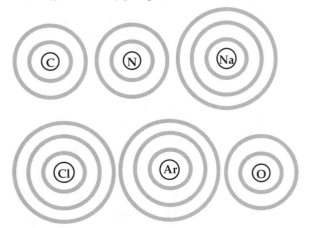

2. Calcium has an atomic number of 20. Draw structures for Ca and Ca^{2+}. What is different about these structures? Why is the most common ion of calcium Ca^{2+}? Why is the most common ion of lithium Li^+?

3. Nitrogen atoms can form triple bonds with each other. What is a triple bond? How many electrons are shared between two N atoms?

4. Water is a polar molecule. This property contributes to cohesion and surface tension. Draw six water molecules. In your drawing, indicate how hydrogen bonding between molecules contributes to cohesion and surface tension. (Be sure to include appropriate covalent bonds in each molecule.)

5. Rank the following solutions in order from the most acidic to the most basic.
   Lemon juice, pH = 2
   Mylanta, pH = 10
   Sprite, pH = 3
   Drain cleaner, pH = 15
   Seawater, pH = 8
   Of the preceding, which has the highest concentration of H^+ ions? Which has the lowest concentration of H^+ ions?

6. If you have 12 moles of a substance, how many molecules do you have of that substance? Suppose the substance has a molecular weight of 342. How many grams of that substance would you have to dissolve in a liter of water to make a 12 M solution?

Answers

Knowledge and Synthesis Answers

1. **b.** The atomic number refers to the number of protons in an atom. The atomic weight can vary with the isotope and is the total of the number of protons and neutrons in an atom.

2. **b.** The first energy shell requires only two electrons to be full; subsequent shells may hold eight or more electrons.

3. **d.** The element with atomic number 18 (Argon) is the most stable because its outermost electron shell is full.

4. **a.** An element is always composed of only one type of atom. Molecules contain many types of atoms that are bonded together.

5. **a.** Covalent bonds, resulting from the sharing of a pair (or more) of electrons, form the strongest types of bonds. Ionic bonds, where electrons are donated or received, are the second strongest. Hydrogen bonds are the weakest bonds and result from the attraction of partial charges to one another. Van der Waals forces are not bonds but attractions between nonpolar molecules.

6. **d.** Hydrophilic molecules are polar and have a partial positive and a partial negative pole. These molecules are said to be hydrophilic because they easily form hydrogen bonds with water.

7. **c.** Hydrogen bonds, though weak individually, are quite effective in large numbers and are responsible for maintaining the structural integrity of many large molecules such as proteins and DNA.

8. **d.** The molarity of the solution is 2. A 1 M solution is made by adding 180 grams to 1 liter of water. Adding the same amount to half a liter would result in a solution with twice the molarity.

9. **b.** The polar nature of water makes it an excellent solvent because it can bind ions and other polar molecules. This prevents solutes from coming out of solution. The other listed properties also result from water's polar nature.

10. **d.** pH is a logarithmic scale. Each increase or decrease in pH value is by a factor of 10. Therefore, a difference between pH 3 and pH 7 would be a difference of $10 \times 10 \times 10 \times 10$, or 10,000.

11. **b.** A solution with a pH of 2 is acidic, and one with a pH of 8 is basic. The lower the pH value, the higher the concentration of H^+ ions.

12. **a.** Each carbon atom weighs 12 atomic mass units, each hydrogen atom is 1 atomic mass unit, and each oxygen is 16 atomic mass units. Adding these yields a molecular weight of 180 atomic mass units. Because one mole is equal to the atomic weight in grams, one mole of glucose equals 180 grams.

13. **c.** Organic molecules must contain carbon.

Application Answers

1.

2.

Calcium (Ca) has two electrons in its outer shell.

Calcium ion (Ca^{2+}) has lost its two outer electrons and therefore has a positive charge of 2 because it has two more protons than electrons. Lithium has only one electron in its outer shell; when that electron is lost, the ion gains a positive charge of only 1.

3. Three pairs of electrons are shared between two N atoms in a triple bond.

4. The partially positive hydrogens of one water molecule are attracted to the partially negative oxygens of another molecule of water. This attraction tends to lead to water molecules "sticking" together and creating surface tension.

5. Lemon juice, pH = 2 (highest concentration of H^+ ions)
Sprite, pH = 3
Seawater, pH = 8
Mylanta, pH = 10
Drain cleaner, pH = 15 (lowest concentration of H^+ ions)

6. You would multiply Avogadro's number (6.02×10^{23}) times 12 to determine the number of molecules. This gives you 7.224×10^{24} molecules of this substance. A 12 M solution would be made by multiplying the formula weight of the molecule times 12 and adding that many grams of the substance to one liter of water. The formula weight of this molecule is 342, so you would need 4104 g.

3 Life and Chemistry: Large Molecules

Important Concepts

Living organisms are arranged into unique biological systems.

- One theory of the origin of life on Earth is based on the idea that water and some molecules characteristic of life may have come to earth from space. A meteorite from Mars, ALH 84001, may have had internal conditions that would allow for living organisms to survive a long space journey.

- Another theory of the origin of life on Earth states that the conditions on Earth billions of years ago led to the emergence of biological molecules. This theory is known as chemical evolution. Stanley Miller and Harold Urey produced a primitive atmosphere containing hydrogen gas, ammonia, methane gas, and water and using electrical charges were able to produce the initial building blocks of life.

Macromolecules are very large polymers with a wide range of biological functions.

- Specific types of macromolecules share similar roles throughout all of life that are dependent on the chemical properties of their component monomers. Macromolecules provide energy storage, structural support, protection, catalysis, transport, defense, regulation, movement, and information storage.

- Polymers are made from the covalent bonding of monomers in a process called condensation. See Figure 3.3a in your text.

- Polymers may be broken down into their constituent monomers through hydrolysis. See Figure 3.3b in your text.

- Both condensation and hydrolysis reactions require an input of energy.

- The size and structure of each polymer is related to its biological function.

Polymers of amino acids are called proteins and function in structural support, catalysis, defense, and movement.

- Enzymes are specialized proteins that catalyze (speed up) chemical reactions in cells. The shape of a protein is integral to its function.

- Proteins range in size from small proteins with 50 to 60 amino acids to large proteins with more than 4,600 amino acids. Proteins may be made of single chains of amino acids or multiple interacting chains.

- Twenty amino acids are found in proteins. Each of the 20 amino acids has a central carbon atom that has a hydrogen atom, amino group, and carboxyl group attached to it. The specific amino acids differ in the unique side chain attached to the central carbon, providing them with specific properties.

- The shape and structure of proteins are influenced by the properties of the side chains of the constituent amino acids. Five amino acids have charged side chains; five have polar, hydrophilic side chains; seven have hydrophobic, nonpolar side chains; and three have special hydrophobic functions.

- Cysteine's terminal –SH can form disulfide bridges with other cysteines and influences how a protein chain folds.

- Glycine is small and fits into folds of proteins.

- Proline has limited bonding ability and often contributes to looping in proteins.

- Amino acids are chained or polymerized together during condensation. The resulting bonds are called peptide linkages. See Figure 3.5 in your text.

- Peptide chains always begin with the N terminus and end with the C terminus.

- The "partial double-bond" character of the peptide bond is relatively rigid, limiting the extent of protein folding. Further, hydrogen bonds are favored within the folded proteins, helping maintain the form and function of proteins.

Proteins have levels of structural organization.

- Primary structure of a protein refers to the sequence of amino acids. All higher levels of structure are due to the specific primary structure of the protein.

- Secondary structure refers to regular repeated patterns of the amino acid chain and is stabilized by hydrogen bonds. The α helix consists of a coiling of a single chain. The β pleated sheet occurs when two chains form hydrogen bonds between themselves.

- Tertiary structure refers to bending and folding of an amino acid chain. This structure is stabilized by disulfide bridges, hydrophobic interactions, van der Waals forces, and ionic interactions. R groups are responsible for this level of folding.

- Quaternary structure refers to the three-dimensional interactions of multiple protein subunits.

Protein shape influences function.

- The ability of proteins to bond noncovalently to other molecules makes them biologically important. This property allows them to function as carrier molecules, receptors, and structural elements.

- Binding sites in proteins are of a specific shape, allowing for selective and specific interactions with other molecules. The particular shape and the ability of chemical groups to interact are a result of the protein's three-dimensional shape and the properties of its constituent amino acids.

- Because a protein's three-dimensional structure is stabilized by relatively "weak" bonds, environmental conditions such as temperature, pH changes, or altered salt concentrations can change the shape of the protein into an inactive form. This is called denaturation.

- Chaperonins are a class of proteins that prevent denatured proteins from bonding inappropriately.

Sugar polymers called carbohydrates serve the roles of primary energy source and carbon skeleton in biological systems.

- Monosaccharides are simple sugar monomers that are used in the synthesis of complex carbohydrates. More complex arrangements include disaccharides (2 monosaccharides), oligosaccharides (3–20 monosaccharides), and polysaccharides (large, complex carbohydrates).

- Simple sugars (monomers) are biologically important as an energy source (glucose) and as a structural backbone for RNA and DNA (ribose and deoxyribose).

- Simple sugars typically have three to six carbon atoms and may exist as different isomers.

- Dehydration synthesis between monomers results in glycosidic linkages.

- Polysaccharides are very large chains of monomers that provide energy storage or structural function. The specific structure of polysaccharides contributes to their function. For instance, starch and cellulose are both chains of glucose, but the glycosidic bonds are in different orientations and yield very different chemical properties.

- Added functional groups, such as a carboxyl group, give carbohydrates altered functions. Sugar phosphates and amino sugars, such as chitin, have significant biological activity.

Lipids are chemically diverse and provide many biological functions. They are all water insoluble.

- The insolubility of lipids in water is due to many nonpolar covalent bonds. Lipids aggregate due to their nonpolar nature and van der Waals forces.

- Fats and oils, composed of fatty acids and glycerol, are also known as triglycerides and function primarily in the storage of energy.

- Each triglyceride contains a glycerol bound to three fatty acids with carbon atoms that are single-bonded (saturated fatty acids) or double-bonded (unsaturated fatty acids). The characteristics of the carbon bonds, saturated versus unsaturated, influence the shape of the molecule and determine its point.

- Phospholipids compose cellular membranes. The replacement of a hydrophobic fatty acid with a hydrophilic phosphate group allows phospholipids to have a "water-loving" head and "water-hating" tails. The hydrophobic tails tend to aggregate with the phosphate heads facing an aqueous environment. This bilayer effect allows the establishment of an inside and an outside aqueous environment separated by a hydrophobic membrane.

- Carotenoids and steroids are synthesized from a series of isoprene units. Carotenoids trap light energy during photosynthesis. Steroids serve as signaling molecules and are synthesized from cholesterol.

- Some vitamins are lipids and are not synthesized in the human body. Vitamins A, D, E, and K are all lipid derived.

- The nonpolar nature of lipids makes them an excellent water repellent. Wax coatings on hair, feathers, and leaves are all lipid derived.

The nucleic acids (DNA and RNA) store, transmit, and use information.

- The primary role of nucleic acids is to store and transmit hereditary information.

- Nucleic acids are made up of nucleotide monomers. Each nucleotide monomer consists of a pentose (five-carbon) sugar, a phosphate group, and a nitrogenous base. DNA has one less oxygen atom in the ribose sugar than RNA and is thus called *deoxy*ribonucleic acid. DNA and RNA also differ in one of their nitrogenous bases.

- The backbone of DNA consists of alternating sugars and phosphates with the bases projecting outward. The sugars bond to the phosphates in phosphodiester linkages.

- In both RNA and DNA the nucleotides are joined by phosphodiester linkages. RNA is a single strand of nucleotides, whereas DNA is a double strand. The two strands in DNA are stabilized by hydrogen bonds.

- Only four types of nucleotides are found in DNA: 2 purines (adenine and guanine) and 2 pyrimidines (thymine and cytosine).

- Nucleotide bases are capable of complementary base pairing. In DNA, adenine and thymine pair, and cytosine and guanine pair. In RNA, thymine is replaced by uracil. Complementary base pairing is facilitated by the size and shape of the molecules and by hydrogen bonding. DNA replication and the production of RNA from a DNA template are both made possible by complementary base pairing.

- The uniformity of the DNA molecules makes it suited for carrying information. The sequence of the nitrogenous bases provides the code. The regularity of the molecule allows it to be "read" in any cell.

- Similarities in base sequences of DNA can be used to help determine evolutionary relationships among organisms because it is a conserved and regular molecule.

- Specific RNA, ribozymes, can function as catalysts and may have been important in the origin of life. RNA can also pass along information due to the bases present.

- Nucleotides have additional biological functions. ATP and GTP serve as energy sources. CAMP serves as a messenger in many cellular processes.

All life on Earth comes from existing life.

- The conditions on Earth are no longer suitable for the generation of new life forms. All life comes from another living organism.

The Big Picture

- Macromolecules provide for specific biological activity and provide energy storage, information storage, structural components, catalytic activity, and other functions within the cell. Biologically important macromolecules fall within four basic classes: proteins, carbohydrates, nucleic acids, and lipids. All macromolecules are synthesized from smaller molecules through condensation and can be broken down through hydrolysis.

- Macromolecules all have regular primary structures held by covalent bonds, but many, especially proteins, can be folded into complex shapes held by ionic or hydrogen bonds or attractive forces.

- Chemically active side chains or functional groups contribute to the specific functions of macromolecules.

Study Strategies

- Make a table of macromolecule characteristics. This will help you see the patterns of similarity and the differences.

- To avoid being overwhelmed, remember that each macromolecule is synthesized from smaller components that give the molecules its specific function.

- Avoid the temptation to memorize structures. If you understand condensation and hydrolysis and know the basic components of the macromolecules, memorizing the structures of the large macromolecules is unnecessary.

- Review the following tutorials and activities on the Student Website/CD:
  Tutorial 3.1 Synthesis of Prebiotic Molecules in an Experimental Atmosphere
  Tutorial 3.2 Macromolecules
  Tutorial 3.3 Pasteur's Experiment
  Activity 3.1 Forms of Glucose
  Activity 3.2 Nucleic Acid Building Blocks
  Activity 3.3 DNA Structure

Test Yourself

Knowledge and Synthesis Questions

1. Which of the following statements concerning polymers is *not* true?
   a. Polymers are synthesized from monomers during condensation.
   b. Polymers are synthesized from monomers during dehydration.
   c. Polymers consist of at least two types of monomers.
   d. Both b and c

2. You are a biochemist and have recently discovered a new macromolecule. Studies of the bond types found in this macromolecule reveal many hydrogen bonds and peptide linkages. You most likely have found what type of macromolecule?
   a. Carbohydrate
   b. Lipid
   c. Protein
   d. Nucleic acid

3. An α helix is an example of which level of protein structure?
   a. Primary
   b. Secondary
   c. Tertiary
   d. Quaternary

4. You have isolated a monomer with the following components: a phosphate group, a sugar, and a nitrogen-containing base. Polymers synthesized from this monomer belong to what class of macromolecule?
   a. Carbohydrate
   b. Lipid
   c. Protein
   d. Nucleic acid

5. Cellulose and starch are composed of the same monomers. Which of the following results in their being structurally and functionally different?
   a. They have different types of glycosidic linkages.
   b. They have different numbers of glucose monomers.
   c. They are held together by different bond types.
   d. None of the above

6. DNA utilizes the bases guanine, cytosine, thymine, and adenine. In RNA, _____ is replaced by _____.
   a. adenine; arginine
   b. thymine; uracil
   c. cytosine; uracil
   d. cytosine; arginine

7. The pairing of purines with pyrimidines to create a double-stranded DNA molecule is called
   a. complementary base pairing.
   b. phosphodiester bonding.
   c. antiparallel synthesis.
   d. dehydration.

8. Lipids, as a complex class of macromolecules, are similar in that they are synthesized from
   a. glycerol.
   b. fatty acids.
   c. steroid precursors.
   d. cholesterol.

9. Amino acids are linked together into proteins by which of the following bond types?
   a. Noncovalent bonds
   b. Peptide bonds
   c. Phosphodiester bonds
   d. Both a and b

10. Which of the following characteristics differentiate carbohydrates from other macromolecule types?
    a. Carbohydrates are constructed of monomers that always have a ring structure.
    b. Carbohydrates never contain nitrogen.
    c. Carbohydrates consist of a carbon bonded to hydrogen and a hydroxyl group.
    d. None of the above

11. Which of the following statements about carbohydrates is *not* true.
    a. Monomers of carbohydrates have 6 carbon atoms.
    b. Monomers of carbohydrates are linked together during dehydration.
    c. Carbohydrates are energy storage molecules.
    d. None of the above

12. What would you expect to be true of the R groups of amino acids located on the surface of protein molecules found within the interior of biological membranes?
    a. The R groups would be hydrophobic.
    b. The R groups would be hydrophilic.
    c. The R groups would be polar.
    d. The R groups would be able to form disulfide.

Application Questions

1. Differentiate between primary, secondary, tertiary, and quaternary structures as they relate to proteins. If a protein is immersed in a pH solution far from its optimum, which structures are most likely to disassociate first, and why?

2. Dietary guidelines encourage people to stay away from saturated fats. What is meant by the term "saturated fat"? Why is this fat type of more concern than unsaturated fats in the diet? How are the two fats structurally different?

3. Draw a phospholipid. What characteristics of phospholipids make them perfectly suited for membranes? What do you think might happen if phospholipids did not form a bilayer? How might they arrange themselves in an aqueous environment?

4. Complex carbohydrates should be a mainstay of your diet. What properties make them excellent food sources?

5. Discuss how a protein's three-dimensional structure makes it perfect for acting as a carrier and receptor molecule. Why are proteins uniquely suited for this function, whereas other macromolecules are not?

6. What properties of DNA structure make it well suited for its function as an informational molecule?

7. Amino acid R groups have a variety of chemical properties. How do these different properties contribute to the final three-dimensional shape of the molecule?

8. You have isolated a protein with the following amino acid sequence: RSCFLA. Using Table 3.2 in your book, draw this protein. In your drawing, label the N terminus and the C terminus and show all peptide linkages. How many water molecules were generated in the synthesis of this protein?

9. Explain how the pattern of charges in a macromolecule can affect its final form and function. You may use any of the four macromolecule classes to illustrate your point.

10. Consider the following triglycerides (*A* and *B*) in answering a–d.

A   *B*

a. In *B*, circle the remnant of the glycerol portion of the triglyceride.
b. Which triglyceride (*A* or *B*) is probably a solid at room temperature? _____ Explain your answer.
c. Which triglyceride (*A* or *B*) is probably derived from a plant? _____ Explain your answer.
d. How many water molecules result from the formation of triglyceride *B* from glycerol and three fatty acids?

11. Use the base-pairing rules for DNA and RNA to label the complementary strand of RNA (right) to the single strand of DNA (left) shown following, where C = cytosine, G = guanine, A = adenine, T = thymine, and U = uracil.

Circle and label an example of a nucleotide and a nucleoside in the figure showing the double-stranded DNA and RNA hybrid. Based on the orientation of the sugar molecules, label the four ends of the molecule as 3′ or 5′ in the figure showing the double-stranded DNA and RNA hybrid.

Answers
Knowledge and Synthesis Answers

1. **d.** Polymers are synthesized via condensation and broken down via hydrolysis. Polymers may be made of a single type of monomer or from multiple types of monomers.

2. **c.** Proteins consist of amino acids joined by peptide linkages. Higher levels of structure are stabilized by hydrogen bonds.

3. **b.** α helices are examples of secondary structure and are maintained by hydrogen bonds between the amino acid residues.

4. **d.** These are all constituents of a nucleotide, the monomer that makes up nucleic acids.

5. **a.** Starch has β-glycosidic linkages, whereas cellulose has α-glycosidic linkages. This leads to structural and functional differences between the two macromolecules.

6. **b.** RNA and DNA differ by one oxygen molecule in their ribose sugar and in the substitution of uracil (RNA) for thymine (DNA).

7. **a.** Complementary base pairing results from the attraction of charges and the ability to pair purines with pyrimidines through hydrogen bond formation.

8. **b.** All lipids are synthesized from fatty acid precursors.

9. **b.** The peptide bond that links amino acids to form proteins is a type of covalent bond.

10. **c.** Carbohydrates always have carbons bonded to hydrogens and hydroxyl groups. They may have a variety of other associate molecules in addition to these.

11. **a.** Carbohydrate monomers may have different numbers of carbon atoms. Five-carbon monomers (pentoses) and six-carbon monomers (hexoses) are both common.

12. **a.** The interior of the plasma membrane is hydrophobic; therefore, an imbedded protein would have hydrophobic residues.

Application Answers

1. See Figure 3.6 in your text for diagrams of primary, secondary, tertiary, and quaternary structures. The quaternary structure is the least stable and breaks down first in nonoptimum conditions. Protein structures continue to denature by unfolding the tertiary structures, then secondary structures. Primary structure is maintained by covalent bonds and is the last to break down. Disruptions in tertiary and quaternary structure are often reversible. Disruptions in primary or secondary structure are generally irreversible.

2. Saturated fats contain only single bonds and are "saturated" with hydrogen. This allows saturated fat molecules to pack together densely. This characteristic is easily seen in that most saturated fats are solid at room temperature. Unsaturated fats contain double bonds that affect the shape of the molecule. They are not "saturated" with hydrogens. This characteristic keeps them from packing together tightly, and they tend to be liquid at room temperature. Saturated fats are dangerous because of this "packing" ability.

3. See Figures 3.20 and 3.21 in your book for drawings of phospholipids. The hydrophilic head and the hydrophobic tail of phospholipids allow them to have an "inside" that resists an aqueous environment and an "outside" that can reside in such an environment. When they exist as a bilayer, the hydrophobic tails aggregate together. If they did not exist in two layers the tails would still try to aggregate. This would result in a spherical aggregation of phospholipids called a micelle in which the tails are arranged toward the center of the sphere, away from the aqueous environment, and the heads are immersed in the aqueous environment.

4. Complex carbohydrates are easily broken down into glucose monomers, which provide nearly all cellular energy. By storing glucose monomers in large carbohydrates, the osmotic strain on any given cell is reduced without sacrificing availability of energy.

5. The three-dimensional nature of proteins allows them to form binding sites. These binding sites are uniquely shaped to interact with other molecules.

6. DNA structure is highly regular and conserved, which makes it usable by all cells, but the different bases allow it to encode specific genetic information.

7. The size of the R group, the charge of the R group, and any special binding properties all contribute to the final orientation of a protein molecule.

8.

9. In proteins, for example, charge distribution affects folding, with hydrophobic regions turning inward and hydrophilic regions turning outward. Another excellent example is phospholipids. See the answer to Question 3.

10. a.

b. Triglyceride *A* is probably solid at room temperature; its fatty acid chains are saturated (no double bonds) and relatively long, both characteristics of solid, animal-derived triglycerides.
c. Triglyceride *B* is probably derived from a plant; its fatty acid chains are unsaturated (double bonds) and relatively short, both characteristics of liquid, plant-derived triglycerides.
d. *Three* water molecules will result. A water molecule results for each of the three fatty acids added to glycerol by a condensation reaction.

11.

4 Cells: The Basic Units of Life

Important Concepts

The cell is the building block of life.

- The cell theory states that the cell is the basic unit of life, that all organisms are composed of cells, and that all cells come from preexisting cells.

- The bubble theory for the origin of the cell states that aggregates of molecules known as protobionts may have been the initial precursor to the cell.

- Cell size is limited by surface area-to-volume ratio. Very large cells are not feasible because as volume increases surface area increases at a slower rate. The capacity of a cell for chemical activity is related to the cell volume. However, cells with large volumes are unable to take up enough material across their surface to support the metabolic activity of the volume. Some cells circumvent this situation by means of long cellular extensions (e.g., nerve cells).

- Because cells are small, either light microscopes or electron microscopes must be used to visualize them. Light microscopes, using lenses and light, resolve objects as small as 0.2 μm. Electron microscopes, using electron beams, can resolve objects as small as 0.5 nm. Staining techniques are used to assist in visualizing cellular components.

- All cells are surrounded by a plasma membrane composed of a phospholipid bilayer with many embedded and protruding proteins. These proteins make the membrane selectively permeable, permitting some small molecules to pass through but preventing others from doing so. The membrane allows for maintenance of the internal cellular environment and is responsible for communicating and interacting with other cells.

Cells are classified as either prokaryotic or eukaryotic.

- Prokaryotic cells do not have internal membrane compartments and are generally smaller than eukaryotic cells.

- Prokaryotic cells are found only in domains Archaea and Bacteria. Prokaryotes are one-celled organisms.

- Prokaryotic cells have a cell membrane and a nucleoid containing the genetic material that is not membrane-enclosed. The cytoplasm of the cell consists of the liquid cytosol and insoluble particles that act as subcellular machinery (i.e., ribosomes for protein synthesis).

- Some prokaryotic cells have special features.

  - Cell walls: This rigid structure is outside the cell membrane and is composed of peptidoglycans. Cell wall type is frequently used as a characteristic to identify bacteria.

  - Outer membrane: This membrane is outside the cell wall. It is polysaccharide rich, and some of the polysaccharides are disease-causing toxins.

  - Capsule: This slime layer is outside the wall and outer membrane. Though it provides protection, it is not necessary for the cell's survival.

  - Internal membranes: Photosynthetic prokaryotes have stacks of folded membranes on which photosynthesis is carried out. Other prokaryotes have membrane folds called mesosomes, which function in energy reactions and cell division.

  - Flagella: These tiny protein machines containing the protein flagellin cause motion of the cell.

  - Pili: These are protein extensions that assist in adherence.

  - Cytoskeleton: Actinlike proteins may help the cell maintain cell shape.

- Eukaryotic cells are larger than prokaryotic cells and are distinguished by having membrane-enclosed internal components such as the nucleus, mitochondrion, endoplasmic reticulum, Golgi apparatus, lysosome, vacuole, and chloroplast.

- Each organelle (membrane-enclosed compartment) has its own internal environment uniquely suited to its function. The membrane assists in regulating that environment. The function and study of organelles can be carried out by microscopy or by cell fractionation and organelle isolation.

Some organelles function to store and process information.

- The nucleus stores DNA and is the site of DNA duplication, DNA regulation of cellular activity, ribosome assembly, and RNA synthesis.

- The nucleus is bounded by a double membrane called the nuclear envelope. Small openings in the nuclear envelope called nuclear pores allow passage of RNA and ribosomes to the cell cytoplasm. The outer membrane of the nuclear envelope is continuous with the endoplasmic reticulum.

- Inside the nucleus, DNA molecules and proteins make up chromatin. At cell division, chromatin condenses to form chromosomes. Chromatin is organized on the nuclear matrix and the nuclear lamina. The lamina helps maintain nuclear shape.
- Ribosomes found in the cytoplasm, the endoplasmic reticulum, and the mitochondria act as information transcription centers and guide the synthesis of proteins from nucleic acid blueprints.

The endomembrane system functions in manufacturing and packaging cellular products.

- The endomembrane system consists of endoplasmic reticulum (ER), Golgi apparatus, and vesicles that shuttle between the two.
- The endoplasmic reticulum is a complex of membrane sacs throughout the cell and is continuous with the nuclear membrane.
- The endoplasmic reticulum is classified into two types based on the presence or absence of attached ribosomes.
  - Rough endoplasmic reticulum (RER) is studded with active ribosomes. Because ribosomes actively aid in the synthesis of new proteins, the RER is important for storage, transport, and modification of the proteins.
  - Smooth endoplasmic reticulum also functions in protein modification and transport but, more importantly, functions in the modification of chemicals taken in by the cell and as the site of hydrolysis of glycogen and synthesis of steroids and lipids.
- The Golgi apparatus, composed of flattened membrane stacks called cisternae and vesicles, is important in protein modification, packaging, and polysaccharide synthesis in plants.
- Different regions of the Golgi have different enzymes and functions. The *cis* region is closest to the nucleus or ER. The *trans* region is closest to the cell surface. Proteins are released from the ER in a vesicle and are transported to the *cis* region, where the vesicle membrane fuses with Golgi membrane to release the contents. Vesicles in turn pinch off of the *cis* Golgi for transport between cisternae or away from the Golgi.
- Lysosomes are "digestion centers" within a cell. They contain a host of powerful enzymes in a slightly acidic environment that break down engulfed molecules and cellular waste. In the process of phagocytosis, the cell takes up material for digestion in lysosomes. Through the process of autophagy, lysosomes digest organelles such as mitochondria, breaking them down to monomers for reuse in new organelles.

Mitochondria and plastids function in energy transformation.

- During cellular respiration, mitochondria convert the potential chemical energy stored in glucose into adenosine triphosphate (ATP), a form usable by the cell.
- Mitochondria have two membranes, an outer smooth membrane and a highly folded internal membrane. The folds are called cristae, and the remaining internal space is called the matrix. The matrix holds ribosomes and DNA. Enzymes used during cellular respiration are embedded in the cristae.

- Plastids are not found in animal cells. Plant cells have several different types of plastids, each with unique functions.
- Chloroplasts are the site of photosynthesis where light energy is converted to chemical energy. Chloroplasts are green because they contain the photosynthetic pigment chlorophyll.
- Chloroplasts have two membranes. The innermost membrane contains circular compartments or thylakoids that are folded into stacks called grana and hold chlorophyll and enzymes for photosynthesis. The liquid content of the chloroplast is called stroma and contains ribosomes and DNA.
- Other plastid types include chromoplasts, which contain pigments involved in flower color, and starch-storing leucoplasts.

The endosymbiotic theory explains the origin of mitochondria and chloroplasts.

- Mitochondria and chloroplasts have their own DNA and are enclosed in a double membrane. They also reproduce within the cell. In other words, they act very much like cells within cells, but they are still under nuclear control.
- The endosymbiotic theory suggests that a small photosynthetic prokaryote was engulfed and retained by a larger cell, with each getting mutual benefit from the other (endosymbiosis). Mitochondria may have origins as an engulfed respiring prokaryote. See Figure 4.18.
- There is evidence to support this theory, including DNA moving between organelles, similarities in DNA and biochemistry between modern-day bacteria and chloroplasts, and the potential for the initial formation of the mitochondria and chloroplasts' outer membrane from the engulfing cells' plasma membrane.

Other specialized organelles are found in some but not all cells.

- Peroxisomes function to break down harmful peroxide by-products.
- Vacuoles, found in plants and protists, function in storage of waste products, structural support, reproduction, food store, or in water regulation, depending on the organism.

The cytoskeleton provides support, shape, and movement for the cell.

- Microfilaments, composed of a double helix of the protein actin, stabilize cell shape and assist with contraction of the cell. These filaments are involved in muscle cell contraction, cell division, cytoplasmic streaming, and cell shape. In muscle cells the protein myosin interacts with actin to produce muscle contraction.
- Intermediate filaments are found only in multicellular organisms and function in stabilizing structure and resisting tension. These make up the lamina, form desmosomes, and anchor the nucleus.
- Microtubules are long hollow tubes of a protein called tubulin that contribute to the rigidity of the cell and act as a framework of movement for motor proteins. They have a very specific structure that can be quickly added to or reduced. Motor proteins use microtubules as tacks from one area of the cell to another.

- Cilia and flagella of eukaryotes are powered by microtubules. Though cilia and flagella differ in size and function, they have the same basic structure: a "9 + 2" arrangement of microtubules. See Figure 4.23.

- Movement of cilia and flagella occurs when the microtubules slide past one another. The sliding is caused by an ATP-driven shape change in molecules of dynein while bound to the microtubules. See Figure 4.24. The motor proteins dynein and kinesin are both involved in the transfer of vesicles along microtubules within the cell.

- Centrioles are also made up of microtubules and are involved in cell division.

Extracellular structures are outside the plasma membrane and provide support, protection, and anchoring for the cell.

- Plants have semirigid cell walls composed of cellulose, which function to support and protect the cell. Plasmodesmata are small holes in the cell walls that allow connections between plant cells.

- The collagen- and glycoprotein-containing extracellular matrix of some animal cells functions to hold cells together, filter materials, orient cell movement, and assist with chemical signaling. Some cells, such as bone cells, secrete an elaborate and rigid matrix.

The Big Picture

- All living things are composed of cells, and all cells come from preexisting cells. There are two basic cell forms: the small prokaryotic cell with no organelles and the eukaryotic cell with many membrane-enclosed organelles.

- The organelles of eukaryotic cells have a variety of functions, including information storage and transmission, modification and packaging of cellular products, and energy transformation. See Figure 4.7 of your text to help you understand the organization, structure, and function of each component.

- The cytoskeleton of the cell provides support, structure, and protection. Cytoskeletal components also provide movement of the cell and within the cell.

Study Strategies

- Terminology is the most difficult part of this material. It is easier to learn if you understand the concepts before you memorize the terms.

- Use Figure 4.7 while learning the organelles and their components. Visualizing the jobs each must do and how their functions relate to the structure makes learning easier.

- Review the following tutorials and activities on the Student Website/CD:
  Tutorial 4.1 Eukaryotic Cell Tour
  Tutorial 4.2 The Golgi Apparatus
  Activity 4.1 The Scale of Life
  Activity 4.2 Know Your Techniques
  Activity 4.3 Lysosomal Digestion

Test Yourself
Knowledge and Synthesis Questions

1. You have found a mass of cells in the sediment surrounding a thermal vent in the ocean floor. The salinity in the area is quite high. Upon microscopic examination of the cells, you find no evidence of membrane-enclosed organelles. How would you classify this cell?
   a. As a eukaryotic cell
   b. As a prokaryotic cell
   c. As a member of domain Archaea or Bacteria
   d. Both b and c

2. Centrifugation of a cell results in the rupture of the cell membrane and the contents compacting into a pellet in the bottom of the centrifuge tube. Bathing this pellet with a glucose solution yields metabolic activity, including the production of ATP. One of the contents of this pellet is most likely which of the following?
   a. Cytosol
   b. Mitochondria
   c. Lysosomes
   d. Golgi bodies

3. Eukaryotic cells are thought to be derived from prokaryotic cells that underwent phagocytosis without digestion of the phagocytized cell. This mutualistic relationship is explained by the
   a. endosymbiotic theory.
   b. cell theory.
   c. evolutionary theory.
   d. parasite theory.

4. Though science fiction has produced stories like "The Blob," we don't see very many large single-celled organisms. Which of the following tends to limit cell size?
   a. The ability to maintain a continuous large membrane
   b. The ability to reproduce a large cell
   c. Surface area-to-volume ratios
   d. All of the above

5. Microscopes are used to resolve images that cannot be seen with the unaided eye. Electron microscopes use _____ to resolve images, whereas light microscopes use _____ to resolve images.
   a. light and lenses; diffraction of electron beams
   b. diffraction of electron beams; light and lenses
   c. lasers; light and lenses
   d. None of the above

6. Match the following cellular functions to the organelle in which they occur.

____ Lysosome	a. DNA synthesis
____ Nucleus	b. Protein folding
____ RER	c. Packaging of materials
____ SER	for export
____ Golgi apparatus	d. Protein synthesis
____ Chromoplast	e. Breakdown of
____ Mitochondrion	phagocytized material
____ Chloroplast	f. Photosynthesis
	g. Cellular respiration
	h. Pigment production

7. Which of the following organelles are enclosed in double membranes?
   a. Nucleus
   b. Chloroplast
   c. Mitochondrion
   d. All of the above

8. Movement of cells is accomplished in both prokaryotes and eukaryotes with which of the following structures?
   a. Cilia
   b. Pili
   c. Dynein
   d. Flagella

9. Which of the following statements is true regarding mitochondria and chloroplasts?
   a. Animal cells produce chloroplasts.
   b. Mitochondria and chloroplasts may be found in the same cell.
   c. Mitochondria and chloroplasts are not found in the same cell.
   d. Chloroplasts can revert to mitochondria in certain conditions.

10. Which of the following best describes ribosomes?
    a. Ribosomes guide protein synthesis.
    b. Ribosomes are found only in the nucleus or on the RER.
    c. There are no ribosomes in the mitochondria.
    d. All of the above

11. Nuclear DNA exists as a complex of proteins called _____ that condenses into _____ during cellular division.
    a. chromosomes; chromatin
    b. chromatids; chromosomes
    c. chromophors; chromatin
    d. chromatin; chromosomes

12. Rough endoplasmic reticulum and smooth endoplasmic reticulum differ
    a. only by the presence or absence of ribosomes.
    b. both in the presence or absence of ribosomes and in their function.
    c. only in microscopic appearance.
    d. None of the above

Application Questions

1. The role of a certain cell in an organism is to secrete a protein. Trace the production of that protein from the nucleus through all necessary organelles to the point of release from the cell.

2. Explain how microtubules and dynein function to make cilia and flagella move.

3. Compare and contrast prokaryotic and eukaryotic cells.

4. Explain the significance of organelles. What are the costs and benefits of having large compartmentalized cells?

5. What is the primary function of a cell membrane? What characteristics of membranes allow them to contribute to metabolic activity?

6. The organelles that contain their own DNA are all enclosed in double membranes. Relate this observation to the endosymbiotic theory.

7. Vacuoles in plants hold waste material. Why do plants tend to retain cellular wastes whereas animal cells do not?

8. Compare and contrast the prokaryotic cell wall with the plant cell wall.

9. There are structural similarities between mitochondria and chloroplasts. Using the idea that form follows function, why are there similarities in these two organelles?

10. The extent of the presence of a cytoskeleton has become known relatively recently. Taking the advances in microscopy as a reference point, discuss how it is possible to better understand the cytoskeleton.

Answers
Knowledge and Synthesis Answers

1. **b.** There are several characteristics that suggest this is a prokaryote. It survives in high salinity and high heat, but the sure indication is that it contains no membrane-enclosed organelles.

2. **b.** The pellet is undergoing cellular respiration, a function that occurs in the mitochondria. You can also assume that if the single membrane of the cell itself is ruptured, other organelles enclosed in single membranes would be ruptured as well.

3. **a.** The endosymbiotic theory is used to explain the presence of DNA in mitochondria and chloroplasts as well as the presence of two membranes around these organelles.

4. **c.** As volume increases, the surface area available for exchange does not increase proportionally. There comes a point at which the surface is not large enough for maintenance of the metabolic activity of the cell.

5. **b.** In electron microscopy, a concentrated beam of electrons is focused on an object, allowing resolution of structures as small as 5 nm. Light microscopy, using lights and lenses, can only resolve objects down to approximately 0.2 μm.

6. **e.** Lysosome
   a. Nucleus
   d. RER
   b. SER
   c. Golgi apparatus
   h. Chromoplast
   g. Mitochondrion
   f. Chloroplast

7. **d.** The nucleus, the mitochondria, and the chloroplasts are the only organelles enclosed in double membranes.

8. **d.** Though the flagella are of different structure, they serve the same role in prokaryotes and eukaryotes.

9. **b.** Mitochondria and chloroplasts may be found in the same cell. Almost all eukaryotic cells contain mitochondria.

10. **a.** Ribosomes, found in the nucleus, the cytosol, and in organelles such as the mitochondria, RER, and chloroplasts, complex with RNA to guide protein transcription.

11. **d.** The complex of proteins and DNA is called chromatin. Chromatin only takes the form of chromosomes during cell reproduction.

12. **b.** Both the structure and the function of RER and SER differ.

Application Answers

1. The protein could be synthesized on ribosomes in the nucleus, or synthesis could take place in the cytosol or on the RER. From there it most likely would be transported along the RER for further modification before being packaged in a bit of ER membrane and transported to the *cis* region of the Golgi. Further modification might take place in the cisternae of the Golgi before the protein is encased in a final vesicle. When the vesicle fuses with the plasma membrane of the cell, its contents are released to the outside.

2. Dynein molecules bind to pairs of microtubules in the flagella or cilia. With the addition of cellular energy, the dynein molecules undergo a conformational change that results in microtubules sliding past one another, resulting in a whiplike motion of the flagella.

3. Prokaryotic Cells
   - Small in size
   - No membrane-enclosed organelles
   - Found only in domains Archaea and Bacteria
   - Survive and metabolize in chemical environments with no oxygen
   - DNA is in nucleoid

   Eukaryotic Cells
   - 10 or more times greater in size
   - Membrane-enclosed organelles
   - Found in all domains other than Archaea and Bacteria
   - Metabolize mainly in oxygen-rich environments
   - DNA is in nucleus

4. Organelles allow different metabolic environments to exist in the same cell. This partitioning of jobs allows for greater specialization but comes at an energy cost. Eukaryotic cells are more energy expensive.

5. A cell membrane exists to form an inside and an outside of a cell. The presence of an inside and an outside allows the establishment of different environments. In addition, membranes hold integral proteins with a variety of chemical properties and activities. This allows for the enzymatic activity associated with membranes. Stacks of membranes, such as those in mitochondria and chloroplasts, increase the amount of chemical activity in an area.

6. See Figure 4.18 for a description of the origin of double membranes from endosymbiosis.

7. The presence of cell walls in plants limits what can be expelled outside the plant cell. Holding wastes in an internal compartment circumvents this problem.

8. Prokaryotic Cell Wall
   - Functions in protection and support
   - Composed of peptidoglycans
   - Located outside the cell membrane
   - Influences cell shape

   Plant Cell Wall
   - Functions in protection and support
   - Composed of cellulose
   - Located outside the cell membrane
   - Influences cell shape

9. Both mitochondria and chloroplasts are involved in energy-transformation activities that require many enzymes. The stacking or folding of membranes provides enzymatic activity centers for these reactions.

10. To visualize the components of the cytoskeleton, electron and transmission microscopy is necessary. With increasing powers of resolution, it is possible to see more and more cellular components and to understand them.

5 *Cellular Membranes*

Important Concepts

Biological membranes have a specific content and structure.

- Biological membranes are composed of lipids, proteins, and carbohydrates. Lipids provide the structure and barrier functions for the membrane. Proteins are involved in creating channels and transporting materials across the lipid barrier. Carbohydrates are involved in signaling and adhesion and are found on the outside of the plasma membrane. The fluid mosaic model describes how lipids interact to produce a fluid membrane on which the proteins and carbohydrates float.

- The majority of a biological membrane is composed of lipids, with phospholipids being the most abundant component. Phospholipids have both hydrophilic ("water-loving") heads and hydrophobic ("water-hating") tails. They arrange themselves into a bilayer with the hydrophobic tails touching and the heads extending into the aqueous environment inside and outside the cell. This arrangement allows for the fluid movement of the two layers on top of one another and sealing of any disruptions in the membrane.

- Though the basic structure of a bilayer is always the same, the inner and outer halves differ in lipid composition and thus have slightly different properties. Phospholipids differ in their length, degree of unsaturation, and degree of polarity. Additionally, cholesterol can make up 25 percent of the lipid content in the membrane. The amount of cholesterol present and the degree of fatty acid saturation influences the fluidity of the membrane.

- Proteins may be embedded in or extend across membranes. Regions or domains of a membrane protein with hydrophobic amino acid side chains tend to be found in the hydrophobic environment of the membrane. The regions of a membrane protein containing hydrophilic amino acid side chains extend out from the membrane. The phospholipids and proteins are independent of one another, allowing for movement throughout the membrane. This happens for the most part because the constituents interact only noncovalently.

- Proteins that penetrate the phospholipid bilayer are called integral proteins. Special types of integral proteins, transmembrane proteins, span the entire bilayer. Those not embedded are referred to as peripheral proteins. Proteins are distributed in membranes asymmetrically "as needed," and the numbers of proteins and their placement varies greatly based on cell type. The "inside" and "outside" of a membrane often have different properties due to the different characteristics of the transmembrane proteins on the two sides.

- Some membrane proteins may be anchored to cytoskeletal components or bound to lipid rafts. Such proteins have very specific functions.

- Membrane carbohydrates serve as recognition sites for other cells and molecules. Carbohydrates are frequently bound to lipids or proteins, forming glycolipids or glycoproteins respectively. Because of the nature of carbohydrates, they contribute to cell adhesion.

Cells must be able to recognize and adhere to other cells to form tissues, organs, and organisms.

- Cells have the ability to recognize and adhere to other cells allowing for the grouping of cells in an organism. Cell adhesions are specific to the organism and to the cell type. A sponge is an animal composed of many cells that exhibits species-specific cell adhesion.

- A cell's ability to recognize and adhere to other cells is due to the presence of recognition proteins on the cell surface. Often the proteins on two cells that adhere to one another are the same; these molecules and the bonds they form are referred to as homotypic. Occasionally the two proteins are different, but have complementary binding sites. In this case the molecules and the bonding are called heterotypic.

Cells must make connections to other cells in order to communicate and stabilize tissues.

- Cell junctions are sites where cell recognition proteins from two cells form a "cement" junction. The three main types of cell junctions are tight junctions, desmosomes, and gap junctions.

- Tight junctions are found specifically in epithelial cells and function to prevent substances from moving through the intercellular space. They also restrict migration of membrane proteins and phospholipids that would otherwise be free to float around in the "fluid" plasma membrane. Tight junctions help tissues move materials in specific directions.

- Desmosomes are structural connections between cells that "reinforce" a tissue. Because a desmosome involves the keratin fibers of the cytoskeleton and crosses the cell membrane, the connections have integrity and are not easily broken.

- Gap junctions facilitate communication between cells. They are made of protein connexons that span adjacent plasma membranes and act as channels.

Membranes are selectively permeable and allow some substances, but not others, to pass through.

- Passive transport is the movement of a substance across a lipid bilayer that does not require an input of energy, either through channel proteins or by carrier molecules. The energy for passive transport is derived from the unequal distribution of the substance on either side of the membrane.

- Diffusion is a process of random movement toward equilibrium and is the net movement of a substance from areas of greater concentration to lesser concentration. The rate of diffusion depends on the size of diffusing substance, electrical charge of diffusing substance, temperature of solution, and concentration gradient.

- Diffusion within small areas such as single cells may occur rapidly, but it occurs more slowly with increasing distance.

- Membranes alter the rate of diffusion. A membrane is said to be permeable to those substances that can pass through it and impermeable to those that cannot. If a substance can pass through a membrane, it will diffuse until concentrations on either side of the membrane are equal. At equilibrium, the molecules of the substance continue to move across the membrane, but the net movement of molecules in both directions is equal. If a substance cannot permeate the membrane, it cannot diffuse across it.

- In simple diffusion, small molecules pass through a membrane. The more lipid soluble the molecule is, the faster it diffuses. Lipid-soluble molecules readily pass through biological membranes. In contrast, charged and polar molecules do not readily pass through a membrane due to the formation of many hydrogen bonds with water and the hydrophobic nature of the internal layer of the membrane. The exception to this rule is water.

- Osmosis is the diffusion of water across a membrane. Water will move across a membrane to equalize solute concentrations on either side of the membrane if the solute cannot move. Water moves from areas of high solute concentration to low solute concentration in an attempt to equalize solute concentrations on both sides of the membrane.

- Solute concentrations separated by a membrane are classified as isotonic, hypertonic, or hypotonic. Isotonic solutions have the same solute concentrations on both sides of a membrane. A hypertonic solution has a higher solute concentration than the solution on the other side of the membrane. A hypotonic solution has a lower solute concentration than the solution with which it is being compared. Environmental solute concentrations dictate the direction of osmosis in living cells.

- Turgor pressure is the pressure within cells and is due to the uptake of water by osmosis. Turgor pressure can build up only if there is a cell wall to limit cell expansion.

- Facilitated diffusion is the passive movement of a substance across a membrane with the help of membrane-bound proteins that act as channels or carriers.

- A substance may cross the membrane by facilitated diffusion through protein channels running through the plasma membrane. The interiors of these proteins have polar amino acids that allow polar molecules and ions to cross the membrane. Most ion channels are "gated." The gating of a channel occurs when the channel protein changes shape in response to some signal and either opens or closes off the channel. Specific channels called aquaporins allow water to cross the membrane. Facilitated diffusion through protein channels is still governed by factors affecting rates of diffusion.

- Facilitated diffusion may be aided by carrier proteins that bind a substance, and transport is across the membrane. Diffusion in this case is not only limited by concentration gradient, but also by the number of available membrane carrier proteins. When all the carrier proteins are bound to the substance, they are saturated, limiting the rate of diffusion.

Active transport is directional, works against a concentration gradient, and requires an input of energy.

- Three types of protein systems are involved in active transport. Uniport transporters move a single solute across a membrane in one direction. Symport transporters move two solutes in the same direction across a membrane. Antiport transporters move two solutes in opposite directions.

- Primary active transport requires the direct participation of ATP to move ions against their concentration gradient. The sodium–potassium pump is an important antiporter that uses ATP to transport two K^+ ions and three Na^+ ions.

- Secondary active transport uses ATP indirectly by coupling solute transport with the ion concentration gradient established by primary transport.

Endocytosis and exocytosis move large molecules into and out of cells.

- Endocytosis is the process by which a cell brings large substances into the cell. This is accomplished by the cell membrane, which folds around the substance to form a vesicle. See Figure 5.15.

- Large substances and even entire cells are engulfed in the process of phagocytosis. Once inside the cell, the vesicle fuses with a lysosome for digestion. The cell takes up liquids in small vesicles from the outside in the process of pinocytosis.

- Animal cells use receptor-mediated endocytosis to capture specific macromolecules from the environment. Receptors for specific macromolecules cluster together on the cell surface in coated pits containing the protein clathrin. Upon binding of the specific molecule to the receptors, the coated pit invaginates to form a vesicle. The resulting vesicle becomes clathrin coated until it is well inside the cell, where it loses its coat and fuses with a lysosome. Animal cells take up cholesterol by the process of receptor-mediated endocytosis.

- Exocytosis moves materials out of the cell. Binding proteins found on the surface of the cell-produced vesicles bind with receptor proteins of the cytoplasmic side of the cell membrane. The vesicle membrane fuses with the cell membrane, and the contents of the vesicle are released outside the cell membrane.

- Membranes have multiple biological functions.
- Though the primary membrane functions are to act as barriers and binding sites, membranes also have roles in information processing, energy transformation, and the organization of chemical reactions.
- In order to carry out these functions, membranes are necessarily dynamic. They are continually being formed, modified, and degraded.

The Big Picture

- Cellular membranes are a dynamic composition of a phospholipid bilayer, integral and peripheral proteins, and carbohydrates. The nature of the constituent phospholipids allows the formation of a barrier that is semi-permeable. Small hydrophilic molecules can traverse the membrane via simple diffusion, water may cross the membrane by osmosis, small charged ions may pass through via protein channels, and some other molecules may be shepherded through via carrier molecules. Transport of other molecules must be active and requires energy input, either directly from ATP or as coupled to ATP-driven transport. Larger substances depend on endocytosis and exocytosis for transport into and out of the cell.
- In multicellular organisms, adhesion and cell-to-cell connections are important for structure and communication. Tight junctions allow multiple cells to form a sheet that directs movement of substances in a single direction. Desmosomes enhance structural support of sheets of cells that are prone to abrasion. Gap junctions allow communication by connecting adjacent cells.

Common Problem Areas

- The biggest challenge when studying membranes is understanding diffusion and osmosis. It is very easy to get the terminology confused, especially the terms "hypertonic" and "hypotonic" when discussing osmosis. Looking at the Latin roots of the words is helpful. "Hyper-" generally means excess, and "hypo-" generally means "less than." "Tonic" refers to solute concentration. Therefore, "hypertonic" means excess solutes and "hypotonic" means fewer solutes.
- Secondary active transport also tends to be confusing. Remember that secondary active transport does not use ATP directly, but is tightly coupled to ion transport that does require ATP.

Study Strategies

- When studying the fluid mosaic model, think about the properties of the constituent molecules. This will make understanding the membrane's structure much easier. Draw a cartoon of the plasma membrane and all the potential components.
- Draw diagrams of the movement of water and solutes when working with diffusion and osmosis. Diagrams help you to visualize what is happening across a membrane.
- Review the following tutorials and activities on the Student Website/CD:
Tutorial 5.1 Passive Transport
Tutorial 5.2 Active Transport
Tutorial 5.3 Receptor-Mediated Endocytosis

Activity 5.1 The Fluid-Mosaic Model
Activity 5.2 Animal Cell Junctions

Test Yourself
Knowledge and Synthesis Questions

1. Which of the following statements regarding cellular membranes is *not* true?
   a. The hydrophobic nature of the phospholipid tails limits the migration of polar molecules across the membrane.
   b. Integral proteins and phospholipids move rapidly and fluidly throughout the membrane.
   c. Membrane phospholipids flip back and forth from one side of the bilayer to the other.
   d. Glycolipids and glycoproteins serve as recognition sites on the cell membrane.

2. Which of the following contributes to differences in the two sides of the cell membrane?
   a. Differences in peripheral proteins
   b. Different domains expressed on the ends of integral proteins
   c. Differences in phospholipid types
   d. All of the above

3. Which of the following cell membrane components serve as recognition signals for interactions between cells?
   a. Cholesterol
   b. Glycolipids or glycoproteins
   c. Phospholipids
   d. All of the above

4. Which of the following types of junctions are responsible for communication between cells?
   a. Tight junctions
   b. Desmosomes
   c. Gap junctions
   d. None of the above

5. You are monitoring the diffusion of a molecule across a membrane. Which of the following will result in the fastest rate of diffusion?
   a. An internal concentration of 5 percent and an external concentration of 60 percent
   b. An internal concentration of 60 percent and an external concentration of 5 percent
   c. An internal concentration of 35 percent and an external concentration of 40 percent
   d. Both a and b

6. If a red blood cell with an internal salt concentration of about 0.85 percent is placed in a saline solution (salt solution) that is 4 percent, which of the following will most likely happen?
   a. The red blood cell will loose water and shrivel.
   b. The red blood cell will gain water and burst.
   c. The turgor pressure in the cell will greatly increase.
   d. The cell will remain the same.

7. In which of the following is solution X hypotonic relative to solution Y?
   a. Solution X has a greater solute concentration than solution Y.
   b. Solution X has a lower solute concentration than solution Y.
   c. Solution X and solution Y have the same solute concentration.
   d. None of the above

8. Which of the following statements regarding osmosis is *not* true?
   a. Osmosis refers to the movement of water along a concentration gradient.
   b. In osmosis, water moves to equalize solute concentrations on either side of the membrane.
   c. If osmosis occurs across a membrane, then diffusion is not occurring.
   d. The movement of water across a membrane can affect the turgor pressure of some cells.

9. Channel proteins allow ions that would not normally pass through the cell membrane to pass through via the channel. What properties of the proteins are responsible for this?
   a. The channels are often composed of charged or polar amino acid groups.
   b. The channels are often composed of hydrophobic amino acid groups.
   c. Both a and b
   d. None of the above

10. Which of the following limits the movement of molecules when carrier-mediated facilitated diffusion is involved?
    a. Concentration gradient
    b. Availability of carrier molecules
    c. Temperature
    d. All of the above

11. Active transport differs from passive transport in that active transport
    a. requires energy.
    b. always requires direct input of ATP.
    c. moves molecules against a concentration gradient.
    d. Both a and c

12. Single-celled animals like amoebas engulf entire cells for food. Which of the following represents the manner in which amoebas "eat"?
    a. The amoeba binds only what it has receptors for, the amoeba's cell membrane surrounds the cell to be digested, a vesicle forms, and the vesicle fuses with a lysosome for digestion.
    b. The amoeba's cell membrane surrounds the cell to be digested, a vesicle forms, and the vesicle fuses with a lysosome for digestion.
    c. The cell is taken into the amoeba's vacuole, a vesicle is formed, and the vesicle fuses with a lysosome for digestion.
    d. None of the above

13. Sodium–potassium pumps are common in many cells. Which of the following are necessary for the pumps to work?
    a. ATP-driven pumping proteins
    b. A signal to activate the pumps
    c. A concentration gradient to work against
    d. All of the above

14. Bacterial cells are often found in very hypotonic environments. Which of the following characteristics keeps them from continuing to take on water from their environment?
    a. The presence of a cell wall allows a buildup of turgor pressure that prevents any more water from entering the cell.
    b. The presence of a cell wall allows a buildup of tonic pressure that prevents any more water from entering the cell.

c. The cell expels water as fast as it takes it up.
d. None of the above

15. Which of the following may affect the rate of diffusion?
    a. Temperature
    b. Molecule size
    c. Concentration gradient
    d. All of the above

Application Questions

1. Draw a diagram of a cell membrane and label the phospholipids, integral proteins, peripheral proteins, and carbohydrates. Using your diagram, describe the fluid mosaic model.

2. One characteristic of cell membranes is their ability to "self-seal." Describe what properties of cell membranes prevent them from rupturing easily.

3. When scientists want to rupture cells to study the contents, they use a detergent and physical force. Why are both force and a detergent necessary?

4. Proteins for export are frequently modified in the endoplasmic reticulum, passed to the Golgi apparatus, and then exported from the cell. Explain the role of membranes in a protein's journey from the ER to the extracellular environment.

5. A marathon runner has just come into your emergency room with severe dehydration. You must decide what type of solution to pump into his veins so that he survives. Your choices are: pure water, 0.9 percent saline, and 1.5 percent saline. You cannot remember which solution is correct, so you treat blood samples that you have placed on microscope slides with each solution and observe what happens. (Hint: Blood cells are approximately 0.9 percent saline.) Describe what you think will happen to the blood cells when exposed to each solution. Which solution you would choose to rehydrate the runner?

6. Compare and contrast active and passive transport.

7. You are interested in the epithelial tissue that lines the intestine. This single sheet of cells must absorb nutrients but prevent many other substances from crossing it. Describe which cell-to-cell connections exist to allow movement of substances into this tissue, resistance to abrasion, and communication between cells.

8. Barrier formation is only one function of the cell membrane. Describe some other functions of the membrane and discuss how the membrane is suited for those functions.

Answers

Knowledge and Synthesis Answers

1. **c.** Because of the hydrophobic tails and hydrophilic head of the phospholipids, it is impossible for them to flip back and forth from one side of the membrane bilayer to the other.

2. **d.** The cell membrane is asymmetric and has different properties and functions on the cytoplasmic side versus the extracellular side. These properties arise from differences in the constituents of the membrane.

3. **b.** Both glycolipids and glycoproteins serve as recognition signals.

4. **c.** Gap junctions are involved in chemical and electrical signaling between cells.

5. **d.** Diffusion may take place in either direction across a membrane and always follows a concentration gradient. The larger the gradient, the faster diffusion will occur.

6. **a.** The cell will lose water as solute concentrations on both sides of the membrane equalize.

7. **b.** Hypotonic solutions have lower solute concentrations than the solution they are being compared to.

8. **c.** Diffusion and osmosis are not mutually exclusive and may take place at the same time.

9. **a.** The charged or polar lining of the channel proteins allows passage of polar and charged molecules.

10 **d.** Anything that affects the rate of diffusion will affect carrier-mediated facilitated diffusion. In addition to those effects, carrier-mediated facilitated diffusion also relies on the availability of carrier molecules.

11. **d.** Active transport works against a concentration gradient and requires energy to do so. That energy does not always have to be directly supplied in the form of ATP.

12. **b.** See Figure 5.15.

13. **d.** Sodium–potassium pumps are forms of primary active transport.

14. **a.** Turgor pressure limits osmosis, and once a cell is turgid, no more water may be taken on.

15. **d.** Temperature, molecule size, molecule charge, and concentration gradients all affect the rate at which diffusion takes place.

Application Answers

1. Refer to Figure 5.1 in your text.

2. The nonpolar regions of the fatty acids tend to aggregate together to the exclusion of water. This property leads to "sealing" of ruptures in the membrane.

3. Detergents disrupt cell membranes because of their interactions with the fatty acids in the membranes. Physical forces allow the detergents to reach the phospholipid tails and break up their aggregations.

4. The protein leaves the ER encased in a membrane vesicle. Signal molecules on the surface of the membrane bind to receptor molecules on the membrane of the Golgi. The membranes fuse, and the protein contents of the vesicle enter the Golgi for modification. Upon leaving the Golgi, the protein is encased in membrane, which again, with the appropriate signals and receptors, binds to the cytoplasmic surface of the cell membrane. The vesicle membrane and the cell membrane fuse, and the contents of the vesicle are released to the extracellular side of the membrane.

5. In pure water, the blood cells will take on water through osmosis, swell, and eventually rupture. In 0.9 percent saline, the cells should neither gain nor lose a significant amount of water and should remain the same. In a 1.5 percent saline solution, the cells should lose water and shrivel. You might jump to the conclusion that for quick rehydration the patient needs a hypotonic solution. But remember that you will be adding this substance directly to the bloodstream—the patient would end up with ruptured cells. You want to use a solution isotonic to the patient's blood cells, 0.9 percent, to avoid rupture of cells.

6. The main difference between active and passive transport is that active transport goes against a concentration gradient and requires energy, whereas passive transport diffuses passively and does not require energy.

7. Tight junctions allow multiple cells to form a sheet that directs movement of substances in a single direction. Desmosomes enhance structural support of sheets of cells that are prone to abrasions. Gap junctions allow communication by connecting adjacent cells.

8. Membranes function in processing energy transformation and in the organization of chemical reactions. Integral and peripheral proteins contribute to these functions. The membrane serves as a holding site for the catalytic enzymes associated with these processes.

6 Energy, Enzymes, and Metabolism

Important Concepts

Energy is the capacity to do work.

- No living cell manufactures energy. The energy needed for life must be obtained from an environmental energy source and transformed to a usable form.

- Energy transformations are linked to chemical transformations in cells.

- Kinetic energy is energy in motion and does work. Potential energy is stored energy. Energy can be stored biologically in chemical bonds of fatty acids and other molecules. Breaking the bonds converts the energy to kinetic energy.

- Metabolic processes either store or release energy in chemical bonds.

  - Anabolic reactions link simple molecules to create complex molecules and store energy in the resulting bonds.

  - Catabolic reactions break down complex molecules and release stored energy.

- First law of thermodynamics: Energy is neither created nor destroyed. Energy can be converted from one form to another.

- Second law of thermodynamics: Not all energy can be used. When energy is converted from one form to another, some becomes unusable.

  - Total energy (enthalpy, H) = usable energy (free energy, G) + unusable energy (entropy, S) × absolute temperature (T).

- Because we cannot measure absolute energy, we can only measure change (Δ) in energy. We determine the change in usable energy as follows: $\Delta G = \Delta H - T\Delta S$

  - If ΔG is negative, free energy is released, and the reaction is exergonic.

  - If ΔG is positive, free energy is required, and the reaction is endergonic.

- With each conversion, entropy (S) tends to increase; therefore, processes tend toward disorder and randomness.

- All biochemical reactions must either release ($-\Delta G$) or take up ($+\Delta G$) usable energy.

- Reactions that are exergonic in one direction are endergonic in the other. The direction of the reaction depends on concentrations of reactants versus products. The point at which the forward reaction occurs at the same rate as the reverse reaction is termed chemical equilibrium.

- The further toward completion a reaction's equilibrium lies, the more free energy is given off (larger $-\Delta G$).

- A larger $+\Delta G$ means equilibrium favors the reverse reaction (equilibrium falls toward the reactant). A ΔG near zero indicates that both products and reactants have free energy.

ATP is the primary energy currency of living cells.

- ATP (adenosine triphosphate) is needed to capture, transfer, and store free energy (usable energy) in cells to do work. It can also be converted into one of the building blocks of DNA and RNA.

- Hydrolysis of ATP to ADP and P_i releases free energy ($\Delta G = -12$ kcal/mol). The equilibrium falls toward ADP production.

- The formation of ATP is endergonic, requiring as much energy as is released by ATP hydrolysis. This results in an "energy-coupling cycle" of the endergonic and exergonic reactions.

- ATP is formed when coupled with the exergonic reactions of cellular respiration. ATP is hydrolyzed to fuel endergonic reactions like protein synthesis. See Figure 6.6.

Enzymes catalyze biochemical reactions by lowering activation energy.

- Catalysts speed up reactions without being consumed in the process. Most biological catalysts are protein enzymes.

- Energy barriers exist between reactants and products. Energy must be added to get past this barrier. You may think of it as a little shove to get the reaction going. This is called activation energy (E_a). See Figures 6.8 and 6.9.

- For reactions to take place, the reactants must be slightly destabilized and turned into "transition-state species" with higher free energy than that of either the reactants or the products. This is accomplished with the help of a catalyst.

- Enzymes are highly specific biological catalysts that act on specific reactants or substrates. They have specific substrate-binding areas on their surfaces called active sites. The enzyme's specificity comes from the shape of its active site. Only specific substrates can fit in and bind with an enzyme's active site. Once bound, the site is referred to as an enzyme–substrate complex (ES).

- The enzyme alters the conformation of the substrate, thus lowering the activation energy for the reaction. See Figure 6.11. From this the product is formed, and the enzyme remains unchanged. Such reactions can be represented as follows:

$$E + S \rightarrow ES \rightarrow E + P$$

- Enzymes do not alter the equilibrium of a reaction. Both the forward and reverse reactions are sped up.

- Enzymes speed up reactions by orienting substrates for maximum chemical interactions, adding charges to substrates, or by inducing strain on the substrate.

The structure of an enzyme determines its function.

- Enzymes are very large macromolecules with small active sites for binding small metabolites.

- Binding of small molecules to the active sites depends on H-bonds, charge interactions, and van der Waals forces.

- Binding of substrate by an enzyme can change the enzyme's shape. The enzyme's shape can be altered by the substrate to produce an "induced fit." The large size of an enzyme helps it position the correct amino acids at the active site and helps regulate shape and allow for the induced fit.

- Cofactors, coenzymes, and prosthetic groups all contribute to an enzyme's activity.

- Cofactors are inorganic ions.

- Coenzymes are organic molecules that bind in the active site and are considered a substrate. Coenzymes do not change form during the chemical reaction.

- Prosthetic groups are bound to the protein itself.

- All three have roles in altering the shape of an enzyme for maximum activity.

- The rate of a reaction increases as the substrate concentration increases until all enzyme active sites are occupied. At that point, no amount of additional substrate will increase the reaction rate because no more enzyme is available for catalysis. Enzyme efficiency is measured in turnover number, or how fast an enzyme can convert substrate to product and free up its active site.

Metabolism is a series of enzyme-catalyzed reactions organized into pathways in which the products of one reaction become the reactants for the next in the series.

- All life must maintain stable internal conditions (homeostasis). The regulation of enzymes helps living organisms maintain homeostasis.

- Metabolism is the sum of all biological reactions that occur in a living organism. Enzymes are regulators of metabolism and are subject to regulation as well. Inhibitors bind to enzymes and slow the rate of reaction.

  - Irreversible inhibition: Inhibitors covalently bond to an enzyme and destroy the active site.

  - Reversible inhibition: Inhibitors bind to and alter the active site but the inhibition is reversible, depending on the concentration of the inhibitor and substrate. Some reversible inhibitors are called competitive inhibitors

because they compete with the substrate for the active site. Others are called noncompetitive inhibitors because they bind elsewhere but alter the active site so that the substrate cannot bind or the rate of binding is reduced.

Allosteric enzymes can be regulated by other molecules.

- Allostery is the regulation of an enzyme by a molecule that binds to the enzyme at a site other than the active site. Noncompetitive inhibitors are an example of inhibitors that work allosterically.

- Most allosteric enzymes exist naturally in an active form and an inactive form, and they can switch back and forth between the two forms. When an allosteric regulator binds to the enzyme, the enzyme becomes locked in either its active form or inactive form.

  - A positive regulator locks an enzyme into its active form.

  - A negative regulator locks an enzyme into its inactive form.

- Allosteric binding sites, like the sites that bind substrates, are highly specific.

- Most allosteric enzymes have multiple "types" of subunits with different jobs.

  - Catalytic subunits bind substrate in the active site.

  - Regulatory subunits bind effectors in one or more allosteric sites.

- Allosteric interactions help control metabolism. The first step in an enzyme-mediated pathway is referred to as the commitment step. Once initiated, the pathway is followed to completion. Frequently, the final product allosterically inhibits the enzyme of the commitment step. This prevents overproduction of the final product. This process is called end-product or feedback inhibition.

Enzyme function is influenced by the cellular environment

- Changes in pH influence charges of amino and other groups on the enzyme or substrate. This may change the folding of the enzyme or its ability to interact with the substrate and drastically alter its catalytic ability. Enzymes typically have an ideal pH at which they function best.

- Increases in temperature may aid in reduction of activation energy by adding kinetic energy; however, this may also result in denaturing of enzymes. Each enzyme tends to have an optimal temperature. Organisms may produce different forms of the same enzyme, called isozymes, which have different optimal temperatures.

The Big Picture

- Cells need energy to carry out their functions. Metabolism is a series of energy-transferring reactions that fuel the processes of the cell. All involved reactions either give off energy (exergonic) or require energy to proceed (endergonic). Energy may be stored in chemical bonds (potential energy) and released when the bonds are broken to do work (kinetic energy). ATP serves as the energy shuttle of many metabolic processes.

- Enzymes aid biological reactions by lowering the activation energy required to start the reaction. Each enzyme has a specific three-dimensional conformation that interacts specifically with the substrate. Interaction between the enzyme and substrate results in optimum orientation for a reaction to take place. Enzymes are proteins and are affected by temperature, pH, concentrations, or reactants and inhibitors.

- Metabolism is regulated by enzymes. Most metabolic pathways are under allosteric control. Enzyme complexes allow for interactions and regulation of adjacent active sites. Often the final product of a specific pathway regulates the commitment step of the pathway itself.

Common Problem Areas

- The chemistry and physics involved in this material are often overwhelming, and you may question why you are learning this in biology. Remember that biological processes are based on physical and chemical properties. To understand the function of organisms, it is necessary to understand the basics of chemistry and physics as they pertain to energy transfer.

- Allosteric regulation can be confusing because the terms "allosteric regulation" and "allosteric enzyme" are related but slightly different. Remember that an allosteric enzyme is an enzyme with more than one binding site: an active site (for binding and acting on substrates) and one or more allosteric sites. When allosteric regulators are bound to the allosteric site, they control whether or not the enzyme can bind substrate.

Study Strategies

- With this chapter, the best strategy is to take "small bites." You may find the concepts to be very unfamiliar; therefore, make sure you understand each section before proceeding. This chapter contains a large amount of terminology. Making a terminology/vocabulary list may be helpful.

- When studying the energy equations, be sure that you understand how any change in either free energy, unusable energy, or absolute temperature will affect the total energy of the system. Also focus on what a larger or smaller ΔG value means in terms of how reactions proceed and whether they require or release energy.

- When thinking of enzyme-mediated reactions, be sure to consider that altering anything on the left side of the equation is going to lead to changes in the amount of product formed. Always remember that enzymes are conserved across the equation.

- Review the following tutorials and activities on the Student Website/CD:
  Tutorial 6.1 Enzyme Catalysis
  Tutorial 6.2 Allosteric Regulation of Enzymes
  Activity 6.1 ATP and Coupled Reactions
  Activity 6.2 Free Energy Changes

Test Yourself

Knowledge and Synthesis Questions

1. ATP is necessary for the conversion of glucose to glucose 6-phosphate. Splitting ATP into ADP and P_i releases energy into what form?
   a. Potential
   b. Kinetic
   c. Entropic
   d. Enthalpic

2. Before ATP is split into ADP and P_i it holds what type of energy?
   a. Potential
   b. Kinetic
   c. Entropic
   d. Enthalpic

3. Which of the following statements concerning energy transformations is true?
   a. Increases in entropy reduce usable energy.
   b. Energy may be created during transformation.
   c. Potential energy increases with each transformation.
   d. Increases in temperature decrease total amount of energy available.

4. A reaction has a ΔG of –20kcal/mole. This reaction is
   a. endergonic, and equilibrium is far toward completion.
   b. exergonic, and equilibrium is far toward completion.
   c. endergonic, and the forward reaction occurs at the same rate as the reverse reaction.
   d. exergonic, and the forward reaction occurs at the same rate as the reverse reaction.

5. ATP hydrolysis releases energy to fuel cellular functions. ATP hydrolysis is
   a. endergonic.
   b. exergonic.
   c. chemoautotrophic.
   d. None of the above

6. Enzymes are biological catalysts and function by
   a. increasing free energy in a system.
   b. lowering activation energy of a reaction.
   c. lowering entropy in a system.
   d. increasing temperature near a reaction.

7. Which of the following contribute to the specificity of enzymes.
   a. Each enzyme has a narrow range of temperature and pH optima.
   b. Each enzyme has a specific active site that interacts with a particular substrate.
   c. Substrates themselves may alter the active site slightly for optimum catalysis.
   d. All of the above

8. Coenzymes and cofactors as well as prosthetic groups assist enzyme function by
   a. stabilizing three-dimensional shape and maintaining active sites.
   b. assisting with the binding of enzyme and substrate.
   c. Both a and b
   d. None of the above

9. Which of the following are characteristics of enzymes?
   a. They are not consumed by the enzyme-mediated reaction.
   b. They are not altered by the enzyme-mediated reaction.
   c. They lower activation energy.
   d. All of the above

10. Ascorbic acid, found in citrus fruits, acts as an inhibitor to catecholase, the enzyme responsible for the browning reaction in fruits such as apples, peaches, and pears. One possibility for its function could be that ascorbic acid is very similar in size and shape to catechol, the substrate of the browning reaction. If this is true, then this inhibition is most likely an example of _____ inhibition.
    a. indirect
    b. competitive
    c. noncompetitive
    d. none of the above

11. Refer to question 10. Suppose further studies indicate that ascorbic acid is not similar to catechol in size and shape but that the pH of the ascorbic acid solution is denaturing catecholase. If this is true, then this inhibition is most likely an example of _____ inhibition.
    a. competitive
    b. irreversible
    c. noncompetitive
    d. none of the above

12. Metabolism is organized into pathways. The pathway is linked in which of the following manners?
    a. All cellular functions feed into a central pathway.
    b. All steps in the pathway are catalyzed by the same enzyme.
    c. The product of one step in the pathway functions as the substrate in the next step.
    d. Products of the pathway accumulate and are secreted from the cell.

13. Which of the following represents an enzyme-catalyzed reaction?
    a. $E + P \rightarrow E + S$
    b. $E + S \rightarrow E + P$
    c. $E + S \rightarrow P$
    d. $E + S \rightarrow E$

14. Which of the following graphs of enzyme-mediated reactions represents an allosteric enzyme?
    a.

    b.

c.

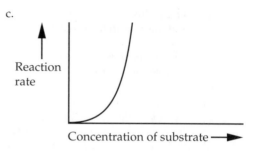

Application Questions

1. It is estimated that approximately 90 percent of energy that passes between levels in a food web is "lost" at each level. Explain the first law of thermodynamics and discuss why this apparent loss of energy does not contradict the law.

2. Amylase is a digestive enzyme that breaks down starch and is secreted in the mouth of humans. Amylase functions well in the mouth but ceases to function once it hits the acidic stomach environment. Explain why amylase does not function in the stomach.

3. The ultimate goal of metabolism is to drive ATP synthesis. ATP is considered the energy currency of the cell. Discuss how ATP couples endergonic and exergonic reactions and why it is so important in cellular functions.

4. Figure 6.19 shows the behavior of an allosteric enzyme that has binding sites for a negative regulator. Describe the behavior of an enzyme with binding sites for a positive regulator instead of a negative regulator.

5. Explain how substrate concentration affects the rate of an enzyme-mediated reaction.

6. You are studying a new species never before studied. It lives in acidic pools in volcanic craters where temperatures reach 100°C. You determine that it has a surface enzyme that catalyzes a reaction leading to its protective coating. You decide to study this enzyme in the laboratory. Under what conditions would you most likely find optimum activity of this enzyme?

7. Explain how free energy, total energy, temperature, and entropy are related.

8. Use a graph to explain how temperature affects enzyme activity.

Answers

Knowledge and Synthesis Answers

1. **b.** The released energy is available to do work; therefore, it is kinetic energy.

2. **a.** Potential energy is energy held within chemical bonds that may be converted to working kinetic energy.

3. **a.** Total energy = Free energy + entropy × temperature. Any increase in entropy is necessarily going to reduce free energy.

4. **b.** A negative ΔG indicates an exergonic reaction with energy being liberated. A large ΔG indicates that equilibrium lies toward completion.

5. **b.** ATP hydrolysis is exergonic, resulting in a ΔG of −12 kcal/mol.

6. **b.** Enzymes reduce activation energy and speed up reactions.

7. **d.** Enzymes are specific to particular substrates that may actually "adjust" the fit of the active site. They also function in specific, narrow optimum ranges of pH and temperature.

8. **c.** Cofactors, coenzymes, and prosthetic groups assist with the maintenance of an enzyme's three-dimensional shape and the conformation of the active site.

9. **d.** Enzymes are not consumed or altered in any way during an enzyme-mediated reaction, and they function to lower the activation energy of a reaction.

10. **b.** Competitive inhibitors compete for the active site with the substrate.

11. **b.** Denaturing an enzyme alters its three-dimensional structure, often irreversibly.

12. **c.** Within a given pathway, the products of the preceding step act as substrates for subsequent steps.

13. **b.** Substrate is converted to product and the enzyme is unchanged.

14. **b.** Allosteric enzymes produce sigmoid curves on reaction rate graphs. See Figure 6.20*b*.

Application Answers

1. The first law of thermodynamics states that energy cannot be created or destroyed, but that it may be converted from one form to another. In the transfer between levels of a food web, approximately 90 percent of the energy is converted to unusable heat energy. There is a net loss of usable energy during each conversion, but the total amount of energy (usable and unusable) remains the same.

2. The pH optimum of amylase is approximately 7. At that pH, the protein has the three-dimensional shape to allow starch to bind to its active site and catalyze its hydrolysis. At stomach pH (approximately 2), the protein is denatured and its three-dimensional shape and active site are lost; therefore, it can no longer catalyze the reaction.

3. The conversion of ATP to ADP and P_i releases approximately 12 kcal/mol of energy. This energy release fuels (endergonic) reactions in the cell. Equilibrium of the reaction is far to the right and favors the formation of ADP. In the converse, the formation of ATP from ADP and P_i is energy intensive and can be coupled to highly exergonic reactions within the cell. Thus ATP functions as an energy shuttle between endergonic and exergonic reactions. The small size of the molecule and its ubiquitousness allow it to be available and move freely within the cell.

4. A positive regulator would stabilize the enzyme in its active form. In the absence of the regulator, the enzyme would alternate between its inactive and active forms. When it encountered substrate, it would bind it only if it happened to be in its active form. When bound to the regulator, the enzyme would be fixed in its active form, and it would bind substrate at a greater rate.

5. Increasing substrate concentration will result in an increased rate of reaction until all available active sites are occupied. At that point, no amount of substrate increase will increase the rate of reaction. See Figure 6.16.

6. Optimum enzyme activity is most likely going to be under acidic conditions at high temperatures. Optimum activity is found when the enzyme is in its most functional shape. In most cases, this occurs under the enzymes' native conditions.

7. Total energy = free energy + unusable energy × absolute temperature.

8. See Figure 6.23.

7 Cellular Pathways That Harvest Chemical Energy

Important Concepts

Metabolic activities of all organisms share similar properties.

- All organisms share similar metabolic processes that are arranged in pathways. Each step of a pathway is catalyzed by a specific enzyme that may act as a key regulator of the pathway. Organelles serve to compartmentalize pathways.

Glucose is the energy source most often used by living organisms.

- All living organisms require a source of energy to survive. Most organisms, with the exception of some bacteria, use glucose ($C_6H_{12}O_6$) as a metabolic fuel source. Cells oxidize glucose for energy: $C_6H_{12}O_6 + 6 O_2 \rightarrow 6 CO_2 + 6 H_2O$ + energy. The energy is typically captured in ATP.

- The metabolism of glucose involves up to three metabolic processes. In the first pathway, glycolysis, glucose is converted to pyruvate with a small net energy release. In the second pathway, cellular respiration, the pyruvate from glycolysis is converted to CO_2 and energy in the form of ATP. Cellular respiration occurs only in the presence of O_2. In the absence of O_2, pyruvate is converted to lactic acid or ethanol with a small net energy release in the process of fermentation. Both fermentation and glycolysis occur in the absence of O_2, making them anaerobic processes. Cellular respiration is an aerobic process requiring O_2.

- Complete conversion of glucose to CO_2 in the presence of O_2 releases 686 kcal/mol of energy. Incomplete breakdown through fermentation produces substantially less energy.

Redox reactions are the energy-shuttling mechanism of metabolism.

- Energy is transferred from compound to compound through transfer of electrons in redox reactions.

- The gain of electrons or hydrogen atoms is called reduction.

- The loss of electrons or hydrogen atoms is called oxidation.

- Oxidation and reduction are always coupled. In order for a material to lose an electron or a hydrogen atom, another material must accept it.

- Oxidizing agents accept electrons and become reduced. In metabolism, O_2 is the oxidizing agent.

- Reducing agents donate electrons and become oxidized. In metabolism, glucose is the reducing agent.

- The net ΔG of a redox reaction is negative; therefore, energy is liberated.

Just as ATP is an energy shuttle in metabolism, NAD is an electron shuttle in the redox reactions of metabolism.

- NAD (nicotinamide adenine dinucleotide) may exist as either NAD^+ or $NADH + H^+$.

- Two electrons are transferred in the reduction of NAD^+ to $NADH + H^+$. The oxidation of NADH by O_2 is exergonic and liberates about 50 kcal/mol.

Glycolysis is the process of converting glucose to pyruvate and usable energy.

- Glycolysis is a fundamental metabolic pathway occurring in the cytoplasm of the cell. It occurs during both aerobic and anaerobic cell respiration.

- Glucose is converted to two molecules of pyruvate with a net production of two ATP and two $NADH + H^+$.

- Though glycolysis results in a net gain of two ATP, some steps are endergonic and require energy input.

- Refer to Figure 7.6 in your text to follow each of the ten enzyme-catalyzed reactions involved in the conversion of glucose to two molecules of pyruvate. The first five steps are endergonic and require energy input. Subsequent steps are exergonic.

- The following are key points to remember about glycolysis:

  - Glycolysis begins with a 6-C sugar that is ultimately converted to two 3-C pyruvate molecules.

  - The initial five energy-intensive steps of glycolysis involve phosphorylating and converting the 6-C sugar into a molecule that is split into two 3-C phosphorylated molecules: glyceraldehyde 3-phosphate (G3P). ATP is used as a source for phosphate groups and energy.

  - The subsequent energy-producing steps occur when phosphate groups are transferred to ADP to make ATP, and NAD^+ is reduced.

  - The conversion of glyceraldehyde 3-phosphate (G3P) to 1,3-bisphosphoglycerate (BPG) is significant because it is an oxidation process and energy is held in two NADH + H$^+$ after this conversion. How this energy is used depends on whether the cell is respiring aerobically or going through fermentation.

- The phosphate groups of BPG are transferred to ADP to produce ATP in the process of substrate-level phosphorylation.

- Two ATPs are used in glycolysis and four ATPs are produced, resulting in a net ATP gain of two.

Aerobic conditions allow the oxidation of pyruvate to acetate, which is combined with Coenzyme A to form acetyl CoA.

- Pyruvate oxidation to acetyl CoA occurs on the inner mitochondrial membrane.

- CO$_2$ is liberated and NADH + H$^+$ and acetyl CoA are produced.

The citric acid cycle oxidizes acetate to CO$_2$ and forms FADH$_2$, NADH + H$^+$, and ATP.

- The citric acid cycle is also known as the Krebs cycle and the tricarboxylic acid cycle. It consists of eight reactions.

- Figure 7.8 details the reactions of the citric acid cycle.

- In the initial step of the citric acid cycle, citrate (citric acid) is formed by the combination of acetyl CoA and oxaloacetate. During this step CoA is liberated and recycled for combination with another pyruvate.

- The concentrations of the intermediate molecules involved in the citric acid cycle are maintained at constant levels.

- In the third and fourth reactions of the cycle, one CO$_2$ molecule and one NADH + H$^+$ are formed at each step. Reaction six yields one FADH$_2$. The last reaction in the cycle results in the formation of one NADH + H$^+$ and oxaloacetate, which reenters the cycle at the beginning. One ATP is also produced with a single cycle. So one turn of the cycle produces two CO$_2$ molecules, three NADH + H$^+$, one FADH$_2$, and one ATP.

- Remember that each glucose molecule yields two acetyl CoA entering the cycle, so yields-per-glucose are doubled.

- The majority of the citric acid cycle takes place in the mitochondrial matrix.

- The citric acid cycle evolved as an efficient method to harvest the energy stored in acetyl CoA in a conservative add-on manner. Many of the intermediates play other roles throughout the cell.

Energy is liberated from reduced NAD$^+$ and FAD through electron transport in the respiratory chain.

- The process of oxidative phosphorylation involves the passing of electrons through a series of membrane-associated electron carriers in the mitochondria to ultimately produce ATP.

- The respiratory chain is composed of four enzyme complexes plus cytochrome c and ubiquinone (Q). The first two complexes shuttle the electrons of NADH + H$^+$ and FADH$_2$ to Q. The third complex moves electrons from Q to cytochrome c. The final complex passes the electrons on to O$_2$, the ultimate acceptor, resulting in H$_2$O

as a by-product. Figures 7.10 and 7.11 detail the respiratory chain.

- The purpose of the electron shuttling is to pump protons across the mitochondrial membrane against a concentration gradient. This movement of electrons by these proton pumps results in the establishment of a proton gradient across the inner mitochondria membrane. This also polarizes the membrane (creates a charge differential). These coupled effects are called the proton-motive force.

ATP is produced as protons move back across the mitochondrial membrane.

- An ATP synthase in the inner mitochondrial membrane couples the movement of protons with the synthesis of ATP in the chemiosmotic mechanism.

- Because the ATP synthase reaction may go in either direction, ATP is removed from the mitochondrial matrix immediately to favor ATP formation. Also, the proton gradient is maintained to favor ATP synthesis.

- If the proton flow is uncoupled from ATP synthase, the resulting energy is lost as heat. This phenomenon is used in thermoregulation.

In the absence of O$_2$, ATP is generated via fermentation.

- Fermentation occurs in the cell cytoplasm. The purpose of fermentation is to reduce pyruvate (or a metabolite) and return NAD for use in glycolysis in the absence of O$_2$. This allows the cell to continue to produce small amounts of ATP by glycolysis.

- Glycolysis occurs at a much higher rate under anaerobic conditions to maintain ATP levels.

- Obligate anaerobes lack the mechanisms to carry out oxidative phosphorylation; they produce energy by fermentation only.

- There are two types of fermentation based on the final end products produced: lactic acid fermentation and alcoholic fermentation.

Fermentation and cellular respiration have different energy yields.

- Fermentation yields a net total of two ATP for every one glucose molecule, regardless of whether it occurs through lactic acid fermentation or alcoholic fermentation.

- Aerobic respiration yields a total of 36 ATP for every one glucose molecule. Two ATP are produced in glycolysis, two from the citric acid cycle, and 32 from the respiratory chain.

- The general formula for aerobic cellular respiration is:

$$C_6H_{12}O_6 + 6\,O_2 \rightarrow 6\,O_2 + 6\,H_2O + 36\,\text{ATP}$$

Glycolysis and cellular respiration occur while other metabolic processes are occurring.

- Components of the metabolic pathways are also components of other metabolic processes including catabolism and anabolism.

- Catabolism is the breaking down of molecules to release energy. Polysaccharides, lipids, and proteins all feed into the metabolic pathways at different points. Polysaccharides are broken down into glucose and enter at glycolysis. Lipids are broken down and can enter at glycolysis or the citric acid cycles as acetyl CoA. Proteins enter glycolysis and the citric acid cycle as amino acids.

- Just as intermediates can be broken down (catabolyzed) to release energy, they can also be synthesized in anabolism for storage or use by the cell. Gluconeogenesis is the anabolic process by which glucose is produced from the intermediates of glycolysis.

- Metabolism, both anabolism and catabolism, is regulated by enzymes to maintain stable concentrations of intermediates. These stable concentrations of intermediates make up the "metabolic pool."

Metabolic enzymes are allosterically controlled to regulate the production of ATP.

- Metabolism responds to both positive and negative feedback provided by intermediates and end products of the process.

- Each process has a specific control point for regulation. The control point for glycolysis is phosphofructokinase, which is inhibited by ATP. This allows glycolysis to speed up during fermentation and slow down during cellular respiration.

- The main control point for the citric acid cycle is isocitrate dehydrogenase. $NADH + H^+$ and ATP inhibit the enzyme to slow down the process, and NAD^+ and ADP act as activators.

- If the citric acid cycle slows due to abundant ATP, glycolysis slows as well.

- If excess acetyl CoA is produced and ATP is abundant, acetyl CoA can be shuttled to fatty acid synthesis for storage.

The Big Picture

- Metabolic processes occur in pathways and are regulated by enzymes. These pathways may be further controlled by compartmentalization into organelles in eukaryotic cells. Because metabolic processes are energy-transferring reactions, ATP and NAD are necessary for shuttling energy and electrons between steps of the pathways. Redox reactions are the basis of metabolism. As one compound is oxidized, another is reduced.

- Glucose provides cellular energy. It may be metabolized in the absence of oxygen through glycolysis and fermentation. In the presence of oxygen, it is metabolized through glycolysis and cellular respiration in the form of the citric acid cycle, electron transport, and oxidative phosphorylation. Anaerobic respiration produces significantly less cellular energy (in the form of ATP) than does aerobic respiration.

- Each step of the catabolic pathway is regulated by the products of subsequent pathways and depends on oxygen being present as the final electron acceptor. This level of control is achieved through allosteric regulation.

Common Problem Areas

- The biggest mistake you can make in learning glucose metabolism is to focus on memorizing the pathways without understanding what is happening. When you begin your study, do not focus on the pathways themselves. Focus on understanding the beginning and end products of each pathway and why the pathway does what it does.

Once you conceptually understand why the pathway is present, move on to learning the pathway itself.

- Students often fail to see how regulation of the pathways occurs. By understanding how the presence or absence of intermediates affects the entire pathway, you can better understand the pathways themselves.

Study Strategies

- Remember that ATP and NAD are energy currencies. They move energy from pathway to pathway. Think about what roles ATP and NAD have in redox reactions to completely understand the pathways.

- Do not memorize pathways. Make sure you *understand* the pathways and how they are connected. Also understand how affecting one pathway can alter another pathway.

- This chapter is very visual in nature. You should spend a significant amount of time with the diagrams.

- Review the following tutorials and activities on the Student Website/CD:
  Tutorial 7.1 Electron Transport and ATP Synthesis
  Tutorial 7.2 Two Experiments Demonstrate the
     Chemiosmotic Mechanism
  Activity 7.1 Glycolysis and fermentation
  Activity 7.2 Energy Pathways in Cells
  Activity 7.3 The Citric Acid Cycle
  Activity 7.4 Energy Levels
  Activity 7.5 Electron Transport Chain
  Activity 7.6 Regulation of Energy Pathways

Test Yourself

Knowledge and Synthesis Questions

1. Which of the following cellular metabolic processes can occur in the presence *or* the absence of oxygen?
   a. The citric acid cycle
   b. Electron transport
   c. Glycolysis
   d. Fermentation

2. Which of the following statements regarding glycolysis is true?
   a. A 6-C sugar is broken down to two 3-C molecules.
   b. Two ATP molecules are consumed.
   c. A net sum of two ATP molecules is generated.
   d. All of the above

3. During which process is most ATP generated in the cell?
   a. Glycolysis
   b. The citric acid cycle
   c. Electron transport coupled with chemiosmosis
   d. Fermentation

4. One purpose of the electron transport chain is to
   a. cycle $NADH + H^+$ back to NAD^+.
   b. use the intermediates from the citric acid cycle.
   c. break down pyruvate.
   d. All of the above

5. Cellular respiration is allosterically controlled. Which of the following act as inhibitors at the various control points?
   a. ATP
   b. $NADH + H^+$
   c. Both a and b
   d. None of the above

6. Which of the following describes the role of the mitochondrial membrane?
   a. The membrane acts as an anchor for the membrane-associated enzymes of cellular respiration.
   b. The membrane allows for the establishment of a proton-motive force.
   c. Both a and b
   d. None of the above

7. In a redox reaction between G3P and NAD^+ yielding BPG and $NADH + H^+$, _____ is oxidized and _____ is reduced.
   a. G3P; NAD^+
   b. BPG; $NADH + H^+$
   c. G3P; $NADH + H^+$
   d. NAD^+; $NADH + H^+$

8. Which of the following is true regarding redox reactions?
   a. Oxidizing agents accept electrons.
   b. A molecule that accepts electrons is said to be reduced.
   c. Redox reactions involve electron transfers.
   d. All of the above

9. Cyanide poisoning inhibits aerobic respiration at cytochrome *c* oxidase. Which of the following are results of cyanide poisoning at the cellular level?
   a. Oxygen is not reduced to water.
   b. ATP cannot be synthesized in the mitochondria because electron transport is never completed.
   c. Cells (with the exception of brain cells) must switch to anaerobic respiration.
   d. All of the above

10. Not all cells can switch between aerobic and anaerobic respiration. If a cell cannot switch to anaerobic respiration in the absence of oxygen, what happens?
    a. The cell concentrates oxygen to continue aerobic respiration.
    b. The cell dies.
    c. The cell uses fermentation.
    d. None of the above

11. The main purpose of cellular respiration is to
    a. convert energy stored in the chemical bonds of glucose to an energy form that the cell can use.
    b. convert potential to kinetic energy.
    c. convert kinetic to potential energy.
    d. create energy in the cell.

12. Which of the following statements concerning the synthesis of ATP in the mitochondria is true?
    a. ATP synthesis cannot occur without the presence of ATP synthase.
    b. The proton-motive force is the establishment of a charge and concentration gradient across the mitochondrial membrane.
    c. The proton-motive force is necessary to drive protons back across the membrane through channels established by the ATP synthase channel protein.
    d. All of the above

Application Questions

1. Why is oxygen necessary for aerobic respiration?

2. Glycolysis yields two molecules of pyruvate, two ATP, and two $NADH + H^+$ regardless of whether oxygen is present or not. What are the fates of these molecules in the absence of oxygen? What would happen if $NADH + H^+$ was not recycled?

3. Explain how the proton-motive force drives chemiosmosis.

4. The fate of acetyl CoA differs according to how much ATP is present in the cell. Explain what happens to acetyl CoA when ATP is limited, and compare that to what happens when acetyl CoA is abundant. How do these processes help regulate metabolism?

5. Cellular respiration occurs simultaneously along with many other cellular processes. Describe generally how cellular respiration interacts with other cellular metabolic events.

6. Compare and contrast energy yields from aerobic respiration and fermentation.

7. One of the by-products of aerobic cellular respiration is carbon dioxide. Assuming you begin with labeled glucose, trace the fate of that molecule until carbon dioxide is released.

8. Identify the commitment steps of glycolysis, the citric acid cycle, and electron transport. What regulators affect each of these commitment steps?

Answers

Knowledge and Synthesis Answers

1. **c.** Glycolysis proceeds during both fermentation and cellular respiration. Only in cellular respiration is oxygen needed as the terminal electron acceptor of the pathway.

2. **d.** During glycolysis, 6-C glucose is broken down into two 3-C pyruvate molecules. In the process, four total ATP are produced but two are consumed, leaving a net production of two ATP molecules.

3. **c.** Most of the ATP produced during cellular respiration is produced as electron transport and chemiosmosis are coupled in oxidative phosphorylation.

4. **a.** The electron transport chain is responsible for oxidizing $NADH + H^+$ back to NAD^+.

5. **c.** Both ATP and $NADH + H^+$ allosterically control metabolism. ATP controls both phosphofructokinase and isocitrate dehydrogenase, which are commitment steps for glycolysis and the citric acid cycle respectively. $NADH + H^+$ controls isocitrate dehydrogenase.

6. **c.** The mitochondrial membrane is necessary for the anchoring of proteins as well as the establishment of a barrier across which a gradient can be established.

7. **a.** A molecule is oxidized when it loses electrons or protons and is reduced when it gains electrons or protons.

8. **d.** Oxidizing agents accept electrons and cause oxidation of another molecule. Reducing agents donate electrons and cause the reduction of another molecule.

9. **d.** Cyanide stops aerobic cellular respiration because cytochrome *c* oxidase loses the ability to reduce oxygen to water. Those cells that can switch to anaerobic respiration (fermentation) do so. Brain cells cannot make that switch and so are killed.

10. **b.** In the absence of oxygen, cells that cannot go through fermentation die.

11. **a.** Cellular respiration is the cell's way of converting potential energy in the chemical bonds of glucose to potential energy that the cell ultimately can use.

12. **d.** The proton-motive force results in a concentration and charge gradient across the mitochondrial membrane. In order for that gradient to equalize, the protons must flow through a channel protein. If this channel protein has an associated ATP synthase, ATP is generated as protons flow through.

Application Answers

1. Oxygen acts as the terminal electron acceptor in the electron transport pathway. Without it, NADH + H$^+$ cannot be cycled back to NAD$^+$. The accumulated NADH + H$^+$ acts as an inhibitor to the citric acid cycle and effectively shuts it down. Therefore, in the absence of oxygen, a cell only can undergo glycolysis.

2. In the absence of oxygen, pyruvate is either reduced to lactate (in lactic acid fermentation) or it is metabolized and its metabolites are reduced to ethyl alcohol (in alcoholic fermentation). In either case, NADH + H$^+$ is the reducing agent, and it is oxidized back to NAD$^+$ in the process. The two ATP would be used as cellular energy. If NADH + H$^+$ was not oxidized to NAD$^+$, there would eventually be no NAD$^+$ available for glycolysis.

3. The proton-motive force results in a concentration and charge gradient across the mitochondrial membrane. In order for that gradient to equalize, the protons must flow through a channel protein. If this channel protein has an associated ATP synthase, ATP is generated as protons flow through.

4. If ATP is limited, acetyl CoA enters the citric acid cycle, and cellular respiration utilizes it to produce ATP. If ATP is abundant, acetyl CoA is shuttled to fatty acid synthesis, thus storing the energy in chemical bonds.

5. Consult Figure 7.17 to see where different metabolic pathways in the cell interact.

6. Fermentation yields only two ATP. Cellular respiration yields 36 ATP. Though this difference is great, organisms can survive quite well relying on fermentation because the rate at which glycolysis occurs is increased nearly tenfold.

7. Follow Figures 7.6 and 7.8. Carbon dioxide is liberated as α-ketoglutarate is oxidized to succinyl CoA.

8. This control point for glycolysis is phosphofructokinase, which is inhibited by ATP. This allows glycolysis to speed up during fermentation and slow down during cellular respiration. The control point for the citric acid cycle is isocitrate dehydrogenase. NADH + H$^+$ and ATP inhibit the enzyme, and NAD$^+$ and ADP are activators. Electron transport is controlled by the amount of NADH + H$^+$ fed in and by NADH-Q reductase.

8 Photosynthesis: Energy from the Sun

Important Concepts

The photosynthetic production of O_2 by green plants is necessary for most organisms to obtain the energy for life.

- Photosynthesis uses CO_2, water, and light to produce carbohydrate and O_2 in the reaction

$$6 CO_2 + 12 H_2O \rightarrow C_6H_{12}O_6 + 6 O_2 + 6 H_2O$$

- In plants, water for photosynthesis must be acquired by the roots and transported to the leaves, where photosynthesis takes place. CO_2, O_2, and water vapor are exchanged through openings or stomata in the leaf's surface. Light is required for photosynthesis to occur. Photosynthesis in eukaryotes occurs in chloroplasts.

- Photosynthesis occurs in a two-pathway process. The light reactions produce ATP from light energy via photophosphorylation. The Calvin–Benson cycle uses ATP and NADPH + H$^+$ from the light reactions to trap CO_2 and produces sugars that act as an energy store. *Neither* pathway operates in the absence of light.

The interaction of light and pigment is fundamental to photosynthesis.

- Light comes in discrete packets called photons and has wavelike properties. Photons can be scattered, transmitted, or absorbed. The properties of light depend on its wavelength. In order for light to be biologically available it must be absorbed by the receptive molecule and contain enough energy to carry out chemical work. When a molecule absorbs a proton, it goes from a ground state to an excited state.

- Pigments are molecules that absorb wavelengths in the visible spectrum. The pigment chlorophyll appears green because blue and red light are absorbed and green light is reflected.

- Compounds have unique absorption spectra depending on which wavelengths of light are absorbed (Figure 8.6). An action spectrum can be measured that relates biological activity to particular wavelengths.

- In plants, photosynthesis relies on chlorophyll *a*, chlorophyll *b*, and accessory pigments. Chlorophylls are more abundant, but they can only absorb photons with blue or orange-red wavelengths. Accessory pigments, such as carotenoids and phycobilins, absorb photons that chlorophylls cannot and transfer energy to the chlorophylls.

In photosynthesis, a pigment molecule in the excited state passes the absorbed energy along.

- Once a pigment absorbs light, two things can happen: The energy can be released as fluorescence, or it can be passed to another pigment.

- Photosynthetic antennae function to pass the energy from light from one pigment to another until the reaction center is reached, where light energy is converted to chemical energy. In plants, the reaction center is chlorophyll *a*.

- Excited chlorophyll acts as a reducing agent and begins a cascade that ultimately drives ATP synthesis.

- Noncyclic electron flow produces ATP and NADPH + H$^+$. Water is oxidized to form O_2, H$^+$, and electrons. Electrons move from water to chlorophyll, through electron carriers, and ultimately to NADP$^+$. Release of free energy drives ATP synthesis through chemiosmosis.

- Noncyclic electron flow requires two photosystems to continuously absorb light. Photosystem II takes slightly higher energy (680 nm) than photosystem I (700 nm).

- Figure 8.11 shows the flow of electrons from the splitting of water to the reduction of NADP$^+$. The electron transport pumps protons actively into the thylakoid compartment. This proton gradient powers the formation of ATP.

- Noncyclic flow produces equal amounts of ATP and NADPH + H$^+$, but the Calvin–Benson cycle requires more ATP than NADPH + H$^+$. Cyclic electron flow supplies the additional ATP by producing ATP without producing NADPH + H$^+$.

- Cyclic electron flow involves only photosystem I and cycles back to chlorophyll (Figure 8.10). The amount of NADPH + H$^+$ regulates whether cyclic or noncyclic electron flow occurs. If NADPH + H$^+$ is abundant, electrons are accepted by plastoquinone rather than NADP$^+$.

- ATP synthesis in either cyclic or noncyclic electron flow is produced via chemiosmosis. Just as in the mitochondrion, a proton-motive force coupled to ATP synthase is established in the chloroplast as electrons are passed along the transport chain. This results in the photophosphorylation of ADP to ATP.

The Calvin–Benson cycle incorporates CO_2 into sugars.

- In the Calvin–Benson cycle, ATP and NADPH + H$^+$ from the light reactions are used to convert CO_2 to sugars. Because ATP and NADPH + H$^+$ are energy-rich coenzymes that cannot be stockpiled, the Calvin–Benson cycle reactions occur only in the light.

- The Calvin–Benson cycle occurs in the stroma of the chloroplasts.

- The cycle was revealed using radiolabeled CO_2 and observing where it was incorporated. Following this process, researchers revealed the pathway by which CO_2 is converted to sugar. See Figure 8.12 to understand how this was done.

- CO_2 is "fixed" in ribulose 1,5-bisphosphate (RuBP), which is split into two molecules of 3-phosphoglycerate (3PG) by an enzyme called rubisco. Rubisco is the world's most abundant protein, and it regulates the Calvin–Benson cycle.

- ATP and NADPH + H$^+$ from the light reactions are then used to convert the fixed CO_2 to carbohydrate (glyceraldehyde 3-phosphate; G3P).

- Finally, ATP is used to regenerate RuBP so that additional CO_2 may be fixed.

- The resulting carbohydrate (G3P) is converted to starch or sucrose, either for storage or for conversion to glucose and fructose for cellular fuel.

Rubisco also mediates the process of photorespiration.

- Rubisco can also fix O_2 at the expense of CO_2 to produce glycolate. Glycolate is ultimately converted to glycine in the peroxisome and then to glycerate and CO_2 in the mitochondrion.

- ATP and NADPH from the light reaction are used in photorespiration, but CO_2 is released instead of used to make carbohydrate.

- Whether CO_2 or O_2 is fixed depends on relative concentrations of CO_2 and O_2. Excess O_2 (abundant, for example, on hot, dry days) forces photorespiration.

Photorespiration is metabolically expensive, so plants have evolved mechanisms to avoid it.

- To avoid photorespiration, the level of CO_2 around rubisco must be high. This is difficult to achieve because O_2 levels in air are much higher than CO_2 levels. If hot conditions force the closing of stomata, CO_2 levels are quickly depleted.

- Plants that fix CO_2 to 3PG are called C_3 plants and are very sensitive to CO_2/O_2 levels. Photorespiration occurs frequently and limits the range of conditions under which C_3 plants can grow.

- Many tropical plants are C_4 plants. They fix carbon through phosphoenolpyruvate (PEP) and oxaloacetate using the enzyme PEP carboxylase. PEP carboxylase has no oxygenase activity and fixes CO_2 at very low levels. The resulting four-carbon compound diffuses to the bundle sheath cells in the interior of the leaf, releases CO_2, and is recycled. The released CO_2 goes through the Calvin–Benson cycle including rubisco as normal. The early fixation process is a CO_2-concentrating mechanism around rubisco.

- Cacti, pineapples, and other succulents undergo crassulacean acid metabolism (CAM) that separates CO_2 fixation and the Calvin–Benson cycle. Plants that carry out CAM open their stomata only at night and store CO_2 for use during the day when light is present. CO_2 is fixed as oxaloacetate and converted to malic acid. CO_2 is stored in malic acid until it is transferred to the chloroplasts or photosynthetic cells when light is available.

Metabolic pathways in plants involve photosynthesis and respiration.

- Because plants are autotrophic, they can synthesize all necessary molecules from CO_2, H_2O, phosphate, sulfate, and NH_4.

- All cells, whether photosynthetic or not, go through respiration, and they do so in the light and the dark.

- The Calvin–Benson cycle and the respiratory pathway are closely linked. G3P can be converted to pyruvate and enter respiration, or it can enter the gluconeogenic pathway to form sucrose. In order for plants to grow, the energy stored must exceed the energy used in respiration.

- Figure 8.18 illustrates how plant metabolism is interconnected.

The Big Picture

- Photosynthesis allows organisms with the appropriate pigments and metabolic processes to convert light energy from the sun to chemical energy that can be used in the cell or stored. This conversion process forms the basis of all food webs and is vital to life on earth. Photosynthetic organisms fix CO_2 into sugars. This process requires water and light energy, and the by-products are O_2 and water.

- The process of photosynthesis occurs in two steps. The first step utilizes light energy to start chemical energy generating the electron transport system. The movement of electrons through the electron transport system generates cellular energy in the form of ATP that can be used in the fixing of CO_2. The fixing of CO_2 is mediated by an enzyme called rubisco and is concentration dependent.

- Rubisco can also act as an oxygenase and allow a plant to go through a metabolically expensive process called photorespiration. Plants that grow in conditions favoring photorespiration have evolved a wide variety of systems to avoid photorespiration.

Common Problem Areas

- There is the temptation to learn pathways without understanding why they occur. Be sure to focus on why the plant has a particular pathway before attempting to learn the pathway itself.

- It is very common to think that the light reactions happen in the light and the Calvin–Benson cycle occurs in the dark. This is not the case! The light reactions must occur simultaneously with the Calvin–Benson cycle to supply the needed energy in the form of ATP and NADPH + H$^+$.

- Remember that plants respire as well as photosynthesize and that photorespiration and respiration are different processes. All plant cells go through cellular respiration. Cellular respiration takes the energy stored in photosynthesis and makes it available to drive other cellular processes. Photorespiration is energy expensive and is of little benefit to the plant. Photosynthesis occurs in specialized photosynthetic cells and happens at the same time as respiration.

Study Strategies

- The process of photosynthesis is easier to understand when you visualize what is happening. Use the figures in your book to understand where energy and carbon are flowing. Remember that what the plant is doing is taking light energy and converting it to chemical energy that can be stored. Watch for where energy is flowing and how.

- The properties of chlorophyll and light itself greatly affect how well a plant photosynthesizes. Take some time to understand the properties of light and how pigments interact with light.

- Rubisco is the most abundant protein on the planet. Focus on how the relative concentrations of O_2 and CO_2 dictate how the enzyme functions. Also understand that C_4 and CAM plants have modifications to avoid photorespiration.

- Review the following tutorials and activities on the Student Website/CD:
  Tutorial 8.1 The Source of the Oxygen Produced by Photosynthesis
  Tutorial 8.2 Photophosphorylation
  Tutorial 8.3 Tracing the Pathway of CO_2
  Activity 8.1 Calvin–Benson Cycle
  Activity 8.2 C_3 and C_4 Leaf Anatomy

Test Yourself

Knowledge and Synthesis Questions

1. The main purpose of photosynthesis is to
   a. consume CO_2.
   b. produce ATP.
   c. convert light energy to chemical energy.
   d. produce starch.

2. Which of the following best represent the components that are necessary for photosynthesis to take place?
   a. Mitochondria, accessory pigments, visible light, water, and CO_2
   b. Chloroplasts, accessory pigments, visible light, water, and CO_2
   c. Mitochondria, chlorophyll, visible light, water, and O_2
   d. Chloroplasts, chlorophyll, visible light, water, and CO_2

3. Chlorophyll is suited for the capture of light energy because
   a. certain wavelengths of light raise it to an excited state.
   b. in its excited state chlorophyll gives off electrons.
   c. chlorophyll's structure allows it to attach to thylakoid membranes.
   d. All of the above

4. Plants give off O_2 because
   a. O_2 results from the incorporation of CO_2 into sugars.
   b. plants do not respire because they photosynthesize.
   c. water is the initial proton donor, leaving O_2 as a photosynthetic by-product.
   d. All of the above

5. Cyclic and noncyclic electron flow are used in plants to
   a. meet the ATP demands of the Calvin–Benson cycle.
   b. avoid producing excess NADPH + H$^+$.
   c. balance ATP and NADPH + H$^+$ ratios in the chloroplast.
   d. All of the above

6. Which of the following statements concerning the light reactions of photosynthesis is true?
   a. Photosystem I can operate independently of photosystem II.
   b. Photosystems I and II are activated by different wavelengths of light.
   c. Photosystems I and II transfer electrons and create proton gradients across the thylakoid membrane.
   d. All of the above

7. ATP is produced during the light reactions via
   a. CO_2 fixation.
   b. chemiosmosis.
   c. reduction of water.
   d. All of the above

8. Because of the properties of chlorophyll, plants need adequate _____ light to grow properly.
   a. green
   b. blue and red
   c. infrared
   d. ultraviolet

9. Which of the following statements concerning the Calvin–Benson cycle is *not* true?
   a. Light energy is not required for the cycle to proceed.
   b. CO_2 is assimilated into sugars.
   c. RuBP is regenerated.
   d. It uses energy stored in ATP and NADPH + H$^+$.

10. Which of the following statements concerning rubisco is true?
    a. Rubisco is an enzyme.
    b. Rubisco catalyzes both the beginning steps of photorespiration and the Calvin–Benson cycle.
    c. Rubisco is the most abundant protein on earth.
    d. All of the above

11. Which of the following begins the Calvin–Benson cycle and is the commitment step that results in the entire pathway being carried out?
    a. 3PG is reduced to G3P using ATP and NADPH + H$^+$.
    b. RuBP is regenerated.
    c. CO_2 and RuBP join forming 3PG.
    d. As a cycle, it can start at any point.

12. The Calvin–Benson Cycle results in the production of
    a. glucose.
    b. starch.
    c. rubisco.
    d. G3P.

13. Which of the following statements regarding photorespiration is true?
    a. Photorespiration is a metabolically expensive pathway.
    b. Photorespiration is avoided when CO_2 is abundant.
    c. Photorespiration reduces the overall CO_2 that is converted to carbohydrates.
    d. All of the above

14. The fixation of CO_2 by PEP carboxylase functions to
    a. concentrate CO_2 for use in photosynthetic cells.
    b. allow plants to close stomata without having photorespiration occur.
    c. allow plants to photosynthesize in the dark.
    d. Both a and b

15. CAM plants differ from C_4 plants in that
    a. CO_2 is stored as malic acid.
    b. photosynthesis can occur at night in these plants.
    c. their stomata close during periods that favor photorespiration.
    d. they use PEP carboxylase to fix CO_2.

16. Which of the following statements is true regarding the relationship between photosynthesis and cellular respiration in plants?
    a. Photosynthesis occurs in specialized photosynthetic cells.
    b. Cellular respiration occurs in specialized respiratory cells.
    c. Cellular respiration and photosynthesis can occur in the same cell.
    d. Both a and c

Application Questions

1. Plants consume CO_2 and give off O_2; how is this possible if plants must also undergo cellular respiration?

2. Why do plants undergo both the light reactions of photosynthesis and the Calvin–Benson cycle? Why don't they simply use the ATP produced in the light reactions of photosynthesis to drive cellular processes?

3. Why do plants undergo both photosynthesis and cellular respiration? Why don't they simply use the ATP produced in the light reactions of photosynthesis to drive cellular processes?

4. Rubisco has both carboxylase and oxygenase activities. These processes compete with one another. What determines which function the enzyme has? What conditions favor photorespiration? What conditions favor photosynthesis?

5. How did the evolution of photosynthetic organisms lead to the evolution of life on land?

6. Compare and contrast C_3 and C_4 plants.

7. The Calvin–Benson cycle was once referred to as the "dark" reactions of photosynthesis. Why is this a misnomer?

8. Explain the differences between cyclic and noncyclic electron flow. Why are both processes necessary?

9. How do accessory pigments enhance photosynthetic activity in plants?

10. Why are plants green?

Answers

Knowledge and Synthesis Answers

1. **c.** Photosynthetic organisms are the only life forms capable of trapping light energy and converting it to chemical energy. Because of this they form the basis of food chains.

2. **d.** Chloroplasts are the site of the photosynthetic reactions; chlorophyll is excited by photons of light and serve as reaction centers for the photosystems; visible light is necessary to excite chlorophyll and accessory pigments; water is the initial electron donor for the pathway; and CO_2 is necessary to make precursor molecules for energy storage.

3. **d.** The "tails" of chlorophyll molecules are associated with the thylakoid membranes of the chloroplasts. This close membrane association assists with establishing the proton-motive force that will drive ATP synthesis. When excited by light, the chlorophyll moves into an excited state and passes electrons to acceptor molecules. This begins to set up the proton gradient across the membrane that will drive ATP synthesis.

4. **c.** Water is split at photosystem II to donate electrons to the reaction center. The resulting protons are moved across the membrane to establish the proton-motive force and O_2 is given off as a by-product.

5. **d.** ATP is required at higher levels in the Calvin–Benson cycle than NADPH + H$^+$ is; therefore, there must be a mechanism for producing additional ATP. Cyclic electron flow provides that mechanism. If noncyclic electron flow were to be sped up to meet ATP needs, an excess of NADPH + H$^+$ would result. Shifting between cyclic and noncyclic flow balances ATP/NADPH + H$^+$ ratios.

6. **d.** Photosystems I and II operate depending on whether electron flow is cyclic or noncyclic. Activity is controlled by the ATP levels in the chloroplast. Photosystem II is activated by light of a higher energy level than photosystem I. Both photosystems transfer electrons and create proton gradients across the thylakoid membranes.

7. **b.** In the light reactions, ATP synthesis occurs when protons flow through an ATP synthase channel protein in the thylakoid membrane. This is a chemiosmotically driven process.

8. **b.** Chlorophyll and accessory pigments absorb light in the blue and orange-red wavelengths of visible light. Green light is reflected; therefore, plants appear green.

9. **a.** Light energy is required for the Calvin–Benson cycle to proceed. ATP synthesis is dependent on light energy, and the Calvin–Benson cycle is dependent on ATP.

10. **d.** Rubisco, the most abundant enzyme on earth, has both oxygenase and carboxylase activities.

11. **c.** The first step of the Calvin–Benson cycle is the fixation of CO_2 into 3PG. This is the regulatory step, and it requires ATP and NADPH + H$^+$.

12. **d.** The Calvin–Benson cycle produces only G3P. G3P can then be metabolized into storage products like sugars and starch.

13. **d.** Photorespiration uses as much ATP as photosynthesis, but results in no energy gains for the plant and reduces net carbon fixation by 25 percent compared with the Calvin–Benson cycle. If CO_2 is abundantly available, rubisco acts as a carboxylase rather than an oxygenase.

14. **d.** Plants do not photosynthesize in the dark. PEP carboxylase allows the fixation of CO_2 and the concentration of it at rubisco.

15. **a.** CAM plants functionally store CO_2 as malic acid.

16. **d.** Photosynthesis occurs only in cells that have the necessary structures, but cellular respiration occurs in *every* living cell.

Application Answers

1. Plant cells undergo cellular respiration in all cells. Therefore, all cells consume O_2. Photosynthesis occurs in specialized cells that consume both CO_2 and O_2. Because atmospheric O_2 levels are high, there is excess O_2 available for the plant to utilize; therefore, O_2 continues to be emitted from the plant.

2. The light reactions of photosynthesis produce ATP. ATP cannot be stored for use later (such as when light is not available); therefore, there has to be a mechanism for that energy to be stored. The Calvin–Benson cycle stores the energy in the chemical bonds of G3P, which can be incorporated into carbohydrates for longer-term storage.

3. Though photosynthesis produces all the necessary energy for a plant, a plant cannot be continuously photosynthetically active. Therefore, a plant stores energy in carbohydrates. Cellular respiration is necessary to break down stored carbohydrates.

4. Whether rubisco acts as a carboxylase or an oxygenase depends on the relative ratio of O_2 to CO_2. At higher CO_2 levels it acts as a carboxylase. At low CO_2 levels it acts as an oxygenase. Photorespiration is favored during hot, dry weather, which forces the closing of stomata and leads to increases in O_2 levels within the leaf. Photosynthesis is favored when stomata can remain open and light intensity is optimal.

5. Increases in photosynthesis led to an oxygen-rich atmosphere. Abundance of oxygen led organisms to evolve aerobic metabolism and the ability to live a terrestrial existence once atmospheric oxygen levels reached a high enough point.

6. Refer to Table 8.1 in your book.

7. Light is required for both the light reactions of photosynthesis and the Calvin–Benson cycle. The Calvin–Benson cycle depends on the ATP generated during the light-dependent reactions.

8. See Figures 8.9 and 8.10. See also the answer to Question 5 under "Knowledge and Synthesis Questions."

9. Accessory pigments allow utilization of light in all wavelengths of the visible spectrum. The energy absorbed is channeled to the reaction centers of the photosystems.

10. The primary pigments in plants are chlorophylls. Chlorophylls absorb blue and orange-red wavelengths of light and reflect green light, thus making plants appear green. See the absorption spectra and action spectra of chlorophyll in Figures. 8.6.

9 Chromosomes, the Cell Cycle, and Cell Division

Important Concepts

Cell division.

- Cell division requires a reproductive signal either from inside or outside the cell, subsequent DNA replication, segregation of the newly replicated chromosomes to opposite poles of the cell, and the division of the cytoplasm (cytokinesis) to form two daughter cells.

- Cells divide asexually to produce two daughter cells with identical genetic information.

- In eukaryotic cells, asexually dividing cells segregate their chromosomes during mitosis. For organisms that reproduce sexually, haploid gametes are formed during meiosis.

Prokaryotes divide by fission.

- Prokaryotic cells divide by increasing in size, replicating their DNA (a single circular chromosome), and dividing into two new cells through the process of fission.

- DNA replication in prokaryotic cells requires a number of different proteins and special sites on the chromosome where replication begins (the *ori* site) and terminates (the *ter* site).

Eukaryotic cells divide by mitosis or meiosis.

- Cell division in multicellular eukaryotes occurs in response to the needs of the entire organism.

- Eukaryotic cells have more than one chromosome, and these chromosomes are linear.

- Eukaryotic chromosomes are located in the nucleus.

- Eukaryotic cells replicate their chromosomes, segregate those chromosomes during mitosis, and divide their cytoplasm during cytokinesis.

- DNA replication, mitosis, and cell division occur in a special sequence during the reproductive cycle of the cell, also known as the cell cycle.

- There are two major phases of the cell cycle: interphase and mitosis. Interphase is divided into three distinct parts, G1 (or Gap1), S (or DNA Synthesis), and G2 (or Gap2).

- G1 is the phase of the cell cycle in which the decision and subsequent preparation for DNA synthesis occurs.

- S is the phase of the cell cycle in which all of the chromosomes are replicated; each replicated chromosome consists of two identical sister chromatids.

- During G2, the cell prepares for mitosis.

- During M (mitosis), the cell segregates the newly replicated chromosomes to opposite poles of the cell, and nuclear division occurs. At the end of M, cytokinesis occurs and cell division is complete.

Both kinases and cyclins play a key role in cell cycle transitions.

- Cdk, a cyclin-dependent kinase that phosphorylates other proteins in the cell, is catalytically active when it is bound to another protein, cyclin.

- Different cyclin-Cdk protein complexes act as checkpoints and signal the transition of G1 to S, the initiation of DNA replication, the condensation of the chromosomes in G2, and the entry into the cell cycle.

- Other proteins, in response to environmental signals, regulate cyclin-Cdk complexes.

- Growth factors are external chemical signals that stimulate cell division.

Eukaryotic chromosome structure

- Eukaryotic chromosomes are composed of protein and DNA, collectively called chromatin. Cohesin is a protein that holds the sister chromatids together along their length; condensin proteins help compact the DNA.

- When eukaryotic chromosomes are replicated, the daughter chromosomes, termed sister chromatids, are still joined at the centromere.

- In a eukaryotic nucleus, DNA molecules are wrapped around beadlike particles of protein called nucleosomes. Nucleosomes are composed of eight histone molecules, two from each class. Histones are very basic proteins.

- The nucleosomes are further condensed around a central axis into a 30 nm fiber, with the assistance of another histone, histone 1. These fibers of DNA and nucleosomes are further compacted into loops. During mitosis more compaction occurs, and the sister chromatids become visible. See Figure 9.7.

Preparation for separation of the chromosomes

- Separation of the sister chromatids into the daughter cells is facilitated by microtubules.

- Microtubules grow from a microtubule-organizing center, the centrosome, which doubles and separates during mitosis. The microtubules form a spindle along which the chromosomes will move. Later in the cell cycle, some of these microtubules (the kinetochore microtubules) attach to the chromosomes and help separate them to opposite poles of the cell.

- Centrosome duplication occurs during S phase. Centrosomes may contain centrioles, which consist of two hollow tubes of microtubules positioned at right angles to one another.

- Duplicated centrosomes migrate to opposite poles of the cell during the G2-to-M transition.

Mitosis

- Mitosis can be divided into five phases: prophase, prometaphase, metaphase, anaphase, and telophase.

- During prophase, the newly replicated chromosomes condense so that the sister chromatids (which are held together by cohesion proteins located at the centromere) are visible. Kinetochores assemble at the centromere of each newly replicated chromosome.

- Prometaphase signals the breakup of the nuclear envelope, the attachment of kinetochore microtubules to the kinetochores of each newly replicated chromosome, and movement of the chromosomes to the equatorial plate.

- Metaphase signals the positioning of all of the centromeres at the equatorial (metaphase) plate.

- At the end of metaphase the cohesion proteins are hydrolyzed by the protease separase. Separase is normally bound to an inhibitory subunit called securin; when all the chromatids are connected to the spindle, securin is hydrolyzed, and separase becomes active.

- During anaphase, the chromosomes move to opposite poles of the spindle assisted by dynein motors acting at the kinetochores, by the shortening of the microtubules at the poles, and by the sliding of polar microtubules, which pushes the poles further apart.

- At telophase, the poles break down, the chromosomes begin to uncoil, and the nuclear envelope and nuclei coalesce, resulting in two identical nuclei.

Cytokinesis

- During cytokinesis in animal cells, the cell membrane contracts due to the interaction of actin and myosin microfilaments at the cell membrane. See Figure 9.10a.

- In plant cell cytokinesis, a cell plate forms between the newly segregated chromosomes. This cell plate is derived from Golgi vesicles located at the equatorial region, which fuse to form a plasma membrane. The plasma membrane secretes plant cell wall materials to complete cell division.

- On completion of cytokinesis, there are two distinct cells, each with a full complement of chromosomes.

Cell reproduction

- There are two kinds of reproduction: asexual, in which the replicated chromosomes are separated by mitosis, and sexual, in which gametes from two different parents fuse to form a zygote. This fusion is called fertilization. Gametes each contain a single set of chromosomes (n) resulting from meiosis, and the zygote contains two sets of chromosomes ($2n$), one set from each parent.

- Somatic cells from diploid organisms contain pairs of homologous chromosomes, one set of chromosomes coming from each parent. Homologous chromosomes are similar in size and appearance and bear corresponding genetic information.

- Haploid cells contain only one set of chromosomes. Gametes are haploid cells.

- Mature organisms can exist primarily in the diploid or haploid state or can alternate between haploid and diploid states during their life cycles.

- Haplontic organisms are haploid as mature organisms. They produce spores that fuse to form a zygote, which is diploid. The zygote then undergoes meiosis to become haploid again. Those cells divide mitotically to become the mature organism.

- Diplontic organisms are diploid in their mature state; only their gametes are haploid.

- Most plants and some protists have mature forms that alternate between diploid and haploid stages (see Figure 9.12).

- Fusion of gametes to form a zygote increases diversity because the zygote's genetic makeup is derived from two different parents, each contributing one haploid set of randomly selected chromosomes.

- Mitotic chromosomes can be characterized by their shape, number, and size. Each organism has a distinctive set of chromosomes, called a karyotype.

Meiosis

- Meiosis is distinguished from mitosis by *two* nuclear divisions (meiosis I and II), but only *one* round of DNA replication, resulting in a reduction in the number of chromosomes from diploid to haploid.

- Each haploid cell has one complete set of chromosomes, and the combinations of alleles on the chromosomes of each gamete are different. The advantage of creating gametes with different combinations of alleles is genetic diversity in the future zygote, potentially increasing its ability to survive.

- There are five phases in each meiotic division: prophase, prometaphase, metaphase, anaphase, and telophase.

- In prophase of meiosis I, the newly replicated chromosomes pair with their homologs. Because there are four chromatids (two from each homolog), these paired chromosomes are called tetrads or bivalents.

- The synaptonemal complex is the aggregate of proteins that promotes the synapsing of the chromatids during prophase. Crossing over between nonsister chromatids on homologous chromosomes occurs during prophase at special sites called chiasmata (see Figure 9.15).

- Prometaphase of meiosis I marks the disappearance of the nuclear envelope, the disaggregation of the nucleoli, the formation of the spindle, and the attachment of the microtubules to the kinetochore. Kinetochores of both chromatids in each chromosome become attached to the same half spindle. As a result, when the chromosomes are separated to opposite poles by microtubules attached to the kinetochore, the entire homolog is pulled to the pole.

- In some organisms the completion of meiosis I is followed by an interphase termed interkinesis, whereas other organisms enter meiosis II immediately after meiosis I.

- No DNA replication occurs before meiosis II begins. The number of chromosomes on the equatorial plate is half the number in the mitotic nucleus (even though each chromosome still consists of two chromatids). Unlike the chromatids in mitosis, each chromatid is different from its sister chromatid, due to crossing over in meiosis I.

- Segregation of each chromatid occurs as in mitosis, with microtubules forming a spindle and the chromatids separating due to the breakdown of cohesion proteins.

- Each of the four haploid nuclei resulting from meiosis is genetically different due to crossing over in prophase I and due to the random segregation of each chromatid to the daughter cells. (In humans there are 23 homologs and thus 2^{23} ways these chromosomes can segregate into the daughter nuclei.)

Abnormal numbers and kinds of chromosomes

- Aneuploidy, an abnormal number of chromosomes, occurs in cells whose chromosomes fail to segregate in meiosis I (both homologs fail to separate and go to the same pole) or meiosis II (both chromatids fail to separate and go to one pole; see Figure 9.18).

- One of the causes of nondisjunction in meiosis I may be a lack of cohesins, which orient chromosomes at the equatorial plate during metaphase I. In the absence of cohesins, homologs may line up at random and wind up in the wrong daughter cell.

- Translocations, another chromosome abnormality, occur when part of one chromosome breaks away and becomes attached to another chromosome.

- Polyploids are cells that have increased numbers of complete sets of chromosomes. Polyploids with an even number of sets of chromosomes ($4n$, $6n$, $8n$) can undergo meiosis successfully, whereas odd-number polyploids ($3n$, $5n$, etc.) cannot, because each homolog needs to pair with is partner.

Cell death

- Cell death can occur in cells in two ways: by necrosis (cells poisoned or starved of essential nutrients) or by apoptosis (programmed cell death). Apoptosis is a normal process during development.

- Signals for cell death include a lack of mitotic signals and recognition of DNA damage. See Table 9.2.

The Big Picture

- In the presence of an internal or external environmental signal, cells undergo a specific set of steps to replicate their DNA, segregate those newly replicated chromosomes, and divide their cytoplasm into two daughter cells. These steps ensure that the daughter cells receive a complete set of genetic instructions.

- Diploid organisms produce haploid gametes, which fuse during fertilization to form new diploid organisms.

- During the generation of these gametes, crossing over between sister chromatids and random segregation of those chromatids create genetically diverse gametes, leading to increased species diversity and survival.

Common Problem Areas

- Be sure you can distinguish the differences between mitosis and meiosis. In mitosis, the replicated chromosomes are segregated to the daughter cells, resulting in each daughter cell's having two sets of chromosomes (diploid), like its parent cell. After every DNA replication, there is one cell division. In meiosis, the replicated chromosomes undergo two cell divisions (for every one round of DNA replication), resulting in cells that have only one set of chromosomes (haploid). Furthermore, the homologous chromosomes pair in prophase of meiosis I (which is not the case in mitosis), and crossing over occurs.

Study Strategies

- Make several sequential drawings that show a cell with four chromosomes and what those chromosomes look like after DNA replication and during each phase of mitosis. Repeat these sequential drawings for meiosis I and II, using different colors for each homologous chromosome. In your drawings be sure to include crossing over in prophase I of meiosis.

- Draw a picture of the cell cycle and indicate the points at which the cyclin-Cdk complexes act during that cell cycle. Refer to Figure 9.4 as well as the information in your text.

- List the five phases of mitosis and describe what happens in each phase. Repeat this process for meiosis.

- Review the following tutorials and activities on the Student Website/CD:
Tutorial 9.1 Mitosis
Tutorial 9.2 Meiosis
Activity 9.1 The Mitotic Spindle
Activity 9.2 Images of Mitosis
Activity 9.3 Sexual Life Cycle
Activity 9.4 Images of Mitosis

Test Yourself

Knowledge and Synthesis Questions

1. Which of the following is true of mitosis?
   a. The chromosome number in the resulting cells is halved.
   b. DNA replication is completed prior to the beginning of this phase.
   c. The chromosome number of the resulting cells is the same as that of the parent cell.
   d. Both b and c

2. Which of the following is true of meiosis?
   a. The chromosome number in the resulting cells is halved.
   b. DNA replication occurs after the beginning of this phase.
   c. The homologs do not pair during prophase I.
   d. The chromosome number of the resulting cells is the same as that of the parent cell.

3. Which of the following is true of kinetochores?
   a. They are located at the centromere of each chromosome.
   b. They are the sites where microtubules attach to separate the chromosomes.
   c. They are organized so that there is one per sister chromatid.
   d. All of the above

4. Which of the following is true of the mitotic spindle?
   a. It is composed of actin and myosin microfilaments.
   b. It is composed of kinetochores at the metaphase plate.
   c. It is composed of microtubules, which help separate the chromosomes to opposite poles of the cell.
   d. It originates only at the centrioles in the centrosomes.

5. Imagine that there is a mutation in the Cdk2 gene such that its gene product is nonfunctional. What kind of effect would this mutation have on a mature red blood cell?
   a. The cell would be unable to replicate its DNA.
   b. The cell would not be able to enter G1.
   c. The cell would be unable to reproduce itself.
   d. There would be no effect, because mature red blood cells do not enter the cell cycle.

6. Imagine that there is a mutation in the Cdk2 gene such that its gene product is nonfunctional. What kind of effect would this mutation have on a cell at the tip of a plant's root?
   a. The cell would be unable to replicate its DNA.
   b. The cell would be unable to enter G1.
   c. The cell would be unable to reproduce itself.
   d. Both a and c

7. Which is true of DNA replication and cytokinesis in *Escherichia coli*?
   a. DNA replication occurs in the nucleus.
   b. Cytokinesis is facilitated by microfilaments of myosin and actin.
   c. Cell reproduction is initiated by reproductive signals, which result in DNA replication, DNA segregation, and cytokinesis.
   d. The *E. coli* chromosome is linear.

8. Which of the following is true of chromatids?
   a. They are replicated chromosomes still joined together at the centromere.
   b. They are identical in mitotic chromosomes.
   c. They are identical in meiotic chromosomes.
   d. Both a and b

9. Histones are positively charged because
   a. the majority of the ions in the nucleus of the cell are negatively charged.
   b. histones interact with acidic residues of proteins found in the nucleus.
   c. the basic side chains of histone proteins interact with the negatively charged DNA.
   d. histones have a majority of acidic residues in their protein sequence.

10. Chromosome movement during anaphase is the result of
    a. the molecular motors at the kinetochores that move the chromosomes toward the poles.
    b. molecular motors at the centrosome that pull the microtubules toward the poles.
    c. shortening of the microtubules at the centrosome that pull the chromosomes toward the poles.
    d. Both a and d

11. Programmed cell death (apoptosis)
    a. occurs in cells that have been deprived of essential nutrients.
    b. occurs only in cells that have damaged DNA.
    c. is a natural process during development.
    d. is signaled by the initiation of mitosis.

Application Questions

1. How is cell division different in prokaryotic cells and eukaryotic cells?

2. By administering a drug that inhibits cytokinesis, you have created peaches that are tetraploid. How many sets of chromosomes do these peaches have? (What is the ploidy of these chromosomes?) Will these peaches produce gametes that are fertile? What if the peaches were triploid?

3. How does cytokinesis differ in animal and plant cells?

4. Describe how two meters of DNA in a typical human cell can fit into the nucleus, which is 5 μm in diameter.

5. Describe two ways that the genetic diversity of organisms is increased during meiosis.

6. What is the fate of cells that exit the cell cycle?

Answers

Knowledge and Synthesis Answers

1. **d.** Mitosis occurs after DNA replication and results in cells having the same number of chromosomes as the parent cell.

2. **a.** Meiosis occurs after one round of DNA replication. Homologous chromosomes pair during prophase I of meiosis, and the resulting cells have half the number of chromosomes as the parent cell.

3. **d.** Kinetochores, one per sister chromatid, are assembled at the centromere of each chromosome and are the sites in which microtubules attach to segregate the chromosomes.

4. **c.** The mitotic spindle is composed of microtubules, not actin and myosin filaments. The spindle originates from the centrosome, which may or may not have centrioles.

5. **d.** Many cells, such as red blood cells, muscle cells, and nerve cells, lose their ability to divide as they mature.

6. **d.** A cell at the tip of a plant's root needs to be able to divide to contribute to the growth of the root. Dividing cells do enter the cell cycle, and cyclin-Cdk complexes signal transitions in the cell cycle. The cyclin E-Cdk2 complex acts in the middle of G1, and the cyclin A-Cdk2 complex acts in S1 and stimulates DNA replication. Without functional Cdk2, no catalytically active cyclin-Cdk complexes can form. The cell will be unable to replicate its DNA and will not progress through the cell cycle to reproduce itself.

7. **c.** *Escherichia coli* is a prokaryote and lacks a nucleus, has a circular chromosome, and does not synthesize actin or myosin proteins. Cytokinesis in *E. coli* is a result of a reproductive signal that causes the DNA to be replicated and segregated, and finally causes the cell to divide.

8. **d.** Chromatids are highly condensed, newly replicated chromosomes, which will be segregated to the daughter cells. After DNA replication, chromatids are still attached to one another at the centromere. Meiotic sister chromatids are different from one another due to crossing over in prophase of meiosis I. Mitotic sister chromatids are identical.

9. **c.** The positive charges on histone proteins are due to the large number of basic amino acid residues found in these proteins. These positive charges interact with the nega-

tively charged phosphate sugar backbone of DNA during assembly of the DNA on the nucleosome.

10. **d.** Chromosomes are attached to the microtubules at their kinetochores. There are molecular motors at the kinetochores, which help move the chromosomes to opposite poles. Chromosomes are also pulled toward the poles by the shortening of the kinetochore microtubules.

11. **c.** Programmed cell death occurs during the development of many organisms (for instance, tadpoles lose their tails to become adult frogs). One of the stimuli for programmed cell death is DNA damage, but it is not the only cause of death. Necrosis (cell death that is not programmed) occurs when cells have been deprived of essential nutrients. The initiation of mitosis is part of the cell cycle, in which cells reproduce, and is not a step in programmed cell death.

Application Answers

1. In prokaryotic cells there is only one circular chromosome, which is attached to the membrane. As the cell enlarges to prepare for division, the newly replicated daughter chromosomes, which are still attached to the membrane, are separated at opposite sides of the cell. During fission, the cell membrane pinches in, and cell wall components are synthesized between the daughter cells. In eukaryotic cells, there are more chromosomes, and they are linear. The cell undergoes a sequential set of steps called the cell cycle, in which the chromosomes are replicated and then separated to opposite poles of the cell. Microtubules are used to segregate the chromosomes equally into the daughter cells, and actin and myosin filaments cause the cell membrane to contract to form two daughter cells.

2. Peaches that are tetraploid have four sets of chromosomes. Because there are an even number of chromosomes ($4n$), each replicated homologous chromosome will be able to find a replicated homolog to pair with at meiosis and will produce fertile gametes. These gametes will be diploid. Triploid cells will not be fertile because one of the three homologs will not find its pair during prophase of meiosis I, and the single homologs will be segregated randomly into the daughter cells.

3. In animal cells, cytokinesis results from the interaction of myosin and actin filaments, which causes the cell membrane to pinch in and divide the cytoplasm into two cells. In plant cells, a cell plate forms between the newly segregated chromosomes, and Golgi vesicles fuse at that site to form the new cell membranes. Cell wall components are then secreted between the plasma membranes to complete cytokinesis.

4. See Figure 9.6.

5. Genetic diversity is increased during crossing over of prophase I of meiosis so that each gamete has chromosomes with different combinations of alleles. During meiosis, each homologous chromosome is randomly segregated to one of the two poles, resulting 2^{23} different possible combinations of homologous chromosomes per gamete.

6. Cells that exit the cell cycle no longer divide. They may live for years without dividing (as nerve and muscle cells do) or they may go through apoptosis after only a few days (as cells in the blood and intestine do).

10 Genetics: Mendel and Beyond

Important Concepts

Genes are units of inheritance.

- Mendel studied the inheritance of characters (such as flower color) and traits (a particular feature, such as a red, purple, or white color) as they were passed from one generation of pea plants to another.

- He used plants that were true-breeding (which meant that the observed trait must be the only form for many generations). To test if plants were true-breeding, Mendel allowed these plants to self-fertilize for many successive generations to determine if the trait was consistently expressed in every generation.

- In genetic crosses, the two individuals in the initial cross are the parents (the P generation) and their offspring are the first filial generation (the F_1 generation). The F_2 (second filial generation) are the offspring from a cross of the F_1 generation.

- Monohybrid crosses are matings between two true-breeding parents, each expressing one different trait (i.e., a cross between one true-breeding parent with a spherical seed and one true-breeding parent with a wrinkled seed). Mendel found that in these crosses, the F_1 generation expressed only one trait (spherical), and in the F_2 generation both the spherical and the wrinkled traits were expressed in ratios of 3:1 (spherical to wrinkled). He termed the F_1 trait (the spherical seed) dominant and the F_2 trait (the wrinkled seed) recessive.

- Mendel's theory stated that particles of inherited traits existed in pea plants in pairs and that these particles separated from each other and were assorted independently during gamete formation. Each of the gametes received one of these particles, whereas the zygote received two particles for a trait, one from each parent. This unit of inheritance is a gene.

- Alleles are different forms of the same gene. The gene for seed shape in peas has a spherical allele and a wrinkled allele.

- True-breeding parents are homozygous for that trait and have two copies of the same allele, one on each homologous chromosome.

- Heterozygous individuals have two different alleles for a gene, one on each homologous chromosome. For example, the F_1 generation of a monohybrid cross between pea plants with spherical seeds and pea plants with wrinkled seeds is heterozygous for seed shape, having one spherical allele and one wrinkled allele.

- The phenotype of an organism is the physical appearance of that organism (i.e., having seeds that are spherical or wrinkled).

- Mendel's theories have been restated as laws. Mendel's first law, the law of segregation, states that when any individual produces gametes, alleles separate so that each gamete receives only one allele for each gene.

Predicting the outcome of genetic crosses.

- Punnett squares can be utilized to analyze the potential outcome of a genetic cross. All of the possible gametes from one parent are lined up on one side of the Punnett square, and all of the gametes of the other parent are lined up on the other side. Genotypes of the progeny are predicted by placing both alleles of the gametes in the square (see Figures 10.4, 10.6, and 10.7).

- Genes are located at particular sites on chromosomes called loci (locus, singular).

- Test crosses are used to determine if a given individual is heterozygous or homozygous for a dominant allele. That individual is crossed with an individual homozygous for the recessive trait (the test cross). If the test individual is heterozygous, then approximately one-half of the offspring from this cross will express the dominant trait and one-half will express the recessive trait. If the individual from this test cross is homozygous for that trait, then all of the offspring will express the dominant trait (see Figure 10.6).

- Mendel's second law, the law of independent assortment, states that all of the alleles of different genes assort independently during gamete formation. When parents that are heterozygous for two traits—for example, seed shape and seed color—are crossed (a dihybrid cross), all the alleles are expressed in the F_2 generation. In the F_2 progeny from this dihybrid cross, both yellow and green seed color are observed as well as wrinkled and spherical seeds. Review the Punnett square in Figure 10.7.

Probability calculations can be used to predict the outcome of genetic crosses.

- The product rule states that when predicting the outcome of two independent events happening together, that outcome is equal to the product of the probability of each of those individual events (see Figure 10.9).

- The sum rule is used to predict the probability of an event that can occur in two or more different ways. The probability that the event will occur in one of these ways is equal to the sum of the individual probabilities (see Figure 10.9).

Pedigrees can be used to predict how traits are inherited from generation to generation.

- If a trait is due to a rare dominant allele:
  - The affected person has an affected parent.
  - On average, one-half of the offspring of an affected parent will be affected.
  - The phenotype is seen equally in both sexes.
- If a trait is due to a rare recessive allele:
  - The unaffected parents have affected children.
  - On average, one-fourth of the children from unaffected parents express the trait.
  - The phenotype occurs in both sexes.

Mutations are changes in the DNA sequence that are stable and passed on to the progeny (heritable).

- "Wild type" refers to traits that occur in most individuals in nature, whereas "mutant" refers to traits that are different from wild type.
- A polymorphic trait is a trait that has many different forms, with each individual phenotype present in less than 99 percent of the population.

There are several extensions of Mendel's laws.

- Multiple alleles: There can be more than two alleles for a particular trait (for example, rabbit fur color; see Figure 10.12).
- Incomplete dominance: Individuals that are heterozygotic for a particular trait appear as intermediate phenotypes of the parents. For example, a cross between a red-flowered plant and a white-flowered plant yields F_1 plants with pink flowers (see Figure 10.13).
- Codominance: Both alleles are expressed in the individual (for example, blood types in humans).
- Pleiotropic: A single allele can have more than one effect in an individual. For example, in the Siamese cat, a single allele can affect both pigmentation and crossed eyes.
- Epistasis: The phenotypic expression of one gene is affected by another gene. For example, an allele that prevents pigmentation in fur can mask other alleles that produce a specific pigment in that individual, producing an individual lacking any color, an albino (see Figure 10.15). Complementary gene action is a form of epistasis in which the expression of one gene depends on another.
- Hybrid vigor or heterosis: Individuals heterozygous for a particular trait are superior to the homozygote. For example, hybrid varieties of corn produce much better yields than their homozygous parents.
- Environmental effects: Environmental conditions such as light, temperature, or nutrition can affect the expression of a particular trait. Both penetrance and expressivity are related to environmental effects.

- Penetrance is the proportion of individuals within a group with a particular genotype that actually express that genotype.
- Expressivity is the degree a particular genotype is expressed in an individual.
- Quantitative or continuous variation: More than one gene (quantitative trait loci) can affect a particular phenotype, with each allele intensifying or diminishing the phenotype (for example, human height).
- Linkage: Different alleles on the same chromosome do not assort independently, as first noted by Morgan in his studies on *Drosophila*.
  - Linkage groups are the full set of loci on a given chromosome.
  - Recombination frequencies between alleles that are linked (due to crossing over during prophase of meiosis I) can be used to map positions of genes on a chromosome. Recombination frequencies are expressed in map units and are equal to the percent of recombinant progeny divided by the total progeny. A map unit (also referred to as a centimorgan) corresponds to a recombination frequency of 0.01 (see Figures 10.20, 10.21, and 10.22).

Sex is determined in different ways in different species.

- Monoecious organisms (earthworms, pea plants, corn, etc.) produce both female and male gametes.
- Dioecious organisms (vertebrates, some insects) produce only male or only female gametes resulting in two separate sexes.
- In bees, sex determination is decided by number of chromosomes and environmental conditions. Diploid bees are female workers unless they are fed royal jelly, in which case they become queens. Haploid bees are males.
- In grasshoppers, sex is determined by the presence of a sex chromosome and its ratio to the other chromosomes (called autosomes). In female grasshoppers, there are two X chromosomes; in male grasshoppers there is only one X chromosome.
- In mammals, there are two X chromosomes in females and one X chromosome and one Y chromosome in males. The *SRY* (sex determining region on the arm of the Y chromosome) determines maleness in males. *SRY* is responsible for the primary sex determination in males and causes the embryo to develop sperm-producing tissue, the testes. Males without the *SRY* gene develop as females. The *DAX1* gene, which encodes the antitestis factor, is found on the X chromosome and inhibits the development of testis tissue. In XY individuals (males), the *SRY* gene product inhibits the effect of *DAX1*, and testes develop.
- Secondary sex determination, the outward manifestation of male and female characteristics, is determined by autosomal and X-linked genes that control levels of testosterone or estrogen. Secondary sex characteristics in humans include body type, breast development, body hair, and voice.
- In *Drosophila*, sex is determined by the ratio of X chromosomes to autosomes: XX is female (as is XXY), and XY is

male (as is XO). The Y chromosome is not necessary for sex determination but is required for male fertility.

- In the worm *Caenorhabditis elegans*, sex is determined by the X to autosome ratio. Hermaphrodites have an X:A ratio greater than 0.67; males have ratios less than 0.67.

- In birds, moths, and butterflies, ZZ (XX) are male and ZW (XY) are female. In these species the egg (rather than the sperm) determines the sex of the offspring.

Sex-linked inheritance is governed by loci on the sex chromosome.

- In mammals and other organisms, including some insects, females have two copies of each X-linked gene, whereas males only have one copy of the X-linked genes.

- Because males are hemizygous for X-linked traits, all their X-linked traits will be expressed.

- If an allele is located on the X chromosome, reciprocal crosses do not have the same result. For instance, a cross between a homozygous red-eyed female *Drosophila* and a white-eyed male *Drosophila* will generate all red-eyed flies. The reciprocal cross between a white-eyed female (who must be homozygous, because the white-eyed allele is recessive) and a red-eyed male will produce an F_1 generation in which all female flies have red eyes and all male flies have white eyes.

- Males with X-linked traits can only pass those traits to their daughters.

- Females who are carriers of an X-linked trait pass that trait on average to one-half of their sons and one-half of their daughters. Because sons always receive the X chromosome from their mother, one-half of those sons, on average, will express the X-linked trait. Daughters who receive the X-linked trait will be carriers because their father donates the other X chromosome.

- X-linked traits can skip a generation; a trait that is expressed in the grandfather will be carried by his daughter and may be expressed in his grandson.

- There are several dozen genes in humans that are Y linked. These traits are always passed from father to son.

Cytoplasmic inheritance differs from the Mendelian pattern.

- Mitochondria and plastids contain genomes that are passed to the zygote from the gamete that makes the largest cytoplasmic contribution (in humans, the egg).

- These organelles are generally so numerous that genes for cytoplasmic traits are polyploid.

- Genes in organelles mutate faster than those in the nucleus, leading to multiple alleles for many nonnuclear genes.

- In plants mutations in plastid genes affect proteins that assemble chlorophyll molecules into photosystems, resulting in white rather than green leaves.

- Mutations in mitochondrial genes affect the electron transport chain and the synthesis of ATP. These mutations have a strong effect on tissues that are sensitive to ATP concentrations, particularly the nervous and muscular systems and kidneys.

The Big Picture

- Genes are units of inheritance that are passed down to the progeny by the fusion of gametes to form a new organism.

- Meiosis generates gametes that contain only one allele for each gene, and those alleles are segregated independently (unless they are linked).

- By studying the expression of particular phenotypes in successive generations, one can begin to understand the alleles that gave rise to those phenotypes (i.e., rare recessive, rare dominant, sex-linked).

- Exceptions to Mendel's rules also help explain the interactions of gene products of particular alleles that lead to the observed phenotype.

Common Problem Areas

- It is easy to be overwhelmed by all the exceptions to Mendel's rules. Initially, make sure you understand Mendel's laws of segregation and independent assortment. Examine the exceptions to these rules by thinking about the molecular role of these alleles. For example, what is the explanation for a codominant phenotype at the molecular level, such as the AB blood type in humans? Both gene products are synthesized by the cell and modify the cell surface glycoproteins, producing both A and B antigens. Or what is the molecular explanation for epistasis? The enzyme required to produce an initial substrate necessary for pigmentation is missing, even though the other gene products (further down the biochemical pathway) for producing a particular color are present in the cell.

Study Strategies

- Review the figures of meiosis in Chapter 9 to see how Mendel's laws of independent assortment and segregation apply. Draw the chromosomes with different alleles (*S* on one chromosome, *s* on the other) and follow how those alleles are segregated to the gametes during meiosis. Repeat this process, but draw two separate pairs of chromosome with each bearing a different allele (for instance *Ss* and *Yy*). Show how these alleles sort independently in the gametes (review Figure 10.8).

- Use the Punnett square to predict the outcome of different monohybrid and dihybrid crosses. Use the probability rules to predict the genotypes and phenotypes of these same crosses and compare the two methods.

- After reviewing Mendel's laws of independent assortment and allele segregation, make a list of the exceptions to those laws with specific examples for each exception.

- Review the linkage problem in Figure 10.22 to familiarize yourself with allele loci on a chromosome and how recombination frequencies affect the genotypes of the gametes.

- Review the following tutorials and activities on the Student Website/CD:
  Tutorial 10.1 Independent Assortment of Alleles
  Tutorial 10.2 Alleles That Do Not Sort Independently
  Activity 10.1 Homozygous or Heterozygous?
  Activity 10.2 Concept Matching I
  Activity 10.3 Concept Matching II

Test Yourself
Knowledge and Synthesis Questions

1. In the beginning of Chapter 10, hemophilia is mentioned as a trait carried by the mother and passed to her sons. What is the pattern of inheritance for this trait?
   a. Hemophilia is an allele carried on one of the mother's autosomal chromosomes.
   b. Hemophilia is an allele carried on the Y chromosome because more males have this genetic disorder than females.
   c. Hemophilia is an allele carried on the X chromosome and can be directly inherited by the son from the father or the mother.
   d. Hemophilia is carried on the X chromosome and can only be inherited by the son if the mother is a carrier.

2. Originally, genetic inheritance was thought to be a function of the blending of traits from the two parents. Which exception to Mendel's rules is an example of blending?
   a. Polygenic inheritance
   b. Incomplete dominance
   c. Codominance
   d. Pleiotropism

3. True-breeding plants
   a. produce the same offspring when crossed for many generations.
   b. result from a monohybrid cross.
   c. result from a dihybrid cross.
   d. result from crossing over during prophase I of meiosis.

4. What is the probability that a cross between a true-breeding pea plant with spherical seeds and a true-breeding pea plant with wrinkled seeds will produce F_1 progeny with spherical seeds?
   a. 1/2
   b. 1/4
   c. 0
   d. 1

5. What is the pattern of inheritance for a rare recessive allele?
   a. Every affected person has an affected parent.
   b. Unaffected parents can produce children who are affected.
   c. Unaffected mothers have affected sons and daughters who are carriers.
   d. None of the above

6. What is the pattern of inheritance for a rare dominant allele?
   a. Every affected person has an affected parent.
   b. Unaffected parents can produce children who are affected.
   c. Unaffected mothers have affected sons and daughters who are carriers.
   d. None of the above

7. What is the pattern of inheritance for a sex-linked allele?
   a. Every affected person has an affected parent.
   b. Unaffected parents can produce children who are affected.
   c. Unaffected mothers have affected sons and daughters who are carriers.
   d. None of the above

8. Penetrance and expressivity are related to
   a. the increased expression of a particular trait when a hybrid species is formed.
   b. quantitative traits that diminish or intensify a particular phenotype.
   c. the influence of environment on the expression of a particular genotype.
   d. the expression of one gene masking the effects of another gene.

9. Sex determination in grasshoppers, humans, and *Drosophila* is similar because
   a. females are hemizygous.
   b. males have one X chromosome and females have two X chromosomes.
   c. all males always have one Y chromosome in all three species.
   d. the ratio of X chromosomes to sets of autosomes determines maleness or femaleness.

10. Linked genes are genes that
   a. assort independently.
   b. segregate equally in the gametes during meiosis.
   c. always contribute the same trait to the zygote.
   d. are found on the same chromosome.
   e. recombine during mitosis.

11. Cytoplasmic inheritance
   a. results from polygenic nuclear traits.
   b. is the result of gametes contributing equal amounts of cytoplasm to the zygote.
   c. is determined by genes on DNA molecules in mitochondria and chloroplasts.
   d. follows Mendel's law of segregation.

Application Questions

1. Draw a pedigree for three generations in which the grandfather has red–green color blindness and his daughter is a carrier. This daughter has four sons. Predict how many of the sons will be color-blind.

2. Draw a sample pedigree with three generations in which the paternal grandfather has a rare dominant autosomal trait.

3. Draw a sample pedigree with three generations in which the maternal grandmother and paternal grandfather are carriers of a rare recessive autosomal trait.

4. Cytoplasmic traits in certain species of trees are passed from the male plant to all of its progeny. Compare this observation to cytoplasmic inheritance in humans.

5. You are a genetics counselor who is working with a 21-year-old pregnant woman who has just discovered that her father has Huntington's chorea, a rare dominant autosomal trait. This disease usually develops in middle-aged individuals, so people carrying this trait do not find out they have this genetic disorder until midlife. What are the chances that the child she is carrying will develop Huntington's chorea? (Assume that her husband's family has no history of Huntington's chorea.) What is the chance that she has Huntington's chorea?

Answers
Knowledge and Synthesis Answers

1. **d.** Hemophilia is an X-linked trait and can only be inherited by the son from his mother's X chromosome. The father contributes the Y chromosome to his son (not his X chromosome) and thus cannot pass any of his X-linked alleles to his son.

2. **b.** Incomplete dominance results in the progeny's expressing an intermediate form of the two parental alleles. (In a cross between red-flowered plants and white-flowered plants, the expression of pink-flowered plants would be a "blend" of the parental traits.) Codominance is not an example of blending because both alleles are fully expressed in the individual.

3. **a.** Monohybrid and dihybrid crosses produce heterozygous individuals; true-breeding individuals are always homozygous.

4. **d.** This is an example of a monohybrid cross. All of the F_1 progeny would have spherical seeds. (The F_1 generation would all have the genotype *Ss*, producing the phenotype of spherical seeds because the spherical allele, *S*, is dominant to the wrinkled allele, *s*.)

5. **b.** Rare recessive alleles can be carried by both parents but not expressed in those parents. The parents will be heterozygous for this allele (*Aa*). Their children will have a one-fourth probability of expressing that recessive allele (*aa*).

6. **a.** If an allele is dominant, every affected individual has at least one dominant allele. An affected individual must have received that allele from one of his or her parents. Because the allele is dominant, that parent must also be affected.

7. **c.** The most common sex-linked alleles are X-linked and are passed from a mother to her son (because the mother always donates one of her X chromosomes to her son and the father always donates the Y chromosome to his son). Daughters can also receive the X-linked allele from their mothers, but the father donates the other X chromosome, so daughters can be carriers.

8. **c.** Penetrance and expressivity are related to the effects the environment has on a particular phenotype. Answer **a** refers to hybrid vigor, answer **b** refers to quantitative traits, and answer **d** refers to epistasis.

9. **b.** In these three species, females have two X chromosomes and males have one X chromosome. The ratio of X chromosomes to autosome sets is important for sex determination in *Drosophila*, but not in grasshoppers or humans. In *Drosophila* male flies can be XO (they are normally XY) and female flies can be XXY (they are normally XX).

10. **d.** Linked genes by definition are on the same chromosome and thus do not assort independently, do not contribute the same trait to the zygote, and do not recombine during mitosis or segregate equally to the gametes during meiosis.

11. **c.** The genes on the mitochondria and chloroplast chromosomes are passed on to all of the progeny from the gamete that contributes the majority of the cytoplasm.

Application Answers

1. See Figure 10.24, generations II, III, and IV. One-half of her sons could be color-blind.

2. See Figure 10.10.

3. See Figure 10.11, generations II, III, and IV.

4. In humans, the gamete with the largest cytoplasmic contribution is the egg, so cytoplasmic inheritance is passed from the female parent to all of her children. In certain tree species, the male gamete contributes the majority of the cytoplasm to the zygote, so that all of the mitochondria and chloroplasts in the zygote are inherited from the male parent.

5. She has a 50 percent chance that she will develop Huntington's chorea. Because the trait is an autosomal dominant allele, one-half of her father's gametes will contain the homologous chromosome carrying that allele and one-half of his gametes will contain the homologous chromosome that carries the wild-type allele. If she received the Huntington's allele, her child has a 50 percent chance of receiving this allele from her. The product rule is used to predict the probability that her child will inherit the Huntington's allele: 1/2 (the probability that she has the Huntington's allele) × 1/2 (the probability her child will inherit this allele from her) = 1/4 (the probability her child has the allele). Her child has a 25 percent chance of carrying the Huntington's chorea allele and thus of developing the disease. She has a 50 percent chance of carrying the Huntington's chorea allele.

11 DNA and Its Role in Heredity

Important Concepts

Specific observations and discoveries revealed that DNA is the genetic material.

- Staining techniques showed that DNA is present in the nucleus.

- Different organisms showed varying amounts of DNA in their nuclei.

- Proportional amounts of DNA were found in somatic and germ cells.

- Griffith's transformation experiments indicated that some transformation factor (later identified as DNA) could cause a heritable change in living cells.

- Avery, MacLeod, and McCarty's work used purified DNA to demonstrate that the transforming factor that contained the genetic information was DNA.

- Hershey and Chase, using a virus that infected bacteria, demonstrated definitively that DNA carried the hereditary information.

A series of related yet independent experiments elucidated the structure of DNA.

- Franklin and Wilkins prepared X ray crystallographic images, which were critical to understanding the structure of DNA.

- Chargaff provided rules for relative ratios of pyrimidines and purines, the building blocks of DNA.

- Watson and Crick coupled results from previous experimentation and model building and revealed the three-dimensional, double-helical structure of DNA.

The form of DNA is tied to its function.

- DNA is a right-handed double helix formed by two antiparallel DNA strands. The two strands are aligned in the helix with the 5′ end of one strand pairing with the 3′ end of its complementary strand.

- The sugar–phosphate backbones of the polynucleotide chains are on the outside of the helix, and the nitrogenous bases point toward the center of the helix.

- Base pairs are complementary; they consist of a pyrimidine pairing with a purine via hydrogen bonding (A pairs with T, G pairs with C). Hydrophobic interactions between the base pairs help them to stack on top of one another in the center of the molecule, stabilizing the molecule and maintaining a constant diameter.

- The specific base sequences in DNA are used to store genetic information. Altering the base sequence in the DNA by mutation may cause a change in the genetic information. Faithful replication of the DNA molecule is precisely generated using complementary base pairing. The expression of genetic information in the DNA results in specific phenotypes.

DNA must be able to be replicated in order to be heritable.

- All of the following must be present for replication to take place: a template strand, all four deoxyribonucleoside triphosphates (dATP, dCTP, dGTP, and dTTP), a primer base-paired to the template DNA, and DNA polymerase.

- Meselson and Stahl's experiments demonstrated that DNA replication is semiconservative: each parental strand of the DNA duplex served as a template for the newly synthesized (daughter) DNA, and each new DNA molecule was composed of one parental (template) and one newly replicated (daughter) strand (see Figure 11.9).

- During replication, DNA is unwound, and the new strand is synthesized by complementary base pairing of each incoming nucleotide to the parental strand via hydrogen bonds. New nucleotides are always added at the 3′ (—OH) end of the elongating strand via covalent bonds.

- Replication complexes assemble at origins on the DNA to begin replication. Prokaryotes have one replication origin; most eukaryotes have many. Replication proceeds in both directions from the replication origin, forming replication forks. Both leading-strand and lagging-strand synthesis occur at replication forks (see Figure 11.16).

- Replication complexes are composed of many proteins: helicases unwind parental DNA, single-stranded binding proteins prevent parental strands from reassociating, primases generate RNA primers, DNA polymerases elongate daughter strands from the 3′ OH end of those primers, and DNA ligases seal the nicks in the sugar–phosphate backbone.

- New daughter strands must begin with a short RNA primer, formed using the parental strand as a template. Once a primer is synthesized, DNA polymerase then adds deoxyribonucleotides to the 3′ (—OH) end of the primer.

Replication proceeds in a specified direction.

- Nucleotides can only be added to the 3′(—OH) end of a strand. Thus, the parental strand is read in the 3′-to-5′ direction, and the daughter strand elongates in the 5′-to-3′ direction.

- The leading strand elongates continuously in the same direction as the movement of the replication fork (5′ to 3′; "forward").

- The lagging strands (Okazaki fragments) are synthesized 5′ to 3′ in a direction opposite to that which the replication fork is moving ("backward"). This process requires many primers, and the resulting Okazaki fragments that are generated are ligated (covalently joined by DNA ligase) to form a continuous strand.

- Telomeres are repetitive sequences found at the ends of eukaryotic chromosomes, which, with the protein telomerase, help maintain the stability of those ends during each replicative cycle.

Many mistakes are made during initial synthesis of DNA; therefore, DNA proofreading and repair mechanisms are essential.

- DNA polymerase proofreads and repairs base mismatches during replication.

- Mismatch repair systems scan for possible errors in newly synthesized DNA.

- Excision repair mechanisms replace abnormal bases with functional bases.

Advanced techniques in molecular biology have allowed us to make multiple copies of DNA from a single strand and to determine the sequence of bases in DNA.

- The polymerase chain reaction (PCR) allows the rapid amplification of short DNA sequences into many identical copies (see Figure 11.20).

- DNA sequencing is now a highly automated process in which modified bases are mixed with the normal substrates for DNA replication to generate a mixture of DNA fragments. These fragments are used to determine the base sequence of the DNA.

The Big Picture

- Knowing that DNA is the genetic material has allowed scientists to understand how hereditary information is passed on at the molecular level and how mutations can alter that hereditary information.

- Advances in our knowledge of how DNA is replicated provide new technologies for understanding genes, their function, their expression, and genetic relatedness in different organisms.

Common Problem Areas

- Students frequently get lost in the history of discovery. Do not focus on *who* and *when,* but instead on *how* the experiments provide evidence to prove that DNA is the genetic material.

- Understanding DNA replication is a highly visual process. Use the figures in your text to "see" what is happening. Make your own diagrams of these processes.

- 5′ and 3′ ends are often confused when comparing parental (template) strands and daughter (newly synthesized) strands. On both strands, the 3′ end corresponds to a site where a hydroxyl group (—OH) is (or was), and the 5′ end corresponds to a site where a phosphate tail is (or was).

- Take advantage of laboratory activities that simulate or actually allow you to experience sequencing of PCR. Nearly all molecular research labs utilize one or both of these techniques.

Study Strategies

- Focus on how scientific evidence proved that DNA was the genetic material and how Meselson and Stahl demonstrated that DNA replication was semiconservative.

- Draw a picture of the replication fork. Position the primers on that fork, and then draw in the leading and lagging strands. Label all of the 3′ and 5′ ends. Use an arrow to indicate where ligase will seal up the ends.

- Make a list of all the proteins involved in DNA replication. Describe the function of each protein.

- Review the following tutorials on the Student Website/CD:
  Tutorial 11.1 The Meselson-Stahl Experiment
  Tutorial 11.2 DNA Replication, Part 1: Replication of a Chromosome and DNA Polymerization
  Tutorial 11.3 DNA Replication, Part 2: Coordination of Leading and Lagging Strand Synthesis

Test Yourself

Knowledge and Synthesis Questions

1. Griffith's experiments showing the transformation of R strain pneumococcus bacteria to S strain pneumococcus bacteria in the presence of heat-killed S strain bacteria gave evidence that
   a. an external factor was affecting the R strain bacteria.
   b. DNA was definitely the transforming factor.
   c. S strain bacteria could be reactivated after heat killing.
   d. All of the above

2. Experiments by Avery, MacLeod, and McCarty supported DNA as the genetic material by showing that
   a. both protein and DNA samples provided the transforming factor.
   b. DNA was not complex enough to be the genetic material.
   c. only samples with DNA provided transforming activity.
   d. even though DNA was molecularly simple, it provided adequate variation to act as the genetic material.

3. Hershey and Chase used radioactive ^{35}S and ^{32}P in experiments to provide evidence that DNA was the genetic material. These experiments pointed to DNA because
   a. progeny viruses retained ^{32}P but not ^{35}S.
   b. presence of ^{32}P in progeny viruses indicated that DNA was passed on.
   c. absence of ^{35}S in progeny viruses indicated that proteins were not passed on.
   d. All of the above

4. X ray crystallography provides information about the _____ of DNA but is limited because of the _____ of DNA. The technique is based on the pattern of _____ off the atoms in the molecule.
   a. structure; difficulty of purification; light absorption
   b. dimensions; molecular weight; diffraction
   c. molecular weight; shape; diffraction
   d. dimensions; linearity; light absorption

5. Chargaff observed that
   a. the amount of purines is roughly equal to the amount of pyrimidines in all tested organisms.
   b. the amount of A is roughly equal to the amount of T in all tested organisms.
   c. the amount of A + T is roughly equal to the amount of G + C in all tested organisms.
   d. Both a and b

6. Watson and Crick's model allowed them to visualize
   a. the molecular bonds of DNA.
   b. how the purines and pyrimidines fit together in a double helix.
   c. that the two strands of the DNA double helix were antiparallel.
   d. All of the above

7. A fundamental requirement for the function of genetic material is that it must be
   a. conserved among all organisms with very little variation.
   b. passed intact from one species to the next species.
   c. accurately replicated.
   d. found outside the nucleus.

8. Evidence indicating that DNA replication was semiconservative came from
   a. DNA staining techniques.
   b. DNA sequencing.
   c. density gradient studies using "heavy" nucleotides.
   d. None of the above

9. Current evidence indicates that replication complexes are attached to stationary nuclear components and that DNA is threaded through these complexes. Which of the following best describes the role of the replication complex?
   a. The complex acts as an enzymatic center for DNA replication.
   b. The complex binds specifically to replication origins, then controls the rate at which replication occurs.
   c. The complex is the initiating site of replication forks.
   d. All of the above

10. The primary function of DNA polymerase is to
    a. add nucleotides to the growing daughter strand.
    b. seal nicks along the sugar–phosphate backbone of the daughter strand.
    c. unwind the parent DNA double helix.
    d. prevent reassociation of the denatured parent DNA strands.

11. The lagging daughter strand of DNA is synthesized in what appears to be the "wrong" direction. This synthesis is accomplished by
    a. synthesizing short Okazaki fragments in a 5´-to-3´ direction.
    b. synthesizing multiple short RNA primers to initiate DNA replication.

    c. using DNA polymerase I to remove RNA primers from Okazaki fragments.
    d. All of the above

12. RNA primers are necessary in DNA synthesis because
    a. DNA polymerase is unable to initiate replication without an origin.
    b. the DNA polymerase enzyme can only catalyze the addition of deoxyribonucleotides onto the 3´ (—OH) end of an existing strand.
    c. RNA primase is the first enzyme in the replication complex.
    d. All of the above

13. Proofreading and repair occur
    a. at anytime during or after synthesis of DNA.
    b. only before DNA methylation occurs.
    c. only in the presence of DNA polymerase.
    d. only in the presence of an excision repair mechanism.

14. DNA replication is an _____ process and _____ energy.
    a. exergonic; does not require
    b. endothermic; does require
    c. endergonic; does require
    d. endodontic; does not require

15. *T. aquaticus* DNA polymerase is not denatured during the heat cycling required to denature DNA. This property led to advances in what technique?
    a RFLP analysis
    b. PCR
    c. Sequencing
    d. EPA

16. Thirty percent of the bases in a sample of DNA extracted from eukaryotic cells is adenine. What percentage of cytosine is present in this DNA?
    a. 10 percent
    b. 20 percent
    c. 30 percent
    d. 40 percent

17. Which of the following represents a bond between a purine and a pyrimidine (in that order)?
    a. C–T
    b. G–A
    c. G–C
    d. T–A

18. Which of the following statements about DNA replication is *false*?
    a. Okazaki fragments are synthesized as part of the leading strand.
    b. Replication forks represent areas of active DNA synthesis on the chromosomes.
    c. Error rates for DNA replication are often less than one in every billion base pairings.
    d. Ligases and polymerases function in the vicinity of replication forks.

19. Which of the following would *not* be found in a DNA molecule?
    a. Purines
    b. Ribose sugars
    c. Phosphates
    d. Sulfur

20. If a nucleotide lacking a hydroxyl group at the 3′ end were added to a PCR reaction, what would be the outcome?
    a. No additional nucleotides would be added to a growing strand containing that nucleotide.
    b. Strand elongation would proceed as normal.
    c. Nucleotides would only be added at the 5′ end.
    d. *T. aquaticus* DNA polymerase would be denatured.

Application Questions

1. Given the following parent strand sequence, what would the daughter strand sequence look like?

   5′ – G C T A A C T G T G A T C G T A T A A G C T G A – 3′

2. Diagram the double helix. Be sure to label those properties that make it most suited as the genetic material.

3. Diagram a replication fork as it would be seen in a replicating segment of DNA. In your diagram label the 5′ and 3′ ends of each parent strand and daughter strand. Indicate which new strand is the leading strand and which is the lagging strand of the daughter DNA.

4. Based on your diagram in Question 3, construct a flowchart that represents what occurs during DNA replication. Divide your chart into three parts: initiation, elongation, and termination of replication. In your chart, indicate the roles of helicase, DNA polymerase, single-stranded DNA binding proteins, nucleotides, parental (template) DNA strands, DNA ligase, RNA primase, and RNA primers.

5. Explain the difference between conservative and semiconservative models of DNA replication. What results supported the semiconservative model? What would the results have looked like had the conservative model of DNA replication been accurate? Are there any other potential hypotheses?

6. Explain the role of Okazaki fragments in the synthesis of the lagging strand.

7. Differentiate between proofreading, mismatch repair, and excision repair. Which of these repair mechanisms is responsible for repairing a mutation that occurs in an adult cell from overexposure to the sun? Explain your answer.

8. Explain how modified nucleotides are used in DNA sequencing.

Answers to Questions
Knowledge and Synthesis Answers

1. **a.** The experiments showed only that something was causing the R strain pneumococcus to develop the S strain virulence. Additional experiments were necessary to provide evidence that the transforming factor was indeed DNA.

2. **c.** These researchers were able to isolate nearly pure DNA samples. It was only these samples that provided transformation activity.

3. **d.** Only DNA incorporates radioactive P. Radioactive S is incorporated into proteins. The fact that progeny viruses retained radioactive P and not radioactive S indicated that DNA is heritable and protein is not.

4. **b.** X ray crystallography allows researchers to measure the distances between atoms and determine the physical dimension of a molecule. The technique becomes quite difficult with molecules that have high molecular weights. Wilkins's contributions were significant because he was able to isolate nearly perfect strands of the molecule. The resulting images are patterns of how the X rays are diffracted off the constituent atoms of the molecule.

5. **d.** Chargaff found that in most DNA sampled, the amount of A equaled the amount of T and the amount of G equaled the amount of C. It follows that the amount of purines (A + G) equals the amount of pyrimidines (T + C). The same does not hold for the amount of A + T versus the amount of G + C.

6. **d.** Model building by Watson and Crick created a three-dimensional visualization of the size, bond angles, base pairings, and overall structure of the DNA molecule. The information contributing to their model came from a variety of experiments performed by other scientists.

7. **c.** Replication of DNA is a fundamental requirement for its function as the genetic material. DNA must be correctly replicated in each cell of an organism and must be passed from parent to offspring or from cell to daughter cell. DNA is not passed from one species to a different species. There is variation in DNA sequences from different organisms. DNA is found outside the nucleus in eukaryotic cells (specifically in the mitochondria and, in plants, chloroplasts), but this DNA is inherited in a non-Mendelian fashion (see Chapter 10).

8. **c.** Density gradient labeling allowed Meselson and Stahl to track parental versus daughter strands of DNA. Their research showed that DNA replication was a semiconservative process with each newly synthesized DNA molecule containing one parental (template) strand base-paired to one daughter (newly synthesized) strand.

9. **d.** Replication complexes bind to replication origins. The enzymes contained in the complex are responsible for unwinding the parental DNA, for initiating replication, and for the actual synthesis of the new strands of daughter DNA.

10. **a.** DNA polymerase adds nucleotides to an existing nucleotide strand.

11. **d.** Okazaki fragments are small segments of newly synthesized DNA that have been added to short RNA primers. The RNA is removed from these small fragments by DNA polymerase I and replaced with DNA. The remaining DNA fragments are ligated (covalently bound) together to form a continuous newly synthesized DNA strand.

12. **b.** DNA polymerase cannot initiate synthesis of a nucleotide strand, it can only add onto an existing strand of RNA primer or DNA. Primers provide DNA polymerase with the 3′ (—OH) end required for the addition of deoxyribonucleotides. The origin is also required for DNA synthesis, but it is the site where the replication complex is initially assembled. Primase is not the first enzyme in the replication complex.

13. **a.** The mismatch repair mechanism operates before the newly synthesized DNA strand is methylated, but for the integrity of DNA to be maintained, repair mechanisms must be active during synthesis, modification, and utilization of DNA.

14. **c.** DNA replication is an energy-consuming process that must have an input of energy to proceed. Energy is provided by the breaking of the phosphodiester bonds of each incoming deoxyribonucleoside triphosphate (see Figure 11.10).

15. **b.** PCR depends on continuous cycles of heating, in which DNA is denatured, followed by cooling, which then allows primer annealing and DNA synthesis to occur. The original limitation of the technique was that DNA polymerases were also denatured at the high temperature along with the DNA itself. Because *T. aquaticus* polymerase is active at high temperatures, it is not denatured during the heating cycle of PCR. This allows the heat cycles of the PCR reaction to be repeated continuously without any addition of extra polymerase enzyme.

16. **b.** If 30 percent of DNA is adenine, then by Chargaff's rule, 30 percent will be thymine. The remaining 40 percent of the DNA is cytosine and guanine. Because the ratio of cytosine to guanine must be equal, the percentage of cytosine in this DNA must be 20 percent.

17. **c.** Guanine is a purine, and its paired pyrimidine is cytosine.

18. **a.** Okazaki fragments are involved in synthesis of the lagging strand.

19. **d.** Sulfur is a constituent of many protein molecules, but is not found in DNA.

20. **a.** A hydroxyl group at the 3′ position of a nucleotide is necessary for the binding of any additional nucleotides. If this hydroxyl group were absent, no other nucleotides could be added to a growing strand.

Application Answers

1. 3′ – C G A T T G A C A C T A G C A T A T T C G A C T – 5′

2. See Figure 11.6 in the text.

3–4. See Figure 11.15 and 11.16 in the text.

5. In the conservative model of DNA replication, the parent DNA remains intact, and a newly synthesized molecule consists of two newly replicated daughter strands. Had this been DNA's method of replication, Meselson and Stahl would not have seen the intermediate density band in the first generation of replication. They would have seen a heavy band corresponding to the parental DNA and a light band corresponding to the daughter DNA. Because they saw an intermediate band, they knew one strand was heavy and one was light; therefore, replication was semiconservative, with each new molecule consisting of one parental strand and one daughter strand. A third hypothesis was the dispersive model, with each new molecule containing bits and pieces of both old and new strands.

6. See Figure 11.17 in the text.

7. Proofreading occurs during synthesis of the DNA molecule, mismatch repair occurs after synthesis, and excision repair removes any abnormalities in "mature" DNA. DNA damage due to UV exposure is generally repaired via excision repair mechanisms.

8. Refer to Figure 11.21 in your textbook for an explanation of DNA sequencing.

12 *From DNA to Protein: Genotype to Phenotype*

Important Concepts

One gene encodes one enzyme.

- Beadle and Tatum studied special mutant (auxotroph) and wild-type (prototroph) strains of *Neurospora*, a bread mold. Auxotrophs are mutant strains that cannot grow on minimal media without the addition of an external mineral, vitamin, or nutrient, because auxotrophs are missing a particular enzyme needed to synthesize these nutrients. Prototrophs can grow on minimal media because in these cells all the enzymes needed to synthesize these nutrients are functional.

- Auxotrophs for a particular nutrient are missing the enzymes needed to synthesize that nutrient from a biochemical intermediate. For the amino acid arginine, for example, each of these enzymes defines a step in the biochemical pathway of arginine synthesis. These mutant strains can be characterized by supplying them with chemical intermediates in the arginine pathway to determine which enzymes are nonfunctional (see Figure 12.1).

- Beadle and Tatum concluded from their experiments that one gene encoded one enzyme. Because enzymes can consist of several different polypeptides, this conclusion has been modified to the theory that one gene encodes one polypeptide.

- Genes also code for some RNAs (such as ribosomal RNA and transfer RNA) that do not become translated into polypeptides.

RNA differs from DNA in three ways.

- DNA consists of two polynucleotide strands, but RNA is generally a single polynucleotide strand. A strand of RNA can base-pair with a single strand of DNA. A strand of RNA may also fold over to base-pair with itself.

- The sugar molecule in DNA, deoxyribose, is replaced by ribose in RNA.

- The bases A, G, and C in DNA also occur in RNA, but T is replaced by U. In RNA, G still base-pairs with C, but A base-pairs with U.

The central dogma of molecular biology says that information flows in only one direction.

- In a eukaryotic cell, a message (the messenger RNA or mRNA) is needed to carry the genetic information (the DNA) from the nucleus to the cytoplasm. The process of synthesizing mRNA from DNA is called transcription.

- An adapter (the transfer RNA or tRNA) is needed to translate the nucleic acid (in the messenger RNA) into proteins. Translation is the process of synthesizing protein from mRNA.

- The information flow is unidirectional, from gene (DNA), to message (RNA), to protein.

- Some viruses use RNA as their genome and during infection make more RNA (using viral RNA as a template). These viruses include poliovirus, tobacco mosaic virus, and influenza virus. Other viruses, such as HIV, make a DNA copy of their RNA and use the DNA to synthesize more RNA. These viruses offer exceptions to the central dogma.

Transcription is the process of synthesizing RNA from a DNA template.

- RNA polymerase is the enzyme that synthesizes RNA using one strand of the DNA as a template. The resulting RNA molecules can be messenger RNAs (which are used to synthesize proteins) or they can become transfer RNA or ribosomal RNA.

- RNA polymerase initiates transcription at special sequences on the DNA called promoters. The promoter tells the RNA polymerase where to start transcription, which strand of DNA to read, and which direction to move on the DNA to begin synthesizing RNA.

- New RNA is synthesized in a 5´-to-3´ direction, antiparallel to the template DNA strand. New nucleotides are added to the 3´ end of the growing strand.

- RNA polymerase unwinds about 20 base pairs in the DNA as it reads and synthesizes the complementary RNA strand.

- RNA polymerase terminates transcription at special sequences in the DNA.

- In prokaryotic cells, translation of the mRNA can begin before transcription terminates.

- In eukaryotic cells, the mRNA is processed in the nucleus and must be transported from the nucleus to the cytoplasm before translation can begin.

Translation is the process of synthesizing proteins from an mRNA template.

- The genetic code provides the key to "reading" mRNA sequences.

- Genetic information in the messenger RNA is read three bases at a time; each three-base sequence is called a codon.

- There are four possible bases that can be used in a codon, so there are 4^3, or 64, different codons.

- Sixty-one codons specify specific amino acids, so the genetic code is redundant, meaning that there is more than one codon for many of the 20 amino acids.

- Three codons, the stop codons UAG, UAA, and UGA, are used to terminate translation.

- The codon used to start translation is AUG, which also specifies the amino acid methionine.

- The genetic code for nuclear genes is almost universally used in all living cells. There are slight differences in the genetic code of one group of protists and in mitochondria and chloroplasts.

- The genetic code was first determined by experiments in which artificial RNAs (poly U, for example) were added to reaction mixtures containing all of the necessary components needed for protein translation (see Figure 12.6). The resulting polypeptides were analyzed to determine what the protein-synthesizing instructions were for these RNA sequences. When the synthetic RNA poly U was used, polyphenylalanine was made. Thus, one of the codons for phenylalanine was UUU.

- Transfer RNA (tRNA) is the link between a codon and its amino acid.

  - tRNA is a small RNA molecule (75–80 nucleotides) that base-pairs with itself to form a characteristic shape, the "cloverleaf," found in all tRNAs (see Figure 12.7).

  - The anticodon of the tRNA (found about midway in the molecule) base-pairs with the codon of the mRNA in an antiparallel fashion. This base pairing occurs on the ribosome during translation.

  - An amino acid is covalently attached to the 3′ end of the tRNA; thus phe tRNA, (with an anticodon of AAA) has phenylalanine covalently attached to its 3′ end. The enzymes that catalyze the covalent attachment of the amino acids to the correct tRNAs are tRNA synthetases (see Figure 12.8).

- The ribosome is the site in the cell where protein translation occurs.

  - Ribosomes are composed of a large subunit and a small subunit; both subunits consist of different ribosomal proteins and RNA molecules (ribosomal RNA).

  - The important sites on a ribosome are the T site (where the incoming charged tRNA enters the ribosome), the A site (where the anticodon of the charged tRNA binds the codon of the mRNA), the P site (where the tRNA with the growing polypeptide chain is located), and the E site (where the tRNA exits from the ribosome). See Figure 12.9.

- Initiation of translation requires the small subunit of the ribosome, the initiator tRNA, the mRNA with the start codon AUG positioned correctly on the ribosome, and the initiation factors.

- When all these components are in place, the large subunit of the ribosome joins the small subunit of the ribosome, and the incoming charged tRNA enters the T site and is positioned at the A site. The anticodon of this tRNA base-pairs with the codon on the mRNA that is at the A site of the ribosome.

  - Base pairing between the anticodon and the codon is always complementary (A pairs with U, C pairs with G) in the first two positions. At the third position of the codon (and the first position of the anticodon) base pairing does not have to be exact. This position is called the wobble position and allows the cell to use one tRNA to base-pair with up to four different codons.

- The large subunit of the ribosome catalyzes the formation of a peptide bond between the amino acid on the tRNA in the P site and the amino acid on the tRNA in the A site. This peptidyl transferase activity can be attributed to the RNA (not to any of the proteins) in the large ribosomal subunit.

- Once the peptide bond is formed, the ribosome shifts down the messenger RNA so that the next codon is placed at the A site. The tRNA that was in the P site is moved to the E site and will exit the ribosome.

- During elongation, the messenger RNA is read 5′ to 3′, with each new codon being moved into the A site during elongation so that the next incoming charged tRNA can try to base-pair with it. Polypeptide synthesis occurs from the N terminus to the C terminus with the polypeptide folding into its tertiary structure as it is synthesized.

  - Elongation factors are proteins that assist in this phase of translation.

- Termination occurs when a stop codon is moved into the A site of the ribosome. Release factor binds the stop codon, and peptidyl transferase hydrolyzes the bond between the amino acid (which will be the C terminus) that is covalently attached to the tRNA in the P site. The polypeptide is released from the ribosome and the small and large subunits of the ribosome dissociate (see Figure 12.12).

- Antibiotics can block various steps in translation in prokaryotic cells, preventing the bacteria from synthesizing proteins and thus killing the bacteria (see Table 12.2). These antibiotics are effective because they specifically target the bacterial translational machinery without affecting the cells of the individual taking the antibiotic.

- One messenger RNA is usually bound by many translating ribosomes, forming a polysome (see Figure 12.13).

- Within the amino acid sequence of some proteins in eukaryotic cells are signal sequences that serve as addresses for the protein's final cellular destination. These signal sequences are recognized by receptor proteins at their destination (in the case of mitochondrial, plastid, peroxisome, and nuclear proteins, see Figure 12.14), or by signal recognition particles in the cytoplasm that direct the partially translated protein to the endoplasmic reticulum for further addressing and completion of translation (see Figure 12.15).

- Proteins that lack signal sequences remain in the cytoplasm.

- Protein modification after translation includes proteolysis, glycosylation, and phosphorylation.

- Proteolysis is the process of cutting proteins into smaller pieces. The removal of the signal sequence from a protein in the ER is an example of proteolysis.
- Glycosylation is the addition of sugars to proteins. Glycosylation in the Golgi apparatus marks proteins as destined for lysosomes, the vacuole (in plants), or the plasma membrane.
- Phosphorylation is the addition of phosphate groups to proteins. Phosphorylation contributes to the tertiary structure of a protein.

Mutations are heritable changes in genetic information.

- Mutations can occur in germ-line cells, and thus be passed on to progeny, or they can occur in somatic cells.
- Some mutations are easy to observe, others are difficult. Conditional mutants are a class of mutations that are only observable under altered conditions, such as high temperature. The inability of certain mutants to grow at high temperature can be related to the unstable tertiary structure of the protein whose gene sequence has been altered.
- Point mutations are changes of one or two base pairs and include silent mutations, missense mutations, nonsense mutations, and frameshift mutations.
- Silent mutations are mutations that do not change the meaning of the codon. For example, the change of a U to an A in the third position of a codon would change a UUU to a UUA, but both codons specify phenylalanine.
- Missense mutations are single-base changes that change one codon to another (UUU, which codes for phe, to CUU, which codes for leu).
- Nonsense mutations are single-base changes that change a codon to a stop codon (UGG, which codes for trp, to UAG, stop).
- Deletions or insertions of one or two bases are frameshift mutations, which alter the reading frame of the genetic message.
- Chromosome mutations are larger deletions, duplications, inversions, or translocations that occur in the chromosome.
- Spontaneous mutations can be caused by tautomerization of the bases, deamination of the bases, errors during DNA replication, and errors during meiosis (leading to unequal crossovers or unequal separation of the chromosomes).
- Induced mutations are caused by chemicals and radiation.
- Mutations for the most part are deleterious, but some mutations may actually benefit the organism by creating an altered gene product that may improve the survival of the organism.

The Big Picture

- Every protein and RNA in the cell has a DNA blueprint (the gene sequence) that specifies the amino acid or nucleotide sequence of that gene product. Those sequences determine how that gene product will fold up three-dimensionally and ultimately function in the cell.
- Alterations in those gene products, caused by mutations in the genes, help us to understand how those proteins and RNAs function in the cell.

Common Problem Areas

- Students often try to understand every detail of gene expression without comprehending the larger picture: accessing genetic information in the DNA and expressing it in the cell so that the cell can function. Familiarize yourself with the details of the central dogma, but then take a step back and understand that all of these details describe the steps required to synthesize gene products.
- Students often confuse transcription and translation. Be able to distinguish between the two processes, and for each one carefully review the template, the product, and the sites of initiation and termination (review Table 12.1).

Study Strategies

- Make diagrams of the processes of transcription and translation. Be sure to orient the 5′ and the 3′ ends of the nucleic acid and the N and the C terminus of the protein.
- Pick a hypothetical gene that encodes a peptide of four amino acids. Draw a flowchart describing how that gene in the chromosome is made into protein. Now follow the details of gene expression, the synthesis of the RNA from the DNA, and the synthesis of the protein from mRNA. Be sure to include the start and stop signals in each of these processes. Alter one of those nucleotides in the DNA sequence. Transcribe and translate the gene. Is the protein sequence changed? Repeat this process using a different mutation in the DNA.
- Review the following tutorials and activity on the Student Website/CD:
Tutorial 12.1 Transcription
Tutorial 12.2 Deciphering The Genetic Code
Tutorial 12.3 Protein Synthesis
Activity 12.1 Genetic Code

Test Yourself
Knowledge and Synthesis Questions

1. Transcription in prokaryotic cells
   a. occurs in the nucleus whereas translation occurs in the cytoplasm.
   b. is initiated at a start codon with the help of initiation factors and the small subunit of the ribosome.
   c. is initiated at a promoter and uses only one strand of DNA, the template strand, to synthesize a complementary RNA strand.
   d. is terminated at stop codons.

2. Which of the following about RNA polymerase is *not* true?
   a. It synthesizes mRNA in a 5′-to-3′ direction reading the DNA strand 3′ to 5′.
   b. It synthesizes mRNA in a 3′-to-5′ direction reading the DNA strand 5′ to 3′.
   c. It binds at the promoter and unwinds the DNA.
   d. It does not require a primer to initiate transcription.

3. Translation of messenger RNA into protein occurs
   a. in a 3′-to-5′ direction and from N terminus to C terminus.
   b. in a 5′-to-3′ direction and from N terminus to C terminus.
   c. in a 3′-to-5′ direction and from C terminus to N terminus.
   d. in a 5′-to-3′ direction and from C terminus to N terminus.

4. If a codon were read two bases at a time instead of three bases at a time, how many different possible amino acids could be specified?
   a. 16
   b. 64
   c. 8
   d. 32

5. Translate the following mRNA:
   3′-G A U G G U U U U A A A G U A- 5′
   a. NH₂ met—lys—phe—leu—stop COOH
   b. NH₂ met—lys—phe—trp—stop COOH
   c. NH₂ asp—gly—phe—lys—val COOH
   d. NH₂ asp—gly—phe—lys —stop COOH

6. What would happen if a mutation occurred in DNA such that the second codon of the resulting mRNA was changed from UGG to UAG?
   a. Nothing. The ribosome would skip that codon and translation would continue.
   b. Translation would continue, but the reading frame of the ribosome would be shifted.
   c. Translation would stop at the second codon, and no functional protein would be made.
   d. Translation would continue, but the second amino acid in the protein would be different.

7. If the following synthetic RNA were added to a test tube containing all the components necessary for protein translation to occur, what would the amino acid sequence be?
   5′-A U A U A U A U A U A U - 3′
   a. Polyphenylalanine
   b. Isoleucine-tyrosine-isoleucine-tyrosine
   c. Isoleucine-isoleucine-isoleucine-isoleucine
   d. Tyrosine-tyrosine-tyrosine-tyrosine

8. What part of the tRNA base-pairs with the codon in the mRNA?
   a. The 3′ end where the amino acid is covalently attached
   b. The 5′ end
   c. The anticodon
   d. The promoter

9. Peptidyl transferase is
   a. an enzyme found in the nucleus of the cell that assists in the transfer of mRNA to the cytoplasm.
   b. an enzyme found in the large subunit of the ribosome that catalyzes the formation of the peptide bond in the growing polypeptide.
   c. an RNA molecule that is catalytic.
   d. Both b and c

10. Termination of translation requires
    a. release factor, initiator tRNA, and ribosomes.
    b. initiation factors, the small subunit of the ribosome, and mRNA.
    c. elongation factors and charged tRNAs.
    d. a stop codon positioned at the A site of the ribosome, peptidyl transferase, and release factor.

11. Which of the following mutations would probably be the most deleterious?
    a. A missense mutation in the second codon
    b. A frameshift mutation in the second codon
    c. A nonsense mutation in the last codon
    d. A silent mutation in the second codon

12. If the DNA encoding a nuclear signal sequence were placed in the gene for a cytoplasmic protein, what would happen?
    a. The protein would be directed to the lysosomes.
    b. The protein would be directed to the nucleus.
    c. The protein would be directed to the cytoplasm.
    d. The protein would stay in the endoplasmic reticulum.

13. Auxotrophs are mutant strains that
    a. can grow on a minimal medium.
    b. require the addition of an essential nutrient to grow on a minimal medium.
    c. behave like wild-type strains.
    d. can only grow if arginine is added to the growth medium.

14. The central dogma of molecular biology states that _____ is transcribed into _____, which is translated into _____.
    a. genes; polypeptides; gene product
    b. protein; DNA; RNA
    c. DNA; mRNA; tRNA
    d. DNA; RNA; protein

15. A gene product can be
    a. an enzyme.
    b. a polypeptide.
    c. an RNA.
    d. All of the above

16. The enzyme that catalyzes the synthesis of RNA is
    a. DNA polymerase.
    b. tRNA synthetase.
    c. ribosomal RNA.
    d. RNA polymerase.

Application Questions

1. What would be the effect of a deletion of the DNA encoding the targeting sequence for that gene product? (Imagine that this protein was targeted to go to the mitochondria and the signal sequence was removed as a result of this deletion.)

2. What would happen if the tRNA synthetase for tryptophan actually added a phenylalanine to the trp tRNAs instead of tryptophan?

3. Mutations can be very harmful to an organism, yet without them life as we know it today would not exist. Explain.

4. Suppose that two different mutant strains of a bacterium are unable to grow on a minimal medium without the addition of the amino acid lysine. Explain how this phenotype might be caused by different mutations in each strain, perhaps on the same gene and perhaps on two different genes.

5. When fed the same biochemical intermediate, two auxotrophic mutants that had been isolated were able to grow. According to the experiments of Beadle and Tatum, the mutations in each of these auxotrophs should be in the same gene, because they were blocked at the same step in a biochemical pathway. Yet, these two auxotrophs had mutations that mapped in different genes. How do you explain this?

6. Antibiotics block different steps in translation in prokaryotic cells. Why are these chemicals so effective in eliminating bacterial infections?

Answers to Questions
Knowledge and Synthesis Answers

1. **c.** Answer **a** describes transcription and translation in eukaryotic cells; answers **b** and **d** describe translation.

2. **b.** RNA polymerase binds at a promoter, unwinds the DNA, synthesizes mRNA in a 5′-to-3′ (not 3′-to-5′) direction and does not require a primer to synthesize the RNA.

3. **b.** Translation of messenger RNA occurs 5′ to 3′ and the polypeptide is synthesized from N terminus to C terminus.

4. **a.** Four possible bases read two at a time gives 4^2, or 16, different codons.

5. **b.** See the codon table, Figure 12.5. Recall that translation occurs in the 5′-to-3′ direction.

6. **c.** UAG is a stop codon, so translation would terminate at that site.

7. **b.** See the codon table, Figure 12.5.

8. **c.** Neither the 3′ end nor the 5′ end of the tRNA is part of the anticodon. The promoter is a DNA sequence, to which RNA polymerase binds to initiate transcription.

9. **d.** Peptidyl transferase is the enzyme that catalyzes the formation of the peptide bond, and it is located in the large subunit of the ribosome. Its catalytic activity is due to ribosomal RNA found in the large subunit of the ribosome.

10. **d.** Termination of translation requires a stop codon to be positioned in the A site of the ribosome, and peptidyl transferase and release factor. Release factor adds water across the last amino acid attached to the tRNA in the P site, creating the C terminus.

11. **b.** The mutation that is potentially the most dangerous is the frameshift in the second codon. All of the other codons would be out of register and no functional protein could be made. A missense mutation in the second codon could be disastrous or minimal, depending on how crucial that second amino acid was to the folding and functioning of the protein. A nonsense mutation at the end of the coding sequence would have less effect on protein function because almost every amino acid in the protein would have been correctly synthesized. Again, this would depend on how important that last codon was to the proper folding and functioning of the protein. A silent mutation would have no effect on protein function because the codons would be the same.

12. **b.** The nuclear sequence would direct this protein to be delivered to the nucleus.

13. **b.** Answer **d** is true only for arginine auxotrophs.

14. **d.** Genes are not transcribed into polypeptides, protein is not transcribed into DNA, and messenger RNAs are not translated into tRNAs.

15. **d.** Gene products can be RNAs (such as rRNA and tRNA) as well as enzymes and other polypeptides. Messenger RNA is translated into a gene product, protein.

16. **d.** DNA polymerase catalyzes the synthesis of DNA, tRNA synthetase covalently attaches amino acids to tRNAs, and ribosomal RNA (in the large subunit) catalyzes the formation of the peptide bond during translation.

Application Answers

1. Because the protein lacked its targeting sequence, it would no longer be moved to the mitochondria and would remain in the cytoplasm after it had been translated.

2. If the tRNA synthetase for tryptophan added phenylalanine to the trp tRNAs, every time a tryptophan codon was read by these trp tRNAs, phenylalanine would be added to the polypeptide. This would create proteins that were nonfunctional, and the cell would die.

3. Mutations very often have damaging effects on a gene, rendering that gene product nonfunctional. Some mutations actually create gene products that are better for the cell or organism and may increase its ability to survive. Mutations over the course of evolutionary time have created new organisms that could survive in different environments or compete more effectively for limited resources. Without mutations evolution does not occur.

4. The mutations in these two strains of bacteria apparently interfere with lysine synthesis. The mutations might both be in the same gene coding for an enzyme necessary for lysine synthesis, but one could be a nonsense mutation in the fifth codon and the other a frame-shift mutation in the twenty-third, for example (the number of mutations that can disable a gene is enormous). If lysine synthesis in this bacterium requires more than one enzyme (as is likely), the two mutations could be in different genes coding for different enzymes. In this case, the phenotypes would not be strictly identical; it should be possible to distinguish the two by trying to grow them on minimal media to which different intermediates in the synthesis of lysine have been added (see Figure 12.1).

5. These two genes must encode different polypeptides that are both subunits for the same enzyme in this biochemical pathway.

6. Many of the infections that cause human disease are of bacterial origin, and when antibiotics are applied to these cells, translation stops and the cells die. Eukaryotic organisms such as humans are not as sensitive to these antibiotics, either because their ribosomes do not bind the antibiotics or because the antibiotics cannot get inside their cells. Protein translation machinery in human cells is unaffected by most antibiotics, but translation in bacterial cells is, leading to the elimination of the bacterial infection.

13 The Genetics of Viruses and Prokaryotes

Important Concepts

Viruses and prokaryotes are useful model organisms for genetic studies.

- Prokaryotic cells and the viruses that infect them (also known as bacteriophage) are easy to grow, have a short generation time, and can produce large numbers of individuals (10^9 E. coli per milliliter of growth medium).

- The genome size of bacterial viruses and bacteria are small, making identification and characterization of genes and gene function in these organisms relatively simple.

- Viruses were first identified in the late 1800s using crude filtration methods. In the 1930s, crystalline preparations of viruses demonstrated that they consisted of protein and nucleic acid. By the 1950s, viral structures were clearly characterized using electron microscopy.

Viruses are intracellular parasites that can reproduce only within living cells.

- Viruses are acellular, do not metabolize energy-rich compounds, and thus generate no ATP. They are obligate intracellular parasites, using the host cell protein translation machinery and energy-rich host compounds to express their gene products, to replicate their genome, and to multiply. Viruses can infect Bacteria, Archaea, animal cells, plant cells, fungi, and protists. Many animal viruses cause disease and are not affected by antibiotics because they use the host cell translation machinery to synthesize viral proteins. (For a review of how antibiotics inhibit bacterial cells, see Table 12.2.)

- Outside the host cell, viruses exist as individual particles called virions. These virions have a central core of DNA or RNA enclosed in a protein capsid. Some animal viruses have a membrane surrounding the protein capsid, acquired from the host's cell membrane during viral release.

- The nucleic acid in viruses can either be DNA or RNA and can either be double stranded or single stranded.

- In prokaryotic cells, viral infection is initiated when proteins on the viral capsid bind receptor proteins on the host bacterium's cell wall. Viral nucleic acid is injected into the cytoplasm through the viral tail assembly.

- Bacterial viruses (bacteriophage) can have two different kinds of life cycles: the lytic cycle, during which the virus multiplies and lyses the cell, and the lysogenic cycle,

during which the viral genetic material is harbored in the host chromosome for many generations. When environmental conditions are altered, the viral genome can excise itself from the host chromosome, multiply, and lyse the cell (see Figure 13.2).

- Some bacteriophage (the virulent viruses) are only able to undergo a lytic cycle. Other bacteriophage (the temperate viruses) can alternate between lytic and lysogenic life cycles.

- During the early phase of the lytic cycle, the phage injects its nucleic acid into the host cytoplasm, and viral genes are transcribed and translated. These early gene products shut down host transcription, degrade host DNA, replicate viral DNA, and stimulate late gene transcription. Late gene products include viral capsid proteins and proteins that lyse the cell at the end of the lytic cycle (see Figure 13.2). The lytic cycle takes about 30 minutes and produces hundreds of bacteriophage per cell.

- Lysogeny occurs when the viral genome inserts itself into the host genome. The virus is called a prophage at this stage. The prophage prevents the host from being infected by other bacteriophage, providing the host immunity to infections by the same virus. Lysogenesis usually occurs when environmental conditions are not limited and the host cell is growing rapidly. When the cell encounters stressful environmental conditions, lysogenic viruses respond with a change in gene expression, resulting in excision from the host chromosome and completion of the lytic cycle.

- The use of bacteriophage to treat bacterial infections has become an active area of research as antibiotic-resistant bacteria are becoming more common.

Animal viruses are diverse.

- Animal viruses infect a wide variety of vertebrates, but among invertebrates only arthropods (crustaceans and insects) are commonly infected. Arboviruses infect insects but do not adversely affect them. Infected insects are vectors for the viruses because they transmit the virus to vertebrates via an insect bite.

- Like bacteriophage, animal viruses consist of a nucleic acid core (either DNA or RNA) surrounded by protein. Some animal viruses (enveloped viruses) have a membrane surrounding the viral protein core, which is derived from the host plasma membrane. The genome of an animal virus usually encodes only a few proteins.

- The lytic cycle of animal viruses, like bacterial viruses, also has early and late stages. Transcription and translation of early genes includes the synthesis of viral proteins that replicate the viral genome and turn off host transcription. Late viral gene products include proteins that package the virus and lyse the cell (see Figure 13.3).

- Animal viruses lacking a membrane are taken into the host cell via the endocytic pathway (refer to the discussion on endocytosis, Chapter 5). Inside the cell, the membrane of the endocytic vesicle is broken down, releasing the virus to the cytoplasm. Enzymes in the cytoplasm degrade the protein capsid of the virus and liberate the viral nucleic acid.

- Enveloped viruses bind host cells using viral glycoproteins (embedded in the viral envelope) that bind specific protein receptors in the host cell membrane. These viruses can be taken into the host via the endocytic pathway and released from the endocytic vesicle into the cytoplasm. Alternatively, some enveloped viruses release the virion into the cytoplasm of the host cell by fusion of their viral envelope with the cell membrane of the host cell. Enveloped viruses leave the host cell by budding out of the plasma membrane, carrying part of that membrane with them.

- The influenza virus, an RNA virus, is taken into the host cell by endocytosis. Inside the cell, the viral envelope fuses with the endocytic vesicle, releasing the RNA into the host cell's cytoplasm. The viral RNA is used to make mRNA for viral proteins and to make more copies of the viral genome (see Figure 13.4).

- HIV is an RNA virus that enters the host cell by fusion of its viral membrane with the host plasma membrane. HIV I is a retrovirus, copying its RNA genome into DNA using reverse transcriptase. The DNA copy of the viral genome is then integrated into the host cell's chromosome, and in this form it is termed a provirus (see Figure 13.5).

Viruses that infect plants need to penetrate the cell wall.

- Plant viruses can infect horizontally (plant to plant) or vertically (plant to offspring).

- In a horizontal infection, the virus is most commonly transmitted from one plant to another by a vector, usually an insect. Virus may also be transmitted by physical contact between a bruised leaf of an infected plant and an uninfected plant. Once inside the cell, the virus reproduces and spreads to other cells in the plant.

- Viral infection of other plant cells is facilitated by the plasmodesmata, which are special plant structures that form cytoplasmic bridges between cells. Viruses travel through these channels with the help of special proteins to infect adjacent plant cells.

Genetic recombination in prokaryotes is not tied to reproduction.

- Unlike viruses, prokaryotes are living cells. They have selectively permeable membranes, perform energy metabolism, and reproduce asexually, generating genetically identical cells (clones).

- New genetic material can be introduced into bacterial cells in several ways, including conjugation, transformation, and transduction.

- Bacteria can pass some of their genetic material to other bacterial cells through the process of conjugation (first characterized by Beadle and Tatum using two different bacterial auxotrophs, see Figure 13.7). Donor cells synthesize a thin extension called a sex pilus, which eventually contacts a recipient cell, forming a conjugation tube (see Figure 13.8). DNA is transferred through the conjugation tube to the recipient cell, where it can recombine with the chromosome of the recipient cell (see Figure 13.9).

- During transformation, external DNA can be introduced to cells, as demonstrated by experiments in which rough nonvirulent pneumococci bacteria were transformed to smooth virulent bacteria (see Chapter 11). Once the DNA enters the living cell it can recombine with the bacterial chromosome (see Figure 13.10a).

- Some viruses can pick up bacterial genes from one cell, pack them in their viral capsid, and transfer them to the next bacterial cell they infect. This process of gene transfer using a viral vector is called transduction (see Figure 13.10b).

Plasmids are small circular DNA molecules.

- Plasmids (sometimes called episomes) are found in the cytoplasm of some bacterial cells. These DNA molecules can move from one bacterial cell to another during conjugation. Plasmids contain a few dozen genes.

  - Some plasmids contain genes that encode enzymes with unusual metabolic functions, such as enzymes that digest oil.

  - Some plasmids contain genes for mating factors, such as the proteins that form the pilus and the conjugation tube. These plasmids are called fertility factors or F factors.

  - Some plasmids contain genes whose products destroy or inactivate antibiotics and heavy metals. These factors are called R factors because they impart resistance to drugs or heavy metals. Some bacteria are resistant to more than one antibiotic due to the presence of multiple R factors.

Transposable elements and transposons carry genes that help them insert into other chromosomes.

- Transposable elements are special DNA sequences that can readily move to other sites on the same chromosome or to another chromosome.

- Transposable elements carry genes whose products can copy or excise the transposable elements and insert them elsewhere in the chromosome. Some transposable elements are copied and inserted at another position, whereas others are simply excised and placed at a different chromosomal site without DNA replication (see Figure 13.12).

- Longer transposable elements that have additional genes are called transposons. Each resistance gene in an R plasmid contains parts of transposons, suggesting that transposition might be the mechanism by which plasmids originally gained antibiotic resistance.

Prokaryotes regulate gene expression in response to changing environments.

- Gene regulation in prokaryotes is primarily transcriptional—gene activity is controlled by allowing or blocking the transcription of different genes.

- Some regulation occurs at the posttranscriptional level, including hydrolysis of messenger RNA, blocking of translation, protein hydrolysis, or protein inhibition.

- Genes can be expressed constitutively (all the time), or they can be expressed in response to specific environmental conditions.

- Enzyme activity can be repressed by end-product feedback (also known as feedback inhibition) or by regulating gene expression (see Figure 13.14).

Repressors block transcription when bound to operators.

- Operons are units of transcription in prokaryotes that consist of structural genes and the DNA sequences that control their transcription. The control sequences include the promoter and the operator.

- The promoter is the first DNA sequence in an operon. RNA polymerase binds to the promoter at the beginning of transcription. If it cannot bind to the promoter, it cannot transcribe the structural genes.

- The operator is a DNA sequence immediately adjacent to the promoter in the operon. If a repressor molecule is bound to the operator, it prevents RNA polymerase from binding to the promoter.

- The *lac* operon is an inducible operon. This means that the repressor is bound to the operator, blocking transcription of the structural genes, unless another molecule (the inducer) binds to the repressor. When the inducer (in this case, lactose) binds the repressor, the repressor becomes inactive and leaves the operator, allowing transcription to proceed.

  - The structural genes for the metabolism of lactose are adjacent to one another in the *lac* operon and are all transcribed from one promoter (review transcription, Chapter 12). There are three structural genes, which code for three enzymes necessary for the cell to use lactose.

  - As long as lactose is present in sufficient quantities, it will bind the repressor and allow production of the proteins that metabolize lactose. As lactose levels drop, lactose molecules separate from the repressor, allowing it to bind the operator again and prevent unnecessary production of proteins for lactose metabolism (see Figure 13.17).

  - The repressor for the *lac* operon is encoded by the *i* gene. The *i* gene is constitutive (it does not have an associated repressor, so it is made at a constant rate). The promoter for the *i* gene does not bind RNA polymerase very well, though, so very few repressor molecules are synthesized. There are generally only about 10 molecules of *lac* repressor per cell, but this is sufficient for proper regulation of the *lac* operon.

- The *trp* operon is a repressible operon. This means that its repressor is *not* bound to the operator unless another molecule, the corepressor (tryptophan in this case), first binds to the repressor. When the corepressor binds to the repressor, the repressor becomes active and binds to the operator, preventing transcription of the structural genes.

  - The *trp* operon has five structural genes (*e, d, c, b,* and *a*) that encode five enzymes necessary for synthesis of the amino acid tryptophan (see Figure 13.18).

- Recall that tryptophan is the corepressor of the *trp* repressor. As tryptophan levels rise in the cell, tryptophan binds to the *trp* repressor, making it active. The repressor then binds to the *trp* operator, preventing transcription of the *trp* operon's structural genes (and thereby preventing the synthesis of additional enzymes for making tryptophan). In this way, the cell avoids making more tryptophan than necessary.

- Inducible systems (like the *lac* operon) control catabolic pathways. The molecules that act as inducers are substrates of these pathways. The presence of the substrate thus turns *on* production of the machinery necessary to catabolize the substrate.

- Repressible systems (like the *trp* operon) control biosynthetic pathways. Their corepressors are the products of these pathways. The presence of the product thus turns *off* production of the machinery necessary to produce more of it.

Transcription can be positively regulated.

- The cAMP receptor protein (CRP), when bound to cAMP, will cause the transcription of certain operons to increase (see Table 13.1 for an outline of positive and negative regulation of transcription).

- The binding site for the CRP protein is just upstream 5′ of the promoter. Glucose affects the levels of cellular cAMP by decreasing the activity of the enzyme that synthesizes cAMP. When glucose levels are high in the cell, cAMP levels are low and the CRP protein is not active. Under these conditions, transcriptional activation by the CRP protein does not occur. As glucose levels fall, cAMP levels increase and the CRP protein binds the available cAMP molecules and activates transcription (see Figure 13.19).

Viruses also need to control transcription.

- In bacteriophage λ (lambda), there are two λ proteins, cI and Cro, which determine whether the phage will complete a lytic or a lysogenic cycle. These two regulatory proteins control a genetic switch.

- Both cI and Cro compete for two operator promoter sites in the λ DNA that control genes involved in lytic and lysogenic functions. If the bacteriophage has infected a host cell that is experiencing rich environmental resources, then lysogeny is favored and the cI-controlled switch is active. If the infected cells are damaged by mutagens or other environmental stress, Cro synthesis is high, the Cro-controlled switch is favored, and the bacteriophage undergoes a lytic cycle (see Figure 13.20).

Many viral and bacterial genomes have been completely sequenced.

- Genomic sequencing and analysis of those sequences is an active area of biological research. The field of functional genomics includes assigning roles to genes (annotation). Information from these analyses includes open reading frames, amino acid sequences of known gene products, and gene regulatory sequences. Signals for the initiation and termination of translation are used to identify unknown proteins.

- As more genome sequences have been completed, comparison of genomes from different organisms has been utilized to understand the physiology of different organisms.

- Genomic analysis for prokaryotes has revealed useful information, including new genes for ATP synthesis in *Chlamydia*, virulence genes in *Rickettsia*, lipid metabolism genes in *Mycobacteria*, new antibiotic resistance genes in *Streptomyces*, and unique genes in pathogenic strains of *E. coli*, *Salmonella*, and *Shigella*. Genomic analysis in simple organisms has also been used to determine the minimal number of genes required for cellular life.

The Big Picture

- Prokaryotic cells and viruses have small genomes. These have been probed by molecular biologists to understand gene function, regulation, and viral life cycles. During viral infection, sequential gene expression occurs early and late to facilitate successful viral production.

- Prokaryotic cells use operons to regulate the expression of several genes. In response to environmental signals (the presence of lactose or the absence of tryptophan for example) the repressor protein releases the operator site, leaving the promoter accessible to RNA polymerase to initiate transcription. In bacteriophage λ, other proteins bind promoters and control whether lysis or lysogeny will occur.

- New genes can be introduced into prokaryotic cells during transformation, conjugation, and transduction. Transposable elements and transposons are mobile genetic elements that move from one part of the chromosome to another. Plasmids are small circular DNA molecules that can contain genes for metabolic factors, drug resistance, heavy metal resistance, and conjugation proteins.

- Genomics is used to study and compare genomes from different organisms and to identify genes, potential genes, and their regulatory sequences. Information from these studies has been used to design new medicines for pathogens and to understand the minimal requirements for a cell to sustain life.

Common Problems

- It is easy to confuse inducible operons (the *lac* operon) with repressible operons (the *trp* operon) because they both use repressors to regulate gene expression. Review the environmental conditions that must exist for each repressor to bind its operator site and what causes each repressor to release the operator site. (In the presence of lactose, the *lac* repressor cannot bind the *lac* operator site; in the presence of tryptophan, the *trp* repressor binds the *trp* operator site.) Review gene expression (Chapter 12) to see how, once transcription is initiated, proteins are produced that will act on those environmental signals.

- Although bacteria reproduce asexually, there are several ways that new genetic material is introduced into bacterial cells, including conjugation, transformation, and transduction. Review transposable genetic elements and plasmids, keeping in mind the advantages that mobile DNA elements provide to their host cells.

- That some viruses insert their chromosome into the host chromosome may seem puzzling at first, because the goal of viral infection seems to be rapid multiplication and infection of adjacent cells. However, lysogeny allows the viral genome to persist while environmental resources

are abundant. When resources are depleted or cell damage occurs, the virus can enter the lytic cycle to infect other cells.

Study Strategies

- Gene expression often occurs in response to environmental signals. Make a list of the environmental signals that a prokaryote receives and then outline the steps the cell initiates to respond to those signals. Include the following signals: lactose in the cell, glucose and lactose in the cell, high levels of glucose in the cell, low levels of tryptophan in the cell, high levels of tryptophan in the cell, bacteriophage λ infection under poor environmental conditions, and bacteriophage λ infection under rich environmental conditions.

- Outline the progress of different animal virus life cycles (Review Figures 13.3, 13.4, and 13.5) from the start of infection through multiplication of virus to viral release. Compare these life cycles to the lytic and lysogenic life cycles of bacteriophage (see Figure 13.2).

- Review the following tutorials and activity on the Student Website/CD:
  Tutorial 13.1 The *lac* Operon
  Tutorial 13.2 The *trp* Operon
  Activity 13.1 Concept Matching

Test Yourself
Knowledge and Synthesis Questions

1. Viruses consist of
   a. a protein core and a nucleic acid capsid.
   b. a cell wall surrounding nucleic acid.
   c. RNA and DNA enclosed in a membrane.
   d. a nucleic acid core surrounded by a protein capsid and in some cases a membrane.

2. Bacterial cells that are resistant to viruses
   a. lack a cell surface receptor that the virus must bind in order to infect the cell.
   b. harbor a prophage in their chromosome, making the bacterial cell immune to further viral infection.
   c. have enzymes (restriction enzymes) that cut up the viral nucleic acid so that the viral replication cannot occur in the host cell.
   d. All of the above

3. Bacterial viruses
   a. infect the cell, replicate their genomes, and lyse the cell.
   b. infect the cell, replicate their genomes, transcribe and translate their genes, and lyse the cell.
   c. infect the cell, replicate their genomes, transcribe and translate their genes, package those genomes into viral capsids, and lyse the cell.
   d. infect the cell, translate their RNA, and lyse the cells.

4. Animal viruses that integrate their DNA into the host chromosome
   a. are RNA viruses.
   b. are prophages.
   c. copy their RNA genome into DNA using reverse transcriptase.
   d. Both a and c

5. During conjugation,
   a. DNA from one bacterial cell is transferred to another bacterial cell using a bacteriophage.

b. mutants that are auxotrophic for one nutrient can be converted to prototrophs when mixed with mutants that are auxotrophic for another nutrient.

c. a pilus is synthesized and DNA is transferred from one bacterium across the conjugation tube to the recipient bacterium.

d. Both b and c

6. Plasmid DNA may contain genes that can
   a. confer drug resistance to the host cell.
   b. regulate conjugation.
   c. confer resistance to heavy metals.
   d. All of the above

7. An operon
   a. is regulated by a repressor binding at the promoter.
   b. has structural genes that are all transcribed from same promoter.
   c. has several promoters, but all of the structural genes are related biochemically.
   d. is a set of structural genes all under the same translational regulation.

8. If the gene encoding the *lac* repressor is mutated so that the repressor can no longer bind the operator, will transcription of that operon occur?
   a. Yes, but only when lactose is present.
   b. No, because RNA polymerase is needed to transcribe the genes.
   c. Yes, because RNA polymerase will be able to bind the promoter and transcribe the operon.
   d. No, because cAMP levels are low when the repressor is nonfunctional.

9. If the gene encoding the *trp* repressor is mutated such that it can no longer bind tryptophan, will transcription of the *trp* operon occur?
   a. Yes, because the *trp* repressor can only bind the *trp* operon and block transcription when it is bound to tryptophan.
   b. No, because this mutation does not affect the part of the repressor that can bind the operator.
   c. No, because the *trp* operon is repressed only when tryptophan levels are high.
   d. Yes, because the *trp* operon can allosterically regulate the enzymes needed to synthesize the amino acid tryptophan.

10. Transcriptional regulation in prokaryotes can occur by
    a. a repressor binding an operator and preventing transcription.
    b. an activator binding upstream from a promoter and positively affecting transcription.
    c. different promoter sequences binding RNA polymerase more tightly, resulting in more effective transcriptional initiation.
    d. All of the above

Application Questions

1. Animal viruses are obligate parasites. What does this mean?

2. Why are antibiotics useless in combating animal viral infections?

3. Genomics has revealed that the bacterium that causes tuberculosis has over 250 genes that metabolize lipids. What does this finding suggest about the bacterium and about medicinal approaches to this disease?

4. A cell has a mutation that deletes the gene encoding the repressor for a certain operon. A plasmid is introduced into the host cell that carries a wild-type copy of the gene for the repressor. Is normal regulation of this operon restored in the presence of this plasmid?

5. A cell has a mutation that deletes the gene encoding the operator for a certain operon. A plasmid is introduced into the host cell that carries a wild-type copy of the operator. Is normal regulation of this operon restored in the presence of this plasmid?

6. Review some of the auxotrophic bacterial strains that Lederberg and Tatum were studying. Describe three different ways that an auxotrophic bacterium can be converted to a phototrophic bacterium.

Answers to Questions
Knowledge and Synthesis Answers

1. **d.** Nucleic acids do not form capsids; cell walls are found in bacterial and plant cells, not viruses. Viruses are organized so that the nucleic acid is surrounded by protein (not membranes), and the protein capsid may be surrounded by a membrane.

2. **d.** Bacteriophage must bind to cell surface receptors to initiate their infective cycle. In bacteriophage λ, a prophage (a viral chromosome inserted into the host chromosome) prevents other λ phage from infecting that bacterium. Some bacterial cells contain restriction enzymes that cut DNA at specific sites to prevent viral DNA from being replicated in the cell.

3. **c.** This sequence includes the most complete details of the viral life cycle.

4. **d.** Answer **b** describes a provirus, which is an animal virus that has inserted its genome into a host chromosome.

5. **d.** Answer **a** describes transduction; **b** describes one possible result of conjugation, and **c** describes what happens during conjugation.

6. **d.** All of these genes can be found on plasmid molecules.

7. **b.** An operon is a set of genes that are all transcribed from the same promoter. Answer **a** is not correct. The repressor binds at the *operator* site, which overlaps the promoter. Answer **d** is not correct because the operon is regulated transcriptionally, not translationally.

8. **c.** If the *lac* repressor is nonfunctional, it cannot bind the operator site, and transcription of the *lac* operon will occur at all times, whether or not lactose is present.

9. **a.** If the repressor can no longer bind tryptophan, then it cannot bind the operator, and transcription of the *trp* operon will always be *on*, whether tryptophan levels in the cell are high or low.

10. **d.** Answer **a** refers to the *lac* and *trp* repressors, answer **b** to the CRP protein, and answer **c** to promoters that have different transcriptional efficiencies.

Application Answers

1. Animal viruses cannot express their genes, replicate their genomes, or multiply unless they are in the cytoplasm of the host cell. They use the host cell's components (ribosomes, ATP) to grow and reproduce.

2. Most antibiotics attack bacterial cells by inhibiting prokaryotic translation. Animal viruses use the host translation machinery, which is different enough from the bacterial translation machinery that it is unaffected by antibiotics. Drugs that would inhibit host translation would kill the host cell (by preventing protein synthesis) as well as the animal virus.

3. The tuberculosis bacterium must use lipids as a source of energy-rich compounds; inhibiting lipid synthesis in this bacterium may inhibit the growth of this bacterium.

4. Yes. The repressor gene can be transcribed and translated from the plasmid DNA and normal regulation will be restored.

5. No. The operator site on the plasmid cannot restore regulation unless it recombines with the host operator site in such a way that it replaces the mutant operator on the host chromosome. The DNA site on the plasmid would bind repressor, but because that site is not adjacent to the promoter or the structural genes on the chromosome, normal regulation of those genes cannot occur.

6. Converting an auxotroph to a prototroph would require the introduction of a wild-type copy of the gene that was nonfunctional in the auxotroph (for example, introducing a gene that encoded an enzyme necessary for leucine biosynthesis to a leucine auxotroph lacking that enzyme). The wild-type copy of this gene could be introduced to the recipient cell by conjugation, transformation, or transduction.

14 *The Eukaryotic Genome and Its Expression*

Important Concepts

Genes in eukaryotic cells differ from those of prokaryotic cells in terms of content and organization.

- The eukaryotic genome is larger (more DNA) and more complex (more genes) than the prokaryotic genome. For example, the size of the human genome is 6 billion base pairs (bp) per cell, whereas bacteria such as *E. coli* have 4.5 million bp per cell, and viral genomes can have as few as 10,000 bp.

- Among eukaryotes, genome size is not related to relative complexity; cells of the lily plant have eighteen times more DNA than human cells.

- Eukaryotic genomes have more regulatory sequences and more regulatory proteins than prokaryotes.

- Much of the eukaryotic genome is noncoding. Some of this noncoding sequence is important for the structural integrity of the chromosome, specifically centromeres and telomeres. Other noncoding regions include introns and highly repetitive sequences, many of which have no known function.

- The eukaryotic genome is located on several linear chromosomes (20–100 million bp per chromosome), whereas the single prokaryotic chromosome (4.5 million bp per chromosome) is circular.

- Each chromosome, consisting of a double helix of DNA, has three essential parts: an origin (the recognition sequence for DNA polymerase to initiate replication), a centromere (which holds the replicated chromosomes together before mitosis and meiosis is completed), and telomeres (the ends of the chromosome).

- Chromosomes are packed tightly into the nucleus (review Figure 9.6) and are separated from the cytoplasm by a nuclear envelope. Transcription occurs in the nucleus; translation occurs in the cytoplasm. Pre-mRNAs have to be processed before they are transported to the cytoplasm for translation.

- During gene expression, transcriptional complexes must be able to access the DNA, including regulatory regions such as regulators, promoters, transcription factor binding sites, enhancers, and silencers. Some of these regulatory sequences (enhancers and silencers) can be located quite far from the promoter.

- Pre-mRNA transcripts include noncoding regions that must be removed from the RNA before the mRNA leaves the nucleus. The 5′ and the 3′ ends of the pre mRNA must also be chemically modified before the mRNA is transported to the cytoplasm.

Model eukaryotes are used to study eukaryotic gene expression and development.

- *Saccharomyces cerevisiae*

  - *Saccharomyces cerevisiae*, budding yeast, is a simple eukaryote that has 12 million bp on 16 chromosomes (per haploid genome).

  - *S. cerevisiae* has 5,600 genes; scientists have identified the probably functions of about 70 percent of these (review "Prokaryotic Genomes," Chapter 13).

  - Within the known genes are genes for metabolism (11%), energy production and storage (3%), DNA replication (3%), protein synthesis (12%), and protein targeting and secretion (6%).

  - The genome of *E. coli* and the yeast genome have the same number of genes for basic functions of cell survival. The genes that are present in the yeast and other eukaryotes but absent from the prokaryotes include histone genes, cytoskeletal genes, and RNA processing genes.

- *Caenorhabditis elegans*

  - The nematode *Caenorhabditis elegans* is used to study development.

  - The worm's body is transparent, and its growth from a fertilized egg to a thousand-celled differentiated adult organism takes just 3 days.

  - The genome of *C. elegans* is eight times larger than that of yeast (97 million bp) and has four times the number of protein-coding genes (19,099 proteins versus 6,000 proteins). Genome comparisons have revealed that 3,000 of these genes in *C. elegans* have direct homologs in yeast.

  - Other gene products in *C. elegans* include proteins used for intracellular communication, cellular differentiation, and cell adhesion.

- *Drosophila melanogaster*

  - The fruit fly *Drosophila melanogaster* has ten times more cells than *C. elegans*. Its genome has 180 million more bp.

  - Genome analysis has revealed that the fly has 13,600 genes (fewer than in *C. elegans*). There are fewer gene families (genes that are related in sequence and function) in *Drosophila* than there are in *C. elegans.* The genome of *C. elegans* also has a larger number of genes devoted to sensing chemicals in the environment and gene families involved in nerve cell formation and development.

- One-third of the genes of *Drosophila* have homologs in the nematode; one-half of fly genes have homologs in mammals.

- The fly genome has 177 genes with sequences known to be similar to human disease genes, including some genes involved in cancer and neurological disease.

- *Fugu rubripes*

  - Puffer fish of the species *Fugu rubripes* have the most compact genome in vertebrates—365 million base pairs encoding 35,000 genes. Humans have about the same number of genes and eight times as much DNA, but human DNA contains 40 percent repetitive sequences.

- *Oryza sativa* and *Arabidopsis thaliana*

  - The genome of rice, *Oryza sativa*, has many genes in common with that of the much-studied *Arabidopsis thaliana*.

  - *Arabidopsis thaliana* is a plant that has a small genome (119 million bp) with 26,000 protein-coding genes. Many of these genes are duplicates of each other, though; when the duplicates are removed, only 15,000 genes remain.

  - *Arabidopsis* contains genes unique to plants including genes involved in photosynthesis, in water transport into the root, in cell wall synthesis, in uptake of inorganic substances, and in defense against plant pathogens.

  - Most of the genes in *Arabidopsis* can be found in various species of rice, corn, and wheat.

  - Rice has many more genes than *Arabidopsis*. These additional genes include genes that make it possible for the plant to grow submerged in water and genes that provide resistance to certain diseases.

There are three types of highly repetitive sequences in the eukaryotic genome: satellites, minisatellites, and microsatellites.

- Satellites are 5–50 bp long, repeated side by side up to a million times in the genome. Satellites are found in the centromeres and appear to bind special proteins that make up the centromere.

- Minisatellites are 12–100 bp long and are repeated several thousand times. The number of repeats of minisatellites varies in individual organisms due to errors DNA polymerase makes during replication of these sequences. This variation provides unique molecular markers that can be used to identify individuals.

- Microsatellites are short 1–5 bp sequences that occur in small clusters of 10–50 copies scattered all over the genome.

Moderately repetitive DNA includes telomeres, tRNA and rRNA genes, *Alu* sequences, and transposable elements.

- Telomeres are moderately repetitive sequences located at the ends of the chromosome (see Chapter 11) that maintain the length and integrity of the chromosome.

- The genes for tRNAs and rRNAs also fall within the class of moderately repetitive DNA.

- Multiple copies of these genes are needed to provide the cell with high concentrations of protein translation machinery.

- In mammals there are four different rRNA molecules: 18S, 5.8S, 28S, and 5S. The 18S, 5.8S, and 28S rRNA genes are all transcribed from one promoter, producing a large precursor RNA, which is enzymatically trimmed to yield the mature rRNA molecules.

- In humans there are 280 copies of these rRNA gene clusters, located on five different chromosomes.

- Transposons are moderately repetitive sequences that can move from place to place in the genome.

- These DNA sequences are not stably integrated in the chromosome and can relocate to another part of the genome.

- There are four kinds of transposable elements: SINEs, LINEs, retrotransposons, and DNA transposons.

- SINEs are *short interspersed elements* of up to 500 bp; they are transcribed but not translated.

- LINEs are *long interspersed elements* of up to 7000 bp; some are transcribed and translated. About 15 percent of the human genome consists of LINEs.

- SINEs and LINEs (greater than 100,000 copies/genome) make an RNA copy of DNA; this RNA is then used as a template for DNA synthesis, which is inserted at a new site in the genome.

- Retrotransposons are transposons that make an RNA copy when they move to a different site in the genome. These transposable elements make up about 17 percent of the human genome.

- *Alu* sequences are the most common retrotransposons, accounting for 11 percent of the human genome. They are about 300 bp long and are found in a million copies throughout the genome.

- DNA transposons do not replicate when they move to a new site on the chromosome (see Figure 14.3).

- Transposons adversely affect the cell by inserting their DNA into different genes and inactivating them, as seen in some forms of muscular dystrophy and hemophilia. Insertions of transposons in genes that control cell division can convert a somatic cell to a cancer cell.

- If the insertion of transposons occurs in a germ-line cell, it can be passed down to progeny.

- Transposons can replicate themselves and adjacent chromosomal genes, resulting in gene duplication. Transposons can also pick up adjacent genes during transposition and move them to a new site on the chromosome, potentially creating a new gene that will be advantageous to the organism's survival.

- Transposons help explain why some mitochondrial and chloroplast genes are located in the nucleus whereas other organelle genes remain inside the organelle.

Many protein-coding genes are single-copy DNA sequences that include structural genes.

- Single-copy genes contain noncoding regions, including the promoter, the transcriptional terminator, and introns.

- Introns are noncoding internal sequences that are transcribed in the pre-mRNA and removed in the nucleus. After intron removal, the exons, which contain the coding sequence for the gene, are spliced together.
- Nucleic acid hybridization experiments clearly demonstrated the presence of introns in pre-mRNA molecules (see Figure 14.6).
  - Denatured β-globin DNA was hybridized to β-globin mRNA, and loops of single-stranded DNA representing noncoding intron sequences were observed.
  - Hybridizations with β-globin DNA and pre-mRNA for β-globin revealed no single-stranded DNA loops, indicating that introns were still present in the pre-mRNA.
- In DNA, introns separate gene-coding sequences into distinct parts called exons. All of the exons are in order on the gene. In some cases, different exons code for different functional domains in the final protein.
- Pseudogenes are inexact copies of genes and are usually positioned near the real gene on the chromosome.
- Gene families are sets of duplicate or related genes (such as the β-globin genes).
- Gene families provide the organism with a functional gene while allowing mutations in the other members of the gene family. Some of these mutations may create new genes that are advantageous to the organism.
- In humans, the globin gene family consists of three α-globin genes and five β-globin genes.
- Different globins are expressed at different times in development. γ-globin, which is expressed in the fetus, binds oxygen more tightly to ensure oxygen transfer from the mother across the placenta to the fetus.
- There are globin pseudogenes, which are nonfunctional because they lack promoters and/or recognition sites for intron removal.

Modifications of pre-mRNA (RNA processing) occurs in the nucleus of eukaryotic cells.

- A cap of modified GTP is added to the 5′ end of eukaryotic mRNA to facilitate binding of mRNA to the ribosome in the cytoplasm and to protect the mRNA from degradation in the cytoplasm.
- At the 3′ end of the mRNA a poly A tail is added.
- The pre-mRNA is first cut by an enzyme that recognizes the sequence A A U A A A at the 3′ end of the mRNA.
- The poly A tail (from 100–300 nucleotides) is then added to the 3′ end.
- The poly A tail assists in the transport of the mRNA from the nucleus to the cytoplasm and increases the mRNA stability in the cytoplasm.
- Splicing removes introns from pre-mRNA.
- Introns are removed from pre-mRNAs by small nuclear ribonucleoprotein particles (snRNPs).
- SnRNPs bind consensus sequences at the 3′ and the 5′ boundaries of the introns using complementary base pairing.
- Other proteins bind the snRNPs to form the spliceosome, which uses the energy of ATP hydrolysis to cut the RNA,

release the intron, and rejoin the RNA to form the mature mRNA.
- Some forms of β-thalassemia (in which there is an inadequate amount of β-globin, resulting in severe anemia) are the result of a mutation in the intron–exon consensus sequence causing inappropriate splicing of β-globin RNA and a nonfunctional gene product.
- Mature RNA is transported from the nucleus to the cytoplasm through the nuclear pores.

The expression of eukaryotic genes must be precisely regulated.

- Multicellular organisms must regulate their genes precisely so that the correct genes are expressed at the appropriate time and development can proceed normally. Differentiated cells have different functions and must express genes specific for those functions.
- Gene regulation in eukaryotes occurs both at the transcriptional and the translational level.
- Some gene products alter chromosome function and structure, making the DNA more accessible for the transcription complex or providing extra copies of certain genes to supply the cell with the necessary gene products.
- All cells must express housekeeping genes, such as proteins involved in glycolysis and protein translation.
- In prokaryotes, operons often include genes that code for several related gene products, all of which are transcribed as a unit. In eukaryotic cells, each gene is usually associated with its own promoter. Eukaryotic genes that code for related gene products (different enzymes in a metabolic pathway, for example) must have similar control sequences near their promoters.
- There are three RNA polymerases in eukaryotic cells.
  - RNA polymerase I transcribes rRNAs.
  - RNA polymerase II transcribes mRNAs.
  - RNA polymerase III transcribes tRNAs and small nuclear RNAs.
- Eukaryotic promoters are diverse and have several kinds of regulatory sequences in front of those genes, including promoters, enhancers, silencers, and regulators. These sequences are bound by proteins that affect transcriptional efficiency.
- Initiation of transcription in eukaryotic cells requires regulatory proteins called transcription factors. These are proteins that bind the promoter and form a transcription complex that RNA polymerase can bind to initiate transcription. This is in contrast to prokaryotic promoters, which have conserved sequences that RNA polymerase binds to initiate transcription.
- Most eukaryotic genes have a TATA sequence in their promoters as well as other recognition sequences that can be bound by other transcription factors.
- TFIID, a transcription factor, binds the TATA box, changing the shape of the DNA and the transcription factor. Other transcription factors then bind the promoter, and finally RNA polymerase binds to initiate transcription.
- Regulator regions, which bind proteins that activate transcription (see Figure 14.13), occur just upstream of the promoter.

- Enhancer sequences are located far away from the gene they affect; activator proteins bind enhancer sequences and stimulate the transcription complex.
- Coordinated gene regulation is achieved by positioning similar regulator sequences near the promoters of each of those genes. Proteins bind to these regulatory sequences and stimulate RNA synthesis. In plants, regulatory elements coordinating several different genes that respond to drought are called stress response elements (SREs; see Figure 14.14).
- Silencer sequences are bound by repressor proteins and negatively regulate transcription.
- Transcription of any eukaryotic gene is determined by the combination of all of these regulatory factors, transcription factors, repressors, and activators.

Proteins that bind DNA (activators, transcription factors, regulators, and repressors) have characteristic protein motifs.

- DNA protein-binding motifs include helix-turn-helix, leucine zipper, zinc finger, and helix-loop-helix. (see Figure 14.15).

In order for gene expression to occur, the DNA must be made accessible to the transcription complex.

- The DNA in eukaryotic chromosomes is associated with nucleosomes and other chromosomal proteins, which can inhibit the initiation and elongation steps of transcription.
- Upstream of the initiation site, protein complexes help the nucleosomes become disaggregated so that the transcription complex can bind the DNA and transcription can be initiated.
- Once transcription has started, other protein complexes bind the DNA to help the transcription complex move through the nucleosomes. This process is called chromatin remodeling.
- Transcriptionally active chromatin (euchromatin) is less condensed and takes up less nuclear stain than the transcriptionally inactive, highly condensed chromatin (heterochromatin).
- The genotype for the sex chromosomes in mammals is XY in males and XX in females. However the expression of X-linked genes is the same in both sexes, due to X-inactivation in females.
  - Early in embryonic development, one of the two X chromosomes is randomly inactivated, producing a highly condensed heterochromatic chromosome, called a Barr body.
  - The number of Barr bodies is equal to the number of X chromosomes minus one (XXY, Klinefelter's syndrome, and XXX both have two Barr bodies). Normal females (XX) have one Barr body.
  - The DNA of the inactive X is hyper-methylated (methyl groups are added to the 5′ carbon of cytosine).
  - Transcription of the *XIST* gene occurs on the inactive X chromosome.
  - The *XIST* RNA (also known as interference RNA or RNAi) binds the X chromosome from which it was transcribed and inactivates it (see Figure 14.18).

DNA sequences can be regulated by moving them to a new site on the chromosome to activate transcription.

- The mating type locus (MAT) in the budding yeast, *Saccharomyces cerevisiae*, is the site in the yeast genome where the mating type (*a* or α) is determined.
- Yeast can be either *a* or α mating type, depending on which allele is in the MAT locus.
- The *a* and α alleles are located in an inactive region of the chromatin, but during cell division, one of these two alleles is copied and placed into the MAT locus.
- If the yeast is an *a* mating type, then a new DNA copy of α is made. The *a* allele at the MAT locus is removed by enzymes, and the α allele is inserted at the MAT locus and transcribed, resulting in a yeast cell of α mating type.
- DNA rearrangements naturally occur in the genes for human antibodies. In some cancers, DNA rearrangements result in the transcriptional activation of inactive genes because they are moved adjacent to an active promoter.

Specific genes can be amplified to meet the needs of the cell, as seen in the amplification of the ribosomal RNA genes.

- Egg cells in frogs and fishes selectively amplify the ribosomal RNA gene clusters so that there are greater than a million copies of ribosomal DNA per egg cell.
- Gene amplification is also seen in cancer cells, specifically in oncogenes (genes that cause cancer) and in genes encoding protein receptors for cancer drugs.

Posttranscriptional modifications to RNA include alternate splicing, increasing mRNA stability, and RNA editing.

- Different RNAs can be made from the same gene by alternate splicing.
- In mammals, tropomyosin pre-mRNA is spliced five different ways to generate five different forms of this gene product, each for a different tissue: skeletal muscle, smooth muscle, fibroblast cells, liver, and brain.
- When an mRNA arrives in the cytoplasm, ribonucleases degrade it. This degradation can be affected by other cytoplasmic factors.
- The sequence of mRNAs can be altered by RNA editing (adding additional U's) or altering individual nucleotides (changing C's to U's).
- Translational and posttranslational regulation occurs in many eukaryotic genes.
- Messenger RNAs in the cytoplasm are not always translated, because other factors regulate the accessibility of these mRNAs to the translational machinery. For example, if the G of the 5′ cap is not modified, as seen in the embryo of the hornworm moth, the mRNA is not translated. After fertilization, this G is modified, and the mRNA is translated.
- Ferritin is a storage protein for iron. The mRNA necessary for making ferritin is present in cells at steady levels, regardless of the concentration of iron ions in the cell. When the iron level in the cell is low, a repressor protein binds to ferritin mRNA, preventing its translation into ferritin. As the iron level rises in the cell, some iron ions

bind to the repressor protein, altering its shape so that it can no longer bind the ferritin mRNA. This permits translation of ferritin from the ferritin mRNA; the new ferritin is then available for storage of iron ions.

- The translation of the globin mRNA increases in the presence of excess heme in the cell. Heme removes a block to translational initiation of globin mRNA.

Protein modification includes glycosylation, phosphorylation, and the removal of signal sequences. Protein degradation regulates the lifetime of a protein in the cell.

- Proteins that are targeted for degradation are first linked to ubiquitin. This 76-amino acid protein binds a huge protein complex called a proteasome. The proteasome is a hollow cylinder that has ATPase activity.

- When a protein-ubiquitin complex enters the proteasome, ubiquitin is removed, the protein is unfolded, and three proteases digest the protein (see Figure 14.22).

- Human papilloma virus targets p53 (a cellular protein that inhibits cell division) for proteasome degradation, leading to unregulated cell division and cancer.

The Big Picture

- Eukaryotic genomes are larger and more complex than prokaryotic genomes. The chromosomes in eukaryotic cells are linear and there are more than one.

- Eukaryotic DNA contains many genes and many noncoding regions, including nongene sequences, moderately repetitive sequences, highly repetitive sequences, nontranscribed regions, and introns. Some of the repetitive DNA sequences, such as centromeres and telomeres, have a structural function for the chromosome.

- DNA is highly condensed in the nucleus due to its association with nucleosomes and other chromosomal proteins. Gene expression is a two-step process: The chromatin must be remodeled to expose the DNA containing the genes so that the transcriptional complexes (which consist of transcription factors, regulatory proteins, activators, and RNA polymerases) can access those DNA sites to regulate transcription of that gene. Transcription factors and other regulatory proteins have particular motifs that facilitate their binding to DNA.

- In eukaryotic cells, transcription is separated from translation by the nuclear membrane. RNAs are processed in the nucleus before being transported to the cytoplasm. Posttranscriptional regulatory elements affect gene expression, including different mRNA stabilities, alternate splicing, RNA editing, translation regulation, protein modification, and protein degradation.

Common Problem Areas

- Sometimes it is difficult to understand why there are so many different ways gene expression is regulated in eukaryotic cells. Examine a picture of a eukaryotic cell to get a feeling for all of the specialized parts of the cell and how gene expression must be tailored for each of these parts. Now think about a multicellular organism and how different genes must be expressed in all of the differentiated tissues.

- Grasping the different events that are involved in the regulation of gene expression in eukaryotic cells may be overwhelming. Begin by examining how transcription of a eukaryotic gene occurs by reviewing chromatin structure, the promoter, RNA polymerase, and transcription factors. Next examine how regulators, enhancers, and silencers, when bound by proteins, can modulate the levels of transcription of a eukaryotic gene. To complete your analysis, review the details of posttranscriptional regulation.

Study Strategies

- Make a list of the sites on the DNA that affect transcriptional initiation. Next, list the names of the proteins that bind those DNA sites and describe how they affect transcription.

- Draw a simple diagram of a eukaryotic gene.

- Include the flanking 5′ and 3′ sequences of the gene that are important for regulation and gene expression.

- Draw a diagram of the pre-mRNA and indicate sites on the RNA where processing occurs.

- Make another diagram showing the mature mRNA as it appears in the cytoplasm. Indicate sites on the mRNA where posttranscriptional regulation might occur.

- Finally, describe ways that proteins can be modified to regulate gene expression.

- Give examples of moderately repetitive and highly repetitive sequences in the DNA and describe their function.

- Review model organisms described in this chapter and describe two advantages for studying gene regulation in each of these organisms.

- Review the following tutorials and activities on the Student Website/CD:
  Tutorial 14.1 RNA Splicing
  Tutorial 14.2 Initiation of Transcription
  Activity 14.1 Eukaryotic Gene Expression: Structural Features
  Activity 14.2 Eukaryotic Gene Expression: Control Points

Test Yourself

Knowledge and Synthesis Questions

1. Eukaryotic chromosomes
   a. are circular and contain origins and terminator sequences.
   b. are linear and have origins and telomeres.
   c. contain coding and noncoding sequences.
   d. Both b and c

2. Model eukaryotic organisms have helped biologists understand
   a. genes involved in development.
   b. gene families.
   c. genes encoding proteins that are essential for all cells.
   d. All of the above

3. Moderately repetitive DNA includes
   a. only coding sequences.
   b. only noncoding sequences.
   c. coding and noncoding sequences.
   d. satellites, minisatellites, and microsatellites.

4. Transposable genetic elements
   a. always affect the cell adversely, because when they move, they inactivate genes.
   b. are retroviruses.
   c. provide a mechanism for moving genetic material from organelle genomes to the nuclear genome.
   d. always replicate their DNA when they more.

5. Introns are DNA sequences that
   a. code for functional domains in proteins.
   b. are removed from pre-mRNA by spliceosomes.
   c. allow one gene to make different gene products, depending on which introns are removed during splicing.
   d. Both b and c

6. Globin genes are
   a. only transcribed in bone marrow cells.
   b. translationally regulated by excess heme.
   c. part of a family of globin genes, with different genes expressed at different times in development.
   d. All of the above

7. Pre-mRNAs must be processed in the nucleus in order to
   a. increase their stability in the cytoplasm.
   b. allow RNA polymerase to initiate transcription.
   c. permit coding sequences to be joined to adjacent noncoding sequences.
   d. facilitate ribosome recognition in preparation for DNA synthesis.

8. Coordinated regulation of genes in eukaryotic cells
   a. is the result of positioning the same regulatory sequence in front of each gene.
   b. results from all of those genes being under the control of one promoter.
   c. occurs because related genes all have the same operons.
   d. occurs because enhancers cause DNA to bend.

9. The transcription complex includes _____ and _____.
   a. transcription factors; promoters
   b. regulator proteins; regulators
   c. repressor proteins; silencers
   d. Both a and b

10. DNA binding proteins
    a. have distinct three-dimensional structures that allow them to bind to the DNA.
    b. can be transcription factors.
    c. can help condense the DNA in the nucleus.
    d. All of the above

11. Chromatin structure must be altered for gene expression to occur because
    a. condensed chromatin is replicated but not transcribed.
    b. condensed chromatin makes most DNA sequences inaccessible to the transcription complex.
    c. decondensed chromatin has more nucleosomes per DNA molecule.
    d. heterochromatin is actively transcribed and euchromatin is not transcribed.

12. When DNA sequences are moved to new sites on a chromosome
    a. new genes can be transcribed.
    b. genes can be inactivated.
    c. new genes can be created.
    d. All of the above

13. rRNA gene copies are amplified
    a. to ensure that gene expression in somatic cells will continue.
    b. to make more copies of DNA origins for replication.
    c. to ensure that there is enough translational machinery for the rapid development of the embryo.
    d. to help heterochromatin become more accessible to the transcription complex.

14. Posttranscriptional regulation can include
    a. binding of repressor on silencer regions.
    b. insertion and alteration of nucleotides.
    c. decreasing mRNA stability in the cytoplasm.
    d. Both b and c

15. Genes can be inactivated by
    a. inaccurate removal of introns.
    b. transposable genetic elements.
    c. movement of genes to heterochromatic regions of the chromosome.
    d. All of the above

Application Questions

1. The cells of the lily plant have eighteen times more DNA than human cells, yet human DNA is more complex. Explain this phenomenon.

2. When normal cells become transformed into cancer cells, gene regulation and chromosome stability in those cells is altered. Describe three different ways that cancer cells are different from normal cells.

3. A mutation occurs in the promoter of a eukaryotic gene, eliminating the TATA box. How will this mutation affect the transcription of this gene?

4. A different mutation eliminates the enhancer sequence on the DNA for this gene. How will this mutation affect the transcription of this gene?

Answers to Questions

Knowledge and Synthesis Answers

1. **d.** Prokaryotic chromosomes are circular and contain termination sequences.

2. **d.** Important developmental genes and gene families have been found in several different eukaryotic organisms. Comparisons of genes in eukaryotes and prokaryotes have revealed many similar genes that provide the same essential functions to those cells.

3. **c.** Moderately repetitive sequences include the ribosomal gene clusters as well as noncoding sequences such as telomeres and *Alu* sequences. Satellites, minisatellites, and microsatellites do not contain genes.

4. **c.** Transposable elements do not always have an adverse effect on genes; sometimes their insertion can create new genes. Retrotransposons are like retroviruses, but lack the genes for the viral capsid proteins. DNA transposons do not replicate their DNA before they move to a different site on the chromosome.

5. **d.** Introns are removed by spliceosomes, and differential splicing can produce different gene products as seen in the tropomyosin example (review Figure 14.20).

6. **d.** Globin genes cannot be transcribed in the human red blood cells in adults because those cells lack nuclei. Globin genes are translationally regulated by heme and they are part of the family of globin genes.

7. **a.** Pre-mRNAs are processed in the nucleus for many reasons (see Figures 14.9 and 14.10), including to increase their stability in the cytoplasm. Transcription by RNA polymerase *precedes* pre-mRNA processing. Noncoding sequences (introns) are removed from the pre-mRNA, not joined to coding sequences (exons). Ribosome recognition does not lead to DNA synthesis.

8. **a.** Each eukaryotic gene is transcribed from a separate promoter, and eukaryotic genes do not have operons to regulate them.

9. **d.** When repressor proteins are bound to silencers, transcription is inhibited.

10. **d.** DNA binding proteins must have three-dimensional structures appropriate for binding DNA. Some DNA-binding proteins are transcription factors, some are telomerases, which bind the ends of chromosomes, and some are nucleosomes, which bind the DNA and help pack it into the nucleus.

11. **b.** Nucleosomes condense the DNA. Heterochromatin is transcriptionally inactive, whereas euchromatin is transcriptionally active.

12. **d.** Transposable elements can be transcribed if they move adjacent to a promoter. Transposable elements can inactivate a gene if they insert into the middle of a coding sequence. Transposable elements can move parts of genes with them to new chromosomal sites, potentially creating new genes.

13. **c.** Ribosomal RNA amplification occurs in the eggs of fishes and frogs to prepare for the rapid development that immediately follows fertilization. Amplified rRNA is not needed in somatic cells.

14. **d.** Repressors that bind at silencer sequences regulate transcription and thus are not posttranscriptional mechanisms of regulation.

15. **d.** Inaccurate removal of introns can create mRNAs that are missing coding sequence or that have extra noncoding sequences. Transposable genetic elements can move into the coding regions of genes, inactivating those gene products. Moving a gene to a heterochromatic (transcriptionally inactive) region of a chromosome results in that DNA being inaccessible to the transcription complex.

Application Answers

1. The DNA of the lily plant must contain more noncoding and repetitive sequences and less coding sequence than human DNA.

2. Genetic changes can make a normal cell into a cancerous cell in the following ways: A transposable element can move into the coding sequence of a gene that regulates cell division, inactivating that gene. Amplification of oncogenes can transform normal cells into cancer cells. Targeting proteins that control cell division for proteasome degradation (as seen in human papilloma virus and P53) will cause cells to become cancerous.

3. If the TATA box of the promoter for a eukaryotic gene is deleted, transcription factor TFIID will not bind the promoter. The transcription complex will not assemble at the promoter and there will be no initiation of transcription.

4. If the enhancer for a gene is deleted, transcription for that gene will still occur, but at a reduced rate.

15 Cell Signaling and Communication

Important Concepts

Cells respond to specific environmental signals by changing their cellular function. This process of receiving a signal and communicating it to the cell is called signal transduction.

- There are three steps in cell signaling: binding of the signal to a receptor, transduction of the signal (communication to the rest of the cell) and cellular response (a change in cellular function).

- Signals must bind to a receptor protein on the cell membrane or in the cell's cytoplasm.

- Binding of a signal molecule to a receptor causes a conformational change in the receptor protein that relays or amplifies the signal within the cell.

- The relay or amplification of the signal results in a change in the function of the cell.

- A signal is not enough to elicit a response. The receptor for a given signal must be present, and the signal must bind the receptor protein.

- Signals may be from the physical environmental (i.e., light, chemical, temperature) or from other cells (hormones, neurotransmitters, other chemicals).

- Autocrine signals are local signals that affect the cells that make them.

- Paracrine signals are local signals that affect nearby cells.

- Hormones are circulatory signals that travel through the circulatory system and affect distant cells.

- Response to osmotic change in *E. coli* is an example of signal transduction (see Figure 15.2). An environmental signal (a change in solute concentration in the inner membrane space) causes a conformational change in a receptor protein (EnvZ), which is a transmembrane protein in the bacterium's plasma membrane. Signal transduction then proceeds along the following pathway: EnvZ picks up a phosphate from ATP and transfers it to OmpR (a protein in the cytoplasm). The OmpR protein changes shape so that it can bind to the promoter for the gene that encodes the protein OmpC, thereby increasing production of OmpC protein. The OmpC protein blocks pores in the outer membrane, preventing solutes from entering the intermembrane space.

Receptors are specific. Not all cells have the same receptors, so a given signal can have different effects (or no effect) on different cell types.

- Binding to the receptor requires a fit of the ligand to the receptor binding site.

- Binding must cause a conformational change on the receptor to have an effect.

- The ligand is not changed in the binding process, and binding of the signal to the receptor is reversible.

- Binding sites may be inhibited by competing chemicals.

- There are two classes of receptors: plasma membrane receptors, which bind large and/or polar ligands that cannot cross the plasma membrane, and cytoplasmic receptors, which bind ligands that diffuse across the plasma membrane.

There are three main types of plasma membrane receptors.

- Ion channels: Ligand binding causes a conformational change to open "gates" that allow ions (Na^+, K^+, Ca^+, or Cl^-) to pass through (see Figure 15.5).

- Protein kinases: Ligand binding stimulates the transfer of a phosphate group from ATP to a specific protein (see Figure 15.6).

- G protein-linked receptors: Ligand binding to this transmembrane receptor protein changes its shape so that a G protein on the cytoplasmic side can bind. The G protein binds GTP and a subunit of the G protein is released from the receptor protein. When the G protein subunit encounters an effector protein, the GTP is hydrolyzed to GDP. The effector protein may be an enzyme or an ion channel that is altered by the G protein. Effector proteins may be inhibited or activated by G proteins (see Figure 15.7).

Cytoplasmic receptors are in the cell cytoplasm.

- Ligand binding causes a conformational change in the receptor that allows passage of the receptor-signal complex into the cell nucleus where it can function as a transcription factor (see Figure 15.8).

Signal transduction may be direct (occurring at ligand binding) or indirect (requiring second messenger molecules).

- Direct signal transduction is a function of the receptor itself.

- Protein kinase receptors are examples of direct signal transduction (see Figure 15.9). They can initiate protein kinase cascades where several different kinases are activated after the initial signal is bound. Protein kinase cascades can result in the transcriptional activation of genes. Different target proteins in protein kinase cascades can mediate different cellular responses.

- Indirect signal transduction utilizes a second messenger to mediate the response between receptor binding and cellular response.

- Secondary messengers are released to the cytoplasm after ligand binding and function to amplify the signal. They allow the cell to have multiple responses to a single signal. Common secondary messengers include the following:

  - cAMP: This secondary messenger can affect ion gate opening in the cell or can activate protein kinases.

  - IP_3 and DAG: These end products of the hydrolysis of PIP2 can interact with G proteins, protein kinase C or Ca^+ channels in the endoplasmic reticulum.

  - Free ions: Calcium ions frequently act as secondary messengers either independently or through calmodulin.

  - Nitric oxide: This gas stimulates the formation of cGMP, which can initiate a protein kinase cascade.

Signal transduction is highly regulated.

- Nitric oxide breaks down rapidly.

- Ca^+ concentration is controlled by membrane pumps and ion channels.

- Protein phosphatases remove phosphate groups from phosphorylated proteins.

- GTPases convert GTP on G proteins to GDP, thereby inactivating G proteins.

- cAMP levels are affected by phosphodiesterase, which converts cAMP to AMP, which has no secondary messenger activity.

Signal effects are changes in cell function.

- Ion channels open (as seen in sensing a particular odor).

- Enzymes are inhibited or activated (as seen with epinephrine stimulation).

- Transcription is stimulated (as seen with lipid soluble hormones).

Signals are often transduced between cells.

- Cell-to-cell signaling can be facilitated by gap junctions (see Figure 15.1). Gap junctions are protein-lined channels that allow the passage of small signal molecules and ions between adjacent cells.

- Plasmodesmata are membrane-lined channels connecting adjacent plant cells. They are filled with endoplasmic reticulum-derived tubules called desmotubules.

The Big Picture

- Signal transduction is the means by which cells receive information from the environment or other cells and react to those signals. Transduction is a highly regulated series of events that depends on the binding of a signal ligand to a receptor protein. This binding must cause a change in the shape of the receptor protein, which results in a cascade of intracellular events that leads to a change in cell function. The effects of signals are often mediated by secondary messengers.

Common Problem Areas

- It's easy to get overwhelmed by the different examples of signal transduction. Try to simplify your studying approach to signal transduction by looking at particular details of the systems. For instance, distinguish between direct and indirect signal transduction. Compare plasma membrane receptors with cytoplasmic receptors. List the three kinds of secondary messengers for signal transduction. Then, pick one of the signal transduction examples and make a table that includes the signal, the receptor, the transduction (responders and amplification) and the effect in the cell. Expand your table to include other examples of signal transduction.

- G protein action frequently gives students trouble. Remember that the G protein interacts with the receptor, that it binds GTP, and then it interacts with the effector protein.

- Gap junctions and plasmodesmata are physical connections between adjacent cells, and although they are similar in function, they are fundamentally different in structure.

Study Strategies

- Make a flowchart or a diagram of the signal transduction pathways.

- There are several different secondary signals. List each one and give an example of how that signal was activated and what it affects.

- Review the figures in your book to clarify the different examples of signal transduction.

- Review the following tutorial and activities on the Student Website/CD:
  Tutorial 15.1 Signal Transduction Pathway
  Activity 15.1 Signal Transduction
  Activity 15.2 Concept Matching

Test Yourself

Knowledge and Synthesis Questions

1. Which of the following does *not* occur during signal transduction?
   a. Ligand binding to receptor
   b. Conformational change to the receptor protein
   c. Conformational change of the signal
   d. Alteration of cellular activity

2. When looking at how *E. coli* cells cope with osmotic changes, the ultimate goal of the *E. coli* cell is
   a. to change the permeability of the membrane.
   b. to change what DNA is transcribed.
   c. phosphorylation of OmpR.
   d. binding of the ligand to EnvZ.

3. What do the following receptors all have in common: G protein-linked receptors, protein kinases, and ion channels?
   a. Ligand binding
   b. Conformational change once the ligand is bound

c. Amplification of the signal
d. All of the above

4. How does signal binding to a receptor differ from enzyme–substrate binding?
    a. The signal is not altered in any way during the process.
    b. The process is reversible.
    c. Cell receptors can become saturated by signal molecules
    d. None of the above

5. Secondary messengers function to
    a. amplify the signal.
    b. bind to the active site of the receptor.
    c. result in multiple effects from a single signal.
    d. Both a and c

6. Caffeine is a stimulant that works because it acts as _____ to the adenosine receptors in a person's brain, and stimulates _____ in that person's heart and liver that increases blood flow and blood glucose.
    a. an alloster; a pathway
    b. an inhibitor; a cascade pathway
    c. an inhibitor; ligands
    d. a signal; inhibitors

7. Cytoplasmic receptors only bind
    a. small signals that can diffuse through the plasma membrane.
    b. secondary messengers like cAMP.
    c. hydrophilic molecules.
    d. All of the above

8. Protein kinase cascades are significant because
    a. amplification can occur at each step in the path.
    b. information at the plasma membrane is communicated to the nucleus.
    c. the multiple steps allow for specificity of the process.
    d. All of the above

9. Signal transduction is regulated in which of the following ways?
    a. The amount of signal present can be regulated.
    b. Enzymes in the pathway convert active forms of proteins to inactive forms.
    c. Signals are denatured.
    d. None of the above

10. Which of the following statements regarding receptors is true?
    a. Only one type of receptor is utilized throughout a cascade.
    b. Cascades may involve many types of receptors including ion channel receptors, protein kinase receptors, and G protein-linked receptors.
    c. Receptors amplify signals in a one-to-one ratio.
    d. All of the above

11. Gap junctions and plasmodesmata differ in that
    a. gap junctions are connected by protein tubules called connexons, and plasmodesmata are connected by extensions of the plant plasma membrane.
    b. plasmodesmata allow much larger molecules to pass through them.
    c. gap junctions have no real physical connection but are the space between adjacent cell membranes.
    d. one is of animal origin and the other is of plant origin; otherwise they are physically the same.

Application Questions

1. Based on your knowledge of prokaryotic and eukaryotic cell structure and function, how might signal transduction differ in prokaryotes and eukaryotes?

2. Describe how protein kinases and G protein-linked receptors may interact in a signal transduction cascade.

3. Diagram the Ras signaling pathway, and identify the components by receptor type. Be sure to identify the signal, the receptors, and the resulting effect on the cell.

4. Discuss the role of secondary messengers in a signaling pathway. How are they different from signals and receptors? What roles do they have in common with signals?

5. True or false: A receptor may act as a signal. Explain your answer.

6. Trace how changes in osmotic pressure in the plasma membrane of *E. coli* result in changes in gene transcription. What would be the outcome if OmpR could not bind to the *ompC* promoter?

Answers to Questions

Knowledge and Synthesis Answers

1. **c.** For signal transduction to take place, the ligand must bind to the receptor, the receptor must undergo a conformational change, and the activity of the cell must be altered.

2. **a.** The ultimate goal of any signaling pathway is to alter the function of the cell in response to the signal.

3. **d.** All types of receptors bind ligands, undergo conformational change, and amplify the signal.

4. **a.** Substrates are altered in enzyme–substrate complexes. Signals are not altered when they bind to receptors.

5. **d.** Secondary messengers both amplify the signal and have multiple effects within a cell.

6. **b.** Caffeine and adenosine bind to the same receptor protein. The binding of caffeine prevents adenosine from binding in the brain. In other organs, a series of cascades begin in response to caffeine binding.

7. **a.** To bind to a cytoplasmic receptor, the ligand must be able to pass through the plasma membrane.

8. **d.** Amplification, regulation, and specificity are all roles of protein kinase cascades.

9. **b.** Active forms of proteins may be converted to inactive forms via enzyme action.

10. **b.** A given cascade may utilize ion channel receptors, protein kinase receptors, and G protein-linked receptors.

11. **a.** Both plasmodesmata and gap junctions provide physical connections between the cytoplasms of adjacent cells.

Application Answers

1. Signal transduction in prokaryotes is limited to only one cell; there is no nuclear membrane to cross, and cascades tend to be simpler with fewer intermediate steps.

2. G protein-linked receptors frequently expose the protein kinase activities of effector molecules.

3. Refer to Figure 15.9 in your text.

4. Secondary messengers function to amplify and spread signals. They do not bind to receptors and do not act like

signals in the cascade. Signals result in a change in cell function; secondary messengers are part of the pathway that results in the change in cell function.

5. Receptors may act as signals for the next step in a given cascade. See Figure 15.8 in your text.

6. Refer to Figure 15.2 in your text. If OmpR could not bind to the *ompC* promoter, then the *ompC* gene would not be transcribed and translated, and the *E. coli* cell would be unable to prevent further solute entry into the inner membrane space.

16 Recombinant DNA and Biotechnology

Important Concepts

Recombinant DNA technology has allowed scientists to combine DNA molecules from different organisms to create useful gene products and to analyze gene expression and function.

- Gene cloning involves
  - Cutting DNA fragments with restriction endonucleases
  - Covalently linking those fragments to vector DNA, creating a new recombinant DNA molecule
  - Introducing the newly recombined DNA into a host cell
  - Growing that cell in conditions that ensure the recombinant molecule will be retained by the cell
- All the key techniques in recombinant DNA technology rely on complementary base pairing, including sequencing, rejoining, amplifying, and locating DNA fragments.

Restriction endonucleases are bacterial enzymes that cut double-stranded DNA at specific sites.

- Bacteria defend themselves against viral infection by synthesizing restriction endonucleases. Restriction endonucleases cut viral DNA but leave their own DNA unharmed because bacterial enzymes have modified that DNA (usually by the addition of a methyl group).
- Restriction endonucleases cut the DNA at specific sites (restriction sites) and leave a 5′ PO_4 and a 3′ OH at the cut ends. For example, *Eco*RI is a restriction endonuclease that cuts the sequence GAATTC, and its complementary strand CTTAAG, between the G and the A on each strand. This type of cutting leaves single-stranded regions on the ends (also called "sticky ends"). These sticky ends can form complementary base pairs with other DNA molecules that have the same sticky ends (see Figure 16.4).
- *Eco*RI cuts on average about 1 in every 4,000 base pairs in prokaryotic genomes, but sizes of the DNA fragments vary, depending on the actual location of these sites in the chromosome.
- Many restriction endonucleases, including *Eco*RI, are active as dimers consisting of two identical subunits.
- A different class of restriction endonucleases cuts between the same bases on both DNA strands, producing blunt ends.

DNA fragments can be separated using gel electrophoresis.

- A gel is a porous molecular sieve that allows smaller DNA molecules to move through it faster than larger DNA molecules. An electric field is applied to the gel, causing the negatively charged phosphates of the DNA to be pulled toward the positively charged electrode.
- DNA fragments of known sizes are also separated on the gel as reference.
- DNA fragments containing specific sequences can be identified on a gel using a blot procedure and subsequent hybridization with a radioactively labeled probe. Denatured DNA fragments are transferred to a nylon filter (the blotting part of the procedure) and then exposed to a single-stranded DNA probe that has been radioactively labeled. The radioactive DNA probe will hybridize to complementary base sequences on the DNA fragments on the filter (see Figure 16.3).
- The identified DNA fragments can then be removed from the gel and purified.

Purified DNA fragments can be mixed with other DNA molecules that have been cut with the same restriction endonuclease or that have complementary sticky ends to form recombinant DNA.

- DNA ligase (the enzyme that covalently joins the Okazaki fragments during DNA replication and mends broken DNA, see Chapter 11) is used to form a covalent bond on each DNA strand of the recombinant molecule (see Figure 16.4).
- Recombinant DNA is inserted (transfected) into a host cell creating a transgenic cell or organism.
- Reporter genes are often added to recombinant DNA to identify host cells that have the clone of interest.

Genes can be cloned into prokaryotes or eukaryotes.

- Prokaryotes have been used to clone many genes. However, they lack splicing machinery and protein modification enzymes that may be required for eukaryotic gene expression.
- Yeast and plants have been used to clone eukaryotic genes.
- Yeast cells have a rapid cell division cycle (2–8 hours), are easy to grow, and have a small genome size (20 million base pairs).

- Many plant cells are totipotent. They can be grown in culture, coaxed to take up recombinant DNA, and manipulated by alterations in the growth medium to form an entire new transgenic plant containing the recombinant DNA molecule.

Once the recombinant DNA has been successfully introduced into the cell, it must be replicated to be maintained.

- Recombinant DNA can be inserted into the host chromosome, where it is replicated when the chromosome is replicated. Such insertion might happen randomly after the DNA is introduced into the cell. Alternatively, recombinant DNA molecules can enter the host cell by being part of a vector (plasmid, virus, or artificial chromosome).

- Vectors must be small and have restriction sites, origins of replication, and reporter genes.

- Plasmid vectors in *E. coli* are small (from 2,000 to 6,000 base pairs) circular DNA molecules. They have a single set of unique restriction sites, where the cut DNA is inserted, and a drug resistance marker (usually ampicillin resistance) allows the investigator to confirm the presence of the plasmid in the *E. coli* cell. Plasmids synthesize their DNA independently from their own origins of replication, making many copies of plasmid DNA per cell. DNA fragments of up to 10,000 base pairs can be cloned into plasmid vectors (see Figure 16.5*a*).

- Virus vectors are used to clone larger DNA sequences, especially eukaryotic genes, which have introns, exons, and flanking sequences. Genetically engineered bacteriophage λ can accommodate up to 20,000 base pairs of inserted DNA. The new recombinant virus is still able to infect *E. coli* and produce many viral particles, each containing the cloned DNA.

- Yeast artificial chromosomes (YACs) can be used to clone genes in a eukaryote, the yeast cell (see Figure 16.5*b*), where normal eukaryotic DNA replication and gene expression occur.

  - YACs have a yeast origin of replication, telomeres, a centromere, marker genes, and restriction endonuclease sites where foreign DNA can be inserted. They are only about 10,000 base pairs in size, but they can accommodate insertion of 50,000 to 1,500,000 base pairs of DNA.

- *Agrobacterium tumefaciens*, a bacterium that causes crown gall in plants, harbors a Ti plasmid. This plasmid contains a transposon, T DNA, that can insert itself into the host cell's DNA.

  - There are restriction endonuclease sites within the T DNA that can be used to clone foreign DNA. Once DNA has been inserted into those sites, the tumor-causing capabilities of this bacterium are eliminated. However, the recombinant plasmid is still able to act as a transposon.

  - When plant cells are infected by this modified *A. tumefaciens*, the recombinant plasmid can move into the host DNA via transposition. Individual plant cells bearing this recombinant DNA in their chromosomes can be cultured to produce a new transgenic plant containing the recombinant DNA molecule in every cell.

Reporter genes and genetic markers (such as drug resistance) can be used to determine if a recombinant DNA molecule has been successfully introduced into a host cell.

- If foreign DNA has been inserted into the *Tet* gene (which confers resistance to tetracycline) of a plasmid, that gene will be inactivated, and the cell bearing this plasmid will be tetracycline sensitive. The plasmid also has an *Amp* gene, which confers ampicillin resistance to the host cell (see Figure 16.6). To determine if the plasmid is present in the host cell, the investigator will have to screen three kinds of cells: cells that did not receive the plasmid (amp^s, tet^s), cells that received the plasmid but without any inserted DNA (amp^r, tet^r), and cells that received the plasmid with the DNA insert (amp^r, tet^s).

- Other marker genes that are commonly used for detection of recombinant DNA molecules in host cells include the *lac* operon, the genes that code for luciferase, and green fluorescent protein (GFP).

- To favor ligation of the foreign DNA into the plasmid (rather than the plasmid religating with itself), the concentration of foreign DNA must be high relative to the plasmid DNA concentration in the ligation mixture.

DNA molecules can be isolated from genetic libraries, which include random pieces of DNA, from DNA copies of messenger RNA (cDNAs), or from artificially synthesized DNA.

- Genetic libraries of DNA are usually cloned into bacteriophage λ because each virus can accommodate 20,000 base pairs of DNA.

- To make cDNA (complementary DNA), the messenger RNA must first be isolated from the cell. Molecules of oligo dT are hybridized to the poly A tail of the messenger RNA and used as primers on the messenger RNA for reverse transcriptase. A complementary DNA molecule is synthesized by reverse transcriptase, and the new molecule can then be cloned.

- cDNA clones are used to compare gene expression in different tissues at different stages of development. One-third of all genes in an animal are expressed only during prenatal development.

- To make synthetic DNA, the investigator must know the amino acid sequence of the gene. Sequences for transcriptional and translational initiation and termination can be added to the gene sequence. Regulatory regions and preferred codons may also be added to the synthetic DNA.

- Mutant genes can be synthesized and compared to wild-type genes to analyze gene function.

- Gene function can be eliminated by creating knockout genes, which are the result of a homologous recombination event in the cell in which a normal gene is replaced by an inactivated form of the gene. The inserted DNA carries a genetic marker so its expression can be followed in the host's cells.

- In constructing transgenic mice, a plasmid containing the inactivated marker gene is transfected into a mouse embryonic stem cell (see Figure 16.9). If recombination occurs, the marker gene will be expressed. The stem cell is then transplanted into an early mouse embryo, and the resulting phenotype is analyzed.

DNA chip technology allows screening of thousands of DNA sequences at the same time.

- DNA chips are small glass chips containing 10 million copies of different DNA sequences per chip. These sequences (greater than 20 base pairs long) are attached to the chip in a precise order and can contain up to 60,000 different DNA sequences.
- DNA chips have been used to study transcription patterns in cells in different physiological states, to study gene expression in different tissues at different times in an organism, and to detect all the possible variants of a particular gene.
- For example, cDNA copies of cellular mRNA can be made and amplified using PCR (see Chapter 11). Those cDNA molecules are coupled to fluorescent dyes and allowed to hybridize to the DNA chips. The chip is then exposed to fluorescent light to determine which sequences formed a hybrid with the cDNA (see Figure 16.10).

Antisense messenger RNA and RNAi can prevent the expression of specific genes.

- Antisense RNA will form a double-stranded RNA molecule in the cell's cytoplasm, which cannot be translated and will be degraded by the cell.
- A DNA copy of antisense RNA can be cloned into a vector and introduced into the cell's cytoplasm where, under the right conditions, it can be transcribed.
- Interference RNAs (RNAi) are short double-stranded RNA molecules that bind specific mRNAs and target them for degradation.

The two-hybrid system can indicate if two proteins interact with each other in the cell.

- The DNA-binding domain of a transcription factor is fused to a target protein. The activating domain of the transcription factor is fused to the protein the investigator wishes to test.
- The DNA sequence for the transcription factor is placed in front of a reporter gene. If the target protein and the test protein interact, the transcription factor will be active, and the reporter gene will be expressed (see Figure 16.12).

Biotechnology is the use of living cells to produce useful materials for people, including food, medicines, and chemicals.

- For cells to produce a cloned gene product, the vector (an expression vector) must have DNA sequences that allow the cloned gene to be expressed.
- Prokaryotic expression vectors require a promoter, a termination site for transcription, and a ribosome-binding site.
- Eukaryotic expression vectors require the same elements as well as a poly A addition site, transcription factor binding sites, and enhancers.
- Modifications of expression vectors include adding inducible promoters (which respond to a specific signal), tissue-specific promoters, and targeting sequences.

- Medically useful products that have been cloned in expression vectors include tissue plasminogen activator, human insulin, and vaccine proteins (see Table 16.1).
- Recombinant DNA offers breeders the opportunity to choose specific genes that will be incorporated into an organism, to introduce any gene into a plant or animal species, and to make plant breeding more efficient.
- Transgenic plants have been created that express toxins for insect larva, have resistance to herbicides, produce extra nutrients, or can grow in high-salt conditions.
- Human genes, such as blood-clotting factors, antibodies to colon cancer, and elastase inhibitors have been cloned in sheep and goats next to the lactoglobulin promoter, causing those animals to produce human proteins in their milk.
- Creating transgenic plants has raised some concerns about these crops being unsafe for human consumption and about transgenes escaping into other noxious plants. Scientists are examining these issues to promote safeguards.
- DNA fingerprinting is a technique that can be used to identify and distinguish different individuals. Some genes have multiple alleles, and each individual expresses different combinations of these alleles.
- DNA probes have been developed to hybridize to short moderately repetitive sequences that are side by side in the human genome. These sequences, termed VNTRs (variable number tandem repeats), vary from individual to individual.
- At least 1 μg of DNA is needed to complete these DNA fingerprint analyses.
- DNA from single cells can be amplified by PCR and used to analyze the fingerprint.
- DNA fingerprint analyses have been used in forensics, paternity issues, genetic analyses, and breeding selection.
- Blood or tissue can be analyzed for infection by a pathogen using primers whose DNA sequence specifically hybridizes to pathogen DNA.

The Big Picture

- The ability to isolate DNA from any organism, ligate it to vector DNA, introduce that DNA into host cells, and propagate those cells has had an enormous impact on our understanding of genetics, molecular biology, and cell function and development. These techniques have been used to elucidate evolutionary relationships between different organisms and to understand gene regulation and function in greater depth. Applications of these techniques have been used to develop new medicines and diagnostics, agricultural products, and powerful forensic tools.

Common Problems

- There are a variety of ways to clone DNA fragments, and trying to remember all the different cloning methods can be challenging. Consider the following as you review the different cloning procedures: size of the cloned DNA, host cell in which the DNA should be cloned, and expression of the cloned DNA in the host cell. Different vectors can be used in different host cells to address each of these considerations.

- There are many different experimental questions that can be answered using cloning. Ask yourself what cloning strategies could be used to answer the following questions, and review your text for answers: What is the sequence of a gene? What sequences are important for regulation of that gene? What sequences are important for targeting that gene to a particular site in a eukaryotic cell? What is the difference in function between a mutant gene product and a wild-type gene? What kinds of genes are expressed during development of an organism? How can genes be expressed in plants or in the milk of mammals?

Study Strategies

- Outline the specific steps needed to clone a gene in a bacterial cell. Include how the gene is initially isolated, what kind of vectors can be used, how to introduce that recombinant DNA into the cell, and how to confirm that the recombinant molecule is in the host cell. Now outline the steps needed to clone a gene in a eukaryotic cell.

- Complementary base pairing is important for many aspects of biotechnology. Describe each technique in gene cloning that uses complementary base pairing, detailing specifically how base pairing is involved.

- Review the following tutorials and activity on the Student Website/CD:
  Tutorial 16.1 Separating Fragments of DNA by Gel Electrophoresis
  Tutorial 16.2 DNA Chip Technology
  Activity 16.1 Expression Vectors

Test Yourself

Knowledge and Synthesis Questions

1. Cloning a gene may involve
   a. restriction endonucleases and ligase.
   b. plasmids and bacteriophage λ.
   c. yeast artificial chromosomes and complementary base pairing.
   d. All of the above

2. Restriction endonucleases
   a. are enzymes that process pre-mRNAs.
   b. are enzymes that degrade DNA.
   c. protect bacterial cells from viral infections.
   d. All of the above

3. DNA fragments are separated using gel electrophoresis
   a. because DNA is pulled through the gel toward the negative end of the field.
   b. because larger DNA fragments move faster through the gel than smaller DNA fragments.
   c. to identify and isolate DNA fragments.
   d. to synthesize DNA for cloning.

4. Complementary base pairing is important for
   a. ligation reactions with blunt-end DNA molecules.
   b. hybridization between DNA and transcription factors.
   c. restriction endonucleases to cut cell walls.
   d. synthesizing cDNA molecules from mRNA templates.

5. In order for a prokaryotic vector to be propagated in a host bacterial cell, the vector needs
   a. telomeres.
   b. centromeres.
   c. drug resistance genes.
   d. an origin of replication.

6. In order for a eukaryotic vector to be propagated in a host eukaryotic cell, the vector needs
   a. telomeres.
   b. centromeres.
   c. an origin of replication.
   d. All of the above

7. Reporter genes
   a. include genes for drug resistance.
   b. include genes for bioluminescence.
   c. include genes for DNA origins.
   d. Both a and b

8. Vectors include
   a. bacterial and plant plasmids.
   b. viruses.
   c. artificial chromosomes.
   d. All of the above

9. Which of the following biotechnology products is made by insertion of recombinant DNA into bacteria?
   a. Insulin
   b. Tissue plasminogen activator
   c. Erythropoietin
   d. All of the above

10. A cDNA clone is
    a. mostly cytosine.
    b. a copy of the DNA identical to the nuclear gene.
    c. a copy of noncoding DNA.
    d. a DNA molecule complementary to an mRNA molecule.

11. Gene expression can be inhibited by
    a. antisense RNA.
    b. knockout genes.
    c. DNA chips.
    d. Both a and b

12. Expression vectors are different from other vectors because they
    a. contain drug resistance markers.
    b. contain telomeres.
    c. contain regulatory regions that permit the cloned DNA to produce a gene product.
    d. contain DNA origins.

13. DNA fingerprinting works because
    a. genes containing the same alleles make it simple to compare different individuals.
    b. PCR allows amplification of proteins from single cells.
    c. there are multiple alleles for some DNA sequences, making it possible to obtain unique patterns for each individual.
    d. DNA in the skin cells is very diverse.

Application Questions

1. Describe three useful products that have been produced using biotechnology. Outline two specific dangers that could result from producing organisms that contain foreign genes.

2. Your lab assistant has cloned gene *x* into yeast and confirmed that the recombinant DNA molecule is present in the yeast cells. However, the yeast cell is unable to synthesize X protein. Suggest why this part of her experiment is not working and what modifications she needs to make in her cloning procedure.

3. You use a radioactively labeled DNA probe to identify a DNA fragment from an *Eco*RI digestion of *Drosophila*

DNA. When you complete your hybridization, you notice that the probe has hybridized to several different fragments in the *Drosophila* DNA sample. How do you explain this result?

4. You are an investigator at a crime scene who is trying to determine if the suspect in custody actually committed the crime. Describe how, using a drop of blood from the crime scene, you could determine whether the suspect was at the crime scene.

Answers

Knowledge and Synthesis Answers

1. **d.** Cloning a gene requires restriction endonucleases to cut the gene of interest and the vector, ligase to covalently join the DNA to the vector, and vectors including plasmids, bacteriophage λ, or yeast artificial chromosomes. Complementary base pairing is important in forming the recombinant DNA molecule if the DNA fragments have sticky ends.

2. **c.** Enzymes that process pre-mRNAs add 5′ caps and 3′ poly A tails to RNAs, and remove introns. Deoxyribonucleases degrade DNA.

3. **c.** DNA fragments migrate toward the positive end of the electric field with the smallest fragments migrating the fastest. Synthesizing DNA for cloning is done in a test tube, not in a gel.

4. **d.** No complementary base pairing can occur between blunt-end cut DNA molecules. Transcription factors are proteins that bind DNA through interactions with their side chains (which are amino acids) and the nucleotides of the DNA. Restriction endonucleases cut double-stranded DNA, not cell walls.

5. **d.** A prokaryotic vector needs an origin of replication to be propagated in a prokaryotic cell.

6. **d.** A eukaryotic vector requires an origin of replication, telomeres, and a centromere. It does not require a drug resistance marker to be propagated in the cell.

7. **d.** Origins of replication are needed to propagate the vector in the host cell but are not used to detect recombinant molecules in host cells.

8. **d.** All of these molecules can serve as vectors for cloned DNA.

9. **d.** All three of these human proteins can be made by transgenic bacteria.

10. **d.** cDNA clones are not clones that contain mostly cytosine. A cDNA clone is generated by making a DNA copy of a particular messenger RNA using reverse transcriptase. The cDNA clone is not identical to the nuclear gene because in the messenger RNA (which served as a template for the cDNA) the introns have been removed, leaving coding sequence and 5′ and 3′ flanking sequences.

11. **d.** Gene expression can be inhibited by antisense RNA, which complementarily base pairs with the target messenger RNA, making it inaccessible to the translation machinery in the cell. Knockout genes are genes that have been inactivated by the insertion of DNA into their coding sequences. DNA chips are used in hybridization experiments and do not inhibit gene expression.

12. **c.** Expression vectors may contain drug resistance markers and telomeres and must contain DNA origins, but none of these features distinguish them from other vectors. Expression vectors are unique because they contain regulatory sequences that allow the cloned gene to be expressed in the host cell.

13. **c.** Multiple alleles are needed to distinguish differences in individuals. PCR is a technique that amplifies DNA (not protein).

Application Answers

1. Useful products include rice grains that produce β-carotene, plants that are resistant to herbicides and insect larvae, sheep that produce human blood-clotting factors and antibodies to colon cancer in their milk, and others (see Tables 16.1 and 16.2). Dangers include creating genetically engineered foods that could adversely affect human nutrition, the transfer of herbicide- and insect-resistant genes from crop plants into noxious weeds, and the introduction into the wild of new organisms that might have unforeseen ecological consequences.

2. The *x* gene can be cloned into a yeast cell, but unless the vector has the appropriate regulatory signals (promoters, poly A addition sites, translational initiation, and termination signals) no expression of gene *x* will occur. Recloning gene *x* in a yeast expression vector will result in the expression of the *x* gene.

3. The DNA sequence of the probe is not unique in *Drosophila* and may be part of a gene family or a repetitive sequence (see Chapter 14).

4. DNA from the blood drop would be isolated and amplified for DNA fingerprint analysis using PCR amplification. The DNA would be digested with a restriction endonuclease and electrophoresed through a gel. DNA from a blood sample from the suspect would be similarly PCR amplified, digested, and run on the same gel. The gel would be blotted to a nylon filter. A radioactively labeled DNA probe containing a VNTR sequence would be hybridized to that filter to determine if the DNA band pattern was the same from the two samples. This process would be repeated with the two DNA samples using several different restriction endonucleases and several different VNTR probes. A match between DNA band patterns from the crime scene and DNA from the suspect would be needed to confirm the suspect's presence at the crime scene.

17 *Molecular Biology and Medicine*

Important Concepts

Many diseases in humans are caused by a specific genetic defect.

- Proteins function in cells in many ways—as enzymes, receptors, transporters, and structures. Mutations in genes encoding these proteins can cause many diseases.

- Both phenylketonuria (PKU) and alkaptonuria are caused by a mutation in genes encoding enzymes in the same biochemical pathway (see Figure 17.1).

- In PKU, the enzyme that converts phenylalanine to tyrosine, phenylhydroxylase, is nonfunctional, leading to buildup of phenylalanine in the blood. As phenylalanine accumulates, it is converted to phenylpyruvic acid. Somehow, high concentrations of phenylpyruvic acid prevent normal brain development in infants, leading to mental retardation. In addition, patients with PKU have lighter skin and hair color because melanin, a skin pigment, is made from tyrosine.

- The molecular alteration in patients with PKU is a change in the 408th amino acid (from arginine to tryptophan) of phenylhydroxylase.

- Many proteins are polymorphic. Protein sequencing has revealed a 30 percent variation in amino acid sequences for the same protein from different individuals. Some mutations in these genes have no effect, producing functional gene products (see Figure 17.2).

- In sickle-cell disease, which is homozygous in 1 in 655 African-Americans, abnormal alleles produce an abnormal protein (β-globin). The resulting hemoglobin protein forms aggregates in the red blood cells, causing the cells to sickle. Sickle cells block narrow capillaries and damage highly vascularized tissues, especially at low blood oxygen concentrations.

- The abnormal allele in sickle-cell disease is caused by a mutation that results in a change at position number 6 (from a glutamic acid to a valine) in the β-globin gene.

- Hemoglobin is easy to isolate and to study. Hundreds of β-globins with alterations in their primary amino acid sequence have been isolated. Some alterations have no effect, whereas others are quite severe. Hemoglobin C disease results from a change at position 6 of β-globin, altering the glutamic acid to a lysine. The result is anemia that is not as severe as sickle-cell disease (see Figure 17.2).

Most common genetic diseases result from altered membrane proteins.

- Familial hypercholesterolemia (FH) is a disease in which elevated levels of cholesterol are present in the blood. The excess accumulation of cholesterol on the inner walls of the blood vessels narrows the diameter of the vessel, increasing the probability that clots will lodge in the narrowed vessels, resulting in heart attack and stroke. FH affects about 1 in 500 individuals, resulting in early heart attacks and strokes before the age of 45.

- Cholesterol is taken in through the diet and travels in the blood in lipoproteins. Low-density lipoproteins (LDLs) carry cholesterol to the liver, where they bind LDL receptors and are taken into the interior of the cell via endocytosis (see Figure 17.3a). Patients with FH lack normal LDL receptors on the surface of their liver cells, causing cholesterol to remain in the blood.

- Cystic fibrosis occurs in 1 in 2,500 individuals and results from a defective version of a membrane protein that is a chloride transporter. In normal individuals, chloride is transported to the exterior of the cell by this protein, and as a result, water leaves the cell osmotically due to the higher concentration of chloride ions outside the cell. Chloride transport keeps the surface of the lungs (and the mucus) moist. When the chloride transporter is defective, thick, dry mucus accumulates in the lungs and the lining of other organs in the body. The dry mucus prevents the lungs from clearing out bacteria and fungal spores, and the patient is subject to recurrent respiratory infections and liver, pancreatic, and digestive failures. Patients with cystic fibrosis usually die in their twenties or thirties.

Defects in structural proteins also lead to severe disease.

- Duchenne muscular dystrophy is caused by a defect in dystrophin, the protein that connects the actin cytoskeleton to the extracellular matrix. This disease occurs in 1 in 3,000 individuals and results in weaker muscles, respiratory failure, and eventual death.

- Hemophilia is caused by a mutation in a gene encoding a clotting factor in the blood. Lacking this protein, patients can bleed to death from minor cuts.

Abnormal protein conformations can cause disease.

- Transmissible spongiform encephalopathy (TSE) is a degenerative brain disease in which the brain develops holes and the brain tissue appears spongelike. This disease has been identified in sheep and goats (scrapie), cows ("mad cow disease"), and humans (kuru).

- TSE is transmitted from one animal to another via brain extracts from the diseased animal.

- The infectious agent is a nonfunctional conformation of a membrane protein (PrP^{sc}) termed a prion (proteinaceous infective particles). Once these proteins fold inappropriately, they can induce a conformational change in the normal protein (PrP^{c}), so that it, too, becomes abnormal (see Figure 17.4). Altered proteins pile up as fibers in the brain tissue, causing cell death.

Most diseases are multifactorial.

- Although some diseases are the result of a mutant allele resulting in a single dysfunctional protein, most diseases are multifactorial, caused by many genes and proteins interacting with the environment.

Human genetic diseases have several patterns of inheritance.

- Autosomal recessive diseases include PKU, sickle-cell disease, and cystic fibrosis. Parents who are normal but both heterozygous for these alleles have a one in four chance of having a child who expresses the trait (homozygous recessive). Individuals who are homozygous recessive for this trait can synthesize only nonfunctional protein and express the disease. Heterozygotes can synthesize normal proteins 50 percent of the time, which is usually enough for the cell to function.

- Autosomal dominant diseases are diseases in which the presence of one mutant allele is enough to produce the disease. Familial hypercholesterolemia is an autosomal disease in which heterozygotes only produce one-half of the normal LDL receptors, which is not enough to clear the blood of cholesterol. The pattern of inheritance for autosomal dominant alleles is direct transmission from parent to offspring.

- X-linked diseases include hemophilia and Duchenne muscular dystrophy. These diseases are more common in males (because males only have one X chromosome) and are passed from mothers (who are heterozygotes and thus carriers) to their sons. The son who inherits the mutant allele from his mother will express that allele; the daughter who inherits that allele from her mother will be a carrier.

- Chromosome abnormalities include an excess or loss of one or more chromosomes (aneuploidy), a loss of a piece of a chromosome (a deletion), or the transfer of a piece of a chromosome to another chromosome (translocation). Chromosomal abnormalities may be inherited or may be the result of a nondisjunctional event in meiosis (see Chapter 9).

- Many spontaneous abortions (about 50%) are due to chromosomal aberrations.

- Fragile-X syndrome (which occurs in 1 in 1,500 males and 1 in 2,000 females) causes mental retardation in some individuals. It is characterized by a constriction in the tip of the X chromosome, which sometimes breaks during preparation for microscopy (see Figure 17.5).

Identifying the genes that cause disease has been accomplished in a variety of ways.

- The β-globin gene was cloned using cDNA generated from mRNA from red blood cells. β-globin mRNA is abundant because it is the major protein in these cells (see Figure 17.6a).

- The dystrophin gene was identified by comparing X chromosomes isolated from several boys with muscular dystrophy to X chromosomes from normal individuals. The X chromosomes from these patients had visible deletions, allowing investigators to position the site of the dystrophin gene on the chromosome.

- Positional cloning has been used to identify genes that are linked to a particular disease. Genetic markers and family pedigrees are compared to determine if a particular marker is different in affected versus unaffected individuals.

- RFLPs (restriction length fragment polymorphisms, see Figure 17.7) are genetic markers that can be used to distinguish individual variations in chromosomes. If a particular RFLP is positioned adjacent to a mutated gene, the RFLP can be used as landmarks to isolate the gene. Comparing RFLP patterns between affected and unaffected individuals can help determine how close these particular markers are to the gene of interest. DNA fragments from those regions thought to contain a gene can be used to probe cellular mRNA. If a DNA–mRNA hybrid is formed, that DNA fragment contains a gene that is being expressed in the cell and can be sequenced. Comparison of these DNA sequences in affected and unaffected individuals can help determine if this gene is mutated in affected individuals.

Different kinds of mutations can cause human disease.

- Point mutations that change one codon to a different codon (as seen in sickle-cell disease and PKU) can lead to an alteration in the three-dimensional structure of a protein, eliminating its function.

- Point mutations that change a codon to a stop codon produce truncated proteins, eliminating protein function if the mutation occurs at the beginning of the coding sequence (as seen in some types of cystic fibrosis).

- Hot spots, which are usually sites on the DNA containing methylated cytosines, are more prone to mutation than average. Loss of the amino group on a cytosine converts that base to uracil, which is removed from the DNA by repair enzymes. Loss of an amino group on a methylated cytosine converts that base to thymine, which is retained in the DNA, resulting in a base change at that site. Repair of the GT base pair may result in a correction to the original GC base pair or the introduction of a new AT mutation in the DNA at that site.

- Larger mutations include deletions, which can vary in size. Small deletions are generally less severe than large deletions, which can remove more than one gene.

- Expanding triplet repeats are found in fragile-X syndrome (a repeat of CGG in the *FMR1* gene), in myotonic dystrophy (a repeat of CTG), and in Huntington's disease (a repeat of CAG). Triplet repeats are more extensive in affected individuals and are thought to expand during DNA replication due to slippage of DNA polymerase in those regions. Triplet repeats can be found within or outside the coding sequence of a gene.

- Genetic imprinting is observed in genes or groups of genes whose phenotypic effects differ depending on the parent from whom those genes were inherited.

Genetic screening is used to identify people who might be carriers of a particular disease.

- Genetic screening can be done prenatally, on newborns, or on asymptomatic individuals who have close relatives with a genetic disease.

- In PKU screening, blood samples from newborn babies are tested to see if they can support growth of bacteria that are phenylalanine auxotrophs (see Chapter 12). Individuals who test positive for PKU are fed special diets low in phenylalanine to prevent abnormal brain development.

- In vitro fertilized human embryos can be screened at the eight-cell stage for mutant alleles. One of these embryonic cells is removed, and its DNA is amplified using PCR primers specific for the disease gene. Once a normal genotype has been confirmed, the seven-celled embryo can be implanted into the mother and will develop normally. Fetuses can be screened using chorionic villus sampling.

- Newborns can be screened for PKU and other disorders using DNA extracted from their cells and PCR amplification of that DNA that targets the mutant allele. Parents who are potential carriers for a particular mutant allele (such as muscular dystrophy or cystic fibrosis) can be screened for that allele using PCR amplification.

- Once the DNA has been isolated, different kinds of genetic screens can be used. Allele-specific cleavage differences can detect changes in restriction endonuclease sites in the DNA containing the gene of interest (see Figure 17.11). Oligonucleotides that are specific for the mutant allele or the normal allele can be used in hybridizations with DNA isolated from different individuals (see Figure 17.12).

Cancer is a genetic disease.

- Cancer is a genetic disease in which there is a loss of normal control over cell division. Cells divide continuously, forming a malignant tumor that metastasizes to other tissues in the body.

- Benign tumors are not cancerous, grow slowly, and do not spread to other parts of the body.

- Metastasis is characterized by angiogenesis, secretion of digestive enzymes as cells begin to relocate, and synthesis of cell surface proteins that allow the cancerous cells to invade new host tissue.

- Carcinomas are cancers of surface tissues including skin and epithelial cells that line organs. Carcinomas include lung cancer, colon cancer, breast cancer, and liver cancer. Sarcomas are cancers of tissue such as bone, blood vessels, and muscle. Leukemia and lymphomas affect cells that give rise to blood cells.

- Some cancers (15%) are caused by viruses (see Table 17.1).

- About 85 percent of cancers are caused by mutations, either spontaneous or induced by natural or synthetic carcinogens and various forms of irradiation.

- Two kinds of genes are altered in many cancers: oncogenes (which stimulate cell division when activated) and tumor suppressor genes (which inhibit cell division). Point mutations, translocations, and gene amplifications in these genes can lead to uncontrolled cell division. Oncogenes can be activated by one mutational hit. Tumor suppressor genes (such as those involved in retinoblastoma, Wilm's tumor of the kidney, and inherited forms of breast and prostate cancer) require the inactivation of both alleles (the "two-hit" hypothesis).

- Women who inherit a mutant allele of the *BRCA1* gene are more prone to develop breast cancer than women with the normal alleles for *BRCA1*.

- Tumor suppressors control the progress of the cell through the cell cycle. The protein encoded by the *Rb* gene acts during G1 by inhibiting transcription factors that are necessary for progress into DNA synthesis and subsequent cell cycle completion. If a mutation inactivates the *Rb* gene, as occurs in retinoblastoma, cells remain in the cell cycle and continue to divide, forming a tumor.

- *p53* is a gene coding for a protein that stops the cells during the G1 phase of the cell cycle. The protein product of *p53* is a transcription factor for proteins that block the interaction of cyclins and protein kinases (see Chapter 10). *p53* is mutant in many types of cancer cells, including lung and colon cancer cells.

- Multiple mutations transform a normal cell into a cancerous cell (see Figure 17.18). These mutations can be screened using oligonucleotide probes for oncogenes and/or tumor suppressor gene alterations.

Treating genetic diseases includes modifying the phenotype and gene therapy.

- There are three ways to treat the genetic disease: restrict the substrate, inhibit a harmful metabolic reaction, or supply the normal version of the gene.

- For individuals with PKU, treatment involves restricting the substrate (phenylalanine) for the missing enzyme (phenylalanine hydroxylase), which prevents the buildup of phenylpyruvic acid.

- For individuals with familial hypercholesterolemia, drugs are administered that block the patient's own cholesterol synthesis.

- Drugs that kill rapidly dividing cells are administered to cancer patients (see Figure 17.19). These drugs also affect noncancerous dividing cells in the body, causing digestive upsets, hair loss, and anemia.

- Patients with hemophilia A are given the missing gene product, which is a clotting factor, in purified form.

- Ex vivo gene therapy is a technique in which the gene of interest is inserted into the patient's cells (white blood cells or skin grafts), and then those cells are reintroduced into the patient. This technique has been used to replace the adenosine deaminase gene and the blood clotting factor in hemophilia.

- In gene therapy, the new gene is inserted into the affected individual. The gene must have all the important regulatory regions necessary for expression and must be delivered effectively to the individual's cells. This requires effective vectors, efficient uptake, precise insertion into host DNA, appropriate expression and processing of mRNA and protein, and selection of the target cells for the recombinant DNA.

- Another approach to gene therapy involves inserting the gene directly into the cells where the defect exists. For instance, in lung cancer patients, functional alleles of tumor suppressor genes, as well as antisense RNAs against oncogene mRNAs, are provided in aerosols for inhalation.

Sequencing the human genome has been a valuable scientific endeavor.

- To sequence human DNA, it must first be cut into smaller fragments (500 base-pair fragments) and then sequenced. Aligning of the different fragments is done using marker sequences along the chromosome.

- Libraries of human DNA have been created using restriction enzymes that recognize 8–12 base-pair recognition sequences, resulting in much larger DNA fragments. These large DNA fragments have been cloned into a bacterial artificial chromosome (BAC), which can accommodate 250,000 base pairs of inserted DNA.

- Hierarchical DNA sequencing involves the careful ordering of DNA sequences.

- These fragments have been arranged in proper order using markers and comparing those sequences to DNA libraries made with different restriction endonucleases (see Figure 17.21a).

- Shotgun sequencing involves cutting the DNA into random small fragments, which are analyzed for overlapping sequences by computer (see Figure 17.21b). Shotgun sequencing was used to sequence the entire 180 million base pairs of the fruit fly genome.

- The sequence of the human genome has revealed that:
  - Less than 2 percent of the DNA contains coding regions (a total of 30,000 to 35,000 genes).
  - The average gene has 27,000 base pairs.
  - Fifty percent of the genome contains highly repetitive sequences.
  - Almost all genes (99.9%) are the same in all people.
  - Genes are not evenly distributed over the genome.
  - The functions of many genes are unknown.

- There are many advantages of sequencing the entire human genome:
  - Comparison of genes from different organisms that have similar counterparts in humans
  - Identifying disease-related genes using positional cloning
  - Identifying genes that will react best to particular medications (pharmacogenomics)
  - Establishing fingerprint patterns of gene expression in cancer cells
  - Searching for genes that predispose certain groups of people to certain diseases

- Now that complete genetic information for each human being is possible, many individuals, including ethicists, psychologists, and geneticists, want to establish guidelines regarding how this information will be used.

The proteome is more complex than the genome.

- Because of alternative splicing and posttranslational modifications, the sum total of proteins produced (the proteome) is more complex than the genome.

- Proteome analysis includes two-dimensional gel electrophoresis and mass spectrophotometry.

- There is little difference in mRNA expression (1.7%) in the cortex portion of the brains of humans and chimps. However, proteome analysis has shown that there is a large difference in the kinds and amounts of proteins expressed in the cortex of humans and chimps.

The Big Picture

- Some diseases in humans can be directly related to a change in the genome, causing the production of an abnormal gene product. Other molecular diseases are affected by the individual's environment. Understanding the molecular alterations in a particular gene product and its effect can lead to therapeutic approaches, new diagnostic tools, and preventative measures in health care.

Common Problems

- Sometimes it is difficult to understand how a mutant allele can be dominant to a wild-type allele (as seen in familial hypercholesterolemia). Remind yourself of the function of the wild-type gene product, and then predict how the individual will be affected with only half of those gene products being expressed.

- RFLP mapping, oligonucleotide hybridization, and allele-specific cleavage analyses are similar techniques used in genetic screening to detect mutant alleles. Individual differences in DNA sequences (including simple base changes and DNA deletions, as well as a difference in the number of repeated sequences) can be utilized to distinguish individuals and to map genes. Review Figures 17.6, 17.7, 17.11, and 17.12 to distinguish these screening techniques.

Study Strategies

- Make a list of the molecular diseases discussed in this chapter. Include in this list the specific proteins that, when altered, cause the disease, and describe the treatments that are currently available for each.

- Review the cell cycle and the signals that are important for causing cells to divide. Predict how mutations in those genes can alter the regulation of the cell cycle and lead to unregulated cell division and cancer. Be sure to include a review of tumor suppressors and oncogenes in this analysis.

- Review the following tutorials and activities on the Student Website/CD:
  Tutorial 17.1 DNA Testing by Allele-Specific Cleavage
  Tutorial 17.2 Sequencing the Genome
  Activity 17.1 Allele-Specific Cleavage
  Activity 17.2 Concept Matching

Test Yourself

Knowledge and Synthesis Questions

1. Fragile-X syndrome is
   a. due to a single base change in the DNA.
   b. caused by changing a valine to a glutamic acid.

c. caused by triplet expansion.

d. caused by a chromosomal translocation.

2. A nonfunctional membrane protein is responsible for
   a. hemophilia.
   b. sickle-cell disease.
   c. cystic fibrosis.
   d. PKU.

3. Triplet repeat expansion
   a. occurs during DNA replication due to slippage of DNA polymerase.
   b. is caused by errors in DNA synthesis with reverse transcriptase.
   c. differences can be identified using RFLP mapping.
   d. Both a and c

4. In sickle-cell disease the
   a. sixth amino acid is changed from a valine to a glutamic acid.
   b. sixth amino acid is changed to a stop codon.
   c. hemoglobin concentration builds up in the red blood cells.
   d. structure of β-globin is altered, and the hemoglobin protein forms aggregates in the red blood cells.

5. Most cancers are
   a. due to the inheritance of mutant genes.
   b. caused by viruses.
   c. caused by mutations in genes whose products help blood clot.
   d. caused by mutations that are spontaneous or induced.

6. Metabolic disorders
   a. are caused by genes that have alterations in their centromeres.
   b. are the result of mutations in enzymes that are required to synthesize particular compounds in the cell (such as an amino acid).
   c. are due to abnormal membrane proteins that transport chloride ions.
   d. are caused by prions.

7. Prions
   a. cause scrapie in sheep.
   b. are caused by abnormally folded proteins interfering with the normal brain cell function.
   c. cause "mad cow disease."
   d. All of the above

8. Human genetic disease can be
   a. due to an autosomal recessive trait.
   b. passed from mothers to their sons (if X-linked) 50 percent of the time and also can be caused by translocations.
   c. due to a mutant allele that is dominant to the wild-type allele.
   d. All of the above

9. Mutations that result in human disease include
   a. triplet expansions that occur during DNA synthesis.
   b. point mutations that do not change the amino acid sequence of the gene.
   c. prionlike diseases such as kuru.
   d. Both a and b

10. Genetic screening has been used to
    a. identify embryos carrying a mutant allele.
    b. identify newborns who have PKU.
    c. identify fathers who are carriers for X-linked diseases.
    d. Both a and b

11. Genes that when mutated cause cancer are
    a. only those genes involved in the regulation of the cell cycle.
    b. transcription factors only.
    c. only those genes inactivated by viruses.
    d. None of the above account for all forms of cancer.

12. In order for gene therapy to be most effective, genes should be inserted in
    a. white blood cells.
    b. red blood cells.
    c. stem cells.
    d. All of the above

13. The human genome allows scientists to
    a. understand regulatory sequences that are important for gene expression.
    b. locate genes that cause disease.
    c. understand evolutionary relationships by comparing human genes to genes in other organisms.
    d. All of the above

Application Questions

1. Familial hypercholesterolemia (FH) and Duchenne muscular dystrophy (MD) are both molecular diseases that are caused by mutations in particular genes, yet FH is easier to treat than MD. Why?

2. Two individuals have a mutation in gene x but at different sites. The disease adversely affects the first individual, and the second individual experiences no effect. Explain this observation.

3. Why are X-linked diseases more common in males than in females?

4. The β-globin gene (which when mutated can lead to sickle-cell disease) was much easier to clone than the gene responsible for one of the inherited forms of breast cancer (*BRAC1*). Why?

5. Explain how mutation in only one of two oncogene alleles can lead to unregulated cell division, whereas mutations in both alleles of a tumor suppressor gene are necessary.

Answers

Knowledge and Synthesis Answers

1. **c.** Fragile-X syndrome is due to a triplet expansion on the X chromosome, not a single base change, a translocation, or an alteration of protein sequence, as seen in sickle-cell disease.

2. **c.** The chloride transporter is a membrane protein that, when nonfunctional, causes cystic fibrosis. Hemophilia is caused by a defect in a clotting factor; sickle-cell disease is caused by a defect in β-globin. PKU is a defect in an enzyme in a biochemical pathway.

3. **d.** Triplet expansion is thought to be due to slippage of DNA polymerase during DNA replication and will result in an increasing number of triplet repeats in those cells. That increase in the DNA sequence can be distinguished from the DNA sequence of a normal individual using RFLP analysis. Reverse transcriptase is an enzyme from HIV that copies RNA into DNA.

4. **d.** In sickle-cell disease, the sixth amino acid is changed from a glutamic acid to a valine (not from a valine to a glutamic acid, and not to a stop codon). The resulting

change causes the hemoglobin protein to form aggregates in the red blood cell. Hemoglobin concentration in the red blood cell is not altered in sickle-cell disease.

5. **d.** About 10 percent of cancers are due to the inheritance of a mutant allele, and 15 percent are caused by viruses. Most cancers are caused by mutations that are either spontaneous or induced (note that cancers may require more than one mutation, so any given cancer may have more than one cause). Mutations in genes that encode blood clotting factors result in hemophilia.

6. **b.** Some metabolic disorders are caused by mutations in genes that encode enzymes in a biochemical pathway (such as the pathway for the synthesis of phenylalanine). Mutations in centromeres and genes for chloride transporters do not result in metabolic disorders. Prions are not the result of a mutation; they are the result of abnormal folding of a protein.

7. **d.** Prions result from abnormally folded proteins that interfere with normal brain cell function. Mad cow disease is caused by prion formation. They are not caused by a mutant allele for this particular protein.

8. **d.** All these mutations and chromosome alterations can cause human disease.

9. **a.** Triplet expansions cause a variety of diseases including fragile-X syndrome, Huntington's chorea, and myotonic dystrophy. Point mutations that do not alter the amino acid sequence of a protein are silent mutations and do not cause any disease (in almost all cases). Prions are not the result of a mutation in a gene but are due to the abnormal folding of a particular protein.

10. **d.** Genetic screening has been used to analyze in vitro–fertilized embryos (at the 8-cell stage) and DNA from individuals who might be carrying a mutant allele. Fathers are not carriers for X-linked diseases; they express all their X-linked alleles.

11. **d.** Although all the examples listed are examples of mutations that can result in cancer, none of these mutations are the *only* reason an individual develops cancer.

12. **c.** The most effective approach for genes to reside permanently in the individual's cells is for them to be inserted in stem cells, thus providing a continual supply of differentiated cells under the correct environmental signals. Red blood cells have no nuclei and therefore cannot be used for gene therapy. White blood cells have been used

for gene therapy, but because they have a set lifetime in the body, they eventually die, destroying the gene they were carrying.

13. **d.** Human genome sequencing provides all these advantages.

Application Answers

1. Familial hypercholesterolemia causes an increase of cholesterol in the patient and can be treated with drugs that reduce the patient's own synthesis of cholesterol. Muscular dystrophy is caused by a defect in a structural protein, dystrophin, and results in patients who have dysfunctional muscle tissue. This structural dysfunction cannot be reversed with the application of any drug.

2. The mutation in gene x in the first individual must have occurred in an essential region of the gene that is required for its function. The mutation in gene x in the second individual may be a silent mutation, or it may be in a region that is nonessential for the function of that protein.

3. X-linked diseases are more common in males because males have only one X chromosome, so every allele on that chromosome is expressed in the individual. Mutant alleles can be carried by females but masked by a wild-type allele on the other X chromosome.

4. β-globin is one of the major proteins found in the red blood cell, and thus isolating mRNA from this cell and cloning a cDNA copy of this gene was relatively straightforward. Neither the function nor the location of the *BRAC1* gene was known. Nearby genetic markers had to be identified that were linked to the *BRAC1* gene, and then DNA sequencing of that region had to be done. DNA sequences between affected and unaffected individuals had to be compared to identify the gene.

5. Oncogenes control cell division by responding to an environmental signal and initiating cell division in response to that signal. A mutation in an allele of an oncogene would result in that protein causing the cell to initiate cell division, even in the absence of any environmental signal. Half of the wild-type proteins would respond appropriately to cell division signals, but the dysfunctional proteins that were now unregulated would still signal the cell to divide. A single wild-type copy of a tumor suppressor gene, however, could still provide enough gene product to regulate cell division.

18 *Natural Defenses against Disease*

Important Concepts

Animal defense systems are based on the distinction of self from nonself and include nonspecific defenses and specific defenses.

- Pathogens are the harmful organisms and viruses that cause disease in animals. Nonspecific defenses or innate defenses are inherited mechanisms that protect the body from many different pathogens. Nonspecific defenses include physical barriers as well as cellular and chemical defenses (see Table 18.1).

- Specific defenses are adaptive mechanisms aimed at specific targets.

- Lymphoid tissues, including thymus, bone marrow, spleen, and lymph nodes, are essential to an animal's defense system. The lymphatic system moves lymph fluid from the tissues to the heart. Both lymph and blood contain white blood cells and platelets. Lymph nodes contain white blood cells and filter the lymph fluid.

- Phagocytes are white blood cells that engulf and digest nonself materials. Macrophages are important phagocytes that are involved in the interaction of nonself materials and T cells.

- Lymphocytes are the most abundant class of white blood cells and include B and T cells. Immature T cells migrate via the blood to the thymus where they mature. T cells participate in specific defenses against foreign or altered cells, including virus-infected cells and tumor cells. B cells circulate through the blood and lymph and make antibodies, which are proteins that bind to nonself or altered self.

- T cell receptors are proteins on the surface of T cells that recognize and bind to nonself substances on the surfaces of other cells.

- Major histocompatibility complex (MHC) proteins protrude from surfaces of most cells in the mammalian body and help in self-identity.

- Cytokines are soluble signal proteins released by T cells, macrophages, and other cells. They can activate or inactivate B cells, macrophages, and T cells or limit tumor growth.

Barriers, local agents, chemicals, and cellular processes act as nonspecific defenses against invaders.

- Nonspecific defenses include skin, mucus-secreting tissues, lysozyme in tears, digestive enzymes, bile salts, and hydrochloric acid in the stomach.

- About 20 complement proteins act as antimicrobial proteins in the vertebrate blood. They function by destroying microbes, activating other immune responses, or tagging microbes.

- Small proteins known as interferons are produced in response to a viral infection.

- Phagocytotic cells ingest cells and viruses. Four types of phagocytotic cells are neutrophils, monocytes, eosinophils, and dendritic cells.

- Natural killer cells are white blood cells that lyse target cells.

- Inflammation defends the body against infectious agents through the release of histamine from mast cells (see Figure 18.4).

- A signal transduction pathway exists in which over 40 genes are activated as an immune response. The receptor that initiates this pathway is known as toll (see Figure 18.5).

Specific defenses of the immune response occur when immune cells (B and T cells) recognize and destroy nonself substances.

- An antigen is a molecule or organism that is recognized by antibodies or the T cell receptors. Immune cells recognize antigenic determinants or epitopes (i.e., specific amino acids of a protein) on invading pathogens. There can be many antigenic determinants on a pathogen surface.

- Immune cells respond to pathogens by activating lymphocytes of the appropriate specificity. Humans can respond to 10 million different antigenic determinants.

- The immune cells can distinguish self from nonself.

- The immune system "remembers" a pathogen and can respond more rapidly and more effectively the next time it is exposed to it. Immunological memory is the basis of vaccination.

- There are two interactive immune responses against invaders. The humoral immune response involves antibodies from B cells that react with antigenic determinants. The cellular immune response involves T cells binding with antigens.

Clonal selection accounts for the characteristic features of the immune response.

- The body contains millions of different B cells, and each B cell is able to produce only one kind of antibody. The differences among B cells are due to changes in DNA. When an antigen that fits the surface antibody binds to the B cell, that cell is activated and divides to produce more B cells, which secrete antibody. T cells are clonally selected in a similar manner.

Immunologic memory and immunity result from clonal selection.

- In the primary immune response, activated B and T cells produce effector cells and memory cells.

- Effector cells attack a pathogen by producing specific antibodies (in the case of B cells) or cytokines (in the case of T cells).

- Memory B and T cells are long lived and will divide rapidly to produce an effective and more powerful immune response if reexposed to a pathogen. This rapid response, the secondary immune response, provides a natural immunity to those diseases.

- Vaccinations and immunizations are administered to give the recipient artificial immunity and are effective because of this secondary immune response. Safe antigens are produced by treating the antigenic molecule with a chemical (attenuation), producing small peptides through biotechnology, or with DNA vaccines.

- Animals are able to tolerate their own antigens due to clonal deletion, a process that removes any B and T cells that recognize self antigens during their differentiation. An immune cell that recognizes self antigens undergoes programmed cell death (apoptosis).

- Clonal anergy is a suppression of the immune response so those T cells that recognize a self antigen on the body do not send out cytokines. This suppression occurs in those body cells that are missing a costimulatory signal, CD28, for the T cell. The costimulatory signal for the T cell is only expressed on antigen-presenting cells.

- Immunological tolerance occurs when a foreign antigen is introduced into an animal during fetal development, which results in the animal's recognizing that antigen as self.

- Continued exposure to self antigens helps maintain tolerance throughout the lifetime of the organism.

B cells and antibodies are the basis of the humoral response.

- Activation of B cells results from the binding of an antibody to a particular antigenic determinant, and the arrival of a signal from a helper T cell that has bound the antigen on an antigen-presenting cell.

- Antibodies (also called immunoglobulins) are proteins that are similar in structure, but can be grouped into five different classes. Antibody molecules consist of two identical heavy polypeptide chains and two identical light polypeptide chains. Each polypeptide consists of a constant region that is similar in amino acid sequence from one immunoglobulin to another and a variable region that forms the antigen-binding site.

- The variable region produces the antibody specificity in the millions of antibodies. The constant region determines whether the antibody will be secreted or remain on the cell surface of the B cell. It also determines what type of action will be taken to eliminate the antigen.

- The five immunoglobulin classes (IgG, IgM, IgD, IgA and IgE) are based on differences in the constant region of the heavy chain (see Table 18.3).

Monoclonal antibodies are produced by hybridomas.

- Normal immune responses are polyclonal, generated by the activation of many different B cells and resulting in a complex mixture of antibodies.

- Monoclonal antibodies are produced from a clone of B cells that makes only one antibody that binds a unique determinant.

- Monoclonal antibodies are generated by the fusion of normal B lymphocytes with tumor cells of plasma cells to produce hybridomas.

- Monoclonal antibodies have been used in immunoassays, in immunotherapy to target cancer cells with radioactive ligands or toxins, and for passive immunization.

T cells direct the cellular immune response.

- T cells have specific glycoprotein surface receptors that are made up of two different polypeptide chains, both with a variable and a constant region. The variable region provides the specificity of the T cell receptor.

- T cell receptors bind antigen fragments that are displayed on antigen-presenting cells.

- Cytotoxic T (T_C) cells recognize virus-infected cells and kill them by causing them to lyse (see Figure 18.14).

- Helper T (T_H) cells assist both the cellular and the humoral immune systems.

The major histocompatibility complex (MHC) encodes proteins that present antigens to the immune system.

- MHC gene products are plasma membrane glycoproteins that present antigens on an antigen-binding site of the MHC protein. These antigens are inspected by T cell receptors. MHC proteins can display antigenic peptides of about 10–20 amino acids.

- Class I MHC proteins are present on the surface of every nucleated cell in an animal. Degraded cellular proteins are presented to T_C cells on MHC I complexes. T_C cells have a surface protein called CD8 that can recognize MHC I proteins.

- Class II MHC proteins are found on the surfaces of antigen-presenting cells, including B cells and macrophages. T helper cells recognize class II MHC proteins using the CD4 cell surface protein.

- Class III MHC proteins include some of the proteins in the complement system that helps lyse foreign cells.

- T cell receptors recognize antigenic fragments when they form complexes with the MHC I or MHC II molecules.

- MHC genes, antibody genes, and T cell receptor genes have similar base sequences, suggesting a common ancestral gene.

The humoral response and cellular immune response involve an interaction between T cells and MHC complexes.

- In the humoral response, both macrophages and B cells bind antigens and present antigen fragments on the surface of their cells on MHC II complexes. T_H cells bind both the antigen and the MHC II complex and release cytokines. Those cytokines activate the T_H cells to divide (the activation phase) and activate B cells to produce antibodies (the effector phase).

- In the cellular immune response, cytotoxic T cells destroy cells that are displaying antigens from virus or mutant proteins on their MHC I complexes, eliminating virally infected cells as well as tumor cells. T_C cells must receive a second signal for activation, which involves an interaction between proteins on the T_C cell and CD28 proteins on the antigen-presenting cell. The T_C cell then proliferates and divides, releasing cytokines.

- Later, T_C cells are inhibited by the cell surface protein CTLA4.

MHC molecules play a key role in the tolerance of self.

- T cells are tested in the thymus to see if they bind to MHC displaying one of the body's own antigens, or if they are unable to recognize the body's MHC proteins. If T cells have either of these characteristics, they are destroyed.

- MHC proteins are specific to each individual; tissue transplants are thus recognized as nonself in the recipient. Suppression of rejection of organ transplants by the recipient's immune system is accomplished by treating a patient with drugs such as cyclosporin that suppress the immune system.

DNA rearrangements and mutations are used to generate all the diverse antibodies within an organism.

- DNA fragments are rearranged and joined during B cell development to generate antibody supergenes.

- The variable region of the light chain originates from two families of genes, and the variable region of the heavy chain originates from three families of genes.

- For the constant and variable regions of the heavy chain in mice there are multiple genes coding for each of four kinds of segments: 100 *V*, 30 *D*, 6 *J*, and 8 *C* (see Figure 18.18). Each B cell randomly selects one gene from each of these multiple genes to make the final heavy chain coding sequence *VDJC*.

- Light chains are similarly constructed from DNA segments.

- Light and heavy chains are combined to create billions of possible antibodies.

- Increased diversity is achieved by imprecise recombination during DNA rearrangements, addition of extra nucleotides to cut DNA fragments as insertion mutations, and an increased mutation rate in immunoglobulin genes.

- Class switching occurs by rearrangements in the DNA that position the *VDJ* segment (the same variable region on the antibody) adjacent to a constant region farther down the DNA (see Figure 18.20), resulting in the production of an antibody with a different constant region. Class switching in B cells is caused by cytokine release from T_H cells.

- Ancestral antibody genes may have developed from transposons that were able to cut out adjacent DNA sequences and move them elsewhere in the genome.

Disorders of the immune system.

- Hypersensitivity occurs when the immune system over-reacts to a dose of antigen and may cause inflammation as seen in the allergic response.

- Immediate hypersensitivity in allergic reactions is caused by IgE binding the foreign antigen and causing mast cells to release histamines.

- Delayed hypersensitivity does not happen immediately following exposure to an antigen, but hours later. Antigen-presenting cells process an antigen and initiate a T cell response. If the response is massive, macrophages may be activated and damage tissues. Such a hypersensitive response occurs when the bacteria that cause tuberculosis infect the lungs.

- Autoimmunity occurs when B and T cells direct their response against self antigens. Examples of autoimmune disease include systemic lupus erythematosis, rheumatoid arthritis, multiple sclerosis, and insulin-dependent diabetes mellitus. Some of these diseases may have genetic components.

- AIDS is caused by HIV, a virus that eventually destroys T_H cells. HIV is transmitted through body fluids, including blood and semen.

- HIV initially infects macrophages, T_H cells, and dendrite cells (one kind of antigen-presenting cell). These cells carry the virus to the lymph nodes, where they preferentially infect activated T_H cells.

- During the first phase of infection, up to 10 billion viruses are made every day, causing mononucleosis-like symptoms. During the second phase, the virus infects T_H cells, and an immune response is mounted, producing HIV antibodies in the blood. Later, the T_H cells are destroyed and the patient becomes susceptible to a wide variety of infections that would normally be eliminated by T_H cells. At this stage Kaposi's sarcoma, *Pneumocystis carinii*, and virally caused tumors persist in the system of the HIV infected individual.

HIV infects and replicates in T_H cells.

- HIV is an enveloped virus and has a protein coat that surrounds two identical molecules of RNA as well as the enzymes reverse transcriptase, integrase, and a protease.

- The viral envelope protein gp120 attaches to T_H cells and macrophages by binding their CD4 proteins. The HIV membrane fuses with the host cell plasma membrane, and the virus enters the host cell.

- A DNA copy of the viral genome is made using reverse transcriptase, which is an error-prone polymerase. Errors made during viral replication create alterations in viral proteins that can now escape the host's immune response.

- Viral cDNA is inserted into the host DNA using the HIV integrase protein.

- When the infected T_H cell is activated, the entire viral genome is transcribed into RNA, where it can be used to make new HIV particles or to synthesize new viral structural proteins. RNA genomes are packaged into core particles, and the mature virus buds from the host cell.

- Therapeutic treatment for HIV includes drug therapies that interfere with the major steps of the life cycle of the virus. Highly active antiretroviral therapy has been used since the late 1990s to combat HIV. The most promising treatment may be the development of HIV vaccines.

The Big Picture

- Animals have special systems to distinguish self from nonself. Some of these systems are nonspecific, such as nonspecific defenses (complement protein, phagocytotic cells). B and T cells can recognize specific antigens on the surface of pathogens and mount an immune response to destroy those pathogens. When the immune system is compromised, as seen in HIV-infected individuals, the individual becomes susceptible to a variety of pathogens.

Common Problems

- The specificity of the immune response may seem complex initially, but there are really only two major components: B cells (which bind antigen) present antigen and secrete antibody, and T cells bind antigen on antigen-presenting cells and help direct the immune response.

- MHC proteins may seem complicated at first, but they have only a few essential functions in the immune system. These proteins are specific for each individual and are essential for presenting antigen on antigen-presenting cells (MHC II) to T_H cells and recognizing one's own cells (MHC I).

- The specificity of the immune response may seem complex, but it can be understood by reviewing cellular communication. Antibodies of B cells bind with antigen on the surface of pathogens; T cell receptors bind antigen presented on MHC I and MHC II proteins on the surface of cells. Once these membrane proteins interact, cytokines are secreted, and cells begin to divide and proliferate. Antibodies are secreted by B cells.

- The rearrangement of DNA in the nucleus of the B cell may seem foreign to students who think the integrity of the genetic information must be preserved. But this change in the genetic makeup of activated B and T cells is essential to provide the organism with a diverse and specific set of antibodies and cell receptors.

Study Strategies

- List the nonspecific (innate) responses that a human body can make in response to an invading pathogen. Next list cells that are part of the specific responses. Include proteins that are crucial to generating that specificity. Review how genetic rearrangements generate a diverse set of antibodies for B cells.

- Diagram an antibody and label the important components. Highlight those areas where mutations and alterations occur that change the amino acid sequence of the heavy and light chains.

- Review the expression and function of MHC I and MHC II molecules.

- Make a list of the functions of T_H cells and T_C cells.

- Draw a diagram of a chromosome and label it with 4 *V*s (*V1–V4*), 2 *D*s (*D1* and *D2*), 3 *J*s (*J1–J3*) and 8 *C*s (*C1–C8*). Imagine that a B cell is maturing to produce antibody on its surface. Select one *V*, one *D*, one *J*, and one *C* segment to make an antibody. Repeat the process using different segments. Predict how many different antibodies can be generated from this set of gene segments.

- List some of the disorders of the immune system and their probable causes.

- Review the following tutorials and activities on the Student Website/CD:
  Tutorial 18.1 Cells of the Immune System
  Tutorial 18.2 Pregnancy Test
  Tutorial 18.3 Cellular Immune Response
  Tutorial 18.4 Humoral Immune Response
  Tutorial 18.5 A B Cell Builds an Antibody
  Activity 18.1 The Human Defense System
  Activity 18.2 Inflammation Response
  Activity 18.3 Immunoglobulin Structure

Test Yourself
Knowledge and Synthesis Questions

1. Phagocytes
   a. are T and B cells.
   b. present antigen on MHC II complexes.
   c. digest nonself materials.
   d. Both b and c

2. B cells
   a. secrete antibodies.
   b. present antigen on MHC II.
   c. ingest antigens.
   d. All of the above

3. Nonspecific responses in the immune system include
   a. macrophages.
   b. natural killer cells.
   c. complement proteins.
   d. All of the above

4. When the receptor of a T_H cell binds to a pathogen being presented on a macrophage, it
   a. activates itself.
   b. secretes cytokines.
   c. activates B cells.
   d. All of the above

5. Part of the normal immune response includes
   a. the production of B memory cells.
   b. the production of memory macrophages.
   c. antibody secretion by eosinophils.
   d. the production of B cells that attack the individual's own cells.

6. Antibody molecules are
   a. produced by B cells and have a variable and a constant region.
   b. secreted by B cells, once a signal (a cytokine) is received from a T cell.
   c. produced by T cells and have a variable and a constant region.
   d. Both a and b

7. Cytotoxic T cells
   a. release cytokines that activate B cells.
   b. attack pathogens by binding to cell surface antigens on those pathogens.
   c. destroy pathogens by lysing them.
   d. destroy host cells that are infected with virus.

8. The humoral response
   a. is due to the secretion of antibodies.
   b. occurs when T cells bind antigen-presenting cells.
   c. is due to T cells secreting their receptors.
   d. occurs when natural killer cells engulf cancer cells.

9. Autoimmunity is
   a. active when organ transplantation is successful.
   b. caused by clonal anergy.
   c. the response of the immune cells attacking the body's own tissues.
   d. Both b and c

10. Patients with HIV are susceptible to a variety of infections because
    a. the virus produces cell surface receptors that bind to pathogens, making it easier for those pathogens to be infective.
    b. synthesizing a DNA copy of the viral genome makes a person feel sick.
    c. HIV attacks and destroys the T helper cells, which are central to mounting an effective immune response, making those individuals more susceptible to other infections.
    d. HIV destroys B cells so that antibodies cannot be made in response to invading pathogens.

11. DNA rearrangements in the B cell
    a. are responsible for generating single B cells that can express many different antibodies.
    b. lead to mutations in T cells resulting in the elimination of essential T cell genes.
    c. occur only in B memory cells.
    d. are responsible for generating many different antibodies, with each B cell expressing only one set of identical antibodies.

12. Major histocompatibility proteins function in the immune system by
    a. presenting antigen to T cells.
    b. presenting self antigens to T helper cells so those T cells will continue to tolerate self throughout the lifetime of the individual.
    c. generating antibodies to different pathogens.
    d. Both a and b

Application Questions

1. Imagine that a particular bacterium has surface molecules that resemble membrane proteins found on the heart valves. What will happen to the individual infected with this bacterium if the infection is not treated?

2. Organ transplants are more successful when the donor is a relative of the recipient. Why?

3. Vaccinations are effective against an array of different pathogens. How do they work? Vaccinations against HIV virus have not been effective in preventing the disease. Why not?

4. If the cytotoxic T cells were eliminated from a person's array of immune defenses, what kinds of disease would he or she be susceptible to?

Answers

Knowledge and Synthesis Answers

1. **d.** Phagocytes are nonspecific cells (not B and T cells) that digest nonself materials and present protein fragments of those nonself materials on their surface. They do not have antibodies.

2. **d.** B cells are antigen-presenting cells; they bind antigen, digest it, present fragments of that antigen on their MHC II proteins, and secrete antibodies.

3. **d.** Macrophages, natural killer cells, and complement proteins are all part of the nonspecific response of the immune system. B and T cells are part of the specific response of the immune system.

4. **d.** When the T_H cell binds antigen being presented on a macrophage, it secretes cytokines, which activate the T_H cell and B cells.

5. **a.** In a normal immune response, B memory cells are made so that the organism can mount a faster and more effective response to the pathogen if the organism ever encounters it again. Memory macrophages do not exist. Eosinophils do not secrete antibodies. In an abnormal immune response, B cells that attack the individual's own cells can be activated, as seen in autoimmune diseases.

6. **d.** T cells have cell surface receptors with a variable and a constant region and do not produce antibody molecules. Antibodies are not produced by macrophages. Only B cells produce antibodies in response to cytokines released from the T_H cell. Each antibody molecule has two identical heavy chains and two identical light chains, and each of those chains has a variable and a constant region.

7. **d.** Cytotoxic cells bind to virus antigen being presented on MHC I protein and destroy those cells. T_H cells release cytokines to activate B cells. The antibodies of B cells bind cell surface antigens on pathogens. Pathogens are lysed by complement proteins (and not by cytotoxic T cells), which are activated by antibodies bound to those pathogens.

8. **a.** The humoral response refers to that part of the specific response that releases antibodies (B cells) in the lymph and blood (the "humors" of the body). The cellular response is T cell receptors binding antigen on antigen-presenting cells. T cells do not secrete their receptors.

9. **c.** Autoimmunity occurs when the immune cells attack the body's own cells. Transplants can be rejected if the transplanted tissue is recognized by the body as nonself, and is not a result of immune cells attacking the body's own tissue. Clonal anergy suppresses the immune response to self antigens and thus works to prevent autoimmunity.

10. **c.** An HIV-infected individual is more susceptible to a variety of infections because the virus destroys T_H cells, which are essential for mounting an effective immune response. HIV does not bind to pathogens; it binds to T_H cells. HIV does not destroy B cells.

11. **d.** DNA rearrangements occur in B cells and result in each B cell expressing a unique kind of antibody. DNA rearrangements also occur in T cells to generate T cell receptors. This rearrangement does not destroy essential T cell genes. Memory B cells are cells in which the DNA rearrangement has already occurred.

12. **d.** MHC proteins present nonself antigens and self antigens to T cells. T cells that react with self antigens are destroyed, preventing autoimmunity. The cytokines of T cells (and not MHC proteins) activate B cells. MHC proteins do not generate antibodies.

Application Answers

1. If the bacterial infection is not treated, the individual's immune response to this pathogen will include secreting antibodies that may be able to cross-react with the individual's own membrane proteins on the cells of the heart

valve. Such antibodies could create an autoimmune response and heart disease.

2. Organ transplants are successful if the MHC proteins and other cell surface markers are similar, making the immune system more tolerant of the foreign tissue. Close relatives have more genes in common with the patient and may have more surface antigens on their cells that are similar. Organ transplants from unrelated individuals require the administration of immunosuppressive drugs so that the patient's own immune system does not reject the tissue as foreign.

3. Vaccinations work against foreign pathogens because of B and T memory cells that are made during every immune response. If these memory cells are reexposed at a later time to the same antigen, they will mount a faster and more efficient immune response to that pathogen. One challenge in creating vaccines against HIV is that the virus mutates frequently (HIV uses reverse transcriptase, an enzyme prone to introducing genetic mutations, in its reproductive cycle). A vaccine that creates memory cells specific to a certain protein in one strain of HIV will not be effective against mutant HIV forms in which that protein has been modified if the memory cells do not recognize the modified protein.

4. Cytotoxic T cells target virally infected cells and some cancer cells. If these cells were eliminated, the individual would be much more susceptible to viral infections and cancer.

19 Differential Gene Expression in Development

Important Concepts

Once fertilization occurs, the zygote undergoes a series of cell divisions and progressive changes that include growth, differentiation, and morphogenesis. This developmental process produces a mature organism.

- Cell differentiation does not result from the loss of DNA. Somatic cells in the mature organism retain all the genes present in the fertilized egg.
- Cell differentiation results from the differential expression of genes caused by transcriptional and translational regulation and by signaling mechanisms.

Development includes three stages: growth, differentiation, and morphogenesis.

- During development, an organism changes forms as it progresses through its life cycle. The embryo is the earliest stage of development.
- Growth is an increase in size due to cell division and cell expansion. The early growth of plant embryos is characterized by cell expansion, whereas early growth of animal embryos is characterized by cell division.
- During differentiation, cells become specialized and provide specific structures or functions for the organism.
- Early embryonic cells are totipotent: they can develop into any of the different kinds of cells in the mature organism. The developmental possibilities narrow later in development as the cell fate becomes determined and the cell becomes fully differentiated.
- Morphogenesis is the creation of form as seen in body shape and organs. It is a result of pattern formation during development. Pattern formation helps direct differentiated tissues to form specific structures.
- Plant cells are nonmotile and constrained by cell walls. Formation of differentiated tissue in the mature plant is due to organized cell division and expansion of cells.
- Animal cells are motile, and these cell movements play a major role in differentiation and morphogenesis.
- Programmed cell death, known as apoptosis, is an important part of development in both plants and animals.

Differentiation and morphogenesis are directed by gene expression responding to extracellular signals.

- A fate map of an embryo shows which cells in an embryo will develop into which adult structures. Transplantation experiments have shown that the environment in which early embryonic cells exist can redirect them to develop down different developmental paths.
- Later embryonic cells become committed to a particular developmental fate, even though they are not yet differentiated. This is because their fate has already been determined. When these cells are transplanted to other locations in the embryo, they continue to develop into the original differentiated tissue, regardless of their environment (see Figure 19.2).
- Cell fate becomes apparent as the cells differentiate.
- Initially, the fertilized egg or zygote is totipotent and can give rise to every type of cell in the adult body. Cell descendants lose their totipotency as they become committed to a developmental path. A cell's fate first becomes determined, and then the cell differentiates to form specialized cells.
- Differentiation is irreversible in cells that have no nuclei, such as mammalian red blood cells and the water-conducting tracheid cells in vascular plants.
- In plants, differentiation is reversible in some cells. Under the appropriate conditions, plant cells in culture can give rise to a new, genetically identical plant (a clone). They do this by producing a callus or mass of undifferentiated cells. Thus, these differentiated cells contain all the genetic information needed to express genes in the right sequence to generate a whole plant.
- The principle of genomic equivalence states that no information is lost from the nuclei of cells as they pass through the early stages of development.
- The totipotency of early embryonic animal cells has been shown in experiments in which nuclei isolated from frog embryos were fused to enucleated eggs, resulting in the development of tadpoles and mature frogs. Such experiments also show that cytoplasm has a large influence on the fate of a cell. Similar studies have been conducted on rhesus monkeys.
- Whole mammals can be cloned by fusing early embryonic cells or somatic cells with enucleated eggs and implanting these eggs in surrogate mothers. The success of cloning mammals from somatic cells has been due to manipulating the donor cells so that they are positioned in G1 of the cell cycle. Upon fusing with the enucleated egg, the donor cells are stimulated by cytoplasmic factors in the egg to enter the S phase. Sheep, mice, and cattle have all been cloned.

- Teratocarcinoma is a type of cancer in which unspecialized cells differentiate to form a tumor that contains kidney, hair, and tooth cells.

Normal development is a complex series of signals and transduction of those signals, resulting in the differentiation of cells to form a new organism.

- The areas of undifferentiated cells in the growing tips of plant roots are known as meristems. These cells can become any cell type found in the root or stem.

- Undifferentiated dividing cells in animals are known as stem cells. Stem cells in adult animals are specific for the kinds of tissue they replace, typically skin, the lining of the intestines, and blood cells. However, environmental signals can induce stem cells from one tissue to replace cells that have been depleted from another tissue. For example, stem cells isolated from the mouse brain can replace bone marrow cells in mice that have been depleted of bone marrow.

- Embryonic stem cells are totipotent. They can be grown in culture for multiple generations, injected back into blastocysts, and then develop normally. In culture, the developmental fate of stem cells can be directed by the addition of environmental signals, such as growth factors or vitamins.

- Future medical applications include the use of embryonic stem cells to replace human cells that have been lost to disease or injury. The source of these cells could be from in vitro fertilized embryos. Therapeutic cloning is a method that would produce stem cells that are compatible with the immune system of the recipient.

Differentiated cells retain all their original genetic content, even though they are expressing only a very small subset of genes.

- Scientists use nucleic acid hybridization techniques to look for the presence of genes in cells. These experiments have shown that even though a specific gene is not expressed (not present as mRNA), it is still present in the nuclear DNA.

- Undifferentiated muscle cells (myoblasts) respond to the MyoD1 protein (*myo*blast *d*etermination gene 1) by fusing to form muscle fibers. This protein is a DNA-binding protein with a helix-loop-helix motif. When MyoD1 binds to promoters of muscle determination genes, transcription of those genes increases. It also binds to its own promoter and positively regulates MyoD1 mRNA production in myoblasts.

- When MyoD1 protein is injected into other precursor cells, such as fat cells, those cells are reprogrammed to develop into muscle cells.

Unequally distributed cytoplasmic factors in eggs and zygotes set up the initial pathways for precursor cells in developing embryos.

- Cytoplasmic segregation of factors in the egg results in an unequal distribution of maternal elements in each of the cells of the embryo.

- Unequal distribution of cytoplasmic determinants plays an important role in directing embryonic development and controlling the polarity of the organism. For example, if a sea urchin embryo is cut in half at the eight-cell

stage and split from left to right by that cut, the resulting larvae are normal, but dwarfed. If the sea urchin is cut so that its upper half is split from its lower half, the cells from the upper half develop into abnormal ciliated larvae and the cells from the lower half develop into misshapen larvae with an oversized gut (see Figures 19.7 and 19.8).

Tissues in developing embryos can induce other tissues to follow a particular developmental path.

- The vertebrate eye lens forms when surface cells make contact with a double bulging of the forebrain, known as the optic vesicles. If a barrier is placed between these two tissues in the developing embryo, no lens develops. If the forebrain is cut out before it can contact the lens surface cells, no lens develops. The optic vesicles produce a signal, causing the surface cells to develop into a lens. The lens tissue produces embryonic inducers that control the size of the optic cups and the differentiation of the cornea (see Figure 19.9).

- In *C. elegans*, a common model organism, the egg develops into a larva in eight hours at 25°C. The animal has a transparent body, making it is easy to follow the developmental fate of the 959 somatic cells. The adult is hermaphroditic and has both female and male reproductive organs. Eggs are laid through a pore called the vulva. The vulva is induced to form from a single cell called an anchor cell. If this cell is laser ablated, no vulva develops.

- The anchor cell determines the fates of six cells on the *C. elegans* ventral surface by producing an inducer that diffuses and interacts with adjacent cells. The closest cell becomes the primary vulval precursor cell. It produces an inducer that causes the next two cells to become secondary vulval precursors, and the final three cells (which are the farthest out from the anchor cell) become epidermal cells (see Figure 19.10).

- The primary inducer for the development of the vulva in *C. elegans* is a growth factor, LIN-3, that is homologous to mammalian epidermal growth factor. LIN-3 binds to a receptor on the surface of the vulval precursor cell, which causes a signal transduction cascade in that cell through the Ras protein and MAP kinase proteins. The end result of this cascade is the differentiation of this cell into a vulval cell.

- Molecular switches such as an inducer allow a developing cell to proceed down one of two alternative pathways.

Pattern development is the spatial organization of tissue that results in the appearance of body form (morphogenesis).

- During morphogenesis, some cells are programmed to die by death genes through the process of apoptosis.

- In *C. elegans*, there are a total of 1,090 cells. During development, 131 of these cells are programmed to die due to the sequential expression of *ced-4* and *ced-3* genes. *Ced-9* inhibits the actions of *ced-3* and *ced-4* during development.

- In human embryos, apoptosis occurs in the webs of skin that initially form between the fingers and toes so that they disappear by birth. Capsase, a protein homologous to *ced-3*, stimulates this apoptosis. The human homolog to *ced-9* is *bcl-2*. Overexpression of *bcl-2* leads to the development of a particular kind of cancer known as follicular large-cell syndrome.

- The genes controlling apoptosis are crucial for proper development in many organisms. These genes have been preserved in organisms separated by 600 million years of evolution (nematodes and humans).

- As a plant develops, it produces organs including leaves, roots, and flowers. Flowers have four types of organs: sepals, petals, stamens, and carpels. Flower organs occur in whorls organized around a central axis and are derived from meristematic tissue on the plant (see Figure 19.12).

- Plant geneticists have studied the development of flower organs in *Arabidopsis*. There are four whorls of organs in *Arabidopsis*. Three genes expressed in the whorls act as organ identity genes to guide differentiation in each whorl. Gene A is expressed in whorls 1 and 2, which form the sepals and petals; gene B is expressed in whorls 2 and 3, which form petals and stamens; and gene C is expressed in whorls 3 and 4, which form stamens and carpels.

- The three genes, A, B, and C, all encode transcription factors that are active as dimers. Gene regulation is combinatorial, and the combination of different dimers determines which genes are activated. A dimer of the transcription factor encoded by gene A will activate genes that make sepals. A dimer of transcription factor A with transcription factor B will result in petals.

- *Leafy* is a transcription factor that regulates the transcription of the genes A, B, and C. Plants bearing a mutation in *leafy* can produce leaves but no flowers.

- These homeotic genes, A, B, and C, as well as *leafy*, may be able to be genetically modified so that the flower organs can produce more fruit and seeds (which come from the carpels).

- Positional information allows developing cells to determine where they are in the developing organism. Morphogens are signals that establish positional information in the organism. Morphogens act directly on the target cell, and differential concentrations within the embryo cause differential effects.

- In the developing vertebrate limb, unique concentrations of the morphogen BMP2 determine the anterior–posterior axis of the developing limb. The action of such a morphogen works by producing differential gene expression in the target cells.

- Genes that encode developmental morphogens have been highly conserved in diverse organisms.

A cascade of gene expression controls development in *Drosophila*.

- The body pattern of *Drosophila* is segmented, consisting of a head (which is formed from fused segments), three thoracic segments, and eight abdominal segments.

- The first step of development in *Drosophila* is to establish anterior–posterior and dorsal–ventral polarity. Polarity is based on the cytoplasmic distribution of mRNAs and proteins produced in maternal genes. These maternal gene products are distributed to the eggs by the mother in a nonuniform manner to nurture the developing embryo. These maternal effects are not affected by the paternal genotype.

- Mutations in maternal genes (*bicoid* and *nanos*) create larvae lacking anterior structures (*bicoid*) or lacking abdominal segments (*nanos*).

- Segmental genes are expressed when the embryo is at the 6,000-nuclei stage and determine the number, boundaries, and polarity of the segments. Three types of segmental genes are involved in finer and finer regulation of segmental development.

- Gap genes are segmental genes whose products organize large areas along the anterior–posterior axis of the developing embryo. Gap mutants produce larvae that are missing certain larval segments. Pair rule gene products divide the larva into units of two segments each. Mutations in the pair rule genes produce larvae that are missing every other segment. Segment polarity gene products determine the boundary and the anterior–posterior organization of each segment. Mutant larvae lacking segment polarity genes produce larvae in which the posterior structures in the segments have been replaced by anterior structures.

- Differences between segments are encoded by homeotic genes, which give each segment an identity. For instance, thoracic segments produce legs.

- Because early embryos do not transcribe genes, unfertilized eggs have large numbers of mRNA molecules that are made prior to fertilization. This mRNA supports the high level of protein synthesis during early stages of development. Some of these mRNAs provide positional information; *bicoid* mRNA, for example, is localized in the anterior portion of the egg.

- In *Drosophila*, the first twelve divisions are only nuclear divisions with no division of cytoplasm until the thirteenth division. This multinucleate single-celled embryo is called a syncytium. *Bicoid* mRNAs are translated early during development, and they diffuse from the anterior (Bicoid protein) and posterior (Nanos protein) end of the egg, creating a gradient. *Bicoid* encodes a transcription factor that positively affects some genes (*hunchback*) and negatively affects others (*Krüppel*). Nanos protein inhibits the translation of gap genes, including *hunchback*. Gap gene products control the expression of pair rule genes, some of which are transcription factors that control segment polarity.

- Homeotic mutants include *Antennapedia* (producing legs in place of antennae, see Figure 19.16) and *bithorax* (producing a fly with an extra set of wings). Homeotic genes are clustered on the chromosome in the order of the segments they determine. The first set of genes in the homeotic cluster are thoracic, the next genes are upper abdominal, and so on.

- The cluster of homeotic genes that determine the segment identity in the front of the fly (*Antennapedia*) probably have risen from duplications of an ancestral gene.

- Nucleic acid hybridization experiments confirm that genes like *Antennapedia* and clusters next to those genes contain DNA sequences common to all homeotic genes. A DNA sequence (the homeobox), which encodes 60 amino acids (the homeodomain), includes a DNA-binding motif (helix-turn-helix) that has also been found in *bicoid*. *Bicoid* encodes a transcription factor that binds the gap promoters, such as *hunchback*. Hunchback is a protein that binds to enhancers to increase transcription of genes involved in head and thorax development.

The Big Picture

- The development of a mature differentiated organism from a fertilized egg involves the differential activation of genes in response to environmental signals. Initially in many organisms, positional determinants may be present in different regions of the egg due to maternal factors. The asymmetric distribution of these factors will activate genes differentially, and as cell division proceeds, that asymmetric gene expression continues. As a result, different cells experience different environmental signals based on their position in the developing embryo, and they respond by activating different genes. This differential gene activation determines the fate of the cell.

- Early embryonic cells are totipotent; they can develop into any structure in the adult organism. As development proceeds, the developmental potential of embryonic cells narrows. In the adult organism, differentiated cells express particular genes that give those tissues a particular structure and function, even though all the genes are still present in the nucleus of those cells.

- The concentration of specific genes and their products are involved in the regulation of polarity during development. These genes are classified as maternal effect genes, and their products are supplied by the mother.

Common Problem Areas

- Sometimes it is difficult to visualize how a single cell can divide and grow to produce a mature functional organism with highly differentiated tissues. At some very early point in development (either before fertilization or in one of the first set of divisions), different genes begin to be expressed in different cells. This gene expression can be in response to cytoplasmic factors or signals from other cells. Early gene expression sets up positional determinants that activate another wave of genes that further divides regions of the embryo into different developmental regions. As cell division proceeds in the embryo, these developmental genes continue to be differentially expressed, resulting in differentiation and morphogenesis in the organism.

Study Strategies

- Review the genes that are important in the development of *Drosophila*. Make a list of those genes and what their effects are on the phenotype of the *Drosophila* larva.

- Transplantation experiments help explain totipotency. Describe the transplantation experiments in frogs using early embryonic tissues and those using adult somatic cells to address the totipotency of those tissues.

- Review the experiments that demonstrated how tissue induction causes the lens in the vertebrate eye to form.

- Review some of the important genes whose products direct development in the organisms discussed in this chapter. Underline homologs from different organisms.

- Review the following tutorials on the Student Website/CD:
  Tutorial 19.1 Early Asymmetry in the Embryo
  Tutorial 19.2 Pattern Formation in the *Drosophila* Embryo

Test Yourself
Knowledge and Synthesis Questions

1. Cell differentiation is a result of
   a. the loss of particular genes from the nucleus of the differentiated cell.
   b. the differential expression of genes that are responsive to environmental signals.
   c. early embryonic cells remaining totipotent in the mature organism.
   d. mutations in genes that control the synthesis of DNA.

2. A totipotent cell is a cell
   a. whose developmental fate has been decided.
   b. that has differentiated into a specialized tissue.
   c. that is expressing morphogens to form a particular structure.
   d. whose developmental potential is extremely broad.

3. Apoptosis is programmed cell death and
   a. only occurs in developing embryos.
   b. only occurs in the mature organism.
   c. is an important process in both development and in mature organisms.
   d. can cause a type of cancer, follicular large-cell lymphoma.

4. Which of the following statements about plant development is *not* true?
   a. Both cell division and cell expansion contribute to growth in early plant embryos.
   b. Differentiation in plant cells is reversible.
   c. Meristems at the tips of the roots and stems consist of undifferentiated cells.
   d. Organ development in a plant continues throughout its life.

5. Mammals can be cloned using somatic cells fused to enucleated eggs if the
   a. donor cell is in the S phase of the cell cycle.
   b. donor cell is in the G1 phase of the cell cycle.
   c. mammalian embryo can be propagated in culture.
   d. somatic cell is a red blood cell.

6. Identify the function of the protein MyoD1.
   a. It is a transcription factor that controls the expression of genes that form muscle.
   b. It controls segment identity in mice.
   c. It is a transcription factor that regulates its own transcription.
   d. Both a and c

7. Tissue induction occurs when
   a. nuclear genes are lost in some tissues.
   b. signals are secreted in one part of the developing embryo and are transported via the circulatory system to another site in the embryo to activate tissue differentiation.
   c. direct contact from one tissue delivers an environmental signal to the tissue that will be induced.
   d. two tissues come into contact with one another, releasing transcription factors.

8. Apoptosis
   a. can cause abnormal development of the nervous system when inhibited in developing *C. elegans*.
   b. can cause webbing to be retained between the fingers of developing human embryos.
   c. is a natural part of development.
   d. Both a and c

9. It may be possible to genetically engineer plants to produce more seeds and fruits by
   a. fertilizing them more in the spring.
   b. manipulating their gap genes.
   c. altering the homeotic genes that control whorl development.
   d. inducing a mutation that eliminates *leafy* gene function.

10. Morphogens
    a. diffuse within the embryo to set up a concentration gradient.
    b. are expressed as positional signals in developing embryos.
    c. direct differentiated cells to form organs.
    d. All of the above

11. Maternal genes
    a. set up positional axes in the egg prior to fertilization.
    b. are mRNA and proteins that are placed in the egg cell.
    c. are masked by the paternal genes.
    d. Both a and b

12. Homeotic genes
    a. encode protein domains that are important in development and that have been highly conserved over evolutionary time.
    b. are found in greatly diverse organisms.
    c. can produce the wrong structure in the wrong place when mutated.
    d. All of the above

Application Questions

1. Nuclear hybridization experiments were done to show that fully differentiated cells still contained all their genes even though they were only expressing a small set of those genes. What kinds of hybridizations demonstrated this?

2. Undifferentiated stem cells are totipotent. Describe the experiment that clearly demonstrated this fact.

3. Transcription factors that control the fate of undifferentiated stem cells can be used to redirect other committed stem cells. Describe the experiment that showed this.

4. In cloning whole organisms by fusing nuclei (as seen in frogs) or whole cells (as seen in sheep, cows, and goats) with enucleated eggs, why is the nucleus removed from the egg?

Answers to Questions

Knowledge and Synthesis Answers

1. **b.** Genes are not lost from differentiated cells. There are no embryonic cells in a mature organism. Mutations in cells that affect DNA replication would result in cells that were unable to replicate their DNA.

2. **d.** A totipotent cell is a cell that can develop in any number of ways; its fate has not been determined, it has not differentiated, and it is not expressing morphogens.

3. **c.** Apoptosis is important for embryonic development and in adult tissues such as skin and blood cells.

4. **b.** In general, it is easier to reverse differentiation in plant cells than it is in animal cells, as the cloning of a carrot from a differentiated storage cell in its root demonstrates. Not all differentiated plant cells are totipotent, however; tracheids (as an extreme example) die before they become functional as water-conducting cells.

5. **b.** The donor somatic cell must be in G1. Mammalian embryos must be grown in surrogate mothers. Red blood cells in mammals have no nuclei and could not be used for cloning.

6. **d.** The gene *MyoD1* encodes a transcription factor (MyoD1) that controls the expression of genes that form muscle in mice and also regulates its own transcription. *MyoD1* does not control segment identity in mice. Homeotic genes control segment identity in *Drosophila*.

7. **c.** Induction involves contact between two different tissues, with one tissue providing some kind of environmental signal that is received by the other tissue. Nuclear genes are not lost in developing cells (with the exception of red blood cells and the immune cells). Tissue induction occurs because of tissue contact, not because a signal has been carried from one part of the embryo's body by circulation to another part of its body. Transcription factors are found in the nucleus and are not released to adjacent tissues.

8. **d.** Apoptosis causes the webbing to be removed (not retained) between the fingers and toes during the development of the human embryo.

9. **c.** Fertilization is not genetic engineering. Gap genes are found in *Drosophila*, not plants. The *leafy* protein transcriptionally activates the homeotic genes that control organ development (and subsequent seed and fruit formation). In its absence, no fruit or seeds would develop.

10. **d.** Morphogens diffuse in the embryo to form concentration gradients that act as positional signals to direct differentiated tissues to form organs.

11. **d.** The paternal genotype does not affect the expression of maternal gene products in the egg.

12. **d.** Homeotic genes encode proteins that are highly conserved and important for directing development in many different organisms. Mutation in these genes can produce an abnormal developmental event in the organism.

Application Answers

1. Hybridizations were done using a specific DNA probe to nuclear DNA and mRNA. If the DNA probe was complementary to a gene that was not being expressed in the cell, it would not hybridize to the mRNA isolated from that cell, but would still hybridize to the nuclear DNA. This indicated that the gene was still present in the nucleus of the cell even though it was not being expressed in the cytoplasm.

2. Undifferentiated stem cells from mouse brain were transplanted to the bone marrow of mice whose bone marrow had been depleted. The brain cells differentiated into bone marrow cells, indicating their totipotency.

3. The MyoD1 transcription factor from muscle cells directs the transcription of several muscle genes, resulting in the fusion of myoblasts to form muscle cells. When MyoD1 was applied to precursors of fat cells, it redirected those cells to form muscle cells.

4. The nucleus of the egg contains a full haploid set of chromosomes. Fusing a diploid somatic cell with an egg cell that still contained a nucleus would result in a cell that had three copies of each chromosome and little chance of developing normally.

20 Animal Development: From Genes to Organism

Important Concepts

Fertilization is the joining of the sperm and the egg.

- Early development of animals has been studied extensively in a few model organisms: sea urchins, frogs, chickens, and humans.

- Development begins when a haploid sperm joins with a haploid egg.

At fertilization, the egg and sperm contribute differently to zygote formation.

- The egg contributes the majority of the cytoplasm and the mitochondria to the zygote.

- The sperm contributes its haploid nucleus and a centriole that becomes the centrosome and eventually forms the mitotic spindles.

- In mammals, both sperm and egg genes are necessary for development to progress normally; some genes are active only if they come from a sperm, and others are active only if they come from an egg. This phenomenon is known as genomic imprinting.

- Upon fertilization in the frog egg, the cytoplasm of the egg undergoes rearrangement. The nutrients concentrate in the vegetal hemisphere of the egg in the lower half, whereas the nucleus is found in the upper animal hemisphere (see Figure 20.2).

- Sperm enters in the animal hemisphere at what will become the ventral side in frogs and causes a rotation of the cortical cytoplasm to create a gray crescent, which will become the dorsal region of the frog. This helps set up the left-to-right axis and the anterior-to-posterior axis of the developing animal (see Figure 20.2).

- During the movement of the cytoplasm, the transcription factor β-catenin is degraded in parts of the egg so that it is concentrated in the dorsal side of the embryo (see Figure 20.3). The protein GSK-3 moves to the ventral side of the zygote where it targets β-catenin for degradation. On the dorsal side, proteins inhibit the action of GSK-3.

Cleavage is the first series of cell divisions in development.

- The first divisions of the cells occur rapidly with little growth or differentiation in a process known as cleavage. During this time DNA replication and mitosis is rapid. The embryo divides into smaller and smaller cells producing a morula.

- The mass of cells that is generated, the blastula, has a central cavity, the blastocoel. Individual cells in the blastula are known as blastomeres.

- The amount of yolk in the egg determines the plane along which cell division occurs. Cleavage furrows make up the site of cell division. Cleavage furrows are impeded within the egg by the yolk. Animals such as sea urchins and frogs with small amounts of yolk have complete cleavage (see Figure 20.4). Fish, reptile, and bird eggs with large amounts of yolk have incomplete cleavage, with the embryo forming a disc of cells sitting on top of the yolk known as the blastodisc. Some insects undergo superficial cleavage.

- The mitotic spindles help to orient the cleavage planes into a radial cleavage pattern in the sea urchin and frog or spiral cleavage in the mollusks.

In mammals, cells separate into two groups between the 16-cell and 32-cell stage.

- Mammals have a unique rotational cleavage pattern. Cleavage in mammals is slow and asynchronous. The zygote produces both the embryo and extraembryonic structures that interact with the mother. Unlike in other species where cleavage is directed by molecules already present in the egg, in mammals genes are expressed during cleavage and play a large role.

- During cleavage from the 16-cell to the 32-cell stage, the cells separate into an inner cell mass that will become the embryo and a trophoblast that will become the placenta and attach to the uterus during implantation. At the 32-cell stage the mass of mammalian cells is called a blastocyst (see Figure 20.5).

- Mammalian fertilization occurs in the oviduct, and the zygote moves down into the uterus where it undergoes implantation. The zona pellucida inhibits early implantation and is lost just prior to implantation in the uterus.

Blastomeres become determined during late cleavage.

- During cleavage, cytoplasm is distributed to the cells of the blastula in such a way that cells in different regions have different levels of nutrients and informational molecules. Specific blastomeres will become specific tissues and organs, and this can be mapped out in the developing cells as a fate map of the blastula (see Figure 20.6). The blastomeres become committed to specific fates at different times for different species.

- Animals with mosaic development have blastomeres that are set to contribute to specific parts of the embryo. In animals with regulative development, the blastomeres can be removed and other cells will compensate for the loss.

Gastrulation involves movement and differentiation of cells.

- During gastrulation, the germ layers of the tissues form and position themselves in the embryo. Inductive interactions occur during gastrulation due to the spatial relationships between cells, allowing for differentiation and organ formation.

- Three germ layers develop during gastrulation. The endoderm inside the embryo will become tissues lining the gut, respiratory, and circulatory system; the ectoderm outside the embryo will become the skin, hair, nails, glands, and the nervous system; and the mesoderm between the two layers will become the tissues of many organs, including blood vessels, muscle, bones, liver, and heart.

Gastrulation in sea urchins involves invagination of the vegetal pole.

- The vegetal pole in sea urchins moves into the cavity of the blastocoel. Some cells from the vegetal pole break free to become the primary mesenchyme cells that make up the mesoderm (see Figure 20.8). The flat portion of the vegetal pole moves inward as an invagination to form the endoderm and primitive gut, known as the archenteron.

- Secondary mesenchyme cells are produced from cells that break off at the tip of the archenteron and enter the blastocoel.

- The opening of the archenteron or blastopore will become the anus, and the place where the tip of the archenteron meets the ectoderm will become the mouth.

The dorsal lip controls embryonic organization during gastrulation in the frog.

- The frog embryo has more yolk than the sea urchin, and gastrulation is more complex (see Figure 20.9). Initially, cells near the gray crescent begin to bulge into the blastocoel. These initial cells (bottle cells) move along the interior of the blastula and pull the outer surface of cells along with them, creating a dorsal lip. The first cells moving in will become endoderm and the archenteron.

- This process of the movement of cells from the surface into the blastopore is known as epiboly. As epiboly continues, the dorsal lip forms a complete circle around a plug of yolk-rich cells. At the end of gastrulation, the cells have become fate-determined and are layered as the ectoderm, endoderm, and mesoderm.

- Hans Spemann and Hilde Mangold examined the timing and fate of cells during gastrulation to determine if they were totipotent or able to direct development (see Figure 20.10). They found that the dorsal lip controlled the ability to form an embryo when it was transplanted to another gastrula, resulting in two embryos attached at the belly. Thus the dorsal lip acts as the primary embryonic organizer (see Figure 20.11).

- Differential expression of transcription factors appears to be involved in controlling organization and left-right axis formation during gastrulation.

- β-catenin is a candidate for initiating the signal cascade involved in organizer activity. When β-catenin activity is knocked out by antisense RNA, the embryo does not go through gastrulation; overexpression of β-catenin can induce formation of a second axis. Primary embryonic organizer activity is started from the vegetal cells below the gray crescent. The transcription factors *goosecoid* and *siamois* are critical in the signal cascade. Tcf-3 proteins act as repressors for *siamois*. When the Tcf-3 protein is blocked by β-catenin, expression of *siamois* occurs, and the goosecoid protein is produced (see Figure 20.12).

- The left-right axis may be regulated by the differential expression of transcription factors early in gastrulation. The TGF-β protein may be important for organization during frog gastrulation whereas cilia and flow produced in the yolk sac cavity are important in mammals.

In reptile and bird gastrulation, the primitive groove is the blastopore, and Hensen's node corresponds to the amphibian dorsal lip.

- Cleavage produces a blastodisc on top of the large amount of yolk in the egg.

- The blastula is a circular layer of cells composed of an outer epiblast layer and an inner hypoblast layer. The fluid-filled space between the two layers is the blastocoel (see Figure 20.13).

- The primitive streak is formed by the movement of cells in the epiblast toward the midline. Along the primitive streak a primitive groove forms and cells move through this blastopore, becoming endoderm and mesoderm in the blastocoel.

- In the chicken there is no archenteron. A group of cells known as Hensen's node act in the same manner as the dorsal lip of the frog blastopore.

Gastrulation in mammals is similar to that in reptiles and birds.

- In mammals, the cells of the embryo segregate into an epiblast and a hypoblast with a blastocoel in between, just as in birds and reptiles. The embryo develops from the epiblast and the extraembryonic membranes develop from the hypoblast.

- Gastrulation in mammals occurs just as in birds, with the formation of a primitive streak.

The nervous system begins to develop during neurulation.

- After gastrulation, the organs begin to develop in the process of organogenesis. Initial development of the nervous system, neurulation, begins early in organogenesis.

- During neurulation, the notochord develops in the blastocoel from chordomesoderm cells derived from the dorsal mesoderm (see Figure 20.15).

- A neural plate forms above the notochord from the ectoderm, which will begin to fold into a cylinder forming the neural tube. The anterior end of the neural tube will become the major parts of the brain, with the rest forming the spinal cord. Failure of this process leads to serious birth defects such as spina bifida and anencephaly.

- The repeating pattern of the vertebrate body plan forms from blocks of somite tissue derived from mesoderm located on both sides of the notochord (see Figure 20.16). The somites will become muscles, ribs, and vertebrae.

- Neural crest cells break from the neural tube and migrate inward to become peripheral nerves.

Homeotic genes control differentiation in the anterior–posterior directions during development.

- Body segments differentiate as the embryo develops. Hox genes control this differentiation of body segments.

- The Hox genes in mice control the process of differentiation in the anterior–posterior directions during development (see Figure 20.17).

- Hox genes are found on different chromosomes and are expressed during development in their order on the chromosomes.

- Dorsal–ventral differentiation is controlled by a separate set of genes. The *sonic hedgehog* gene controls the development of the cells over the neural tube that become part of the ventral spinal cord.

- Pax genes, such as *Pax3*, are also important in development of the nervous system and somite development.

- After segmentation occurs, organs and organ systems develop rapidly.

Birds and reptiles develop extraembryonic membranes.

- Extraembryonic membranes surround the embryos of reptiles, birds, and mammals (see Figure 20.18).

- In birds, the yolk sac is derived from the hypoblast and surrounds the entire yolk to help retrieve nutrients from the yolk and deliver them to the embryo.

- Cells from the ectoderm and mesoderm form the amnion, which helps provide an aqueous environment, and the chorion, which develops just under the shell. The chorion regulates water, oxygen, and carbon dioxide exchanges across the shell.

- The allantois is derived from endoderm and is a membrane that forms a sac to store metabolic waste products.

The placenta of mammals provides exchanges between the mother and the embryo.

- In mammals, the first extraembryonic membrane, the trophoblast, interacts with the endometrium to attach to the uterine wall and begin implantation.

- The hypoblast cells interact with the trophoblast to form the chorion, which along with tissues from the uterine wall forms the placenta (see Figure 20.19).

- The amnion of the developing mammal surrounds the embryo to produce a closed, fluid-filled environment.

- The allantois of mammals has the function of removing nitrogenous wastes, but its function is relatively minor in many mammals. The allantoic tissues help to form the umbilical cord, which carries major blood vessels that provide a route for exchanges of nutrients, carbon dioxide, and oxygen between the mother and fetus.

- Amniocentesis and chorionic villus sampling are two methods used to detect genetic markers for diseases (see Figure 20.20).

Human pregnancy is divided into trimesters.

- In humans pregnancy lasts about 266 days, compared with 21 days in mice and 600 days in elephants.

- During the first trimester of human development, the embryo develops into a miniature version of the adult called the fetus. Cell division and tissue differentiation are rapid during the first trimester. At this stage the developing fetus is most sensitive to environmental disruptors. The corpus luteum continues to produce estrogen and progesterone during this trimester.

- During the second trimester of development, the first fetal movements are felt, the fetus reaches a size of 600 g, and the woman begins to show physical evidence of her pregnancy.

- The third trimester is marked by rapid growth of the fetus and the maturation of the internal organs.

- Growth continues even after birth.

The Big Picture

- Animal development involves a number of steps: fertilization, cleavage, gastrulation, and neurulation. Fertilization is the joining of the egg and sperm to form a zygote.

- Following fertilization, the cells of the embryo begin to divide in a process known as cleavage. Eggs undergo either complete cleavage or incomplete cleavage, depending on the amount of yolk present. Those that undergo complete cleavage form a blastula, and those that undergo incomplete cleavage form a flat blastodisc.

- Cleavage in mammals is different from cleavage in other groups. It is much slower, and the products of embryonic genes play a role in directing its course; in other animals, cleavage is directed entirely by molecules present in the egg prior to fertilization, not the products of embryonic genes. The mammalian egg produces cells that will become the embryo and extraembryonic structures.

- During gastrulation, the germ tissues—endoderm, ectoderm, and mesoderm—form and move into specific positions. In the sea urchin, an archenteron that will become the gut forms, and specific cells form the three germ layers. In frogs, a dorsal lip is formed, and the germ cells move into place. The dorsal lip is considered the primary embryonic organizer. The blastodisc of reptiles, birds, and mammals goes through a much different pattern of gastrulation.

- Neurulation occurs early in organogenesis. The nervous system begins to form a notochord and neural plates. The neural plates form into a neural tube that will become the brain and spinal cord. Somites produce the repeating segments of the vertebrate body plan. A number of genes play a large role in the differentiation of tissues.

- Reptiles and birds have four major extraembryonic membranes: yolk sac, amnion, chorion, and allantois. Mammals produce a special placenta that provides nutrients between the mother and the developing fetus.

Common Problem Areas

- All fertilized cells must divide to grow into adult animals. Sorting out the features of division that are common to all animals from those specific to certain taxonomic groups can be difficult and confusing until you begin to recognize the basic patterns.

- Students are often confused about the role of the placenta and its various membranes and structures. Is it embryonic? Is it maternal? And how do nutrients, gases, and wastes travel between the mother and embryo or fetus?

Study Strategies

- There is a great deal of terminology in this chapter, particularly with respect to the reproductive anatomy. Yet there are also many terms that are common to worms and wolves alike. Create a list of new terms from this chapter. Then determine which are the more general terms applicable to numerous animals (e.g., "somites," "gastrulation") and learn these first. Then tackle those referring to reproduction in specific animals.

- A key point in understanding animal development is that all specialized tissues arise from the three germ tissues: endoderm, ectoderm, and mesoderm. More important than just committing these three tissues to memory is learning the specialized tissues that derive from the germ tissues. This will better enable you to make predictions about tissue functions and to compare body plans among animals.

- Review the following tutorials and activity on the Student Website/CD:
  Tutorial 20.1 Gastrulation
  Tutorial 20.2 Tissue Transplants Reveal the Process of Determination
  Activity 20.1 Extraembryonic Membranes

Test Yourself

Knowledge and Synthesis Questions

1. The sperm and the egg make different contributions to the zygote. Which statement about their contributions is *false*?
   a. The sperm contributes most of the mitochondria.
   b. The egg contributes most of the cytoplasm.
   c. Both the sperm and the egg contribute a haploid nucleus.
   d. All of the above are true.

2. In which animal is the cleavage pattern rotational?
   a. Frog
   b. Mammal
   c. Sea urchin
   d. All of the above

3. Why do some species have complete cleavage and others incomplete cleavage?
   a. Incomplete cleavage occurs in species with small volumes of cytoplasm.
   b. Complete cleavage, found in mammals, is a more evolved characteristic.
   c. Incomplete cleavage occurs in species with large amounts of yolk.
   d. Complete cleavage occurs only in eggs that have been fertilized by two sperm.

4. A fate map can be used to map out the tissues and organs that the germ layers will become in a frog blastula. What is the fate of the dorsal lip of the blastopore in amphibians?
   a. The lining of the gut
   b. The notochord

c. The epidermal layer of skin
d. The heart

5. Whereas the location of the _____ determines the anterior–posterior axis of the embryo, the _____ determines the embryo's bilateral symmetry.
   a. primitive streak; epiblast
   b. blastopore; neural plate
   c. vegetal hemisphere; notochord
   d. hypoblast; blastopore

6. What is the order of the germ layers from the inside to the outside?
   a. Mesoderm, ectoderm, endoderm
   b. Endoderm, ectoderm, mesoderm
   c. Ectoderm, mesoderm, endoderm
   d. Endoderm, mesoderm, ectoderm

7. Birds develop from a blastodisc whereas frogs develop from a blastula. During gastrulation in frogs, a dorsal lip forms that will eventually form the notochord. What is the equivalent in the chicken blastodisc?
   a. Epiblast
   b. Hypoblast
   c. Hensen's node
   d. Bottle cells

8. Birds develop extraembryonic membranes during development. Which statement about avian extraembryonic membranes is *false*?
   a. The yolk sac surrounds the yolk and provides nutrients.
   b. The amnion and chorion are derived from ectoderm and mesoderm.
   c. The allantois stores nutrients.
   d. The chorion exchanges gases and water between the embryo and the environment.

9. Which of the following statements about the mammalian blastocyst is *false*?
   a. The trophoblast gives rise to the embryo proper.
   b. Maternal genes are expressed during cleavage.
   c. The blastocyst implants in the mother's uterus.
   d. Early mammalian development is slow.

10. The primary embryonic organizer is most likely initiated by _____.
    a. the yolk
    b. TCF-3 protein
    c. β-catenin
    d. None of the above

11. The _____ eventually develop into vertebrae, ribs, and muscles and are found along the sides of the _____.
    a. somites; notochord
    b. neural tube cells; notochord
    c. blastopore cells; dorsal lip
    d. neural crest cells; dorsal lip

12. Why did Hans Spemann call the dorsal lip of the blastopore the embryonic organizer?
    a. It is the point where gastrulation begins.
    b. It becomes part of the nervous system.
    c. It becomes part of the notochord.
    d. It leads to the establishment of the embryonic axes.

13. During its development, the human embryo is contained within a fluid-filled chamber bounded by the membranous
    a. yolk sac.
    b. amnion.
    c. chorion.
    d. allantois.

Application Questions

1. In the following diagram of a sea urchin gastrula, label the ectoderm, endoderm, primary mesenchyme, secondary mesenchyme, blastopore, blastocoel, and archenteron.

2. Label the listed structures in the following longitudinal cross section of a late frog gastrula: blastopore, area where brain will form, yolk plug, notochord, neural plate, archenteron, ectoderm, mesoderm, and endoderm. Also indicate the (A) anterior, (P) posterior, (D) dorsal, and (V) ventral areas of the embryo.

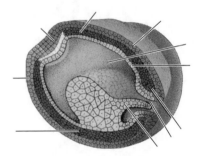

3. Compare cleavage in birds and in mammals. How do differences in cleavage between these two groups influence gastrulation?

4. How does the development of the extraembryonic membranes of birds and mammals differ and how are they similar? What is the main reason for the differences?

Answers

Knowledge and Synthesis Answers

1. **a.** The mitochondria of the sperm degenerate. The egg donates all the mitochondria to the zygote. This fact is used by some scientists to determine how species are related.

2. **b.** Cleavage in mammals occurs in a rotational pattern.

3. **c.** Incomplete cleavage occurs because the cleavage furrows cannot completely penetrate the yolk. Incomplete cleavage occurs in bird and reptile species with large amounts of yolk.

4. **b.** The dorsal lip of the blastopore in amphibians will form the notochord.

5. **b.** The blastopore, which will eventually become the anus, marks the posterior of the embryo; the embryo is bilaterally symmetrical with respect to the neural plate.

6. **d.** The germ layers that are formed during gastrulation are the inner layer of endoderm, the middle layer of mesoderm, and the outer layer of ectoderm.

7. **c.** Chickens develop a primitive streak and groove along the top of the blastodisc. At one end of the primitive streak is the Hensen's node, which is the equivalent of the dorsal lip in the developing frog.

8. **c.** During development, an avian embryo produces wastes, but because they develop in a shell, the wastes must be stored in the egg. The allantois forms a sac that is used for the storage of metabolic wastes.

9. **a.** In the mammalian blastocyst, the trophoblast forms the fetal part of the placenta. A disc-shaped portion of the inner cell mass surrounded by the trophoblast becomes the embryo.

10. **c.** β-catenin is thought to be the initiator of organizer activity during early development.

11. **a.** Somites are located along the notochord in the developing vertebrate. These cells will develop into the vertebrae, ribs, and muscles.

12. **d.** The dorsal lip leads to the establishment of the embryonic axes.

13. **b.** The amnion makes up the membrane that surrounds the developing mammalian fetus.

Application Answers

1.

2.

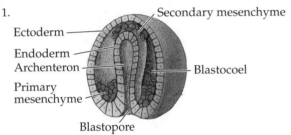

3. Due to the large amount of yolk in the avian egg, cleavage in birds is incomplete, and the result is a blastodisc that sits on top of the yolk (see Figure 20.4). Unlike bird embryos, mammalian embryos undergo complete cleavage in a rotational pattern, resulting in a structure called a blastocyst (see Figure 20.5); the pace of cleavage is also slower in mammals than in birds, and embryonic genes

play a role in directing cleavage in mammals, whereas in birds, cleavage is directed by molecules already present in the egg. In spite of these differences, gastrulation is very similar in birds and mammals. The avian blastodisc develops into a flattened blastula with epiblast and hypoblast surrounding a fluid-filled blastocoel; the mammalian inner cell mass also splits into an epiblast and hypoblast. Although the details vary (see the next question), the developmental fates of these structures are fundamentally the same in both groups: The hypoblast will give rise to extraembryonic membranes, and the embryo proper will develop from the epiblast.

4. Both the birds and mammals develop extraembryonic membranes. The extraembryonic membranes of the birds are used to obtain food from the large reserve of yolk (yolk sac), to exchange water and gases with the environment across the shell (amnion and chorion), and to store nitrogenous wastes (allantois). The mammals have extraembryonic membranes that form a placenta involved in exchanging nutrients, oxygen, and carbon dioxide with the mother. The hypoblast cells that form the yolk sac in birds along with the trophoblast cells and the chorion all contribute to the placenta. An allantois also develops, but has limited function in humans.

21 Development and Evolutionary Change

Important Concepts

Evolutionary developmental biology has grown out of classic genetic and developmental studies.

- Descent with modification results in related species having similarities due to common ancestry and differences due to natural selection. These similarities and differences can be viewed in the tree of life.

- Evolutionary developmental biologists study how heritable change in an organism results in change in development. To do this they examine the genes responsible for the control of development.

- Many of the genes involved in controlling development are conserved. Thus, these genes can be very similar in organisms that are very different in form. The similarity in the genes involved in eye development of the fruit fly and the mouse and the homeobox genes involved in controlling animal body plans are two examples of conserved genes.

- Homologous genes, while providing a common set of instructions, still harbor important differences that result in the differential development of the variety of organisms.

Mutations and changes in the timing of gene expression of homologous genes have led to evolutionary changes in organisms.

- The developing embryo can successfully change because its many parts are made up of modules. A module such as the heart can change independently of the other modules in the developing embryo.

- Changes in the morphology of developing embryos occur because of mutations in the genes controlling development. These gene mutations result in a new phenotype for the organism. A mutation in the insect *Ultrabithorax* (*Ubx*) homeotic gene results in insects having no legs in their abdominal segments. Other arthropods, such as millipedes, do not have this mutation and have abdominal legs. The presence of webbed feet in ducks is another example of mutation acting on a single module.

- Evolutionary changes in morphology have also occurred due to the change in the timing or location of regulatory gene expression during development. Heterochrony occurs when the expression of genes regulating the development of one module differs in timing from other modules in different species.

- Structural changes can also occur due to gene duplication.

Genes regulating plant development appear to be shared by many species.

- Plant development differs from animal development in a number of ways.

- Plant cells do not move in relation to each other during development. In animals, cell movement is an important part of the developmental process.

- The organs involved in plant reproduction can be produced from meristems at any point in the plant's life. In animals, the reproductive organs are produced only once during the animal's lifetime.

- The ability to change development depending on surrounding environmental conditions is called developmental plasticity. Plants exhibit greater developmental plasticity than animals.

- There are important regulatory genes shared by both plants and animals. MADS box genes and homeobox genes are found in both plants and animals.

The environment can have a large influence on the development of both plants and animals.

- Organisms have the ability to exhibit different phenotypes based on a single genotype. The phenotype is dependent on the environment in which the organism develops and lives.

- Environmental signals can be either accurate predictors of the future environment or poorly correlated with the future environment. Environmental signals such as day length, temperature change, and extent of rainfall are accurate predictors of future conditions. Temperature can influence the coloration of the adult butterfly. The *distal-less* gene is involved in the development of wing eyespots. Soil temperature at which the pupae develop dictates whether the *distal-less* gene is expressed. Another example involves butterflies that produce two generations during the summer. Each generation must be ready for a different environment. Environmentally controlled development allows organisms to adapt to the environmental conditions that will be present when they become adults.

- Some species must have specific organisms present in their environment for them to survive. This can be seen in mice that die when raised in microbe-free environments.

- The environment of organisms can also change in an uncertain manner. In the presence of predators, many species of *Daphnia* exhibit developmental plasticity by growing large helmets that are thought to increase their survival. The induction of a large helmet has a trade-off in a decreased number of eggs. In a similar fashion, spadefoot toad tadpoles speed up development when their water source begins to dry.

- In plants, low light levels result in elongation of cells. This allows a plant to grow faster and potentially reach a sunny patch.

- Some environmental signals are poorly correlated with the future environment that an organism will experience. Plant seeds tend to remain constant in size and not change with the environment because of the unpredictable year-to-year variation in the environment.

- Organisms will most likely be unable to respond to environmental changes to which they have never been exposed. Today the normal development of organisms in the wild is being disturbed by environmental changes brought about by humans.

The Big Picture

- Evolutionary developmental biology is the study of how organisms evolved by examining the differences in developmental gene expression and regulation between species. Animals have similar conserved genes, such as homeobox genes, that control development. Heterochrony describes how differences in organisms have arisen because of differences in the timing of different developmental processes. The timing of gene expression can differ between organisms because of the modular makeup of animals.

- Plants share similar regulatory genes with animals but also have a large number that are specific to controlling their development. Plants differ from animals in a number of ways that dictate developmental processes.

- Development can be influenced by environmental signals. These signals can either provide an accurate prediction of the future or be poorly correlated with future conditions.

Common Problem Areas

- The concept of heterochrony can be difficult for students to grasp. Try to remember that when we discuss heterochrony, we are talking about differences between species, not within species.

Study Strategies

- This chapter intorduces many concepts. Try to learn the concepts involved in evolutionary developmental biology first. Then go back and see how the examples fit with the concepts.

- Review the following tutorial and activity on the Student Website/CD:
Tutorial 21.1 Modularity
Activity 21.1 Plant Development

Test Yourself

Knowledge and Synthesis Questions

1. Genes that regulate development are conserved. This means that they
   a. have evolved large differences among multicellular organisms.
   b. have changed very little over the course of evolution.
   c. are always turned on.
   d. None of the above

2. _____ genes are involved in the development of the anterior–posterior axis in both humans and insects.
   a. Developmental plasticity
   b. Heterochrony
   c. Homeobox
   d. All of the above

3. Heterochrony
   a. relates development in at least two species.
   b. involves the relative timing of developmental processes.
   c. has been extensively studied in salamanders.
   d. All of the above

4. Module development has allowed for evolutionary changes to occur while still resulting in a viable organism because
   a. all modules must change together.
   b. modules can change independently of each other.
   c. modules are unimportant for development.
   d. None of the above

5. How do plant cells differ from animal cells?
   a. Plant cells do not migrate during development.
   b. Plants produce undifferentiated cells throughout their lives.
   c. Plants exhibit greater developmental plasticity.
   d. All of the above

6. Developmental plasticity is
   a. the ability to express different genotypes under different environmental conditions.
   b. found only in plants.
   c. the ability to express only one phenotype under different environmental conditions.
   d. None of the above

7. Which of the following is *not* a developmental environmental signal?
   a. Temperature
   b. Light
   c. The presence of predators
   d. All of the above are developmental environmental signals.

8. Which of the following genes are found only in plants?
   a. Homeobox genes
   b. MADS box genes
   c. *gremlin*
   d. None of the above

9. Spadefoot toad tadpoles respond to changes in their aquatic habitat by
   a. laying more eggs.
   b. slowing down development when their pond begins to dry.
   c. speeding up development when their pond begins to dry.
   d. None of the above

10. Yearly difference in rainfall is unknown to a plant producing seeds for the next year. Because of this, plants produce
    a. seeds of a constant size.
    b. seeds of varying sizes.
    c. no seeds.
    d. None of the above

Application Questions

1. There are morphological differences between chickens and ducks that are related to the duck's aquatic lifestyle. Why were these adaptations able to occur without influencing the development of the rest of the duck's morphology?

2. Day length and temperature are both potential environmental signals that may influence development. What do changes in these two environmental factors tell an organism about future conditions?

3. Why do insects have only six legs, whereas millipedes have legs on each body segment?

Answers

Knowledge and Synthesis Answers

1. **b.** Conserved genes are genes that are found in many organisms and have undergone very little change.

2. **c.** Homeobox genes regulate development along the anterior–posterior axis in animals.

3. **d.** Heterochrony is the process by which the timing of gene expression differs between two or more species. Heterochrony has been studied extensively in salamanders.

4. **b.** Embryos are composed of self-contained units called modules that can change independently of one another. This allows for the evolution of one module without disrupting the other modules of the animal.

5. **d.** Plant cells differ from animal cells in a number of ways; all of the choices are ways in which plants and animals differ.

6. **a.** Developmental plasticity is found in both plants and animals and allows a developing organism to express different phenotypes, depending on the environment.

7. **d.** Environmental signals are physical or biological properties of the environment that help organisms anticipate changes. Light, predators, and temperature are all environmental signals that some organisms respond to by altering their development.

8. **d.** Homeobox and MADS box genes are found in both plants and animals, whereas the *gremlin* gene is found only in animals.

9. **c.** The change in water volume or concentration is an environmental signal for the spadefoot toad tadpole. In response to this environmental signal, the tadpole's development into a toad will speed up.

10. **a.** Because current rainfall is an inaccurate predictor for the following season's rainfall, plants will produce seeds each year that are fairly constant in size.

Application Answers

1. Organisms are made up of modules. Change can occur in a module independent of any other module. The webbing found in duck feet is due to changes in the timing of *gremlin* gene expression in the foot module.

2. Both day length and temperature are accurate predictors of the future environment. Changes in day length act as an environmental signal that accurately predicts when future conditions will change. Shorter days signify that fall and winter are approaching. Similarly, changes in temperature occur with changes in the seasons and can signal to an organism that a change in the season will occur soon.

3. Leg development in arthropods involves the homeotic gene *Ultrabithorax* (*Ubx*) and the gene *distal-less* (*dll*). The expression of *dll* in a segment results in the development of a pair of legs in that segment. In insects, the *Ubx* gene has a mutation that suppresses the expression of *dll* in the abdomen. As a result, no legs will develop in the abdomen. Millipedes do not have this mutation, therefore legs develop in all body segments.

22 The History of Life on Earth

Important Concepts

Biological evolution is a change over time in the genetic composition of a population.

- The development of the science of biology, particularly Darwin's theory of evolution by natural selection, depended on evidence supplied by geologists that Earth is very old.

- Short-term evolutionary changes occur rapidly enough to be studied directly.

- Long-term evolutionary changes involve the appearance of new species and evolutionary lineages. The fossil record provides evidence of such changes.

Several types of evidence have been used to estimate the age of rocks, of fossils, and of Earth itself. All the evidence indicates that both Earth and life on Earth are very old.

- Fossils, the remains of ancient organisms, are useful for establishing the relative ages of the sedimentary rocks in which they occur.

- The regular pattern of decay of radioactive isotopes provides a means of estimating the absolute ages of fossils and rocks. The ratio of radioactive ^{14}C to its stable isotope, ^{12}C, is used to date fossils less than 50,000 years old. The decay of ^{40}K to ^{40}Ar is widely used to date ancient evolutionary events.

- Sedimentary and igneous rocks can also be aged because they preserve a record of Earth's magnetic field at the time they were formed.

- Earth's geological history is divided into eras, which are subdivided into periods. The boundaries between these divisions are marked by changes in the types of fossils found in successive layers of sedimentary rock.

- The first forms of life evolved early in the Precambrian era, which lasted for more than three billion years and was marked by stupendous physical changes on Earth.

The history of Earth has been marked by unidirectional and oscillatory changes, as well as by occasional collisions with extraterrestrial objects.

- Earth's crust consists of solid plates floating on a fluid mantle. Convection currents in the mantle result from heat emanating from Earth's core. The currents cause continental drift, which is the gradual shift in the position of the plates and the continents they contain. The drifting of continents has had profound effects on climate, sea level, oceanic circulation, and the distributions of organisms.

- One major unidirectional change has been the increase in concentration of oxygen in Earth's atmosphere. No free oxygen was present in the atmosphere until certain prokaryotes evolved the ability to use water as a source of hydrogen ions for photosynthesis. The gradual increase in oxygen concentration in the atmosphere resulted in the dominance of organisms using aerobic metabolism and in the evolution of larger eukaryotic cells and of multicellular organisms.

- The climate of Earth has alternated between hot/humid and cold/dry conditions. Though most major climatic changes have been gradual, some have occurred within periods of 5,000 to 10,000 years or less.

- The climatic shifts caused by massive volcanic eruptions associated with continental drift are implicated in several mass extinctions, notably the one that occurred in the late Permian period at the end of the Paleozoic era.

- The mass extinction that occurred in the late Cretaceous period at the end of the Mesozoic era was probably caused by the collision of Earth with a large meteorite.

The fossil record reveals patterns in life's history.

- The earliest life on Earth appeared about 3.8 billion years ago, but the fossil record of organisms that lived in the Precambrian is fragmentary.

- The first organisms were unicellular prokaryotes. The first eukaryotes evolved about 1.5 billion years ago.

- Because an oxygen-rich environment favors rapid decomposition, organisms that become fossils are likely either to have lived in a poorly oxygenated environment or to have been transported to such a site soon after death.

- The 300,000 known fossil species represent only a tiny fraction of the species that have ever lived. Some groups, such as hard-shelled marine animals, are much better represented in the fossil record than others.

- The fossil record shows that an organism of any specific type can predictably be found in rocks of a particular age.

- Living organisms resemble recent fossils more closely than they resemble fossils from more ancient periods.

- According to the fossil record, all species eventually become extinct.

- The best Precambrian fossil deposits, dating from about 600 million years ago, contain diverse soft-bodied invertebrates, some of which may represent lineages with no living descendents.

Beginning with the Cambrian period, the Paleozoic era (543–245 mya) was marked by a dramatic increase in the diversity of life.

- During the Cambrian explosion, all of the major groups of animals that have living representatives appeared. Several continental plates began to come together to form the southern land mass known as Gondwana.

- The Ordovician period was marked by a proliferation of marine filter feeders living on the sea floor. Small plants appeared in moist terrestrial habitats. At the end of this period, massive glaciers formed over Gondwana, the sea level and ocean temperatures dropped, and the majority of animal species became extinct.

- The Silurian period witnessed the aggregation of northern continents, the evolution of swimming marine animals, and the appearance of the first terrestrial arthropods and vascular plants.

- During the Devonian period, all major groups of fishes evolved. On land, the first insects and amphibians evolved, and forests of club mosses, horsetails, and tree ferns appeared. A mass extinction of approximately three-quarters of all marine species occurred at the end of this period, possibly caused by the collision of two large meteorites with Earth.

- In the Carboniferous period, swamp forests largely consisting of giant tree ferns and horsetails were widespread. The fossilized remains of these plants formed coal. The period also saw the appearance of the first reptiles, which evolved from an amphibian ancestor, and of the first winged insects.

- The Permian period was marked by the formation of a single supercontinent called Pangaea. As the climate cooled, reptiles became the dominant land vertebrates, and the lineage of reptiles from which mammals later evolved became distinct. The occurrence of the most extensive of all mass extinctions brought the Permian period to a close.

During the Mesozoic era (245–65 mya), the continents that formed Pangaea slowly separated to form Laurasia and Gondwana, and distinct assemblages of plants and animals evolved on each continent.

- During the Triassic period, seed plants (seed ferns and conifers) were the dominant forms of terrestrial vegetation, and reptiles were the dominant vertebrates. A mass extinction occurred at the end of the Triassic.

- The Jurassic period witnessed the radiation of the dinosaurs, the evolution of flying reptiles, and the first appearance of mammals. Bony fishes became dominant in the sea.

- The Cretaceous period was marked by the complete separation of the northern and southern continents. The first flowering plants evolved from gymnosperm ancestors, and the dinosaurs continued to diversify and dominate, despite the presence of many groups of mammals. At the end of this period, the collision of a large meteorite with Earth caused the extinction of the dinosaurs and of many other animal and plant lineages.

Modern groups of plants and animals evolved during the Cenozoic era (65 mya–present).

- During the Tertiary period, the continents drifted toward their present positions. As the climate became cooler and drier, extensive grasslands appeared. Birds and mammals radiated extensively.

- The current geological period, the Quaternary, is divided into the Pleistocene and Holocene (recent) epochs. Modern humans evolved during the Pleistocene, which was a time of severe climatic fluctuations, including four major periods of extensive glaciation. As humans spread geographically, many species of large birds and mammals became extinct.

Three major faunas have dominated life on Earth.

- Life on Earth has been dominated by three major animal assemblages resulting from periods of rapid evolutionary diversification. The first was the fauna stemming from the Cambrian explosion; the second fauna dominated the later Paleozoic; and the third originated from the Triassic explosion and resulted in our present-day fauna. Organisms with new body plans appeared in the Cambrian explosion, but not in the others.

Rates of evolutionary change have been uneven.

- The fossil record shows that no single pattern characterizes evolutionary rates within lineages.

- Relatively unchanging environments favor evolutionary stability in the species of organisms ("living fossils") that are adapted to them.

- Many series of fossils show gradual change in lineages of organisms over time.

- Conversely, changes in the physical or biological environment of a species often trigger rapid evolution of new traits.

- Though species have become extinct throughout the history of life, extinction rates have varied widely, being relatively low during normal times and very high during periods of mass extinction. Some mass extinctions have hit both terrestrial and marine organisms, whereas others have primarily affected either one group or the other.

The causes of evolution that have operated in the past continue to operate, but humans are a new and dominant evolutionary agent.

- By changing Earth's vegetation, importing species into regions where they did not previously live, and manipulating the evolution of certain valuable species, humans are having an enormous influence on the future of life on Earth.

The Big Picture

- The evidence provided by nineteenth-century geologists that Earth is very ancient was crucial to Darwin's theory of evolution by natural selection.

- Radiometric dating is an excellent example of how a discovery in one field of science (nuclear physics, in this instance) can lead to advances in other fields (geology and paleontology). Prior to the development of this technique, no absolute dates could be accurately assigned to any of the important geological and evolutionary events described in this chapter.

- The concept of continental drift, first widely accepted in the late 1960s, has revolutionized our understanding of geological processes and of the history of life on Earth.

- The accumulating evidence of the sudden and catastrophic effect of the impact of large meteorites and of other rapid changes (e.g., widespread volcanic eruptions) on the environment of Earth has given paleontologists a new perspective on the causes of mass extinctions and on the dynamics of the evolutionary process in general.

Study Strategies

- This chapter contains many names of geological eras and periods and many dates. Instructors vary with respect to the importance they place on memorization of this information. In general, it is best to focus on broad evolutionary patterns and trends in studying this material. Most instructors will also place more importance on your learning phylogenetic sequences (e.g., mammals evolved from reptiles, which evolved from amphibians) than on your memorizing the dates at which these events occurred.

- Making a table or chart to summarize and organize the events that occurred in each geological period is an excellent study strategy.

- Review the following tutorial and activity on the Student Website/CD:
  Tutorial 22.1 Evolution of the Continents
  Activity 22.1 Concept Matching

Test Yourself

Knowledge and Synthesis Questions

1. The half-life of an isotope is best defined as the
   a. time a fixed fraction of isotope material will take to change from one form to another.
   b. age over which the isotope is useful for dating rocks.
   c. ratio of one isotope species to another in a sample of organic matter.
   d. None of the above

2. Mountain ranges are ultimately the result of
   a. plates in Earth's crust that move against one another on top of a fluid mantle.
   b. climate changes and the movement of glacial ice sheets.
   c. leftover debris from ancient collisions with an asteroid or meteor.
   d. the breakup of Laurasia and Gondwana.

3. Which of the following would likely be the best setting for fossilization to occur?
   a. The surf zone along a sand beach
   b. A shallow, cool swamp with good deposition rates of mud sediments
   c. The bottom of a hot, dry cave with no running water
   d. A fast-running mountain stream

4. Despite being incomplete as a whole, the fossil record is rather detailed for
   a. soft-bodied insects.
   b. cnidarians and sponges.
   c. most terrestrial animals.
   d. hard-shelled mollusks.

5. When biologists say that Earth's biota became provincialized, they mean that
   a. mass extinctions took place.
   b. continental drift did not occur.
   c. distinctive assemblages of plants and animals arose on different continents.
   d. reductions in species diversity took place.

6. All but one of the following occurred during the Ordovician period. Which statement is *not* correct?
   a. Marine filter feeders flourished.
   b. The number of classes and orders increased.
   c. Modern mammals appeared.
   d. Many groups became extinct at the end of the period.

7. Most of human evolution has occurred during the
   a. Paleozoic era.
   b. Devonian period.
   c. Quaternary period.
   d. Carboniferous period.

8. For terrestrial animals and plants, the most recent mass extinction event that occurred prior to the evolution of humans took place approximately
   a. 10 mya.
   b. 65 mya.
   c. 220 mya.
   d. 400 mya.

9. One of the main factors that distinguishes the Cambrian explosion from all others is that
   a. evolutionarily, it was the most recent explosion.
   b. many new major groups of animals appeared at this time, but not during other explosions.
   c. it was the time when the dinosaurs became extinct.
   d. there was a dramatic drop in species diversity, especially among marine organisms.

10. During which of the following geological times did the most new kinds of body plans appear?
    a. Paleozoic
    b. Triassic
    c. Jurassic
    d. Cambrian

11. The sudden disappearance of the dinosaurs some 65 mya may have been the result of
    a. Earth's collision with a large meteorite.
    b. slow climate changes due to planetary cooling.
    c. competition from better-adapted organisms.
    d. the rise of birds and mammals.

12. Which of the following statements about evolution is *not* true?
    a. Some species living in stable environments appear to have changed very little over many millions of years.
    b. Species have become extinct throughout the history of life.
    c. The fossil record of trilobites provides a good example of gradual evolutionary change within a lineage of organisms.
    d. There is no evidence in the fossil record for periods of rapid evolutionary change.

13. Which of the following statements about patterns or processes in the evolution of life is *not* true?
    a. ^{14}C can be used to date the age of dinosaur bones.
    b. The supercontinent Pangaea formed during the Permian period.
    c. All of the major groups of animals that have species living today appeared in the Cambrian period.
    d. Rates of evolutionary change can be rapid during times of dramatic change in physical environments.

14. Which of the following pairs of organisms were *not* present on Earth in their living forms at the same time?
    a. Tree ferns and bony fish
    b. Amphibians and birds
    c. Gymnosperms and insects
    d. Humans and dinosaurs

Application Questions

1. ^{14}C decays to ^{12}C with a half-life of about 5,700 years. Suppose you find a fossil in which the amount of ^{14}C is only one-sixteenth of what one would find in a living organism with the same carbon content. Approximately how old is the fossil?

2. For each geological period, make a chart summarizing its major geological and evolutionary events.

3. Describe several ways in which the dramatic increase in Earth's human population is affecting the evolutionary process.

Answers to Questions

Knowledge and Synthesis Answers

1. **a.** Radioactive decay is measured as the time it takes for one-half the amount of a substance to spontaneously convert into another substance.

2. **a.** As the crustal plates are pushed together, one may move underneath the other, pushing up mountain ranges.

3. **b.** Fossilization occurs best in areas of low oxygen concentration and rapid sedimentation, where scavengers cannot destroy the body.

4. **d.** Hard-shelled animals such as mollusks are good candidates for fossilization because their shells can withstand decay for long enough to become buried. Many also tend to live in quiet, shallow waters.

5. **c.** Coming from the same root word as *province*, which is a specific area within a country or state possessing unique characteristics, provincialized biotas are geographically unique assemblages of organisms produced by a variety of physical and biological isolating factors.

6. **c.** The modern mammals did not appear until the Tertiary, more than 350 million years after the Ordovician.

7. **c.** The Pleistocene epoch, which is part of the Quaternary, was the time of most hominid evolution.

8. **b.** The mass extinction at the end of the Cretaceous, 65 mya, was the most recent affecting terrestrial life before the evolution of humans.

9. **b.** The Cambrian explosion produced many novel groups of animals characterized by distinctive body plans. Later explosions caused an increase in diversity only within already existing major lineages.

10. **d.** For the reasons outlined in the previous answer, the Cambrian period is the correct answer.

11. **a.** Paleontologists believe that a large meteorite's collision with Earth so altered conditions that the dinosaurs rapidly succumbed during what we know as the great Cretaceous mass extinction.

12. **d.** In some lineages, evolutionary rates apparently have changed through time. Long periods of little or no change have been interrupted by periods of rapid evolution.

13. **a.** Carbon dating is generally only reliable for fossils less than 50,000 years old.

14. **d.** Dinosaurs disappeared at the end of the Cretaceous, over sixty million years before humans evolved.

Application Answers

1. If only $\frac{1}{16}$ of the ^{14}C remains, then four ^{14}C half-lives have passed since the fossil was formed ($\frac{1}{2} \times \frac{1}{2} \times \frac{1}{2} \times \frac{1}{2} = \frac{1}{16}$). Because the half-life of ^{14}C is roughly 5,700 years, the fossil must be about 22,800 years old ($4 \times 5,700 = 22,800$).

2. See Table 20.1 in the text.

3. By dramatically altering Earth's vegetation from forests and grasslands to crops and pastures, humans are causing the extinction of many species, both large and small. By transporting species around the world, humans are reversing the independent evolution of Earth's biota on separate continents. Lastly, by artificial selection and biotechnology (including gene transfer between species) humans are directly determining how certain species are evolving.

23 *The Mechanisms of Evolution*

Important Concepts

Darwin's main contribution to biology was to develop the theory of evolution by natural selection.

- The theory of evolution by natural selection rests on two facts:
  - All populations have the capacity to increase in number exponentially. Because populations rarely increase rapidly, high death rates must balance the potential growth rates.
  - All populations show variations among individuals with regard to numerous inherited characteristics, some of which may affect their chances of surviving and reproducing.
- From these two facts, Darwin made the following inference: Those individuals whose heritable traits enable them to survive and reproduce more successfully than others will pass those traits on to greater numbers of offspring.
- In short, natural selection is differential reproductive success among individuals of a population.
- An adaptation is a trait that helps its bearer survive and reproduce; the term also refers to the evolutionary process that produces such traits.
- The rediscovery of Mendel's laws of inheritance led to the development of population genetics, which applies these laws to entire populations to understand the genetic basis of evolutionary change.

Evolution is any genetic change in a population from generation to generation.

- A genotype is the genetic constitution of a particular trait of an individual. Evolution occurs when individuals with different genotypes survive or reproduce at different rates.
- Agents of evolution actually act on the phenotype, which is the physical expression of the genotype.
- The features of a phenotype are its *characters*. A *trait* is a specific form of a character.
- The gene pool of a population is the sum of all the alleles (alternative versions of a gene) found within it. Although a single diploid individual can have at most two different alleles for any gene, a population may contain many alleles for that gene. The gene pool contains all the variations that result in individuals with differing phenotypes on which agents of evolution act.

- Studies of both domestic and wild species show that virtually all populations contain a certain amount of genetic variation.
- Genotypes do not uniquely determine phenotypes. For example, if one allele is dominant to another, two genotypes (e.g., AA and Aa) can determine the same phenotype. Conversely, one genotype can produce different phenotypes because of the interaction of genetic and environmental factors during development.
- A Mendelian population is a locally interbreeding group within a geographic population. Such groups are often the subject of evolutionary studies.
- Allele frequencies measure the amount of genetic variation in a population, whereas genotype frequencies show how this variation is distributed among its members. Populations that have the same allele frequencies may nevertheless differ in their genotype frequencies.
- The sum of all allele frequencies at a locus is equal to 1, as is the sum of all genotype frequencies. For a locus with two alleles, p and q are typically used to represent the frequencies of the dominant and recessive alleles, respectively, and thus $p + q = 1$. Biologists estimate allele frequencies by measuring numbers of alleles in a sample of individuals from the population.
- The genetic structure of a population is described by the frequencies of different alleles at each locus and the frequencies of different genotypes.

A population that is not changing genetically is at Hardy–Weinberg equilibrium.

- To be at Hardy–Weinberg equilibrium, a population must meet five conditions:
  - Mating must be random.
  - The population must be very large.
  - There must be no migration into our out of the population.
  - There must be no mutation.
  - Natural selection must not be acting on the alleles in question.
- If these conditions are met, allele frequencies at a locus remain the same from generation to generation. Moreover, after one generation of random mating, the genotype frequencies will remain in the proportions $p^2 + 2pq + q^2 = 1$, where p^2, $2pq$, and q^2 represent the

frequencies of the homozygous dominant, heterozygous, and homozygous recessive genotypes, respectively.

- Though populations in nature rarely meet the conditions of the Hardy–Weinberg equilibrium, this equation is important because deviations from it may show that evolution is occurring. Moreover, the pattern of deviations is useful in identifying the agents of evolutionary change operating on the population.

Evolutionary agents are forces that change the allele and genotype frequencies in a population.

- The ultimate source of genetic variation in a population is mutation. Because mutations are random changes in genetic material, most are harmful or neutral, but it is the environment that determines whether a particular mutation is disadvantageous or adaptive.

- Though mutations cannot be totally prevented, their frequency is usually so low as to cause little deviation from Hardy–Weinberg expectations.

- Gene flow occurs when individuals migrate from one population to another and breed in their new location. Gene flow can add new alleles to a population's gene pool, change the frequencies of alleles already present, or both.

- Genetic drift is caused by chance events that alter allele frequencies in a population. It has its greatest impact on small populations, in which it may even cause harmful alleles to increase in frequency.

- During a population bottleneck, when a large population is severely reduced in size, allele frequencies may shift drastically, and genetic variation may be reduced as a result of genetic drift.

- The founder effect occurs when a few individuals originate a new population. As in the case of a population bottleneck, some alleles found in the source population will be missing from the founding population, and others will occur with altered frequencies.

- Nonrandom mating occurs when individuals mate preferentially with others either of the same genotype or of a different genotype. The effect of nonrandom mating is to cause either fewer or more heterozygotes to be found in a population than would be expected in a population in Hardy–Weinberg equilibrium. The effect of self-fertilization, another form of nonrandom mating, is to reduce the frequency of heterozygotes. In any type of nonrandom mating, the allele frequencies remain the same despite the changes in genotype frequencies.

- Fitness is the contribution of a phenotype to the composition of later generations, relative to the contribution of alternative phenotypes. The fitness of a phenotype is determined by the average rates of survival and reproduction of individuals with that phenotype, as compared to individuals with other phenotypes.

- Differences in fitness among phenotypes (and hence in the controlling genotypes) lead to changes in allele frequencies in the gene pool of later generations. Changes in allele frequencies from one generation to the next constitute evolution.

- Natural selection, unlike other agents of evolution, adapts organisms to their environment. In cases in which the distribution of a phenotype approximates a bell-shaped curve because it is controlled by many gene loci, selection can produce any one of three results:

  - Stabilizing selection reduces variation in the population by favoring average individuals.

  - Directional selection changes the mean value for a character by favoring individuals that vary in one direction.

  - Disruptive selection, by favoring both extremes, leads to a population with two peaks in the distribution of the character.

- Sexual selection favors traits that benefit their bearers (generally males) in the competition for access to members of the other sex, or make their bearers more attractive to members of the other sex. Sexual selection often results in sexually dimorphic species, in which males and females differ in appearance.

- The evolution of adaptations depends on the trade-off between costs and benefits.

Most populations maintain substantial genetic diversity.

- Genetic drift, stabilizing selection, and directional selection tend to lessen the genetic variation of a population.

- By generating new combinations of alleles, sexual reproduction amplifies existing genetic variation.

- Neutral mutations—that is, mutations that do not affect the fitness of an individual—tend to accumulate and thereby increase the genetic variation in a population.

- Frequency-dependent selection, in which the less-common genotype or phenotype is favored by natural selection, preserves variation as a polymorphism.

- Geographically distinct subpopulations of a population often vary genetically because they are adapted to different environments.

There are constraints on the evolutionary process.

- Evolutionary changes must be based on modifications of previously existing traits.

Cultural evolution involves learning rather than heredity.

- In cultural evolution, individuals acquire new traits by learning them from other individuals. Cultural evolution occurs most rapidly in human beings but has been observed in other species, notably apes.

Studies of short-term evolutionary processes do not allow us to predict the long-term course of evolution.

- Long-term evolutionary changes are strongly influenced by events that occur so infrequently or so slowly that they are rarely observed during short-term evolutionary studies. To understand the long-term course of evolution, biologists seek evidence of rare events and search for trends in the fossil record.

The Big Picture

- Most biologists rank Darwin's *The Origin of Species* as the most significant book ever written about their subject. Its importance is twofold. First, it presents a vast amount of evidence for the fact of evolution. Second, it describes a mechanism—natural selection—to explain evolutionary change. Evolution through natural selection remains perhaps the most important unifying concept in biology.

- Population genetics unites the concepts of Darwin with the insights into the mechanisms of heredity provided by Mendel. It enables biologists to study evolution with quantitative rigor.
- The loss of genetic variation that occurs during a population bottleneck is a major concern for biologists working to preserve endangered species. As the discussion of the greater prairie chicken makes clear, genetic uniformity in a population renders it more vulnerable to extinction.

Common Problem Areas

- A common mistake made by students when attempting to solve Hardy–Weinberg problems is to use the wrong equation. For example, if you are given phenotypic frequencies for a trait that is at genetic equilibrium and shows dominance, remember that the frequency of the recessive phenotype (and genotype) is equal to q^2, not q. By taking the square root of q^2, you can obtain the frequency of the recessive allele and can then determine the frequency of the dominant allele by subtraction ($p = 1 - q$).
- It is important to bear in mind that for any gene locus with two alleles, the allele frequencies must total unity (i.e., it is always the case that $p + q = 1$). The equation that specifies genotype frequencies ($p^2 + 2pq + q^2 = 1$) is only valid for a population at genetic equilibrium.

Study Strategies

- Population genetics is learned best by doing problems, so be sure you can solve problems similar to those presented in the "Test Yourself" section of this chapter of the *Study Guide*. Your instructor may provide a worksheet with additional examples.
- Review the following tutorials on the Student Website/CD:
Tutorial 23.1 Natural Selection
Tutorial 23.2 Hardy–Weinberg Calculator
Tutorial 23.3 Producing and Maintaining a Chemical is Costly

Test Yourself

Knowledge and Synthesis Questions

1. Evolution occurs at the level of
   a. the individual genotype.
   b. the individual phenotype.
   c. environmentally based phenotypic variation.
   d. the population.

2. What does natural selection act on?
   a. The gene pool of the species
   b. The genotype
   c. The phenotype
   d. Multiple gene inheritance systems

3. Which choice would form a statement about Mendelian populations that is *not* true? A Mendelian population must
   a. consist of members of the same species.
   b. have members that are capable of interbreeding.
   c. show genetic variation.
   d. have a gene pool.

4. In comparing several populations of the same species, the population with the greatest genetic variation would have the
   a. greatest number of genes.
   b. greatest number of alleles per gene.
   c. greatest number of population members.
   d. largest gene pool.

5. The ability to taste the chemical PTC (phenylthiocarbamide) is determined in humans by a dominant allele T, with tasters having the genotypes Tt or TT and nontasters having tt. If you discover that 36 percent of the members of a population cannot taste PTC, then according to the Hardy–Weinberg rule, the frequency of the T allele should be
   a. 0.4.
   b. 0.6.
   c. 0.64.
   d. 0.8.

6. A gene in humans has two alleles, M and N, that code for different surface proteins on red blood cells. If you know that the frequency of allele M is 0.2, according to the Hardy–Weinberg rule, the frequency of the genotype MN in the population should be
   a. 0.16.
   b. 0.32.
   c. 0.64.
   d. 0.8.

7. If the frequency of allele b in a gene pool is 0.2, and the population is in Hardy–Weinberg equilibrium, the expected frequency of the genotype $bbbb$ in a tetraploid ($4n$) plant species would be
   a. 0.0016.
   b. 0.04.
   c. 0.08.
   d. 0.2.

8. Random genetic drift would probably have its greatest effect on which of the following populations?
   a. A small, isolated population.
   b. A large population in which mating is nonrandom.
   c. A large population in which mating is random.
   d. A large population with regular immigration from a neighboring population.

9. Allele frequencies for a gene locus are *least* likely to be significantly changed by
   a. nonrandom mating.
   b. the founder effect.
   c. mutation.
   d. gene flow.

10. Select all of the following evolutionary agents that produce nonrandom changes in the genetic structure of a population.
    a. Self-fertilization
    b. Population bottlenecks
    c. Mutation
    d. Natural selection

11. Suppose a particular species of flowering plant that lives only one year can produce red, white, or pink blossoms, depending on its genotype. Biologists studying a population of this species count 300 red-flowering, 500 white-flowering, and 800 pink-flowering plants in a population. When the population is censused the following year, 600 red-flowering, 900 white-flowering, and 1,000 pink-flowering plants are observed. Which color has the highest fitness?
    a. Red
    b. White
    c. Pink

12. The following graph shows the range of variation among population members for a trait determined by multiple genes.

If this population is subject to *stabilizing selection* for several generations, which of the distributions (a–d) is most likely to result?

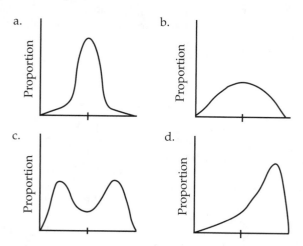

13. In areas of Africa in which malaria is prevalent, many human populations exist in which the allele that produces sickle-cell disease and the allele for normal red blood cells occur at constant frequencies, despite the fact that sickle-cell disease frequently causes death at an early age. This is an example of
    a. the founder effect.
    b. a stable polymorphism.
    c. mutation.
    d. nonrandom mating.

Application Questions

1. Discuss the main use of the Hardy–Weinberg rule in evolutionary biology.

2. The Hardy–Weinberg expression is an expansion of a binomial or $(p + q)^2 = p^2 + 2pq + q^2$. In this equation, the left-hand side represents _____ frequencies, and the right-hand side represents _____ frequencies.

3. In a population with 600 members, the numbers of individuals of three different genotypes are $AA = 350$, $Aa = 100$, $aa = 150$.
    a. What are the genotype frequencies in this population?
        1. $AA =$
        2. $Aa =$
        3. $aa =$
    b. What are the allele frequencies in this population?
        1. $A =$
        2. $a =$
    c. What would be the expected genotype frequencies if this population were in genetic equilibrium?
        1. $AA =$
        2. $Aa =$
        3. $aa =$
    d. Is this population in genetic equilibrium? Explain.

4. The following data were collected in a study in which dark and light moths of the peppered moth (*Biston betularia*) were released and recaptured a few days later in several different areas (*a*, *b*, *c*, and *d*). Based on these data, in which area is the light phenotype most advantageous? (Assume that recapture rates reflect the relative survival of the moths, not the ability of the investigators to find and recapture the moths.) What does this tell us about the fitness of light moths in this area?

Area	Moth type	Released	Recaptured
a	dark	100	25
	light	200	50
b	dark	100	25
	light	200	100
c	dark	200	50
	light	400	100
d	dark	200	100
	light	300	75

5. As peppered moths rest on bark surfaces during the day, they are subject to predation by birds. Assuming that moths that are more camouflaged will be preyed on less, what can you say about the color of bark surfaces in area *d*?

6. Describe what is meant by sexual selection and how the sexual selection of tail length in male widowbirds was studied.

Answers

Knowledge and Synthesis Answers

1. **d.** Evolution is defined as changes in the genetic structure of a population over time, so evolution occurs at the level of the population.

2. **c.** Natural selection acts on phenotypes, not genotypes. For example, a harmful recessive allele is "invisible" to natural selection when it occurs in a heterozygote, where its harmful effect is masked by the dominant allele.

3. **c.** Although most populations show genetic variation, this condition is not part of the definition of a Mendelian population.

4. **b.** Genetic variation is related to the number of different alleles per gene. Regarding choice **a**, recall that all species members have the same number of genes. Population size per se has little to do with genetic variation, so choices **c** and **d** are also incorrect.

5. **a.** Recall that $q^2 = 0.36$ is the frequency of the tt genotype, so q (0.6, the square root of 0.36) is the frequency of the t allele. Assuming there are only two alleles for this trait, $T + t = 1$. The frequency of the T allele is therefore $1 - t = 1 - 0.6 = 0.4$.

6. **b.** Because $p = 0.2$ and therefore, $q = 1 - 0.2 = 0.8$, the frequency of the MN genotype is $2pq$, or $2 \times 0.2 \times 0.8 = 0.32$.

7. **a.** The probability of one allele b in a genotype is equal to its frequency, or 0.2, so the probability that all four of the alleles in a tetraploid organism will be b would be $(0.2)^4$, or 0.0016.

8. **c.** Genetic drift is most significant in small populations.

9. **a.** Unlike the founder effect, mutation and gene flow, all of which change allele frequencies in a population, nonrandom mating only causes a deviation from the frequency of heterozygotes predicted by Hardy–Weinberg equilibrium.

10. **a and d.** Self-fertilization leads to increased numbers of homozygous individuals, and natural selection is also nonrandom.

11. **a.** Fitness measures the relative contribution of a genotype or phenotype to subsequent generations. The red-flowering plants, which doubled in number, had the greatest percentage increase of any of the plants and thus had the highest fitness.

12. **a.** Stabilizing selection results when individuals that are intermediate in phenotype make a larger contribution to future generations than individuals of more extreme phenotype. This leads to reduced variation for the trait and causes the curve to be higher and narrower. Curve b shows greater variation, curve c would result from disruptive selection, and curve d from directional selection.

13. **b.** Polymorphism in a population is the existence of two or more alleles at a particular gene locus at frequencies greater than mutations can produce. If phenotypes are stable through time, then the underlying alleles will also be constant. In this instance, the polymorphism is stable because malaria is a significant cause of mortality in some parts of Africa, and heterozygotes have greater resistance to this disease than individuals with a "normal" phenotype. Thus the superior fitness of the heterozygotes maintains both alleles in the population.

Application Answers

1. Most populations do not meet the Hardy–Weinberg conditions for genetic equilibrium. However, by determining how a population deviates from the expectations of Hardy–Weinberg equilibrium, evolutionary biologists can identify what evolutionary agents are affecting the population.

2. allele; genotype

3. a.
   1. $AA = 350/600 = 0.58$
   2. $Aa = 100/600 = 0.1$
   3. $aa = 150/600 = 0.25$

   b.
   1. $A = (700 + 100)/1200 = 0.67$
   2. $a = (300 + 100)/1200 = 0.33$

   c.
   1. $AA = p^2 = (0.67)^2 = 0.45$
   2. $Aa = 2pq = 2 \times 0.67 \times 0.33 = 0.44$
   3. $aa = q^2 = (0.33)^2 = 0.11$

   d. No. The observed genotypic frequencies differ from those predicted by the Hardy–Weinberg rule using the observed allele frequencies for p and q. The population is *not* in genetic equilibrium.

4. Area b. In areas a and c recaptures were 25% of releases for all moths, but in area b, 50% of the light moths were recaptured, indicating that moths with this phenotype survived best in this area. This suggests that fitness will be higher for light moths in this area, but it won't be possible to determine this for certain until the moths reproduce because fitness, by definition, depends on the relative contribution to subsequent generations (it is possible, for example, that light moths will have a harder time finding mates in this area).

5. Dark moths seem to survive best in area d. If the increased survival of the moths is due to cryptic coloration, then the bark surfaces in that area should also be dark.

6. Sexual selection is the spread of a trait that improves the reproductive success of an individual as determined by mating preferences of individuals of the opposite sex. By artificially lengthening or shortening the tails of male widowbirds, behavioral ecologists were able to show that increased tail length attracted more females but did not confer an advantage in interactions with other males.

24 *Species and Their Formation*

Important Concepts

Species are the fundamental units of evolutionary diversification and of the biological classification system.

- The morphological concept of species used by early biologists grouped organisms into species on the basis of their appearance. For some groups of organisms, morphological similarity is still the major criterion.

- The biological species concept widely used by modern biologists regards a species as a genetic unit, the members of which can interbreed among themselves but are unable to interbreed successfully with members of other such units because of their genetic differences.

- Determining whether two populations constitute different species may be difficult because speciation is frequently a gradual process.

Speciation is the process by which one species splits into two species.

- Evolutionary change can occur without speciation. A single lineage may change through time without diverging into two species.

- Speciation requires that the gene pool of the original species divide into two isolated gene pools. After separation, the agents of evolution cause sufficient genetic differences to accumulate in the populations for them to be unable to interbreed successfully if they come together again.

- In allopatric speciation, the population is initially divided by a geographic barrier. If the barrier results from a geological or climatic change, the two isolated populations are often large and genetically similar but diverge because the environments in which they live are, or become, different. Alternatively, separation may result from a founder event, in which a relatively small group of individuals crosses a barrier and founds a new population. Because it is small, the founding population will usually lack many of the alleles found in the parent population. Natural selection and other agents of evolution will then cause additional genetic divergence of the populations.

- The effectiveness of a barrier in preventing gene flow depends on the size and mobility of the species.

- Evidence suggests that allopatric speciation is the most common mechanism of speciation among animals.

- Sympatric speciation occurs without geographic subdivision of the gene pool of the original species. Polyploidy, an increase in the number of sets of chromosomes, produces new species because the polyploid organisms cannot interbreed with members of the parent species. Polyploid species that have a single ancestor are called autopolyploids, whereas those that have resulted from the hybridization of two species are referred to as allopolyploids.

- New species arise by polyploidy much more easily among plants than among animals because plants of many species can reproduce by self-fertilization.

Speciation requires that reproductive isolating mechanisms evolve to prevent previously allopatric populations from exchanging genes if and when they become sympatric.

- Prezygotic barriers prevent members of different species from mating.

  - Species may simply mate in different areas or different parts of a habitat (spatial isolation).

  - Species may not be able to interbreed because they are fertile at different times (temporal isolation).

  - Differences in reproductive organs may prevent interbreeding (mechanical isolation).

  - The sperm and egg may be chemically incompatible (gametic isolation).

  - The two species may not recognize or respond to each other's mating behaviors, or in flowering plants, the behavioral preferences of the pollinating animals may prevent interbreeding (behavioral isolation).

- Postzygotic barriers can prevent effective gene flow between species, even if mating occurs.

  - Hybrid zygotes may not mature normally (hybrid zygote abnormality).

  - Hybrids may be infertile (hybrid infertility).

  - Hybrids may survive less well than either parent species (low hybrid viability).

- The evolution of more effective prezygotic reproductive barriers is known as reinforcement. It may occur if the hybrid offspring of two species survive poorly.

- Because allopatric speciation occurs gradually, we can observe various stages of the process in present-day populations.

If two populations reestablish contact before reproductive isolation is complete, several results are possible.

- If hybrid offspring are not at a selective disadvantage, they may spread through both populations with the result that the gene pools of the populations combine. Thus no new species would result from the period of isolation.

- If hybrid offspring are less successful, reinforcement may strengthen prezygotic reproductive barriers.

- If hybrid offspring are at a disadvantage but reinforcement fails to occur, a stable, narrow hybrid zone may form.

- The genetic differences that separate closely related species are primarily among the small number of genes involved with reproductive isolation.

Several factors affect the rate of speciation of different evolutionary lineages.

- Characteristics of a species that make it prone to speciation include membership in a large evolutionary lineage, poor dispersal ability, ecological specialization, passage through a population bottleneck, complex behavior, and pollination by animals.

- Periods of environmental change that fragment formerly continuous habitats favor speciation.

In an evolutionary radiation, the speciation rate greatly exceeds the extinction rate within a lineage, and many daughter species arise.

- The native biota of the Hawaiian Islands illustrates that populations colonizing environments that have relatively few species are particularly likely to produce evolutionary radiations.

The Big Picture

- Speciation is the process that has produced the millions of life forms—each adapted to a particular environment and way of life—that constitute life on Earth.

- Archipelagos such as the Galápagos and Hawaiian Islands have been called natural laboratories of evolution. Studies of the evolutionary radiations of the Galápagos finches and of several Hawaiian groups, such as *Drosophila* and the silverswords, have provided crucial insights to evolutionary biologists since the time of Darwin.

- Chapter 57, "Conservation Biology," discusses the reasons for the rapid pace of human-caused species extinctions, including the disappearance of many members of groups that have undergone evolutionary radiations, such as Australian marsupials and Hawaiian honeycreepers. It also describes some strategies for the preservation of Earth's precious biological diversity.

Study Strategies

- Because most instructors consider the heart of this chapter to be the discussions of allopatric (geographic) speciation and reproductive isolating mechanisms, you should pay particular attention to these topics.

- Review the following tutorials and activity on the Student Website/CD:
Tutorial 24.1 Speciation Mechanisms

Tutorial 24.2 Founder Events Lead to Allopatric Speciation
Activity 24.1 Concept Matching

Test Yourself

Knowledge and Synthesis Questions

1. Select all the correct answers that complete the following sentence: It is difficult to apply the biological species concept to groups of organisms that
   a. are asexual.
   b. produce hybrids only in captivity.
   c. show little morphological diversity.
   d. exist only in the fossil record.

2. Which of the following statements about allopatric speciation is *not* true?
   a. Allopatric speciation can sometimes involve small populations.
   b. Allopatric speciation only occurs in species that are widely distributed.
   c. Allopatric speciation always involves a physical barrier that interrupts gene flow.
   d. Allopatric speciation can sometimes involve chance events.

3. A long, narrow hybrid zone exists between the ranges of the fire-bellied toad and the yellow-bellied toad in Europe. Which of the following is a factor contributing to the persistence of this zone?
   a. Reinforcement strengthens the prezygotic barriers between the two species.
   b. Hybrid offspring have the same fitness as nonhybrid offspring.
   c. Both species travel long distances over the course of their lives.
   d. Individuals from outside the hybrid zone regularly move into the hybrid zone.

4. Which type of speciation is most common among flowering plants?
   a. Geographic
   b. Sympatric
   c. Allopatric

5. Which of the following would *not* be considered an example of a prezygotic reproductive isolating mechanism?
   a. One bird species forages in the tops of trees for flying insects while another forages on the ground for worms and grubs.
   b. The males of one species of moth cannot detect and respond to the sex attractant chemicals produced by the females of another species.
   c. Sperm of one species of sea urchin are unable to penetrate the egg plasma membrane of another species.
   d. Mosquitoes of one species are active in foraging and searching for mates at dusk, whereas those of another species are active at dawn.

6. Which of the following factors would *not* be expected to increase the rate of speciation in a group of organisms?
   a. Fragmentation of populations
   b. Poor dispersal ability
   c. High birthrates
   d. Increased behavioral complexity

7. Which of the following is *not* a suggested reason for the evolutionary radiation of silverswords on the Hawaiian archipelago?
   a. Water is an effective barrier for many organisms.
   b. Because islands are small compared with mainland areas, you would expect more species to develop there.
   c. Competition is frequently reduced on islands.
   d. More ecological opportunities exist on islands that have not been colonized by many species.

8. Studies of species of Hawaiian *Drosophila* show that
   a. sympatric speciation via polyploidy is common in insects.
   b. few genes need to be involved in establishing reproductive isolation.
   c. species with nonoverlapping ranges can be very similar.
   d. a geographic barrier is not always necessary for the establishment of reproductive isolation.

9. Which of the following statements about speciation is *not* true?
   a. A small founding population can be involved in speciation.
   b. Speciation always involves interruption of gene flow between different groups of organisms.
   c. The rate of speciation can vary for different groups of organisms.
   d. Speciation cannot occur in a single generation.

10. Which of the following observations constitutes conclusive evidence that two overlapping populations that had been geographically separated have *not* diverged into distinct species?
    a. Matings between members of the two populations produce viable hybrids.
    b. A stable hybrid zone exists where their ranges overlap.
    c. Interbreeding is common between members of the two populations.
    d. None of the above

Application Questions

1. Discuss the conditions on the Galápagos Islands that led to the evolution of the birds known as Darwin's finches.

2. In autopolyploidy, a new species of plant can arise by the doubling of chromosome number in a single individual of one species (provided that the individual is capable of self-fertilization). Why is it virtually impossible for such a tetraploid plant to interbreed successfully with diploid individuals of the "same" species?

3. Discuss what is known about the evolutionary radiations on islands, based on studies of Hawaiian silverswords and tarweeds.

4. Suppose that members of two populations are separated by a geographic barrier and begin to diverge genetically. Many generations later, when the barrier is removed, the two populations can interbreed, but the hybrid offspring do not survive and reproduce well. Explain how natural selection might lead to the evolution of more effective prezygotic barriers in these species.

Answers

Knowledge and Synthesis Answers

1. **a** and **d.** The key criterion of a biological species is that its members be reproductively isolated from other such groups. This criterion is impossible to evaluate in asexual and fossil species.

2. **b.** A wide distribution is not a requisite for allopatric speciation.

3. **d.** These toads provide an example of related species in which reinforcement does not strengthen prezygotic barriers, even though hybrid offspring are only half as fit as nonhybrid offspring. The reason for this is that toads from outside the hybrid zone (not subject to the selective pressure against hybridizing) regularly move into the hybrid zone and mate with members of the other species. The hybrid zone remains narrow because toads do not travel long distances; natural selection removes hybrids from the population before they can disperse very far.

4. **b.** Sympatric speciation is most common among flowering plants. It has been estimated that about 70 percent of all flowering plant species are polyploid.

5. **a.** Provided that the two species are active in the same locality at the same time, a difference in the habitat in which they forage would not in itself be a barrier to interbreeding (though seeking mates in different habitats might well be a barrier to interbreeding). All of the other choices describe reproductive barriers that would act prior to fertilization.

6. **c.** Birthrates per se do not seem to affect the rate of speciation in organisms. All other factors have been shown to increase speciation rates in the lineages of some organisms.

7. **b.** Actually, biogeographers have found that larger land masses tend to have more species than smaller land masses, so you might expect the reverse effect.

8. **b.** Studies of the Hawaiian *Drosophila* show that these species have only slight genetic differences, despite being morphologically diverse.

9. **d.** New species formed by polyploidy can arise in a single generation.

10. **d.** Interbreeding, production of viable hybrids, and establishment of a hybrid zone do not necessarily mean that speciation is not complete. If, on the other hand, the hybrids were successful, fertile, and bred freely with members of both original populations, their gene pools would merge, and you would conclude that speciation had not taken place.

Application Answers

1. The relatively great distance between the Galápagos Islands and the South American mainland and also between each of the islands in the archipelago ensured that once immigrants had arrived on an island, they would be genetically isolated for a substantial period of time. Also, because the islands differ greatly in climate and vegetation, the resident birds were subject to different selection pressures. This, in combination with reduced gene flow between the islands, led to a rapid evolutionary radiation of finches.

2. Recall from Chapter 9 that pairs of homologous chromosomes synapse during prophase and metaphase of the first division of meiosis. The homologs then separate so that each cell resulting from meiosis I is haploid, as are the products of meiosis II. In tetraploids as in diploids, meiosis is normal because pairing of homologs can occur.

Any offspring of a cross between tetraploid and diploid individuals would be triploid, however, and therefore sterile because correct synapsis of homologs could not occur. Because the tetraploid product of autopolyploidy is reproductively isolated from its diploid relatives, it is a new species.

3. Studies of the silverswords of the Hawaiian archipelago show that taxa that evolve on islands frequently show great morphological diversity because of the reduced competition that immigrants encounter on islands. Thus Hawaiian silverswords have evolved tree- and shrublike species because there were few resident tree and shrub species with which they had to compete, unlike their mainland tarweed relatives.

4. Recall that natural selection tends to remove from a population traits that reduce survival or reproductive success. Individuals that interbreed between populations will have lower fitness (they will contribute fewer offspring to future generations) than those who breed within their own population. If the tendency to avoid interbreeding is heritable (and not just the result of chance), the frequency of alleles that prevent interbreeding will increase in each population. How might such a trait be heritable? Any of the prezygotic barriers to interbreeding might be heritable traits. For example, if the species in question are frogs, and if their mating calls started to diverge while the populations were separated, the following traits might be heritable: a tendency to make a call that is more distinct from that of the other population, or the ability to distinguish between the existing calls of the two populations (coupled with a preference for the call of one's own population). As alleles for these traits increased in frequency, they would contribute to the behavioral isolation of the two populations and perhaps eventually to more complete speciation.

25 Reconstructing and Using Phylogenies

Important Concepts

Systematics is the scientific study of the diversity of organisms.

- Taxonomy, a subdivision of systematics, is the theory and practice of classifying organisms.

Phylogenetic trees show the evolutionary patterns of life on Earth.

- A phylogeny is a hypothesis that describes how a group of organisms has descended from a common ancestor. Phylogenetic trees display the order in which lineages are hypothesized to have split. They are constructed by analyzing evolutionary changes in the traits of organisms.

- Homologous traits are keys to reconstructing phylogenetic trees because they are features that are shared by members of a lineage owing to their descent from a common ancestral trait. A trait that differs from its ancestral form is called a derived trait.

- Homoplastic traits create confusion when reconstructing the evolutionary history of a lineage because they are features that are similar for some reason other than descent from a common ancestral trait.

- Two processes generate homoplastic traits: convergent evolution and evolutionary reversals. In convergent evolution, features that were independently evolved become superficially similar. In an evolutionary reversal, a character reverts from a derived state to an ancestral one.

- The ingroup is the assemblage of organisms whose phylogeny is to be determined. One method of distinguishing ancestral and derived traits is to compare the ingroup to an outgroup—a lineage that is closely related to the ingroup but that splits off from the lineage of the ingroup below its base on the evolutionary tree. Ancestral traits are expected to be found in both the ingroup and the outgroup, whereas derived traits should occur only in the ingroup.

Reconstructing phylogenies accurately involves several steps.

- The first step in reconstructing a phylogeny is to choose the ingroup and the appropriate outgroup.

- Any feature of the organisms in the ingroup that will be used to analyze its evolutionary history is called a character (e.g., flower color); a trait is a particular form of the character (e.g., white flowers). To determine the group's evolutionary history, systematists must distinguish between ancestral and derived traits and between homologous and homoplastic traits. The more characters that are measured, the more likely it is that the phylogeny will be accurate.

- Morphology—the sizes and shapes of body parts—is an important source of traits for phylogenetic analysis. Structures in early developmental stages sometimes reveal evolutionary relationships that are not evident in adults.

- The morphology of fossils is particularly useful in helping to distinguish ancestral and derived traits. The fossil record also reveals when lineages diverged.

- Like morphological characters, molecules are heritable characteristics that may diverge among lineages. Thus, comparisons of the amino acid sequences of homologous proteins or of the base sequences of comparable regions of DNA provide excellent evidence of evolutionary relationships.

- In reconstructing a phylogeny, systematists assume that organisms that share unique derived traits have a common ancestor not shared with organisms lacking those traits.

- A phylogenetic tree does not describe the ancestors or date the splits between lineages; it simply shows the sequential order of the splits.

- In reconstructing phylogenies, systematists use the parsimony principle, which states that one should prefer the simplest hypothesis capable of explaining the known facts. In practice, this means that the phylogenetic reconstruction that assumes the smallest number of evolutionary changes is the best hypothesis. Complex computer programs are now employed by systematists to determine the shortest possible phylogenetic tree—that is, the one with the fewest homoplasies. In some cases, several phylogenies may be equally likely on the basis of the parsimony principle. A consensus tree based on all the most parsimonious trees retains the branch points (nodes) common to all the trees and leaves the others "unresolved."

- The maximum likelihood method uses computer analysis for the reconstruction of phylogenies based on molecular data.

Biological classification systems are designed to express evolutionary relationships among organisms.

- The system of binomial nomenclature developed by Linnaeus in 1758 assigns each species two names, one identifying the species itself and the other the genus to which it belongs. The generic name (always capitalized) is followed by the species name (lowercase), and both are italicized.

- A taxon is a group of organisms that is treated as a unit in a biological classification system.

- In the Linnaean system of classification, species are grouped into higher-order taxa. The hierarchy of taxa, ranked from most to least inclusive, is kingdom, phylum, class, order, family, genus, species. Thus, a genus includes one or more species, a family includes one or more genera, and so forth.

- Classification systems are important for several reasons. They improve our ability to explain relationships among organisms and provide a basis for predictions about characteristics to be found among members of a group. In addition, they aid precise communication by providing unique, universally used names for organisms.

- Most present-day taxonomists believe that evolutionary relationships should be the basis for classifying organisms.

- A monophyletic taxonomic group (also called a clade) contains all the members of a lineage descended from a common ancestor and no other organisms. Most taxonomists believe that all taxa should be monophyletic.

- A polyphyletic group consists of members that do not all share a common ancestor. Because they do not reflect phylogenetic relationships, such taxonomic units are eliminated by taxonomists as soon as they are discovered.

- A paraphyletic group includes some, but not all, descendants of a particular ancestor. Some taxonomists believe that certain paraphyletic groups should be retained to emphasize that members of some lineages have undergone rapid evolutionary change and diversification and evolved unique derived traits. A group that evolves such distinctive characters is called a grade.

Phylogenetic trees provide important information to biologists trying to answer many kinds of questions.

- One important use of phylogenetic trees is to determine how many times a particular trait has evolved within a lineage.

- Phylogenetic trees can also help to determine when lineages diverged and how evolutionary radiations came about.

The Big Picture

- The parsimony principle, which holds that one should prefer the simplest hypothesis capable of explaining the known facts, is fundamental not only to the reconstruction of phylogenies but also to every other field of scientific research.

- The concept of evolution revolutionized taxonomy. Prior to Darwin's time, Linnaeus and others had developed "natural" systems of classification based primarily on similarity of morphology. Modern biologists realize that similarities of organisms (other than homoplasies) are the result of descent from a common ancestor and believe that a truly natural system of classification should reflect evolutionary relationships.

- Knowing that organisms are evolutionarily related enables biologists to make predictions about their charac-

teristics. As this chapter points out, this knowledge can provide important hints in the search for organisms with valuable properties, such as the ability to produce medically useful drugs.

Study Strategies

- Here is a mnemonic for the hierarchy of taxa (kingdom, phylum, class, order, family, genus, species) in the Linnaean system of classification: "**K**indly **P**rofessors **C**annot **O**ften **F**ail **G**ood **S**tudents." Note that the plural of genus is genera; the words *general* and *generic* come from the same root and can help you remember that the genus is the more general taxon in the genus/species binomial. *Species* and *specific* also come from the same root, and the species is, of course, the most specific taxon.

- Review the following activities on the Student Website/CD:
  Activity 25.1 Constructing a Phylogenetic Tree
  Activity 25.2 Types of Taxa

Test Yourself
Knowledge and Synthesis Questions

1. Which of the following would *not* be expected to result in homoplasy?
   a. Convergent evolution
   b. The independent evolution of similar structures in different lineages
   c. Selection for traits that perform similar functions
   d. The inheritance of ancestral traits

2. A derived trait is one that
   a. differs from its ancestral form.
   b. is homologous with another trait found in a related species.
   c. is the product of an evolutionary reversal.
   d. has the same function, but not the same evolutionary origin, as a trait found in another species.

3. Which of the following statements about reconstructing phylogenies is *false*?
   a. Traits found in the outgroup as well as in the ingroup are likely to be ancestral traits.
   b. Shared traits are generally assumed to be homoplastic until they can be proven to be homologous.
   c. Phylogenies do not include ancestors of modern groups or date the splits between lineages.
   d. Nodes (branching points) in phylogenetic trees have only two branches because during speciation a lineage normally splits into only two daughter species.

4. The *most* important attribute of a biological classification scheme is that it
   a. avoids the ambiguity created by using common names.
   b. reflects the evolutionary relationships among organisms.
   c. helps us remember organisms and their traits.
   d. improves our ability to make predictions about the morphology and behavior of organisms.

5. Suppose you are writing a scientific paper about a unicellular green alga called *Chlamydomonas reinhardtii*. What would be the proper way to refer to this species after you had used the full binomial earlier in the same paragraph?
   a. *Chlamydomonas reinhardtii*
   b. *Chlamydomonas* spp.

c. *Chlamydomonas* sp.
d. *C. reinhardtii*

6. The organisms that make up a class are _____ diverse and _____ numerous than those in a family within that class. The organisms that make up a phylum all diverged from a common ancestor _____ recently than did the organisms in an order within that phylum.
   a. more; more; less
   b. more; more; more
   c. more; less; less
   d. less; less; more

7. Which of the following incomplete lists of taxonomic categories ranks them properly from most inclusive to least inclusive?
   a. Phylum, order, family, genus
   b. Class, phylum, order, species
   c. Order, class, family, genus
   d. Family, order, class, kingdom

Application Questions

1. Based on the following phylogenetic tree showing the evolutionary relationships of five species (A–E) relative to five traits (1–5), fill in the table using 1 to indicate the presence of a derived trait and 0 to indicate the presence of an ancestral trait.

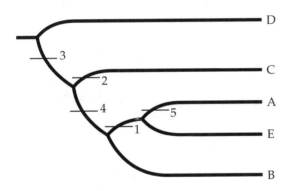

SPECIES	TRAIT				
	1	2	3	4	5
A					
B					
C					
D					
E					

2. In the phylogenetic tree shown in Question 1, which species is considered the outgroup? Explain your answer.

3. Based on the following table showing the ancestral and derived traits of five species (A–E), construct a phylogenetic tree that represents the evolutionary relationships of this group using conventions presented in the textbook. In this table, the ancestral state of each trait is indicated by 0, and the derived state is indicated by 1.

SPECIES	TRAIT				
	1	2	3	4	5
A	1	1	1	0	0
B	0	0	0	0	0
C	1	1	0	1	0
D	0	1	0	0	0
E	1	1	0	1	1

4. Discuss the application of the parsimony principle in the construction of phylogenetic trees.

5. Discuss the implications the following statement has for systematics: "DNA is the genetic material for all eukaryotes."

6. Why are paraphyletic groups (taxa) acceptable to some taxonomists whereas polyphyletic groups are not?

Answers

Knowledge and Synthesis Answers

1. **d.** Homoplasy is the appearance of similar structures in different lineages that were not present in the common ancestor.

2. **a.** Derived traits are those that have undergone a change during evolution from the ancestral (original) character state.

3. **b.** Most shared traits, especially in species with a recent common ancestor, are likely to be homologous, not homoplastic. Therefore it is more consistent with the parsimony principle to assume that traits are homologous until proven homoplastic than to assume the reverse.

4. **b.** Although all of the statements listed are important attributes of biological classification schemes, the most important attribute is that biological classification reflects evolutionary relationships.

5. **d.** After a scientific name is referenced once in a paragraph, typically the genus is abbreviated, but the species name is given in full.

6. **a.** Organisms in a higher taxon are *less* similar, have diverged from a common ancestor *less* recently, and include *more* species than organisms in a lower, included taxon.

7. **a.** The complete hierarchy of taxonomic categories, from most to least inclusive, is: kingdom, phylum, class, order, family, genus, species.

Application Answers

1.

SPECIES	TRAIT				
	1	2	3	4	5
A	1	0	1	1	1
B	0	0	1	1	0
C	0	1	1	0	0
D	0	0	0	0	0
E	1	0	1	1	0

2. Species D. The outgroup has the ancestral form for all traits.

3.

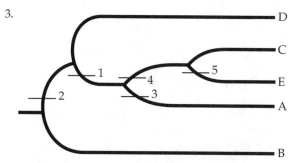

4. In the construction of a phylogenetic tree, you initially assume that derived traits appear only once and never disappear. Given a set of traits for a group of species, sometimes these restrictions must be relaxed to produce a phylogenetic tree for the group. Parsimony involves arranging the species so that you minimize the number of required reversals and multiple origins. Generally, the simplest explanation is most likely to be correct.

5. Some of the implications of this statement are that DNA evolved as the genetic material before eukaryotes had diverged from prokaryotes, that DNA is an ancestral and general homologous trait, and that all surviving eukaryotes use DNA as the genetic material.

6. Paraphyletic taxa are tolerated as a way to emphasize groups that have changed rapidly and evolved distinctive traits. For example, birds are put in their own class, Aves, even though most biologists agree that the birds share a more recent common ancestor with turtles and crocodilians than any of these groups does with other reptiles such as lizards and snakes (see Figure 25.8). This makes the class Reptilia (which traditionally includes lizards, snakes, crocodilians, and turtles, but excludes birds) paraphyletic. Polyphyletic groups are not acceptable because they fail to include multiple groups that share the same recent common ancestor (see Figure 25.7). This is antithetical to the purpose of constructing phylogenies.

26 *Molecular and Genomic Evolution*

Important Concepts

The genome of an organism is its entire set of genes.

- Because meiosis and fertilization shuffle genes in every generation, a successful gene must be able to function in a wide variety of genetic backgrounds. Among the genes of an individual, there are both divisions of labor and strong interdependencies.

The field of molecular evolution is the study of the evolution of macromolecules, particularly nucleic acids and proteins.

- Nucleic acids and proteins are investigated to determine how rapidly and why they have changed. The answers to these questions may be crucial to an understanding of the evolutionary history of groups of organisms.

- Molecular evolution differs from phenotypic evolution in that its rate and direction are strongly influenced by random genetic drift and mutation.

- A synonymous (silent) mutation replaces a nucleotide base in a codon but does not change the amino acid specified by the codon. Because synonymous mutations are unlikely to be affected by natural selection, they accumulate in populations at rates determined by rates of mutation and genetic drift.

- A substitution is the replacement of a nucleotide or longer sequence throughout an entire population or species.

- A nonsynonymous mutation changes the amino acid specified by the codon. Though such mutations are likely to be harmful, they are sometimes selectively neutral, or nearly so.

- The neutral theory of molecular evolution postulates that most evolutionary change in macromolecules, as well as much of the genetic variation within species, is the result of random genetic drift, rather than natural selection. The rate of fixation of neutral mutations is theoretically constant and equal to the mutation rate.

The structures of macromolecules can be determined and compared.

- The polymerase chain reaction (PCR) technique makes it possible for biologists to determine the base sequence of regions of DNA from both living tissues and preserved remains.

- By first aligning amino acid sequences of homologous proteins or nucleotide sequences of homologous regions of DNA and then constructing a similarity matrix, biologists can determine the similarity of these macromolecules in different species. The assumption is that the longer the molecules have been evolving separately, the more differences they will have.

- The rate of evolution of macromolecules is extremely variable. In general, the more essential a molecule is for cell functioning, the slower the rate of its evolution. Regions of a molecule where any change is likely to affect function tend to evolve much more slowly than areas where changes are functionally insignificant. By the same token, nucleotide substitutions occur much more rapidly in pseudogenes—duplicate copies of genes that are never expressed—than in functional genes.

- Because the rate of nucleotide or amino acid substitutions is reasonably constant in many macromolecules, differences in these molecules can be used as a molecular clock to estimate the approximate time of divergence of lineages of organisms.

Duplicated proteins can acquire new functions during evolution via mutation and natural selection.

- The many thousands of genes found in modern organisms have evolved by gene duplication from a relatively small number of ancestral genes.

- Polyploidy, in which the entire genome is duplicated, is one source of duplicate genes.

- After a gene has been duplicated, one copy may retain its original function whereas the other accumulates mutations until it can perform a different task.

- The evolution of lysozyme shows how a protein can become adapted to a new function through mutation and natural selection. While serving in almost all animals as an important first line of defense against invading bacteria, lysozyme also has taken on an essential role in the digestive process of several groups of foregut-fermenters. The independent evolution in these groups of a type of lysozyme adapted to its new environment and function also shows that convergent evolution can occur at the molecular level.

There is tremendous variation in the size and composition of the genomes of various species.

- The size of the coding portion of the genome is larger in more complex organisms. Thus, eukaryotes have many times more genes than prokaryotes; vascular plants and invertebrates have more genes than single-celled organisms, and vertebrates have more genes than invertebrates.

- Most of the variation in genome size of various organisms is due not to differences in the number of functional genes, but in the amount of noncoding DNA.

- Most of the noncoding DNA appears to be nonfunctional, but some of it may have undetected functions.

- When a gene is duplicated, three evolutionary outcomes are possible:

  - Both copies may retain the gene's original function.

  - One copy may become a functionless pseudogene.

  - One copy may retain its original function, whereas the second mutates so extensively that it can perform a different function.

- Successive rounds of gene duplication and mutation can result in a gene family—a group of homologous genes with related functions.

- The globin gene family illustrates that gene diversification can produce molecules with different functions (hemoglobin and myoglobin) as well as functionless pseudogenes.

Molecular genomic information has many uses.

- In lateral gene transfer, genes are passed from one evolutionary lineage to another. Hence the phylogenies of genes often differ from those of organisms.

- Molecules that have evolved slowly, such as ribosomal RNA, are useful for determining ancient lineage splits. Molecules that have evolved rapidly can be used to determine relationships among organisms with more recent common ancestors.

- For ancient lineage divisions, molecular data are often the only evidence available for reconstructing phylogenies. In other cases, molecular data are used in combination with morphological and fossil evidence.

- A gene tree depicts the evolutionary relationships of members of a gene family. Orthologs are genes found in different organisms that arose from a single gene in their common ancestor. Paralogs are related genes that have resulted from gene duplication in a single lineage.

- Biomedical scientists are using molecular data to discover new strategies in our struggle against infectious diseases.

- The AIDS epidemic shows that molecular data cannot solve all disease problems. Because of HIV's high mutation rate, it is able to develop strains resistant to highly active antiretroviral therapy (HAART) in the blood of most patients receiving this treatment.

The Big Picture

- The growing importance of the study of macromolecules in the reconstruction of phylogenies is an example of the far-reaching impact that advances in molecular biology have had in the decades since the discovery of the structure of DNA by Watson and Crick in 1953.

- The emergence of strains of HIV resistant to HAART drug treatment is a dramatic illustration of the power of natural selection to bring about quick evolutionary change in a highly variable, rapidly reproducing population subjected to strong selection pressure.

Study Strategies

- Study Figure 26.2 carefully to understand how a similarity matrix is constructed. Similarity matrices are essential to reconstructing phylogenies based on molecular data.

- Review the following activities on the Student Website/CD:
  Activity 26.1 Amino Acid Sequence Alignment
  Activity 26.2 Similarity Matrix Construction
  Activity 26.3 Gene Tree Construction

Test Yourself
Knowledge and Synthesis Questions

1. Which of the following statements about mutations is *false*?
   a. A silent mutation results in no change in the amino acid sequence of a protein.
   b. Because most substitution mutations are selectively neutral, most evolutionary changes in macromolecules are the result of random genetic drift.
   c. A base substitution mutation in the third codon position is more likely to be neutral than a substitution at the first or second codon position.
   d. Nonsynonymous mutations are virtually always deleterious to the organism.

2. Which of the following statements about gene families is *false*?
   a. Gene families evolve via gene duplication.
   b. Pseudogenes are quickly removed from gene families by deletion.
   c. Members of a gene family can include several functional genes.
   d. Examples of gene families include the homeotic gene complex in *Drosophila* and the α hemoglobin gene family in humans.

3. Which of the following is *not* a valid conclusion based on studies of the structure of proteins, such as cytochrome *c*, in different species?
   a. The rate of accumulation of neutral substitutions is based on the mutation rate.
   b. Many nucleotide substitutions result in neutral changes in the amino acid sequence of the protein.
   c. Fewer differences were observed in the amino acid sequences of a protein if the organismal sources of the proteins were closely related.
   d. Functionally important regions of a protein can be discovered by identifying the regions with the most amino acid substitutions.

4. Which of the following statements about molecular clocks is *false*?
   a. The molecular clock of cytochrome *c* ticks more slowly than that of rRNA.
   b. The tick rate of a clock for a molecule with great functional constraints would be very slow.
   c. The tick rate of a molecular clock that mostly reflects neutral changes is determined by the mutation rate.
   d. The tick rate of a molecular clock that mostly reflects neutral changes is directly related to the slope of the line showing the relationship between the time of divergence of different taxa and amino acid sequence differences for the protein.

5. Which of the following statements about the enzyme lysozyme is true?
   a. A small group of closely related mammals has evolved a special form of lysozyme that functions in digestion.
   b. The lysozymes found in the foregut fermenters resulted from convergent evolution.
   c. Lysozyme could not have evolved a secondary function if it had been an enzyme with a vital primary function.
   d. A higher mutation rate in the foregut fermenters allows their lysozymes to evolve rapidly.

6. Which of the following sequences of organisms ranks them from least to most, in terms of the expected total amount of coding DNA in their genomes?
   a. Bacterium, single-celled eukaryote, roundworm, insect, bird
   b. Bacterium, roundworm, bird, insect, single-celled eukaryote
   c. Single-celled eukaryote, bacterium, roundworm, insect, bird
   d. Roundworm, single-celled eukaryote, insect, bird, bacterium

7. Which of the following molecules would be best to use in a study of the domains of the phylogenetic tree of life?
   a. Lysozyme
   b. Mitochondrial DNA
   c. rRNA
   d. Myoglobin

Application Questions

1. Discuss the importance of the polymerase chain reaction (PCR) to studies of molecular evolution.

2. Although scientists have acquired detailed knowledge of the genes and proteins that govern the different stages in the life cycle of HIV (the AIDS virus), they have been unable to develop a totally effective drug therapy against AIDS. Why?

3. Determine the amino acid sequence for an eight-residue section of a small protein from five different species (1–5).

Position	1	2	3	4	5	6	7	8
Species 1:	Arg	Cys	Leu	Leu	Ser	Thr	Asn	Met
Species 2:	Arg	Cys	Phe	Leu	Leu	Ser	Thr	Asn
Species 3:	Arg	His	Leu	Leu	Ser	Thr	Asn	Met
Species 4:	Arg	Cys	Leu	Ser	Ser	Thr	Asn	Met
Species 5:	Arg	His	Leu	Leu	Ser	Gln	Asn	Met

Complete the following similarity matrix using these sequences:

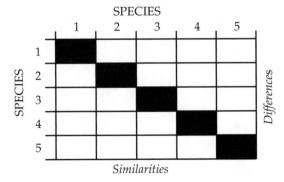

Which species differ from Species 1 because of an amino acid insertion?

Which species differ from Species 1 because of amino acid substitutions?

Answers

Knowledge and Synthesis Answers

1. **d.** A nonsynonymous substitution mutation can be selectively neutral if, as sometimes happens, it results in an amino acid change that has no significant effect on the shape (and hence the functional properties) of the protein.

2. **b.** Pseudogenes (nonfunctional DNA) may be removed via deletion, but numerous pseudogenes persist in many gene families.

3. **d.** Functionally important regions of a protein can be discovered by identifying the regions with the fewest amino acid substitutions. Because the sequences in these regions have been optimized by natural selection to accomplish the function of the molecule, random changes in these areas are not tolerated.

4. **a.** The molecular clock of cytochrome *c* would tick more rapidly (i.e., the rate of neutral changes-per-unit time would be greater) than the clock of rRNA. rRNA has changed very little because it is a molecule with great functional restraints.

5. **b.** Because the animals that evolved a similar lysozyme do not share a recent common ancestor, the mechanism involved is convergent evolution. All other statements are false.

6. **a.** Recall that there is a rough relationship between the amount of coding DNA and organismal complexity.

7. **c.** Unlike the other molecules, rRNA is found in all living organisms and also has evolved very slowly.

Application Answers

1. Because the study of molecular evolution depends on obtaining fairly large amounts of pure DNA for sequencing purposes, PCR is commonly used to amplify and purify DNA. This is especially important in studying DNA from fossil organisms in which the amount of DNA available may be very small and degraded.

2. Drugs that are effective against HIV remove only variants of the virus that are not drug resistant. Because of the exceptionally high mutation rate of the virus, resistant strains arise in the blood of most patients sooner or later. In the absence of drugs to which the virus cannot adapt, the best hope for reducing the incidence of AIDS is to change human sexual behavior.

3.

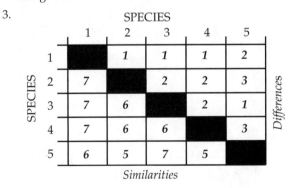

Species 2 shows an amino acid insertion (Phe) at position 3. There are no deletions.

Species 3, 4, and 5 all show amino acid substitutions.

27 Bacteria and Archaea: The Prokaryotic Domains

Important Concepts

Domains attempt to reflect evolutionary relationships among organisms.

- All living things can be placed into one the following three domains: Archaea, Bacteria, and Eukarya. This system reflects evolutionary relationships among these monophyletic groups.

- All organisms have similarities; only derived similarities (synapomorphies) indicate evolutionary relationships. Eukarya have nuclei, membrane-enclosed organelles, and cytoskeletons. All organisms that possess these are descended from the most recent common ancestor of the Eukarya. The absence of these features in Archaea and Bacteria is a retained characteristic inherited from the common ancestor of all life on Earth (a prokaryote). Unique and distinguishing features of Archaea and Bacteria have been discovered through biochemistry and genetic studies.

- All three domains arose from a common ancestor possessing a circular chromosome and machinery for transcription and translation of the DNA code to protein. Divergence from this ancestor occurred billions of years ago. The domain Archaea is more closely related to the Eukarya than to the Bacteria.

Prokaryotes have many features in common.

- Though they are members of two different domains, Archaea and Bacteria have several structural similarities that are indicative of evolutionary adaptation to similar environments.

- Prokaryotes are prolific and are the most numerous of all organisms. Because they are single celled (there are a few exceptions) and very small, they generally escape human notice. They play extremely important roles in the biosphere, for instance in nutrient cycling. They can also metabolize many substances that eukaryotes cannot.

- Prokaryotes generally exist in one of three shapes; spherical (cocci), rod-shaped (bacilli), and spiral-shaped. All of the different shapes may aggregate to form chains or plates of cells. Advanced associations may result in filaments that are encased in a protective sheath.

- Prokaryotes do not have membrane-enclosed organelles; therefore, the organization and replication of DNA proceeds differently in prokaryotes than it does in eukaryotes. Also, the processes of cellular respiration and photosynthesis are carried out differently and rely on infoldings of the cell membrane. Finally, prokaryotes lack a cytoskeleton; therefore, true mitosis does not occur. Prokaryotes go through a process called fission after replicating their DNA.

- Many prokaryotes are mobile and move by gliding, rolling, regulating gas vesicles, or flagella. Bacterial flagella are whiplike and are composed of a single protein called flagellin. In contrast, the flagellum of eukaryotes is complex.

- The cell wall of prokaryotes is thick and rigid and composed of peptidoglycans (bacteria) or protein (archaea). This cell wall is very different from that of plant and fungal cell walls. Cell wall composition is one of many ways of differentiating bacteria and archaea.

- Cell wall structure within the bacteria also serves as a diagnostic characteristic among different bacteria. This difference is used in the Gram staining technique. Whether a cell is Gram-negative or Gram-positive relates to the amount of peptidoglycan in the cell wall.

  - Gram-positive cells stain purple and have a thick layer of peptidoglycan.

  - Gram-negative cells do not retain stain and appear red due to a thin layer of peptidoglycan.

- Some antibiotics work by interfering with the synthesis of peptidoglycan-containing cell walls. Eukaryotic cells are not harmed because their cell walls lack this material.

- Prokaryotes reproduce asexually but may exchange genetic information through transformation, conjugation, or transduction.

Metabolic processes in prokaryotes are varied and diverse.

- Obligate anaerobes are oxygen-sensitive and can only survive in oxygen-poor conditions. Facultative anaerobes can shift between aerobic and anaerobic modes of respiration. Some facultative anaerobes do not respire aerobically but are not harmed by an oxygen-rich atmosphere as obligate anaerobes are. Obligate aerobes require oxygen and cannot survive in the absence of it.

- Photoautotrophs are photosynthetic and convert light energy to chemical energy. Some photoautotrophs use chlorophyll a and give off oxygen as a by-product. Others use bacteriochlorophyll and produce sulfur as a by-product.

- Photoheterotrophs use light for energy but also require outside carbon sources.

- Chemolithotrophs oxidize inorganic substances for energy. Some chemolithotrophs fix carbon dioxide into carbohydrate. Others fix nitrite or ammonia, hydrogen gas, hydrogen sulfide, and other materials.

- Chemoheterotrophs obtain energy and carbon from consuming organic molecules.

- Some bacteria carry out respiratory electron transport using nitrogen or sulfur compounds as electron acceptors. These organisms are extremely important in nutrient cycling.

Prokaryotes make diverse contributions to their environments.

- The nitrogen cycle depends on nitrification by bacteria. Nitrifying bacteria use nitrogen to make ATP, which is used to fix carbon dioxide into glucose and other food molecules. Through this process, unusable nitrogen sources are converted into biologically available nitrogen sources.

- Cyanobacteria, using photosynthesis, enriched the atmosphere with oxygen. Oxygen levels became too high for many existing organisms; many others evolved to fill those niches, and new organisms arose.

- Prokaryotes live in many mutualistic associations. Many of your digestive processes rely on the bacteria that inhabit your intestine.

- A small minority of bacteria are pathogenic. Bacteria cause disease because they produce toxins that are harmful to host tissues. Endotoxins are released when bacterial cells burst, but are rarely fatal. Exotoxins are secreted by bacteria and may be fatal.

- The German physician Robert Koch came up with criteria for determining if a bacterium was a disease-causing agent.

  - The organism must always be found in the individual with the disease.

  - The organism must be taken from the host and grown in pure culture.

  - A sample of the culture must produce disease if injected into a healthy individual.

  - The newly infected individual must yield a new pure culture of the same organism.

- In order to be pathogenic an organism must be able to reach a new host, invade the host, evade the host's defenses, multiply, and infect a new host. Relatively few organisms can do this.

Evolutionary relationships among prokaryotes are determined at the genetic level.

- Physical characteristics are used for classification for identification purposes, but to understand the true evolutionary relationships among very small organisms such as prokaryotes, it is necessary to look at genetic sequences and biochemistry.

- Ribosomal RNA gives us information about relationships among organisms. It is useful for this purpose because it is evolutionarily ancient, is found in all organisms, has the same function in all organisms, and is highly conserved (it evolved slowly). Scientists compare signature sequences to make evolutionary inferences.

- rRNA sequences have not provided clear data. Lateral gene transfer—genetic transfer between species—makes these data hard to understand.

- Mutations are important in prokaryotic variation. Because prokaryotes have single circular DNA pieces, mutations cannot be masked and are passed on through fission. This is in contrast to diploid organisms in which mutation plays a small role in genetic variation.

- Prokaryotes evolve comparatively rapidly because of rapid reproduction, persistent mutation, and selection. Mutations leading to antibiotic resistance are of extreme concern and can occur in a matter of just a few years.

The majority of known prokaryotes are bacteria.

- Bacteria can be subdivided into a series of monophyletic groups. The composition (membership) of these groups is not considered to be the same by all scientists, and this is an active area of research.

- Proteobacteria contain the largest number of known species. This group is believed to be the origin of the eukaryotic mitochondria. The early ancestor of this group was probably a photoautotroph, but today, members of the group have multiple metabolic schemes. This group contains important nitrification bacteria and several pathogens. Another member, *Agrobacterium tumefaciens,* has been important in understanding genetic transformation in plants.

- Cyanobacteria undergo photosynthesis similar to eukaryotic photosynthesizers. They contribute to the current oxygen-rich environment. These organisms frequently exist in colonies with distinct divisions of labor. Some cells are vegetative and photosynthesize. Others form resting stages called spores that can withstand harsh environmental conditions. Still others, called heterocysts, fix nitrogen.

- Spirochetes are corkscrew-shaped prokaryotes characterized by the presence of axial filaments that run through their periplasmic space. Many spirochetes are parasites of humans.

- Chlamydias are the smallest bacteria and exist only as parasites. They have a two-stage life cycle and are responsible for several human diseases.

- Most, but not all, firmicutes are Gram-positive. They produce endospores that are capable of surviving in extremely harsh conditions. Many medically important bacteria fall within this group, including *Clostridium botulinum* (botulism), *Staphylococcus* (many skin infections), and *Mycobacterium tuberculosis* (tuberculosis). However, one of our most potent antibiotics comes from this group. Streptomycin is produced by the actinomycete *Streptomyces.* Most currently used antibiotics come from the actinomycetes, which are firmicutes. Another subgroup, the mycoplasmas, may represent the smallest life form.

Archaea live in some of the harshest known environments.

- Very little is known about the phylogeny of the Archaea, and much research still needs to be done, but there are shared characteristics that distinguish them from the Bacteria. Archaea lack peptidoglycan cell walls and have very distinctive lipids in their cell membranes. These are linked by ether linkages. No other organisms possess this type of lipid.

- Crenarchaeota live in hot, acidic environments and are thermophilic and acidophilic. They require temperatures above 55°C and can survive in an environment with a pH as low as 0.9 while maintaining an internal pH of 7.

- Some members of Euryarchaeota produce methane gas and are obligate anaerobes. These organisms are prevalent in the guts of herbivores. Others live near volcanic vents. Another group of Euryarchaeota live in extremely salty conditions and are referred to as halophiles.

The Big Picture

- All organisms fall within one of three domains. This chapter focuses on the domains Archaea and Bacteria. Subsequent chapters focus on the Eukarya. Archaea and bacteria are prokaryotic organisms and retain many features of the earliest forms of life on Earth. They have no membrane-enclosed organelles, circular chromosomes, and cell walls of peptidoglycans or proteins. Their structures are simple, and they exist as single cells. Some prokaryotes are stationary, others move by means of flagella. Their metabolic processes are highly varied, as are their modes for procuring energy. Many prokaryotic organisms make valuable contributions to their environments through nitrogen fixation, photosynthesis, and other metabolic processes. The majority of prokaryotes are free-living, but some live in mutualistic relationships, and yet others are pathogenic to other organisms.

- The evolutionary relationships among prokaryotes are still highly disputed and best understood at the gene level. The most ubiquitous extant prokaryotes are the bacteria. They are grouped according to physical and biochemical characteristics, and these groupings do not necessarily reflect their evolutionary relationships. The Archaea are more closely related to the Eukarya, and they exist in some of the harshest known environments.

Study Strategies

- This chapter is an overview of two domains of organisms. A single chapter can only cover a few points of prokaryotic diversity. Focus your study on what makes each domain distinct (its synapomorphies) and the unique features that indicate a closer evolutionary relationship between archaea and eukaryotes than between archaea and bacteria. Move on to learning about sample organisms from each domain and groups within each domain.

- The temptation is to focus on the pathogenic organisms, but remember that only a very small percentage of prokaryotes are pathogenic.

- Review the following tutorial and activity on the Student Website/CD:
  Tutorial 27.1 The Evolution of the Three Domains
  Activity 27.1 Gram Stain and Bacteria

Test Yourself
Knowledge and Synthesis Questions

1. Which of the following characteristics is unique to prokaryotes?
   a. Lack of membrane-enclosed organelles
   b Presence of cell walls
   c. Presence of plasma membranes
   d. All of the above

2. A soil sample contains large numbers of rod-shaped bacteria. These bacteria are referred to as
   a. bacilli.
   b. cocci.
   c. spiral.
   d. helici.

3. Which of the following statements concerning prokaryotes is true?
   a. Because prokaryotes do not contain organelles they cannot photosynthesize or carry out cellular respiration.
   b. Prokaryotes have no chromosomes and therefore lack DNA.
   c. Prokaryote flagella are similar in structure to eukaryote flagella.
   d. None of the above

4. Gram-negative bacteria stain pink because
   a. they have specialized lipids in their cell walls.
   b. their peptidoglycan layer is thin.
   c. their peptidoglycan layer is thick.
   d. they are receptive to antibiotics.

5. Archaea are more closely related to _____ than to any other monophyletic group.
   a. bacteria
   b. eukaryotes
   c. bacteria and eukaryotes
   d. fungi

6. The dense films laid down by many prokaryotes are
   a. endotoxins.
   b. denitrifiers.
   c. biofilms.
   d. pathogens.

7. Which of the following groups of bacteria has the highest proportion of pathogens?
   a. Proteobacteria
   b. Cyanobacteria
   c. Chlamydias
   d. None of the above

8. The mitochondria of eukaryotes were derived from _____ by endosymbiosis.
   a. proteobacteria
   b. chemoheterotrophs
   c. eukaryotes
   d. archaea

9. Autoclaves must pass a "spore test" in many states to demonstrate that they are able to sterilize equipment. The spore-producing bacteria they use for this test are most likely from which of the following groups of bacteria?
   a. Firmicutes
   b. Proteobacteria
   c. Cyanobacteria
   d. Chlamydias

10. Archaea often live in harsh environments. Archaea that live in extremely salty conditions are referred to as
    a. thermophiles.
    b. halophiles.
    c. salinophiles.
    d. None of the above

11. Methane gas contributes to the greenhouse effect that is raising atmospheric temperatures. A large portion of all methane emission is from grazing cattle. This is because cows harbor methane-producing archaea from which of the following groups?
    a. Crenarchaeota
    b. Euryarchaeota
    c. Anarchaeota
    d. Proteobacteria

12. One of the major diagnostic characteristics that distinguishes archaea from bacteria is the lack of
    a. peptidoglycan cell walls in archaea.
    b. peptidoglycan cell walls in bacteria.
    c. ribosomes in archaea.
    d. chemoautotrophic bacteria.

13. A bacterium that requires a carbon source other than carbon dioxide, yet can convert light energy to chemical energy, is called a
    a. photoautotroph.
    b. photoheterotroph.
    c. chemoautotroph.
    d. chemoheterotroph.

14. A bacterium that *cannot* live in the presence of oxygen is called a(n)
    a. obligate aerobe.
    b. facultative aerobe.
    c. obligate anaerobe.
    d. facultative anaerobe.

Application Questions

1. You work for a wastewater treatment plant. An organism is getting past your treatment process, and fish are dying in a stream that the runoff enters. You are able to isolate the organism. Using a microscope, how can you tell if the organism is a prokaryote or a eukaryote?

2. You work with the Centers for Disease Control in Atlanta. You have been called to a remote area of Uganda to study a mysterious disease that is causing respiratory ailments in a small village. You isolate a bacterium from several patients that seems like a good candidate for the pathogen. How can you determine if this bacterium is causing the illness?

3. As a U.S. Department of Agriculture field representative, you counsel a young farmer to plant alfalfa in fields with soils that have high nitrite levels. You know alfalfa roots are hosts for nitrogen-fixing bacteria. How will the alfalfa and its associated bacteria help "fertilize" this farmer's soil?

4. A patient comes to your medical practice with a bacterial infection. You are terribly concerned because the bacteria she is harboring produce an exotoxin. Why are exotoxins often more dangerous than endotoxins?

5. Describe how Gram staining depends on bacterial cell wall structure. Would Gram staining work for archaea?

6. Differentiate between the following terms:
    a. obligate anaerobe/facultative anaerobe
    b. photoautotroph/photoheterotroph
    c. chemolithotroph/chemoheterotroph

Answers
Knowledge and Synthesis Answers

1. **a.** Prokaryotes are differentiated from eukaryotes by lacking membrane-enclosed organelles. They also have a single circular piece of genomic DNA and lack a cytoskeleton.

2. **a.** A rod-shaped bacterium is known as a bacillus (plural, bacilli). Cocci are roughly spherical.

3. **d.** Bacteria do carry out both photosynthesis and multiple forms of cellular respiration. All life forms have DNA. The flagella of bacteria consist of a single protein called flagellin rather than multiple proteins as found in eukaryotes.

4. **b.** Gram-negative bacteria have thin peptidoglycan layers that do not retain crystal violet stain.

5. **b.** The Archaea and the Eukarya share a more recent common ancestor with each other than either does with the Bacteria. The evidence for this is, for example, a signature sequence from rRNA that has been found in all archaea and eukaryotes tested so far, but in none of the bacteria.

6. **c.** Biofilms are gel-like polysaccharide matrices that are laid down by prokaryotes and trap other bacteria.

7. **c.** Chlamydias are all parasitic and cause several human diseases.

8. **a.** Endosymbiosis of proteobacteria gave rise to the mitochondria of eukaryotes.

9. **a.** Among the groups listed, only firmicutes and cyanobacteria produce spores. Firmicute spores can withstand extreme environmental conditions, so they would make good spores for testing an autoclave's ability to sterilize equipment.

10. **b.** The term "halophile" means "salt loving."

11. **b.** Methanogenic bacteria that reside in the guts of cows belong to the Euryarchaeota.

12. **a.** Archaea lack peptidoglycans. Instead they have a unique lipid in their cell walls.

13. **b.** Photoheterotrophs require an outside carbon source other than carbon dioxide, yet are able to harvest light energy.

14. **c.** Oxygen gas is toxic to obligate anaerobes.

Application Answers

1. The most immediately apparent difference between a prokaryote and a eukaryote would be the absence of membrane-enclosed organelles in prokaryotes.

2. You can use Koch's postulates. These postulates stipulate that the organism must always be found in individuals with the disease, the organism must be taken from the host and grown in pure culture, a sample of the culture must produce disease if injected into a healthy individual, and the newly infected individual must yield a new pure culture of the same organism.

3. Nitrogen-fixing bacteria can convert nitrites that plants cannot use into nitrates that plants *can* use.

4. Exotoxins are continuously produced by the infecting bacteria. Endotoxins are produced only at cell death. There are a limited number of endotoxin molecules released. Exotoxins can be released until the host dies.

5. Gram staining results depend on the arrangement of peptidoglycans in the bacterial cell wall. This technique does not work with archaea because they do not have peptidoglycan in their cell walls.

6. Obligate anaerobes die in the presence of oxygen gas. Obligate aerobes die without oxygen gas for cellular respiration. Photoautotrophs convert light energy to chemical energy using carbon dioxide as their carbon source. Photoheterotrophs rely on an outside carbon source other then carbon dioxide but still can convert light energy to chemical energy. Chemolithotrophs can produce what they need from inorganic molecules. Chemoheterotrophs depend on nutrients from external sources.

28 Protists and the Dawn of the Eukarya

Important Concepts

Protists are a polyphyletic group of very diverse organisms.

- Protists are eukaryotic organisms that do not fit into the kingdoms Animalia, Plantae, or Fungi; they are a polyphyletic group. Many are unicellular, but some are multicellular.

The origin of the eukaryotic cell was pivotal in evolution.

- The origin of the eukaryotic cell is still largely debated by biologists. The changes leading to evolution of the eukaryotic cell were vast and occurred at a time of great environmental change.

- The eukaryotic cell contains a flexible cell surface, a cytoskeleton, a nuclear envelope, digestive vesicles, and organelles.

- The first step in the path toward a eukaryotic cell may have involved loss of the rigid cell wall. Loss of this rigidity allows infoldings of the cell surfaces to increase the surface area-to-volume ratio. This allows cells to become larger. It also allows vesicles to form from bits of infolded membrane that pinch off. Such infolding may have been the origin of the nuclear envelope (see Figure 28.2).

- New structures had to form during the evolution of the eukaryotic cell. Ribosomes became associated with internal membranes to form the endoplasmic reticulum, a cytoskeleton of actin and microtubules formed, and digestive vesicles developed. The cytoskeleton opened up modes of locomotion with actin/myosin-based flagella. How these structures formed is currently conjecture.

- It is thought that these early motile cells may have been phagocytes that engulfed prokaryotic cells. Because the atmosphere was shifting from anaerobic to aerobic, it is possible that these phagocytes engulfed prokaryotes that could metabolize the toxic by-products of the oxygen-rich atmosphere. These prokaryotes were retained as symbionts and may have led to modern peroxisomes.

- Subsequent endosymbiotic events may have led to the evolution of mitochondria and chloroplasts.

- See Figure 28.3 for a diagram of events in the evolution of the eukaryotic cell.

- An understanding of the evolution of eukaryotes is complicated by lateral gene transfer. The bacterial genes in many prokaryotes are so numerous that lateral transfer cannot explain them all, nor can endosymbiosis. Additional hypotheses are being tested.

Protists have diverse ways to accomplish the same functions.

- Most protists are aquatic. Even those that seem terrestrial live in soil water or other extremely moist environments.

- Mechanisms of metabolism are relatively uniform among protists, but nutritional acquisition varies. Some protists are autotrophic, some are heterotrophic, and some have the ability to switch between the two modes. Many protists are involved in symbioses from mutualism to parasitism.

- Most protists are motile and move via pseudopodia, flagella, or cilia. Cilia and eukaryotic flagella are identical in cross section; they differ only in length.

- Many protists are microscopic, though some algae literally form aquatic forests of very large, multicellular organisms. Single-celled protists are able to increase their cell size with the presence of vesicles, which effectively increase surface area. Vesicles assist with water regulation (contractile vacuoles) and nutrient procurement (food vacuoles).

- Many protists have cell walls or "shells" to protect their cell membranes.

- Many protists contain endosymbionts. Some harbor other protists that photosynthesize or carry out other metabolic processes.

- Most protists undergo both sexual and asexual reproduction, but some lack the ability to reproduce sexually. Sexual and asexual reproductive practices are as diverse as the protists themselves.

The euglenozoans are a monophyletic group.

- Euglenozoans are single-celled flagellates that reproduce asexually by binary fission.

- This group consists of two subgroups, the euglenoids and the kinetoplastids.

- Euglenoids have anterior flagella. Many are photosynthetic and are characterized by a complex cell structure (see Figure 28.11). They are flexible in nutritional requirements and may switch between autotrophism and heterotrophism. Euglenoid chloroplasts have triple membranes.

- Kinetoplastids are parasitic and are characterized by having a single large mitochondrion. The mitochondrion is unique in having a kinetoplast housing multiple circular DNA molecules and associated proteins. Many tropical human diseases are caused by kinetoplastids, including African sleeping sickness.

The Alveolata are a monophyletic group.

- The synapomorphy that defines the Alveolata is the possession of cavities called alveoli just below their cell surfaces.

- This group consists of dinoflagellates, apicomplexans, and ciliates.

- Dinoflagellates are mostly marine organisms with two flagella. They are yellowish in color due to photosynthetic and accessory pigments and are primary producers in marine environments. Many live in symbiosis with other organisms. They are common endosymbionts in corals, for example. Dinoflagellates are responsible for "red tides" in warm marine waters and can damage fish. Toxins produced by these organisms can accumulate in shellfish and be fatal to people eating the shellfish. Dinoflagellates are also bioluminescent and produce light when disturbed.

- Apicomplexans are parasitic protists. Their name comes from the cluster of organelles at the apex of their spores, which assist with invasion of their host organism. The life cycles of these parasites are complex. Apicomplexans in the genus *Plasmodium* cause malaria, a disease that claims more than a million lives annually (see Figure 28.14).

- Ciliates move via cilia and are characterized by multiple macronuclei and micronuclei. Most ciliates are heterotrophic. Paramecia are common and well-studied ciliates.

  - Paramecia move by cilia, their membranes are protected by a pellicle, and they protect themselves with trichocysts.

  - Paramecia reproduce asexually by binary fission. Genetic recombination is accomplished through conjugation, which involves the exchange of equal amounts of genetic material (see Figure 28.17). The resulting recombined paramecia then go through binary fission.

Stramenopiles, most likely a polyphyletic group, are referred to as the brown algae.

- All members are either flagellated or have lost their flagella over the course of evolution.

- This group consists of diatoms, brown algae, oomycetes, and a few smaller groups.

- Diatoms are yellowish brown and store chrysolaminarin and oils as photosynthetic products. They are most noted for their silica-containing cell walls. The cell walls occur in two halves and fit together like a petri plate. All diatoms are symmetrical. Diatoms reproduce both asexually and sexually (see Figure 28.19). Diatom cell walls have significant human uses from filters to reflective paint.

- Brown algae are multicellular and are the largest of the protists.

  - Brown algae are almost exclusively marine. They produce leaflike growths called thalli. The giant kelp that grow in large oceanic "forests" are brown algae.

  - Brown algae are brown because of chlorophylls *a* and *c* and fucoxanthin. They have specialized regions called holdfasts, which anchor them to a substrate and allow them to resist water movement.

- Brown algae are commercially important for the presence of alginic acid, used as a binder in many food and cosmetic products.

- Like many other protists and all plants, brown algae go through alternation of generations in which a diploid sporophyte gives rise to haploid spores through meiosis to form the haploid gametophyte, which, in turn, produces haploid gametes (see Figure 28.22). Heteromorphic alternation of generations occurs when the two generations differ morphologically, and isomorphic alternation of generations occurs when the two generations are morphologically similar.

- Oomycetes are a nonphotosynthetic group of stramenopiles. These are commonly known as water molds and downy mildews, but they are not fungi. Oomycetes are coenocytic, having multiple nuclei per cell. Many are saprobic and feed on dead material. Others are infectious to plants. The Irish potato famine was caused by an oomycete.

The red algae are mostly multicellular and are part of a monophyletic group.

- The red algae are characterized by the pigment phycoerythrin, which gives them their characteristic red color. Depending on light intensity, red algae alter the ratio of pigments in the chloroplast. In high light conditions, red algae may appear very green.

- Most red algae are large marine organisms and have specialized structures.

- Red algae store the products of photosynthesis as floridean starch and have no motile cells at any point in their life cycles.

- Red algae are the source of agar, which is important in food products as well as in the study of biology.

The clade that includes the green algae also includes plants.

- The green algae are not a monophyletic group. This is because plants are also descended from the most recent common ancestor of all green algae, but plants are not included among the green algae. Chlorophyta, the largest lineage of green algae, is a monophyletic group.

  - Chlorophytes contain chlorophylls *a* and *c*, store photosynthetic products as starch in plastids, and are mainly aquatic.

  - Morphology of the chlorophytes is diverse and ranges from simple single-celled organisms to very complex multicellular organisms. Colonial and aggregate forms exist. Colonial forms show the first indications of cell specialization without multicellularity. Individual cells range from single-nucleated to coenocytic.

  - Life cycles within the chlorophytes are diverse. Figures 28.26 and 28.27 illustrate the life cycle of *Ulva* and *Ulothrix*, respectively. Some chlorophytes are isogamous, indicating that their gametes are identical. Others are anisogamous with female gametes much larger than the male gametes. Some chlorophytes have isomorphic life cycles, and others are heteromorphic. In one variation of the heteromorphic life cycle—the haplontic life cycle—a multicellular haploid individual (gametophyte) produces gametes that fuse to form a zygote. Some chlorophytes exhibit diplontic life cycles much like animals, in which meiosis of sporophytes produces gametes directly.

Choanoflagellates are thought to be the closest relatives of the kingdom Animalia.

- Choanoflagellates are flagellated, and the majority are colonial.
- Choanoflagellates are structurally similar to sponges, the most basal animal lineage.

Endosymbiosis is the origin of chloroplasts.

- Some chloroplasts are enclosed in double membranes and others are enclosed in triple membranes. This phenomenon can be explained by endosymbiosis.
- All chloroplasts can be traced to the engulfment of an ancestral cyanobacterium with a single membrane. Its engulfment in a vesicle of the phagocyte's membrane resulted in a double membrane. This initial event is called primary endosymbiosis. Primary endosymbiosis gave rise to the green and red algae.
- Photosynthetic euglenoids, with their triple membranes, arose from secondary endosymbiosis. In this case, a chlorophyte was ingested with its housed chloroplast, resulting in three membranes. Ultimately all of the constituents of the chlorophyte except the chloroplast were lost. Another line is thought to have arisen by means of a red algal endosymbiont.

Some body plans recur in several protist lineages.

- The amoeboid body plan is seen in several groups of protists. It is characterized by pseudopods, cytoplasmic extensions that are used in locomotion. This body plan is highly effective in nutrient-rich environments. Some organisms have an amoeboid body plan at different stages of their life cycle.
- Actinopods have very thin pseudopods reinforced by microtubules. The pseudopods increase surface area, help the organisms float, provide locomotion, and are used in feeding. Radiolarians, a specific type of actinopod, secrete glassy endoskeletons. Other types, called, heliozoans, lack an endoskeleton.
- Foraminiferans secrete shells of calcium carbonate and live either as plankton or on the sea bottom. Their pseudopods are used to trap food. Limestone deposits are often the result of discarded foraminiferan skeletons.
- Slime molds occur in three distinct groups. All groups are motile, feed by endocytosis, form spores on erect fruiting bodies, and undergo drastic changes during their life cycle.
  - Acellular slime molds form multinucleate masses. During the vegetative stage, an acellular slime mold is a wall-less mass of cytoplasm with multiple diploid nuclei. It oozes over a network of strands called plasmodium. This phenomenon is called cytoplasmic streaming and is used to move and engulf food. If exposed to adverse environmental conditions, the slime mold will form a dormant sclerotium from which it can turn back into a plasmodium when environmental conditions are right. It may also transform into a fruiting structure called a sporangiophore.

- Sporangiophores form rigid walls between the nuclei. The walled nuclei go through meiosis to become spores. Spores germinate into swarm cells, which can divide to produce more swarm cells. They can also release flagellated swarm cells, which fuse to form a zygote, which can then form a new plasmodium.
- Cellular slime molds are made of large numbers of cells called myxamoebas with single haploid nuclei. This life cycle can continue as long as conditions are favorable. Once conditions become unfavorable, myxamoebas secrete cAMP, aggregate, and form a motile slug that eventually produces fruiting bodies. Spores from the fruiting bodies germinate to form more myxamoebas. Sexual reproduction occurs with the fusion of two myxamoebas, which then undergo meiosis to release haploid myxamoebas.

The Big Picture

- The many lineages of protists as well as all fungi, animals, and plants are eukaryotic. Eukaryotic cells are characterized by membrane-enclosed organelles, the presence of a cytoskeleton, and a nuclear envelope. The evolution of the eukaryotic cell is a major evolutionary milestone. The eukaryotic cell allows for compartmentalization of processes and increased adaptability. Though the evolutionary lineage is still debated, it is thought that eukaryotes evolved via multiple symbiotic events.
- Protists as a group are not monophyletic. This is because the most recent common ancestor of protists gave rise to the plants, animals, and fungi, and these groups are not included in the Protista. Protists are diverse, ranging from microscopic single-celled organisms to complex multicellular organisms. Several subgroups of protists are covered in this chapter. Some groups are monophyletic, others are not.
- Certain body plans and modes of nutritional acquisition are repeated throughout the protist groups and are good examples of form following function. Protists as a group play a vast role in the environment and medicine and are economically valuable.

Common Problem Areas

- There is a tendency for students to become overwhelmed with organism names and characteristics. Focus on the trends outlined in the chapter and use lab opportunities to understand the organism groupings.

Study Strategies

- This chapter has a lot of information in it. It is best not to try to learn it all at once, but to break it up into smaller pieces and study it over several sessions.
- In order for you to learn the organism groups, it is best to create a chart with key characteristics. This will help you to compare and contrast the groups.
- Review the following tutorials and activities on the Student Website/CD:
  Tutorial 28.1 Food Vacuoles Handle Digestion and Excretion
  Tutorial 28.2 Family Tree of Chloroplasts
  Activity 28.1 Anatomy of a *Paramecium*
  Activity 28.2 An Isomorphic Life Cycle
  Activity 28.3 A Haplontic Life Cycle

Test Yourself
Knowledge and Synthesis Questions

1. Which of the following modes of reproduction can be found in at least some protists?
   a. Binary fission
   b. Sexual reproduction
   c. Spore formation
   d. All of the above

2. During the evolution of eukaryotes from prokaryotes, which of the following did *not* occur?
   a. Infolding of the flexible cell membrane
   b. Loss of the cell wall
   c. A switch from aerobic to anaerobic metabolism
   d. Endosymbiosis of once free-living prokaryotes

3. Which of the following statements about protists is *false*?
   a. Apicomplexans are the only protist group without parasitic representatives.
   b. Foraminiferans and radiolarians are shelled protists.
   c. Ciliates have great control over the direction their cilia beat.
   d. Although they appear structurally simple, amoebas are not primitive organisms.

4. Ciliates, as represented by paramecia, have defensive organelles called _____ in their pellicles.
   a. trichonympha
   b. trichocysts
   c. trichomes
   d. throclea

5. A major difference between the vegetative states of cellular and acellular slime molds is that acellular slime molds
   a. have haploid nuclei, and cellular slime molds have diploid nuclei.
   b. produce fruiting bodies, and cellular slime molds do not.
   c. use cAMP as a chemical cue for aggregation, and cellular slime molds do not.
   d. exist as a coenocytic mass, and cellular slime molds exist as individual myxamoebas.

6. Organisms that exhibit isogamy have
   a. similar male and female gametes.
   b. female gametes that are significantly larger than male gametes.
   c. male gametes that are significantly larger than female gametes.
   d. a diplontic mode of reproduction.

7. Red tides often cause massive fish kills and human illness in those eating shellfish. Which group of protists is responsible for red tides?
   a. Parabasalids
   b. Red algae
   c. Euglenozoans
   d. Dinoflagellates

8. Holdfasts and alginic acid are characteristic of which group of protists?
   a. Parabasalids
   b. Red algae
   c. Brown algae
   d. Stramenopiles

9. A water mold is discovered growing on the surface of a freshwater green alga. Which two groups of protists are involved?
   a. Choanoflagellates and red algae
   b. Oomycetes and chlorophytes
   c. Apicomplexans and chlorophytes
   d. Apicomplexans and ciliophora

10. Which of the following statements regarding conjugation in *Paramecium* is *false*?
    a. Conjugation results in genetic recombination.
    b. Conjugation results in clones.
    c. Conjugation results in offspring.
    d. Both b and c

11. Which of the following statements regarding protists in general is *false*?
    a. Protists are always parasitic.
    b. Protists are all single celled.
    c. Protists are all heterotrophic.
    d. All of the above

12. According to Figure 28.9, which of the following statements is true?
    a. The alveolates are more closely related to the stramenopiles than to any other group.
    b. The most recent common ancestor of the parabasilids and euglenoids also gave rise to the animals.
    c. The red algae are more closely related to the brown algae than they are to any other taxon.
    d. The opisthokonts are not monophyletic.

Application Questions

1. Explain alternation of generations. Why are the protists the first group in which this process could be seen?

2. Differentiate between asexual and sexual reproduction. Select one mode of protist sexual reproduction and explain it fully.

3. You have identified a new organism. What would the criteria be for placing it with the protists?

4. Explain one line of thought regarding the origin of the eukaryotic cell. Why are scientists unsure of the evolutionary relationships among the protists?

5. Most protists are motile. Describe some of the means of motility.

6. Describe the origin of double membrane-enclosed and triple membrane-enclosed chloroplasts.

7. Many protists are significant human pathogens. Select one pathogenic protist, describe its life cycle, identify the protist group to which it belongs, and discuss the disease it causes.

8. Draw the phylogeny of the three domains of life and add the various lineages of protists to it. Now add the plants, animals, and fungi.

Answers
Knowledge and Synthesis Answers

1. **d.** Methods of reproduction are quite varied among the protists.

2. **c.** At the time the first eukaryotes evolved, the environment was becoming oxygen-rich. There was a switch from anaerobic to aerobic metabolism, not vice versa.

3. **a.** Malaria is caused by *Plasmodium*, which is an apicomplexan. In fact, all apicomplexans are parasitic.

4. **b.** Trichocysts are defensive barbs ejected from ciliates when they are disturbed.

5. **d.** The vegetative (feeding) state of an acellular slime mold is called a plasmodium; it consists of multiple diploid nuclei enclosed in a single membrane. The vegetative state of a cellular slime mold is a myxamoeba with a single haploid nucleus. Although the myxamoebas of cellular slime molds do aggregate to form fruiting structures, individual myxamoebas never fuse into multinucleated structures.

6. **a.** *Isogamy* means *same gametes*. The gametes appear identical.

7. **d.** Dinoflagellates are the cause of red tides.

8. **c.** The brown algae are multicellular protists notable for their organ and tissue differentiation and the presence of alginic acid in their cell walls.

9. **b.** Water molds are oomycetes and green algae are chlorophytes.

10. **d.** Conjugation does result in genetic recombination but does not result in the production of clones or offspring.

11. **d.** Protists do not have a unifying characteristic other than being eukaryotic organisms that do not fit into the kingdoms Plantae, Animalia, or Fungi.

12. **b.** Scientists are still working out the evolutionary relationships among the protists, and Figure 28.9 represents a conservative simplification of current ideas. This phylogenetic tree takes no stand on the relationships among the alveolates, stramenopiles, red algae, green algae and plants, and opisthokonts, so answers a and c are incorrect. The opisthokonts, which include animals and fungi in addition to choanoflagellates, appear to be monophyletic, so answer d is incorrect. Although the parabasilids and euglenoids appear next to each other in the tree, their most recent common ancestor gave rise to all the groups listed.

Application Answers

1. In alternation of generations an organism exists in a haploid gamete-producing form and a diploid spore-producing form. Prokaryotes have a single chromosome, so only in eukaryotes, with multiple copies of chromosomes, is a diploid stage of a life cycle possible.

2. Asexual reproduction results in clones of the original organism, and there is no genetic recombination or variation associated with the creation of offspring. Sexual reproduction allows for new genetic combinations. Protists undergo varied types of sexual reproduction, from fusing haploid myxamoebas in cellular slime molds, to haplontic alternation of generations, to a diplontic type of reproduction. Conjugation between paramecia is an example of sexual recombination without reproduction.

3. The organism must be a eukaryote and not fit the criteria for plant, animal, or fungus.

4. See Figure 28.3. The evolutionary relationships are difficult to understand due to limited fossilization, lateral gene transfers, and the contributions of symbiotic prokaryotes.

5. Protists move by flagella, cilia, and pseudopodia. Because many are aquatic, additional modifications allow for floating and movement along currents.

6. Double membrane-enclosed chloroplasts most likely arose from the endosymbiosis of a cyanobacteria in a process called primary endosymbiosis; triple membrane-enclosed chloroplasts most likely arose from the endosymbiosis of a chlorophyte or a rhodophyte (see Figure 28.29).

7. You may select from a variety of pathogenic protists: malaria caused by one of the apicomplexans, African sleeping sickness caused by a kinetoplastid, or others. See Figure 28.14 for the life cycle of *Plasmodium*, the protist that causes malaria.

8. This takes practice! Begin by drawing just the Bacteria, Archaea, and Eukarya. Now "add" the branches representing groups of protists, animals, plants, and fungi. Check your final tree against Figure 28.9.

29

Plants without Seeds: From Sea to Land

Important Concepts

The plant kingdom is monophyletic.

- Plants are photosynthetic eukaryotes that utilize chlorophylls *a* and *b* to produce and store carbohydrates as starches.
- All plants develop from an embryo that is protected by tissues of the parent plant.
- Plants undergo alternation of generations.
  - Haploid gametophytes produce haploid gametes through mitosis, and diploid sporophytes produce haploid spores through meiosis.
  - The diploid sporophyte arises from the fusion of gametes to produce a diploid zygote that develops into the mature sporophyte.
- The extent of the development of the two life stages is characteristic of the different phyla.
  - Nontracheophytes typically have a dominant gametophyte. The small sporophyte is nutritionally dependent on the gametophyte.
  - In the nonseed tracheophytes, the sporophyte is dominant, but the gametophyte, though small, is nutritionally independent of the sporophyte.
  - In the most recently evolved tracheophytes, the sporophyte is dominant; the gametophyte is greatly reduced and nutritionally dependent on the sporophyte.
- There are ten extant (surviving) groups of plants (see Table 29.1). Plants as a whole are monophyletic, but the nontracheophytes, as a group, are paraphyletic.
- Molecular and fossil evidence indicates that plants arose from a green algae ancestor. Plants most likely arose from a group of green algae called charophytes that lived at edges of marshes and ponds between 400 and 500 mya.

Survival on land poses a series of challenges.

- To survive on land, an aquatic organism must evolve ways to reduce its dependence on water.
- Aquatic organisms can float in their environment. Terrestrial plants need to support themselves against gravity if they are to grow upward.
- Terrestrial plants need a mechanism to move water to all their cells.
- Terrestrial plants need a mechanism to prevent loss of water to evaporation.

- The sperm of aquatic organisms can swim in search of an egg. Terrestrial plants need some other mechanism to move sperm from one plant to another.
- Terrestrial plants need a mechanism to prevent desiccation of the developing embryo.

Many characteristics that distinguish plants from green algae are adaptations to a terrestrial environment.

- The earliest plants developed the following modifications for terrestrial life: a waxy cuticle, gametangia, embryos, protective pigments, thick spore walls, and mutualistic associations with fungi.
  - Waxy cuticles, though minimal in early plants, did prevent water loss from tissues as they were exposed to dry air. The extent of the cuticle varies among plant groups and is most well-developed in the more recently evolved gymnosperms and angiosperms.
  - Multicellular gametangia mark the transition from algae to plant. Gametangia enclose the gamete-producing tissue and protect the developing gametes.
  - Plant embryos are significant in that they contain the developing sporophyte and are housed within protective tissues. Recently-evolved angiosperms have a complex protective mechanism.
  - Water protects aquatic organisms from damaging ultraviolet radiation. To compensate for the loss of this protection, plants have evolved pigments that screen out UV radiation.
  - Spore walls protect spores from desiccation during dispersal. This is particularly important in the nonseed plants.
  - Mutualistic associates with fungi help with the absorption of nutrients from the soil that might otherwise be inaccessible to plants.

The development of vascular tissue was key to terrestrial existence.

- Vascular tissue provides for the transport of water and food through a plant. The fundamental unit of a vascular system is the tracheid, a strawlike cell that conducts water in the plant. Some mosses have cells that are precursors to tracheid development, but these are not true tracheids.
- Plants without vascular tissue depend on abundant water sources, thin tissues that can absorb water easily, mechanisms for capturing water vapor in the air, and

capillary action to move water through the plant. They lack the support of lignin and closely hug the ground.

- Not all land plants have vascular tissue; therefore, it is not an absolute requirement for terrestrial life. Plants that lack vascular tissue, however, cannot grow very large or live far from a water source. The evolution of vascular tissue has allowed plants to occupy many niches in the biosphere that would otherwise be inaccessible.

Nontracheophytes were the first plant colonizers of land.

- Most nontracheophytes live in dense mats that hug the ground in moist environments.

- The life cycle of the nontracheophyte is dominated by the gametophyte generation. The sporophyte is very tiny and is completely dependent on the gametophyte. Figure 29.5 illustrates the life cycle of a moss as an example of the nontracheophyte life cycle. You should be able to follow this life cycle and understand which stages are haploid, which are diploid, and where meiosis takes place.

- Liverworts (Hepatophyta) may be the oldest of all plant lineages.

  - The liverwort gametophyte is usually a flat plate of cells (see Figure 29.7). Gametophytes produce antheridia and archegonia on their upper surfaces (sometimes on stalks) and rhizoids (rootlike structures for anchoring and water absorption) on their lower surfaces.

  - The liverwort sporophyte has a stalk that, in many species, is elongated to aid in spore dispersal. Some sporophytes also have structures that forcefully eject spores.

  - Liverworts can reproduce asexually by shedding small clumps of cells called gemmae.

- Hornworts (Anthocerophyta) evolved simple stomata, pores that help with the exchange of gas and water vapor. Unlike stomata in other plants, however, hornwort stomata are unable to close.

  - Hornworts are distinguished from other nontracheophytes by two characteristics: the presence of a single, platelike chloroplast in each cell, and sporophytes that are capable of indeterminate growth. The hornwort sporophyte resembles a long, slender horn (see Figure 29.8).

  - Hornworts often have symbiotic relationships with cyanobacteria that are able to fix atmospheric nitrogen and make it available to the hornwort.

- Mosses (Bryophyta) are the most familiar nontracheophytes.

  - Mosses have specialized cells called hydroids. These hydroids are thought to be precursors to the tracheid seen in tracheophytes, but they lack lignin for structural support. Like tracheids, the hydroid cell dies and leaves a tiny channel that can transport water through the plant.

  - Mosses also differ from other nontracheophytes in that they exhibit apical cell division. This characteristic is seen in "higher" plants and allows for regular, upright growth.

  - The moss life cycle is presented in Figure 29.5.

The tracheophytes have specialized vascular tissues.

- All tracheophytes are characterized by the presence of water-conducting tracheids. Tracheids were a significant advance in the evolution of plants, providing a means of water transport and support for growth.

- Tracheophytes fossilize much better then nontracheophytes. Thus, our morphological knowledge of early tracheophytes is much more extensive than our knowledge of nontracheophytes.

- Tracheophytes are also characterized by branching, independent sporophytes.

- Figure 29.10 illustrates the relationships between the different groups of tracheophytes.

- The first tracheophytes were members of a now-extinct group of plants called the rhinophyta. *Rhynia* had characteristics of extant tracheophytes and showed the earliest features of this group, including an independent, branched sporophyte and tracheids, but they lacked more advanced characteristics such as true leaves and roots (see Figure 29.12).

Significant new features arose in early tracheophytes.

- Roots are believed to have evolved over time from branches that grew underground. Because branches above ground and below ground would have been exposed to different selective pressures, underground branches might have given rise to what we know today as roots.

- True leaves are flattened photosynthetic structures with vascular tissue. Tracheophytes have evolved two different kinds of leaves: microphylls and megaphylls (see Figure 29.13).

  - Microphylls are present in the club mosses. They probably evolved from sterile sporangia.

  - Most familiar leaves are megaphylls. These are thought to have evolved when photosynthetic tissue developed between the ends of small, lateral branches.

- Heterospory appears to have evolved several times in the tracheophytes (see Figure 29.14).

  - Most early tracheophytes exhibit homospory. In homospory, spores (and the gametophytes that grow from them) are all of the same type; gametophytes produce both archegonia and antheridia. In heterospory, one spore type called a megaspore gives rise to the female, egg-producing megagametophyte; another spore type called a microspore develops into a male, sperm-producing microgametophyte.

  - Plants typically produce many more microspores than megaspores.

Extant nonseed tracheophytes are varied and abundant.

- Club mosses (Lycophyta) diverged from the tracheid lineage relatively early and are the most "primitive" of the extant nonseed tracheophytes. They have simple leaves arranged spirally on the stem. Their sporangia are held in apical strobili, which are simple branching structures of fertile sporangia and sterile leaves (see Figure 29.15).

- The horsetails, whisk ferns, and ferns were once treated as separate phyla, but recent evidence suggests that they form a clade, the Pteridophyta.

- Horsetails are represented by few extant species. They have true roots, their sporophytes are large and independent, and their gametophytes are highly reduced but also independent. The leaves of horsetails are simple, forming whorls around the stem (see Figure 29.16). This group is one of the few plant groups that exhibit basal growth.

- Whisk ferns once were thought to be the "missing link" to *Rhynia*, but molecular evidence indicates that they evolved much later from more highly complex plants. Their relatively simple body plan works well and demonstrates that evolution does not necessarily move toward greater complexity.

- Ferns are the largest surviving group of nonseed tracheophytes. Ferns are characterized by large complex leaves with branching vascular strands. Like all other nonseed tracheophytes and nontracheophytes, ferns continue to be dependent on water to carry motile sperm. Ferns have advanced vascular structures and may reach great heights, but they do not produce true wood. Their root systems are poorly developed.

The fern life cycle has independent sporophytes and gametophytes.

- The fern life cycle is dominated by the sporophyte, but the gametophyte is an independent photosynthetic structure (see Figure 29.20).

- Most ferns are homosporous, but some fern groups are heterosporous. A few genera of ferns produce gametophytes that depend on a mutualistic fungus for nutrition.

The Big Picture

- Plants are photosynthetic eukaryotes that utilize chlorophylls *a* and *b*, undergo alternation of generations, and develop from multicellular embryos that are protected by the parent plant. The ten extant groups can be classified as nontracheophytes (those without highly-developed vascular tissue) and tracheophytes (those with highly-developed vascular tissue). This chapter focuses on the nontracheophytes and the nonseed tracheophytes.

- In order to colonize land, plants had to evolve strategies for dealing with desiccation and gravity. This involved mechanisms for extracting water from soil, means of transporting water throughout the plant, methods of ensuring fertilization, and modes of protecting developing embryos.

- This chapter introduces and describes the liverworts, hornworts, and mosses, which are all considered nontracheophytes, and the club mosses, horsetails, whisk ferns, and ferns, which are nonseed tracheophytes.

Common Problem Areas

- It is very tempting to memorize plant types and characteristics, but it is more important to focus on the trends and relationships among the plants. Your goal is to understand the evolutionary relationships among the plants.

Study Strategies

- Focus on relationships among the plants and how to differentiate one group from another. Many of the important distinctions between groups are related to adaptations for terrestrial life.

- When studying life cycles, be sure to note whether the sporophyte or the gametophyte is dominant and make note of where mitosis and meiosis take place.

- Review the following tutorial and activities on the Student Website/CD:
  Tutorial 29.1 Life Cycle of a Moss
  Activity 29.1 Homospory
  Activity 29.2 Heterospory
  Activity 29.3 Fern Life Cycle

Test Yourself

Knowledge and Synthesis Questions

1. A plant may be reliably distinguished from a green alga by which of the following characteristics?
   a. Chlorophyll type
   b. The presence of an embryo protected by parent tissue
   c. The presence of roots
   d. Swimming sperm

2. Which of the following characteristics was (were) necessary for plants to colonize land?
   a. Vascular tissue for moving water throughout the plant
   b. A mechanism to prevent desiccation of tissues
   c. The ability to screen ultraviolet radiation
   d. Both b and c

3. The main difference between nontracheophytes and tracheophytes is
   a. lack of gametophytes.
   b. spore production.
   c. the presence of tracheids.
   d. All of the above

4. In alternation of generations, the sporophyte generation is _____ and the gametophyte generation is
   _____.
   a. haploid; diploid
   b. diploid; haploid
   c. haploid; haploid
   d. diploid; diploid

5. Which of the following characteristics helps to differentiate between liverworts and mosses?
   a. The presence of hydroids
   b. Gametophyte dominance
   c. Sporophyte dominance
   d. Swimming sperm

6. During a plant's life cycle, meiosis takes place in the _____ to produce haploid _____.
   a. gametophyte; gametes
   b. sporophyte; gametes
   c. sporophyte; spores
   d. gametophyte; spores

7. Asexual reproduction in liverworts is accomplished by
   a. gametophytes.
   b. spores.
   c. gemmae.
   d. tracheids.

8. Which of the following limits the size of a hornwort's sporophyte?
   a. The sporophyte's ability to distribute water to all its cells
   b. The gametophyte's ability to produce enough nutrients for the sporophyte
   c. Developmental genes that prevent growth of the sporophyte beyond a certain size
   d. All of the above

9. You are walking along a roadside and find a plant with the following characteristics: a very thin, waxy cuticle, stomata, simple leaves in whorls around a central stem, independent sporophytes and gametophytes, and sporangia in strobili. This plant is most likely a member of which of the following phyla?
   a. Bryophyta
   b. Pteridophytes
   c. Anthocerophyta
   d. Lycophyta

10. Match the phyla with their diagnostic characteristics.
    ____ Mosses   a. No stomata
    ____ Liverworts   b. Descended from a more
    ____ Hornworts   complex ancestor
    ____ Club mosses   c. Complex branching leaves
    ____ Horsetails   d. Unique chloroplasts
    ____ Whisk ferns   e. First strobili of sterile
    ____ Ferns   leaves and sporangia
                         f. Hydroids
                         g. Basal growth

Application Questions

1. Explain why the largest mosses are less than a meter tall.

2. Diagram the evolutionary relationships among the nontracheophytes and nonseed tracheophytes. Label the major differentiating characteristics at each branch of your tree.

3. Discuss how the relationship between sporophytes and gametophytes changes as you move from the first nontracheophytes to later nonseed tracheophytes.

4. Compare and contrast the moss life cycle and the fern life cycle.

5. Discuss the challenges of terrestrial life that plants needed to address for survival.

6. Compare and contrast homospory and heterospory. What reproductive structures result from meiosis in each type of life cycle, and what structures result from mitosis?

Answers

Knowledge and Synthesis Answers

1. **b.** By the definition your textbook uses, all plants produce embryos that are protected by tissue of the parent plant. Green algae and plants use the same types of chlorophyll. Not all plants have roots, so a plantlike organism lacking roots will not necessarily be a green alga.

2. **d.** There are several successful groups of terrestrial plants that lack vascular tissue.

3. **c.** Tracheids are found only in the tracheophytes.

4. **b.** Meiosis occurs in the diploid sporophyte to produce haploid spores that develop into the haploid gametophyte.

5. **a.** Mosses are the only group with hydroids, the precursors to true vascular tissue.

6. **c.** The outcome of meiosis is four cells, each of which has half the genetic material of the parent cell. A haploid cell

already has only half the normal number of chromosomes of most eukaryotes, so a cell must be diploid (or have higher ploidy) to undergo meiosis. In all plant life cycles, the sporophyte is diploid and the gametophyte is haploid; therefore, only the sporophyte can undergo meiosis. Spores are the products of sporophyte meiosis.

7. **c.** Liverworts have specialized asexual reproductive structures called gemmae.

8. **a.** The sporophyte of a hornwort is nutritionally dependent on the gametophyte, but its growth is indeterminate—the sporophyte will continue to grow as long as water is able to reach all its cells.

9. **b.** You have come across a horsetail. These plants are very common along roadsides in damp ditches, particularly in the Midwest.

10. **f.** Mosses
    a. Liverworts
    d. Hornworts
    e. Club mosses
    g. Horsetails
    b. Whisk ferns
    c. Ferns

Application Answers

1. Mosses have very rudimentary water transport cells called hydroids. Hydroids lack the waterproofing and support molecule lignin. Because of this, they can carry water only short distances and cannot support tall growth.

2. See Figure 29.4 and Table 29.1.

3. Early nontracheophytes have reduced sporophytes that are highly dependent on the gametophyte for nutrition. Nonseed tracheophytes, which have evolved more recently, have independent sporophytes and gametophytes, and the gametophyte is highly reduced.

4. See Figures 29.5 and 29.20. Pay particular attention to the relative dominance of the sporophyte versus the gametophyte.

5. Terrestrial life poses problems of support, water conduction, UV radiation, water loss, and embryo protection. The tracheid was an important evolutionary innovation that helped with both water conduction and support. Special pigments protect against UV radiation. A waxy cuticle prevents water loss from cells. Plant embryos are protected in early development by parental tissue.

6. See Figure 29.14. In homospory, meiosis results in a single type of spore, but in heterospory, meiosis results in megaspores and microspores. Mitosis occurs throughout both types of life cycle as cells divide and plants grow, but the reproductive cells that result from mitosis in both homospory and heterospory are sperm and eggs. In homospory, the antheridium produces sperm by mitosis, and the archegonium produces eggs; in heterospory, the microgametophyte produces sperm and the megagametophyte produces eggs.

30

The Evolution of Seed Plants

Important Concepts

Seed plants have characteristics that set them apart from nonseed plants.

- Seed plants consist of two groups of tracheophytes: the gymnosperms and the angiosperms. Gymnosperms include the following phyla: Cycadophyta, Ginkgophyta, Pinophyta, and Gnetophyta.

- Seed plants have highly reduced gametophyte generations that are dependent on the sporophyte for survival. When you look at a seed plant, what you see is the sporophyte.

- Very few seed plants have retained swimming sperm. Most have evolved other means for dispersing male gametes.

- All seed plants are heterosporous. Microspores and megaspores develop in specialized cones or in flowers.

- In the megasporangium, meiosis produces four megaspores, but in seed plants only one of the four megaspores is retained. This megaspore divides by mitosis to form the multicellular (but tiny) megagametophyte, which produces the eggs.

- When the eggs are ready to be fertilized, they are surrounded by megagametophyte cells, which in turn are still within the megasporangium. The megasporangium is surrounded by sterile sporophyte tissues (the integument).

- Microspores develop into pollen grains, the male gametophytes. Pollen grains consist of sperm and supporting cells; they are dispersed by wind and animals.

- Because the female gametophyte is retained within sporophyte tissue, pollen grains do not have direct access to gametophytes and their eggs. The sporophyte housing a gametophyte creates tissue for receiving pollen grains; upon reaching this tissue, pollen produces pollen tubes that deliver the sperm through the sporophyte tissue to the eggs for fertilization.

- The embryo resulting from fertilization grows to a certain size and then becomes dormant within the surrounding tissues. This dormant, protected embryo and its surrounding tissues constitute the seed.

Seeds are complex structures that may contain tissue from three plant generations.

- Recall that alternation of generations involves a multicellular sporophyte generation that alternates with a multicellular gametophyte generation. The embryo within a seed represents the beginning of a new sporophyte generation. It is still surrounded by its parent, the gametophyte, which will provide the embryo with nutrients to begin growth (particularly if it is a gymnosperm). The tough coat that surrounds the seed consists of tissue provided by the embryo's sporophyte grandparent.

- Seeds protect the embryo until conditions are right for germination. Many seeds have adaptations that allow dispersal by wind or another vector.

Gymnosperms are naked-seed plants.

- Gymnosperms are seed plants but they do not form flowers or true fruits. They are diverse and live in a variety of habitats, from sparse deserts to vast forests. The Pinophyta are the most abundant of the gymnosperms.

- Gymnosperms have significant vascular tissue and are capable of woody secondary growth.

- The gymnosperm life cycle can be illustrated by that of the pine, a conifer (see Figure 30.6). Conifers differ from other gymnosperms in that they produce cones and strobili, which are specialized structures for reproduction. Cones house the megasporangia and produce megaspores, megagametophytes, and eggs; strobili house the microsporangia and produce microspores, microgametophytes (pollen), and sperm.

- Pines do not have swimming sperm, and their pollen is modified for wind dispersal. Pollen lands on the female cone and lodges in the pollen chamber. The pollen tube grows through the maternal sporophyte tissue to the female gametophyte, where it releases two sperm. Only one of the two sperm will fertilize an egg. The resulting zygote develops into an embryo that remains encased in the tissues of the megasporangium and gametophyte. The seed is also protected by the scale of the cone (sporophyte tissue) until it is mature and ready for dispersal. Dispersal is aided by modifications of the seed coat.

Angiosperms are characterized by flower formation.

- Angiosperms are a diverse group of plants that all produce flowers and are thus called flowering plants. They are presently the dominant plant form on Earth.

- Angiosperms have the most reduced gametophyte generation and the most highly developed sporophyte generation of all plant lineages.

- Angiosperms differ from all other plants in that they produce triploid endosperm, they have double fertilization, their ovules are enclosed in a carpel, they produce flowers and fruits, and their xylem and phloem have multiple modified cell types including vessel elements, fibers, and companion phloem cells.

Flower structure is varied among plants but consists of similar structural elements.

- All flower parts are modified leaf structures. Although flowers are very diverse, they are constructed from the same small set of structures (see Figure 30.7).

- The male structures in a flower are stamens. Each stamen consists of a filament and an anther; the anther contains microsporangia, which produce pollen. Flowers often have several stamens.

- The female structure in a flower is the pistil, which consists of a stigma, a style, and an ovary. The stigma is modified to receive pollen. The style is a stalk that separates the stigma from the ovary. The ovary contains one or more ovules, each of which houses a megasporangium.

- Nonreproductive floral structures include petals, sepals, and the receptacle. Petals and sepals are often modified to attract animal pollinators. Sepals also serve to protect the developing flower bud. The receptacle is the attachment site for the sepals, petals, stamens, and carpels.

- Flowers that have both megasporangia and microsporangia are called perfect. If either structure is missing, the flower is called imperfect. Species that have both megasporangiate flowers and microsporangiate flowers on the same plant are called monoecious. If they are on separate plants, they are called dioecious.

- Many plants and animals have coevolved in such a way that the nutrition of the animal and pollination of the plant are interdependent. Some plants have evolved methods of limiting pollination to a single species of insect, but most can be pollinated by a range of species.

The angiosperm life cycle is dominated by the sporophyte.

- The angiosperm life cycle differs from that of all other plants in that double fertilization occurs (see Figure 30.11). In double fertilization, one sperm unites with the egg to produce a diploid zygote; the other sperm unites with two other haploid cells of the female gametophyte to produce a triploid cell. This triploid cell divides mitotically to create the (triploid) endosperm tissue. The endosperm provides nutrition for the developing embryo.

- Angiosperm embryos have one or two seed leaves called cotyledons. The number of cotyledons distinguishes the two main lineages of flowering plants: monocots have one cotyledon; eudicots have two. A few other relatively small groups of angiosperms do not fit into these two lineages (see Figures 30.13 and 30.14).

- Fruits develop from the ovary and its supporting tissues. Fruit types depend on the number of carpels associated with the fruit and the extent of support tissue incorporated into the fruit structure.

The evolutionary relationships among angiosperms have not yet been resolved.

- Different phylogenetic methods have led to different conclusions regarding angiosperm phylogeny. Investiga-

tors are still working on the question of how angiosperms first arose and whether they are sister to one gymnosperm phylum.

- Evidence now points to *Amborella* as the living species most similar to the first angiosperms. *Amborella* has a variable number of carpels and stamens, and it lacks vessel elements.

- Botanists recently discovered fossils of two species (both in the genus *Archaefructus*) that may belong to the sister taxon of all angiosperms. The flowers of *Archaefructus* had ovules enclosed in carpels, but no petals or sepals.

The Big Picture

- Seed plants can be characterized into two main groups: the gymnosperms and the angiosperms. Seed plants have developed complex ways of protecting embryos, dispersing gametes, and moving water, nutrients, and food throughout the plant. Both gymnosperms and angiosperms are characterized by the presence of seeds for protecting the embryo until conditions favor germination. The main difference between gymnosperms and angiosperms is the presence of flowers and ovaries in angiosperms. Both groups have highly developed vascular tissue, may exhibit secondary woody growth, and show significant diversity.

- There is great variation among flower types, but the basic flower consists of the structures shown in Figure 30.7. Many plants and animals have coevolved in such a way that the flower structure and the animal are integrally linked for nutrition of the animal and pollination of the plant. The angiosperm life cycle is different from that of all other plants in that double fertilization occurs.

- Both the angiosperm and gymnosperm life cycles are dominated by the sporophyte. Both groups are heterosporous. The roles of megasporangia and microsporangia are similar in both groups.

Common Problem Areas

- Not all gymnosperms are pines. The examples used in this chapter focus on the pine, but gymnosperms are a varied group, and cones are not a diagnostic characteristic of this group. Cones define the conifers only.

- Angiosperms are extremely varied as well. Not all flower parts are found on all plants, and some are highly modified. Also note that cultivated flowers are often sterile, have aberrant parts, and may be altered through selective breeding and treatment.

Study Strategies

- Much of the terminology in this chapter was introduced in Chapter 29, and it is worth reviewing that chapter in light of the material presented here. The details of heterospory can be particularly confusing; reviewing Figure 29.14 can help you to sort them out.

- Be sure that you truly understand what sets angiosperms and gymnosperms apart. Use Figures 30.6 and 30.11 to compare their life cycles.

- Review the following tutorial and activities on the Student Website/CD:
  Tutorial 30.1 Life Cycle of a Conifer
  Activity 30.1 Life Cycle of a Pine Tree
  Activity 30.2 Flower Morphology

Test Yourself
Knowledge and Synthesis Questions

1. You are enjoying a stroll in the botanical gardens. You notice a plant with a beautiful flower. Upon closer inspection you find that the flower has a pistil but no stamens. You look at several more flowers on the same plant, but you are unable to find any that have stamens. This plant is _____ and its flowers are _____.
   a. monoecious; perfect
   b. monoecious; imperfect
   c. dioecious; perfect
   d. dioecious; imperfect

2. Why are gymnosperms referred to as naked-seed plants?
   a. They lack ovules.
   b. They lack ovaries.
   c. They don't protect their embryos.
   d. They don't have seed coats.

3. Which of the following functions make seeds useful to plants?
   a. They provide a mechanism for dispersal.
   b. They protect the embryo.
   c. An embryo may remain dormant in a seed until optimum growth conditions are available.
   d. All of the above

4. Many angiosperms and animals have coevolved. What roles do animals play in the life cycle of plants?
   a. They act as pollinators.
   b. They assist in dispersal of seeds.
   c. They promote fertilization.
   d. All of the above

5. In gymnosperms, fertilization results in _____. In angiosperms, fertilization results in _____.
   a. a zygote; a zygote
   b. a zygote; an endosperm nucleus
   c. a zygote; a zygote and an endosperm nucleus
   d. a zygote and an endosperm nucleus; a zygote

6. The diagnostic characteristic of an angiosperm is the presence of
   a. multiple carpels.
   b. woody growth.
   c. a flower.
   d. All of the above

7. Which of the following statements regarding gymnosperms is true?
   a. All gymnosperms produce cones.
   b. Gymnosperms are heterosporous.
   c. Gymnosperm seeds have no protection.
   d. All gymnosperms are woody.

8. Match the following flower structures with their function.
   _____ Ovule   a. Assists in attracting pollinators
   _____ Anther   b. Secretes sticky material to help
   _____ Stigma   pollen adhere
   _____ Style   c. Site of ovule development
   _____ Ovary   d. Protects the immature flower bud
   _____ Petal   e. Houses the megasporangium
   _____ Sepal   f. Holds the stigma in position
                        g. Houses microsporangia

9. What is the purpose of a fruit?
   a. It aids in dispersal of seeds.
   b. It protects seeds until they are mature.
   c. It attracts pollinators.
   d. Both a and b

10. Vascular tissue in angiosperms is highly developed. The purpose of this vascular tissue is to move
    a. water.
    b. food.
    c. nutrients.
    d. All of the above

Application Questions

1. Discuss the significance of double fertilization in angiosperms.

2. Refer back to the obstacles encountered by plants when they began to colonize terrestrial environments. How have gymnosperms surmounted those obstacles?

3. Pollen is found only in seed tracheophytes. What is the significance of pollen in tracheophyte evolution? Is pollen a sporophyte or gametophyte?

4. Where would you find a female gametophyte in an angiosperm? Explain your answer.

5. What is a cotyledon? How is it used to classify angiosperms into two monophyletic groups?

6. What is the significance of the angiosperm fruit? What types of tissues does the fruit arise from?

Answers
Knowledge and Synthesis Answers

1. **d.** If, upon inspection, you find only female flowers, then you are most likely looking at a dioecious plant in which male and female flowers appear on separate plants. The absence of stamens makes these flowers imperfect.

2. **b.** Gymnosperms lack ovaries and thus lack the ability to produce fruit. Their embryos are protected, and they have ovules and seed coats.

3. **d.** Seeds protect the embryo, provide a mechanism for dispersal, and allow an embryo to remain dormant.

4. **d.** By acting as pollination vectors, animals promote fertilization of the plant. Animals also assist with dispersal of seeds.

5. **c.** Angiosperms differ from gymnosperms by having double fertilization that results in a triploid endosperm nucleus in addition to the zygote.

6. **c.** Only angiosperms have flowers. It is true that only angiosperms can have multiple carpels, but multiple carpels are not a necessary characteristic of an angiosperm.

7. **b.** All gymnosperms are heterosporous. Only conifers are cone producers.

8. **e.** Ovule
   g. Anther
   b. Stigma
   f. Style
   c. Ovary
   a. Petal
   d. Sepal

9. **d.** By the time fruit forms, a flower has already been pollinated, and there is no need to attract additional pollinators.

10. **d.** Vascular tissues in angiosperms move food, water, and nutrients throughout the plant.

Application Answers

1. Double fertilization allows for a diploid embryo and a triploid endosperm. The endosperm provides nutrition to the developing embryo at the time of seed germination.

2. Gymnosperms have a sophisticated vascular system that allows movement of nutrients and also support for large plants competing for sunlight. Wind-dispersed pollen removes the dependence on water for fertilization and increases the range of gymnosperms. Seeds protect embryos from desiccation and provide a mechanism for dispersal.

3. Pollen is the male gametophyte of seed tracheophytes; in other words, it is the male multicellular haploid structure of the seed tracheophyte's life cycle (see Figure 29.14*b*).

   As a gametophyte, pollen's most fundamental function is to create gametes (sperm, in this case), but it also plays a role in helping those gametes reach and fertilize female gametes. Pollen is significant in that it provides a means of dispersal that eliminates the need for water in fertilization.

4. The female gametophyte is highly reduced and is part of the ovule of the angiosperm.

5. A cotyledon is the first "seed leaf" produced in an angiosperm embryo. Monocots have only one seed leaf whereas eudicots have two.

6. Angiosperm fruit arises from tissues of the sporophyte and the gametophyte. How that fruit comes to develop depends on fruit type. Different fruits encompass more or less of the parent sporophyte tissues.

31 Fungi: Recyclers, Pathogens, Parasites, and Plant Partners

Important Concepts

Fungi are heterotrophs with chitin in their cell walls.

- Fungi absorb the nutrition needed for their survival from dead matter (saprobes), from living hosts, and from mutually beneficial symbiotic relationships with other organisms.

- Fungi can reproduce either asexually or sexually.

- The cell walls of fungi are made up of the polysaccharide chitin. The ability to produce chitin indicates that fungi share a more recent common ancestor with choanoflagellates and animals than with any other group.

- Most fungi belong to one of four phyla: Chytridiomycota, Zygomycota, Ascomycota, and Basidiomycota (see Table 31.1). These phyla are traditionally distinguished by structures and behaviors involved in sexual reproduction. Sexual stages have not yet been identified in about 25,000 species of fungi, so these have not yet been assigned to any of the four phyla. Such fungi are known as the imperfect fungi, or deuteromycetes.

- All of the fungal phyla contain unicellular fungi, commonly called yeasts.

- Yeasts reproduce by budding, by fission, or sexually.

The body of a fungus is a mycelium composed mainly of hyphae.

- Hyphae are tubular filaments that grow rapidly into a substrate. The hyphae in one mycelium may collectively grow as much as 1 km per day.

- Most fungi are multicellular, but multicellularity in fungi is significantly different from the familiar multicellularity of plants and animals. In plants and animals, each cell is usually enclosed in its own membrane and has only one nucleus. In multicellular fungi, cells are separated by porous septa that do not completely block the movement of organelles (see Figure 31.3). In many species, each cell has two nuclei; in some, even nuclei can pass through the septa.

- Chytrids and zygomycetes have coenocytic hyphae—hyphae with no septa between nuclei.

- Rhizoids are hyphae that anchor chytrids and some other fungi to the substrate.

- In fungi that are plant parasites, hyphae enter a leaf from a spore on the surface and form mycelium within the leaf (see Figure 31.4).

- Hyphae provide a large surface area-to-volume ratio and thus a large surface area for absorption of nutrients from the substrate.

- Fungi are tolerant of hypertonic environments and temperature extremes.

Fungi may be saprobes, pathogens, predators, or symbionts.

- Saprobic fungi secrete enzymes into the environment that help in absorption of dead matter and the decomposition and recycling of elements used by living organisms.

- Many fungi can use both nitrate and ammonium as a source of nitrogen.

- Parasitic fungi are either facultative (possessing the ability to grow independently) or obligate (dependent on their living host for growth).

- Some predatory fungi trap microscopic animals in a sticky substance; others form a constricting ring around their prey (see Figure 31.5).

- Fungi form two types of symbiotic relationships: Mycorrhizae are associations between fungi and the roots of plants in which a fungus obtains the products of photosynthesis from the plant and provides minerals and water to the plant. Lichens are associations between fungi and unicellular photosynthetic protists (green algae) or cyanobacteria.

Fungi reproduce sexually and asexually.

- Fungi have several means of asexual reproduction: They can produce spores enclosed within sporangia or naked spores known as conidia. Unicellular fungi can reproduce by fission or budding. Just about any part of a mycelium is capable of living independently of the rest; therefore, simply breaking a mycelium into two or more parts is a method of reproduction.

- Sexual reproduction in fungi involves two or more mating types rather than morphologically distinct males and females.

- Self-fertilization is prevented because individuals of the same mating type cannot mate with one another.

- The first step in sexual reproduction is fusion of two hyphae of different mating types. Eventually, two nuclei (one from each mating type) will fuse to create a zygote nucleus. The zygote nucleus may be the only diploid nucleus in the life cycle of a fungus.

- A dikaryon stage occurs in some fungi as the first stage of sexual reproduction. The cytoplasms of two individuals fuse, but the nuclei do not fuse right away, and the resulting hyphae contain genetically different haploid nuclei. A fruiting structure is eventually formed, and the two nuclei fuse to form a zygote. This form of reproduction has no gamete cells (only gamete nuclei). The hyphae are not truly diploid ($2n$), but dikaryotic ($n + n$).

Some fungi cause human diseases, but many more are plant pathogens.

- People with compromised immune systems, particularly people with AIDS, are susceptible to diseases caused by fungi, such as certain types of pneumonia, diarrhea, and candidiasis.

- Less-threatening fungal diseases include ringworm and athlete's foot.

- Most plant pathogens are fungi. Fungal crop diseases cause billions of dollars in damage annually.

Chytrids are aquatic fungi with flagellated gametes.

- Chytrids probably resemble the common ancestor of all fungi more closely than any other fungal group.

- Chytrids are aquatic, have chitinous cell walls, and are the only fungi with flagellated gametes. They can be either parasitic or saprobic.

- The genus *Allomyces* displays alternation of generations. A haploid zoospore becomes a small organism (the gametophyte) with both female and male gametangia that produce gametes (see Figure 31.7). When the gametes fuse, they form a diploid zygote and a diploid organism (the sporophyte). The sporophyte produces diploid zoospores that grow into other diploid sporophytes. Eventually, a sporophyte produces sporangia that give rise to haploid zoospores via meiosis.

Zygomycetes have coenocytic hyphae and (usually) no fleshy fruiting body.

- Over 900 species of zygomycetes have been described. The hyphae of a zygomycete spread randomly over a substrate, periodically producing stalked sporangiophores bearing sporangia.

- The zygomycetes include black bread mold and several species that form mycorrhizal associations with plant roots.

- Sexual reproduction in zygomycetes occurs between adjoining individuals with different mating types (see Figure 31.9). Branches from each individual grow toward each other until they fuse to produce gametangia. A thick-walled zygosporangium with a zygospore forms at the fusion site. A sporangium sprouts from the zygospore.

Ascomycetes include baker's yeast, truffles, and *Penicillium*.

- The ascomycetes are distinguished by their ascus, a sexual reproductive structure. There are approximately 30,000 species of ascomycetes, divided into two groups: The euascomycetes have an ascocarp, whereas the hemiascomycetes lack this structure.

- The hyphae of ascomycetes have regularly spaced septa that nonetheless do not prevent movement of nuclei and other organelles.

- Most hemiascomycetes are microscopic.

  - Unicellular hemiascomycetes can reproduce asexually by fusion or budding.

  - Sexual reproduction in hemiascomycetes occurs between cells of different mating types and involves the forming of a diploid cell population in some species or direct meiosis of the zygote in others. Meiosis results in the whole cell becoming an ascus with either eight or four ascospores.

  - The species of yeast used to make bread and alcoholic beverages, *Saccharomyces cerevisiae*, is a hemiascomycete.

- The euascomycetes produce a dikaryon during sexual reproduction.

  - Euascomycetes include molds and the cup fungi and have a number of uses for humans. Mold from the genus *Penicillium* produces the antibiotic penicillin; other euascomycete molds are important in making cheese. Brown molds are used in brewing sake (a Japanese alcoholic beverage) and soy sauce.

  - Asexual reproduction in euascomycetes involves the production of conidia at the tips of hyphae.

  - A dikaryon is produced during sexual reproduction by the fusing of two mating structures (see Figure 31.13). From the dikaryon hyphae ($n + n$), dikaryotic asci form in which the nuclei fuse. The zygote ($2n$) undergoes meiosis and then produces ascospores (n). The ascospores germinate and grow into new mycelia.

The fruiting structures of basidiomycetes are familiar as mushrooms.

- Puffballs and mushrooms make up some of the 25,000 species of fungi in the phylum Basidiomycota.

- In basidiomycetes, meiosis and nuclear fusion occur in the basidium, found at the tips of hypha (see Figure 31.15). After meiosis, four basidiospores are formed on tiny stalks. The basidiospores are released into the environment and germinate to form haploid hyphae. The hyphae grow and fuse with different mating types to form the dikaryotic mycelium. The fruiting structure, or basidiocarp, grows from the dikaryotic mycelium. The cap of a mushroom is the fruiting structure of a basidiomycete. Basidia form on the underside of the cap along gills and discharge their spores, which will form the new developing basidium.

Fungi form mutually beneficial symbioses with photosynthetic organisms.

- Almost all tracheophytes form associations with fungi. The root hairs of these plants become infected with fungi, forming associations called mycorrhizae. The fungus helps to increase the water and nutrient uptake of the plant, while obtaining nutrients such as sugars and amino acids from it.

- In ectomycorrhizae, fungal hyphae surround roots but do not enter root cells. In endomycorrhizae, hyphae penetrate into root cells (see Figure 31.16).

- The ability to form mycorrhizae may have been the most important step in the colonization of land by plants.

- Lichens are associations between fungi and single-celled photosynthetic organisms, either cyanobacteria, protists from a green algae lineage, or both. Ascomycetes and basidiomycetes form the fungal components of lichens.

- Lichens reproduce by the formation of a vegetative body called a thalus or by producing algae surrounded by hyphae in soredia (see Figure 31.18).

- Lichens are hardy enough to grow in terrestrial habitats inaccessible to plants, but they are very sensitive to environmental pollutants.

The Big Picture

- Fungi are heterotrophic organisms that absorb nutrients from the environment. Fungi can be unicellular or multicellular. The mycelium is the body of a multicellular fungus and is composed of many tubular filaments known as hyphae. Many fungi form symbiotic relationships with photosynthetic organisms, creating mycorrhizae and lichens.

- Reproduction in the fungi occurs both sexually and asexually. Asexual reproduction involves the production of spores, breakage, fission, or budding. Sexual reproduction in multicellular fungi requires the union of hyphae with two different mating types.

- The four phyla in the kingdom Fungi are the Chytridiomycota, Zygomycota, Ascomycota, and Basidiomycota.

  - The Chytridiomycota are aquatic fungi with flagellated gametes. Many chytrids display alternation of generations.

  - The Zygomycota have coenocytic hyphae. Sexual reproduction occurs when hyphae of two mating types join to form a zygosporangium from which a sporangium will eventually grow.

  - The Ascomycota produce an ascus as their reproductive structure. The ascomycetes are separated into the euascomycetes and the hemiascomycetes. The hemiascomycetes reproduce by budding, fission, or the production of an ascus. The euascomycetes have a dikaryotic ($n + n$) stage during reproduction.

  - In the Basidiomycota, the basidium is the reproductive structure. The fusion of haploid hyphae forms a dikaryotic mycelium that produces the fruiting body, a basidiocarp.

Common Problem Areas

- Students are often confused by the numerous ways that fungi reproduce. Create flowcharts to help you understand the ploidy level at each stage of the life cycle in the different phyla of fungi. Make sure to include both the sexual and asexual stages in your charts. Comparisons of your flowcharts for the different phyla will help you learn the differences between the groups.

Study Strategies

- Fungi are placed in particular systematic groupings because they share unique characteristics. Try to learn these features, along with the names of the groups, and look for patterns that distinguish major groups from each other. Most fungi are grouped according to the reproductive structures that they produce.

- Review the following tutorial and activities on the Student Website/CD:
  Tutorial 31.1 Life Cycle of a Zygomycete

Activity 31.1 Fungal Phylogeny
Activity 31.2 Euascomycete Life Cycle

Test Yourself
Knowledge and Synthesis Questions

1. Fungi are placed into one of the four main phyla based on
   a. methods of sexual reproduction.
   b. nucleic acid sequences.
   c. the presence or absence of septa in the hyphae.
   d. All of the above

2. Which of the following would you *not* expect to find in the life cycle of a typical fungus?
   a. Haploid nuclei
   b. Diploid nuclei
   c. Spores
   d. Chloroplasts

3. All fungi are absorptive heterotrophs. Which of the following is an adaptation shown by many fungi that greatly aids this mode of nutrient procurement?
   a. Dikaryosis
   b. A large surface area-to-volume ratio
   c. Conjugation
   d. A complex life cycle

4. Assume that two normal hyphae of different fungal mating types meet. After a period of time, the cell walls between these hyphae will dissolve, producing a
   a. mycelium.
   b. fruiting body.
   c. zygote.
   d. dikaryotic cell.

5. Which of the following is the best way to represent the ploidy of dikaryotic hyphae?
   a. $2n$
   b. n
   c. $n + n$
   d. n/n

6. Which of the following statements about sexual reproduction in fungi is *false*?
   a. Motile gametes are present in all fungal species.
   b. Water is not required for fertilization to occur in most fungi.
   c. There is no true diploid tissue in the life cycle of most sexually reproducing fungi.
   d. Sexual reproduction often begins with contact between hyphae of different mating types.

7. A mycorrhiza is
   a. an imperfect fungus.
   b. the fruiting structure of a basidiomycete.
   c. a symbiotic association between a fungus and cyanobacterium or green alga.
   d. a symbiotic association between a fungus and a plant.

8. Suppose a scientist investigating the classification of an imperfect fungus (a fungus that has never been observed to reproduce sexually) discovers that it has DNA sequences characteristic of basidiomycetes. If this scientist could coax fungi of this species to reproduce sexually, which of the following would he or she most likely observe?
   a. Dikaryotic hyphae segmented by septa
   b. Dikaryotic hyphae without septa
   c. Asymmetrical cell division
   d. A dikaryotic ascus

9. Which of the following fungi have coenocytic hyphae and stalked sporangiophores?
   a. Chytrids
   b. Zygomycetes
   c. Ascomycetes
   d. Basidiomycetes

10. Which of the following fungi display alternation of generations?
    a. Chytrids
    b. Zygomycetes
    c. Ascomycetes
    d. Basidiomycetes

Application Questions

1. Early taxonomists considered fungi to be members of the plant kingdom. What evidence indicates that they are in fact more closely related to animals?

2. Complete the following diagram, which shows the generalized life cycle of a basidiomycete, by adding the following terms in their appropriate locations: spores, dikaryotic hyphae, meiosis, mitosis, fertilization. Also, circle and label the portions of the life cycle corresponding to the haploid (n), diploid ($2n$), and the dikaryotic ($n + n$) stages.

conjugation by opposite mating types

Answers

Knowledge and Synthesis Answers

1. **d.** Traditionally, a fungus's method of sexual reproduction was the primary criterion in its classification, but all of the traits listed are useful in classifying fungi.

2. **d.** Fungi have haploid and diploid nuclei at different stages of their life cycle. They also produce spores. The chloroplasts are not found in the fungi, but in plants.

3. **b.** The large surface area-to-volume ratio of the hyphae increases the ability of a fungus to absorb nutrients.

4. **d.** When two hyphae of different mating types fuse, they form a dikaryotic hypha.

5. **c.** Dikaryotic hyphae are neither truly diploid ($2n$) nor haploid (n). Because dikaryotic hyphae include genetic material from two haploid nuclei that remain separate, the best way to represent their ploidy is as $n + n$.

6. **a.** Not all of the phyla in the kingdom Fungi have mobile gametes. Only the gametes of chytrids are motile.

7. **d.** Mycorrhizae are associations between fungi and the roots of plants. Lichens are symbiotic relationships between fungi and cyanobacteria or green algae.

8. **a.** If this fungus is indeed a basidiomycete, when hyphae of different mating types fuse, they will most likely create dikaryotic hyphae that are segmented by septa.

9. **b.** Coenocytic hyphae are characteristic of both the chytrids and the zygomycetes, but only the zygomycetes regularly produce sporangiophores.

10. **a.** Among the fungi, only the chytrids have a multicellular haploid stage and a multicellular true diploid stage.

Application Answers

1. At the cellular level, fungi bear very little resemblance to plants. Fungi don't even particularly resemble the surviving green algae that are most like the common ancestor of all plants. They do not produce any sort of chlorophyll, nor do they have plastids for storing reserves of photosynthetic products (see the section on the chlorophytes and green algae in Chapter 28). Fungi and plants both have cell walls, but the cell walls of fungi are constructed with chitin, a polysaccharide that plants do not produce. Chitin *is* found in the animal kingdom and in the protist group most closely related to the animals. It is unlikely that this complex molecule evolved more than once, so fungi and animals probably shared a chitin-producing common ancestor more recently than either group did with the plants.

2.

conjugation by opposite mating types

32 Animal Origins and the Evolution of Body Plans

Important Concepts

Animals are descendants of a common ancestor.

- With the possible exception of sponges, the animal kingdom is a monophyletic group.

- In addition to molecular and developmental similarities, all animals share certain characteristics important for multicellularity: special types of cell–cell junctions (tight junctions, desmosomes, and gap junctions) and extracellular matrix molecules (collagen and others).

- The common ancestor of modern animals may have been a colonial flagellated protist such as a choanoflagellate (see Figure 28.28).

- Animals are multicellular organisms that must take organic molecules in with their food. They also have differentiated cells that coordinate their activities.

Animals are classified based on morphology and genetics.

- The classification of animals has typically been based on an animal's developmental pattern and morphology. With the advent of modern molecular techniques, small subunit rRNA and mitochondrial genes have become common tools used for classification.

- The sponges, cnidarians, and ctenophores diverged from the rest of the animal lineage early in evolutionary history (see Figure 32.1). The remaining animals are categorized as either protostomes or deuterostomes.

- Diploblastic animals have two cell layers during development (ectoderm and endoderm), and triploblastic animals have three (ectoderm, mesoderm, and endoderm). Protostomes and deuterostomes are triploblastic; all other animal groups are diploblastic.

- Most protostomes undergo complete radial cleavage, but one lineage has spiral cleavage. The blastopore of the protostome embryo becomes its mouth.

- Deuterostomes have indeterminate radial cleavage. The blastopore of the deuterostome embryo becomes its anus.

Every animal has a body plan.

- The overall organization of an animal's body is known as its body plan. An animal's body plan includes symmetry, type of skeleton, position of major organs, and presence or absence of body cavities.

- An animal that can be divided along at least one plane into similar halves is said to be symmetrical. Asymmetrical animals, such as sponges, have no plane of symmetry.

- The sea anemone is radially symmetrical; any plane running along a sea anemone's main axis will divide it into roughly equal halves (see Figure 32.2).

- Animals exhibiting bilateral symmetry can be divided into two mirror images by only one plane. Bilateral symmetry is common among animals that are able to move quickly.

- Bilaterally symmetrical animals often have sense organs and nervous tissue concentrated at one end; this type of organization is known as cephalization.

- Animals have three different types of body cavities (see Figure 32.3). Acoelomates have no enclosed body cavity, pseudocoelomates have a liquid-filled space known as the pseudocoel in which many of the internal organs are located, and the coelomates have a true body cavity, a coelom, developing within the mesoderm. In coelomates, the internal organs are in pouches of the peritoneum.

- Animals have evolved a skeleton for support that can be internal or external and rigid or flexible.

Sponges have differentiated cells but no true organs.

- About 5,500 marine species and 50 freshwater species of sponges are found in the phylum Porifera. Sponges are groups of cells loosely organized around a water canal system. The Porifera are the only animals that do not have distinct cell layers.

- Although the sponge body plan is relatively simple, sponges have cells that are differentiated for specific functions. This characteristic distinguishes them (and all other animals) from colonial protists.

- Sponges filter small organisms and nutrient particles from water as it flows through pores in the walls of their inner cavity. Choanocyte cells use flagella to divert water through the pores and filter out food particles. The water exits the sponge through large openings called oscula (see Figure 32.4).

- Sponges have a supporting skeleton composed of branching spines (spicules) or an elastic network of fibers.

- Sponges can reproduce sexually by producing both egg and sperm, and asexually by budding and fragmentation.

The cnidarian life cycle has two stages: the polyp and the medusa.

- All cnidarians are diploblastic and have radial symmetry.

- The cnidarian gastrovascular cavity is a blind sac: There is only one opening, which serves as both mouth and anus. The gastrovascular cavity functions in food digestion, respiratory gas exchange, and circulation; it also lends support as a hydrostatic skeleton.

- A layer of jellylike material, the mesoglea, is found between the two cell layers of all cnidarians.

- Cnidarians have low metabolic rates and can catch large prey; this combination of characteristics makes it possible for them to live in environments with few prey. Cnidarian tentacles have specialized cells, cnidocytes, that inject toxins into their prey with the help of stingers called nematocysts (see Figure 32.7).

- The cnidarian life cycle includes a sessile polyp stage and a motile medusa stage (see Figure 32.8). The polyp stage usually reproduces asexually, and the medusa stage reproduces sexually, with the fertilized egg becoming a planula larva and eventually a polyp.

- Much of the difference in appearance between polyps and medusae is due to the mesoglea. In medusae, the mesoglea often constitutes the bulk of the animal; in polyps, it is usually thin.

- The 11,000 or so species in the phylum Cnidaria are composed of three classes: Anthozoa, Hydrozoa, and Scyphozoa (see Figure 32.6).

  - Members of the class Anthozoa (about 6,000 species of sea anemones, sea pens, and corals) lack a medusa stage altogether, with the polyps reproducing sexually and asexually. Sexual reproduction produces a planula that will develop into a polyp. Sea anemones are solitary and occasional motile; corals are sessile and colonial. The coral skeleton is composed of calcium carbonate and varies greatly between species. Corals live in symbiosis with dinoflagellates that provide nutrients for the coral colony through photosynthesis, contributing to the success of corals in clear tropical waters.

  - Species from the class Hydrozoa are found in freshwater and are typically dominated by the polyp stage of the life cycle. The polyps tend to be colonial, with many polyps sharing a common gastrovascular cavity (see Figure 32.10).

  - Scyphozoans include the marine jellyfish, which have medusae with thick mesoglea. These species spend most of their lives in the medusa stage. Polyps are produced sexually by adult medusae. The polyps produce young medusae by budding (see Figure 32.11).

Ctenophores superficially resemble cnidarian medusae.

- The marine comb jellies (phylum Ctenophora) have bodies similar to cnidarians. An important distinction between the two phyla is that ctenophores have a complete gut with a mouth and two anal pores.

- Ctenophores get their name from the eight rows of comb-like plates of cilia known as ctenes. The cilia are used to propel the animal through the water. Prey is caught on sticky filaments on the tentacles or body (see Figure 32.12).

- Ctenophores have a simple life cycle and reproduce sexually. The externally fertilized egg hatches into a miniature ctenophore.

Protostomes and deuterostomes evolved bilateral symmetry.

- Animals with bilateral symmetry form a monophyletic clade; their common ancestor is known as an urbilaterian. All animals with bilateral symmetry share intercellular signaling systems and development regulated by homologous homeobox genes.

- The urbilateria appear to have been complex animals with circulatory systems, antagonistic muscles, and a fluid-filled body cavity.

- The protostome and deuterostome lineages split during the Cambrian period. Both lineages are characterized by triploblastic, bilaterally symmetrical, cephalized body plans.

  - In protostome development, the blastopore becomes the mouth. Other traits that protostomes share are a ventral nervous system and a brain that surrounds the opening to the digestive tract.

  - In deuterostome development, the blastopore becomes the anus. Deuterostomes have dorsal nervous systems.

- Most animals are either protostomes or deuterostomes.

The protostomes belong to two lineages: lophotrochozoans and ecdysozoans.

- Soon after the protostome lineage split from the deuterostome lineage, the protostomes split again into two lineages.

  - Ecdysozoans, which are the subject of Chapter 33, grow by molting their exoskeletons.

  - Lophotrochozoans grow by adding skeletal elements, use cilia for locomotion, and may have free-living larvae known as trochophores (see Figure 32.14).

Flatworms exhibit bilateral symmetry.

- The simplest lophotrochozoans are the bilaterally symmetrical flatworms (phylum Platyhelminthes).

- Flatworms lack respiratory organs and have only simple cells for waste removal. The flat shape of the animal helps in oxygen transport and waste removal (see Figure 32.15).

- The digestive tract has only one opening, but it has many branches that aid in nutrient absorption.

- Many of the 25,000 species of flatworms are parasites of other animals (see Figure 32.16). The parasitic flatworms have nutrients brought to them and disperse their eggs in the feces of the host. Many parasitic flatworms have more than one host species in their life cycle. Some common flatworms are tapeworms and flukes.

Rotifers have a pseudocoel.

- The phylum Rotifera contains 1,800 species that are very small (some are smaller than single-celled protists), bilaterally symmetrical, unsegmented animals (see Figure 32.17). They are pseudocoelomates and use their pseudocoel as a hydrostatic skeleton.

- Rotifers have a complete gut with a mouth and anus. The corona on the head of many rotifers has cilia that direct water and organic matter into their mouth and a structure called the mastax. Food is ground in the mastax for digestion.

Lophophorates are named after the lophophore, a food-collecting and gas exchange structure.

- The lophophorates belong to three phyla: Phoronida, Brachiopoda, and Ectoprocta.

- The lophophorates have a lophophore surrounding their mouth with two rows of ciliated tentacles used for food collection and gas exchange. The phoronid gut is shaped like a U, with the anus close to the mouth.

- The 20 species in the phylum Phoronida are sedentary worms that live in chitinous tubes.

- The Ectoprocta are colonial; individuals in each colony are connected by strands of tissue, and in some species individuals are specialized for feeding, reproduction, defense, and support. Colonies grow via asexual reproduction of the founding members. New colonies are founded by larvae that result from sexual reproduction. Ectoprocts are the only lophophorates that can completely retract their lophophores.

- Brachiopods are marine lophophorates surrounded by a divided shell with two halves (see Figure 32.20). The lophophore is located in the shell, and cilia help draw water and food into the shell. Gas exchange occurs across the body.

The spiralian lineage contains wormlike phyla.

- The spiralians are wormlike organisms with bilateral symmetry and soft bodies. They include the ribbon worms, annelids, and mollusks.

Ribbon worms are unsegmented, dorsoventrally flattened spiralians.

- Nemerteans, or ribbon worms, have a complete digestive tract with two openings. Small ribbon worms use their cilia for movement and large ribbon worms use muscle contractions to move.

- The feeding organ of the ribbon worms is a proboscis that is kept in a rhynchocoel, or fluid-filled cavity (see Figure 32.21). The proboscis has a sharp stylet and can be forcefully ejected from the body to catch prey.

The spiralian lineage also contains the segmented worms.

- Segmentation has evolved a number of times in the spiralians. Segmentation allows for greater control of body movements. Your textbook describes only one segmented spiralian group, the annelids.

- The phylum Annelida is composed of approximately 16,500 species of segmented worms living in marine, freshwater, and moist terrestrial environments. Each segment of the worm has an isolated coelom and a ganglion nerve center connected to a nerve cord. Gas exchange occurs across the thin moist surface of the animal.

- The class Polychaeta contains more than half of all annelid species. Polychaetes tend to be found in marine environments. Outgrowths used in gas exchange, the parapodia, extend from segments over much of the body. Setae extending from the parapodia help to attach the animal to the substrate and aid in movement. The joining

of male and female gametes produces trochophore larvae (see Figure 32.23).

- Unlike polychaetes, oligochaetes live mainly in freshwater and terrestrial environments and are hermaphroditic, containing both male and female reproductive organs in the same individual. The earthworm is a typical oligochaete.

- The class Hirudinea, or leeches, are all hermaphroditic species that live either in freshwater or on land. Parasitic leeches attach externally and feed on a host's blood.

- Vestimentiferans lack a digestive tract. They are able to take up nutrients quickly from the environment. Species from this class are found in the deep ocean near hydrothermal vents and use hydrogen sulfide to obtain energy.

All mollusks share three main morphological features.

- Mollusks have evolved into a morphologically diverse group, with all classes including the same three basic parts in their body plan. All species have a large muscular foot and a tissue fold known as the mantle covering the visceral mass of internal organs. The mantle is extended in some species to create a mantle cavity holding the gills used in respiration; in many species the mantle secretes a hard, calcareous shell. Mollusks have a reduced coelom.

- Only a limited number of monoplacophorans survive today, though they were the most abundant mollusks during the Cambrian period. They are unlike other mollusks in that they have repeated body segments containing gills, muscles, and excretory structures.

- Chitons (class Polyplacophora) are marine mollusks that feed on algae using their radulae. A chiton's shell consists of several plates. They have simple internal organs, multiple gills, and bilateral symmetry.

- Bivalves include clams, oysters, scallops, and mussels. They create a current of water flowing into their two-part shell through a siphon. Large gills inside the shell extract food and also function as respiratory organs.

- Gastropods use their foot either to crawl or swim. The land-dwelling slugs and snails are the only terrestrial mollusks.

- Cephalopods include the octopuses, squids, and nautiluses. They have evolved the ability to control the water movement into the mantle, allowing them to use ejected water as a jet for propulsion. They are also able to control the gas movement in the mantle, which helps in buoyancy control. The gills are found within the mantle cavity.

The Big Picture

- There are a number of morphological and genetic traits that are shared by all animals, indicating that the kingdom Animalia is a monophyletic group. All animals are multicellular organisms that cannot synthesize organic molecules.

- Animals are distinguished by the number of cell layers found in their embryos, their type of body cavity, and their body symmetry.

- Your textbook discusses three phyla that are neither protostomes nor deuterostomes. The phylum Porifera contains the sponges, which are loosely organized groups of cells. The phylum Cnidaria consists of diploblastic animals that have evolved a gut. The cnidarian life cycle has an asexually reproducing polyp stage and a sexually reproducing medusa stage. The three classes of Cnidarians are the Hydrozoa, the Scyphozoa, and the Anthozoa. Animals in the phylum Ctenophora are similar to the cnidarians, except they have a complete gut.

- The lophotrochozoans are all protostomes with a anterior brain, ventral nervous system, and free-floating larva. The simplest lophotrochozoans are the Platyhelminthes and the Rotifera, which both have bilateral symmetry.

- The lophophorates, which include the Phoronida, Brachiopoda, and Ectoprocta, all have a lophophore. A lophophore is an anatomical feature used in food collection and gas exchange.

- The spiralian lineage contains the wormlike animals, including the unsegmented phylum Nemertea and the segmented phylum Annelida. The annelids are further divided into the Polychaeta, Oligochaeta, Hirudinea, and Vestimentifera. The evolution of a segmented body allowed for better control of movement.

- The five classes in the phylum Mollusca share three morphological characters: a muscular foot structure, a mantle, and a visceral mass.

Common Problem Areas

- The difference between deuterostomes and protostomes is often difficult to remember. This chapter, along with the next, is concerned with the protostomes, whereas Chapter 34 discusses the deuterostomes. Create a table to be used with the upcoming chapters containing all of the characteristics of a protostome and a deuterostome. To this table you can add the different phyla that are included in each group. Use this table as a tool to help you organize and remember the differences between the groups.

Study Strategies

- The phylogenies that are presented in your book are good skeletons on which you can add more information about the different phyla. Doing so will provide you with a framework and a point of reference when you are studying the differences between phyla. Remember that groups of animals have a set of traits they share with earlier lineages and a new set of traits that distinguish them from related lineages.

- Review the following tutorial and activities on the Student Website/CD:
  Tutorial 32.1 Life Cycle of a Cnidarian
  Activity 32.1 Animal Classification
  Activity 32.2 Animal Features

Test Yourself

Knowledge and Synthesis Questions

1. Which of the following is not a derived trait that is shared by all animals?
   a. Similarities in 5S and 18S rRNA sequences
   b. The extracellular matrix molecule collagen
   c. Tight junctions, desmosomes, and gap junctions
   d. Bilateral symmetry

2. Which of the following traits do the Ctenophora share with the Lophotrochozoans?
   a. A complete gut
   b. Embryos with three cell layers
   c. Spiral cleavage
   d. An internal skeleton

3. An animal is known to be a deuterostome. Which of the following does not apply to this animal?
   a. Three distinct layers of tissue were present during development.
   b. If a coelom is present, it formed within the embryonic mesoderm.
   c. Its early embryonic cleavage pattern was radial.
   d. It is diploblastic.

4. An important factor contributing to the evolution of diversity among animals is the considerable variation in animal
   a. multicellularity.
   b. cell junctions.
   c. methods of food acquisition.
   d. symmetry.

5. Which of the following statements about the body cavity of animals is true?
   a. The body cavity of coelomates develops from the embryonic ectoderm.
   b. The acoelomate's body cavity is filled with liquid.
   c. The pseudocoel of the pseudocoelomates has a peritoneum.
   d. The acoelomates do not have an enclosed body cavity.

6. Sponges have a very simple body plan. Which of the following statements about sponge structure or function is false?
   a. Choanocytes are flagellated cells that play a role in feeding.
   b. Large species are found in areas of heavy wave action, where food is most abundant.
   c. Individual sponges are both male and female.
   d. Water enters a sponge through pores and exits via one or more oscula.

7. Which of the following traits is not shared by both sea anemones and jellyfish?
   a. A medusa as the dominant stage in the life cycle
   b. Possession of a gastrovascular cavity
   c. Sexual reproduction
   d. Nematocysts on the tentacles

8. Which of the following traits is not shared by ctenophores and cnidarians?
   a. Both are diploblastic.
   b. Both have radial symmetry.
   c. Both have complete guts.
   d. Both have feeding tentacles.

9. The rhynchocoel is a body plan feature found in which of the following groups?
   a. Trematoda only
   b. Rotifera and Nematoda
   c. Rotifera only
   d. Nemertea only

10. Which of the following phyla is not a lophophorate lineage?
    a. Ectoprocta
    b. Phoronida
    c. Annelida
    d. Brachiopoda

11. Which of the following statements about the Rotifera is *false*?
    a. They have a complete gut with an anterior mouth and posterior anus.
    b. They are coelomates.
    c. The corona is a ciliated organ used in acquiring food.
    d. They have a hydrostatic skeleton.

12. A meal containing snails, clams, and octopuses includes which of the following classes of mollusk?
    a. Bivalvia only
    b. Gastropoda and Cephalopoda
    c. Bivalvia and Gastropoda
    d. Bivalvia, Gastropoda, and Cephalopoda

13. The combination of a true coelom and repeating body segmentation allowed the annelids (unlike the anatomically "simpler" worms) to do which of the following?
    a. Evolve complex body shapes and control movement more precisely
    b. Move through loose marine sediments
    c. Be hermaphroditic
    d. Inject paralytic poisons into their prey

14. Which class of cnidarians contains species that live symbiotically with dinoflagellates?
    a. Hydrozoa
    b. Ctenophora
    c. Scyphozoa
    d. Anthozoa

Application Questions

1. The mollusks are a very diverse group morphologically. However, they all share three common morphological traits. Describe the traits and their functions.

2. The following are all characteristics of earthworms: They have segmented bodies, a spiral cleavage pattern, a terrestrial lifestyle, a complete gut, a blastopore that develops into a mouth, tight junctions between cells, and bilateral symmetry. In what order did these characteristics evolve?

3. Scientists have not found a fossilized body of the urbilaterian species that is the common ancestor of all bilaterally symmetrical animals. Is there much hope that they will? How have scientists been able to learn about this species?

Answers

Knowledge and Synthesis Answers

1. **d.** Not all animals have bilateral symmetry. Sponges, cnidarians, and ctenophores do not exhibit bilateral symmetry in their body plan.

2. **a.** Ctenophores and lophotrochozoans have complete guts; ctenophores do not have embryos with three cell layers or spiral cleavage. Neither group has internal skeletons.

3. **d.** Deuterostome embryos have three layers, the ectoderm, the mesoderm, and the endoderm, making deuterostomes triploblastic, not diploblastic.

4. **c.** Animals eat foods from all kingdoms of life. Thus, they have evolved a wide variety of methods for acquiring diverse types of food.

5. **d.** The body cavity of coelomates develops from the mesoderm and contains a peritoneum. The acoelomates lack a body cavity.

6. **b.** Because they are not structurally robust, heavy wave action would destroy large, upright sponges.

7. **a.** Although the jellyfish have both a medusa and polyp stage in their life cycle, the sea anemones have lost the medusa stage and spend their life cycle as polyps.

8. **c.** Ctenophores have evolved a complete gut with a mouth and two anal pores; cnidarians have a blind gut with only one opening.

9. **d.** The rhynchocoel is a fluid-filled cavity containing a proboscis feeding organ found in the phylum Nemertea.

10. **c.** Annelids belong to the spiralian lineage, not the lophophorates.

11. **b.** Rotifers are pseudocoelomates and have a pseudocoel.

12. **d.** Snails are in the Gastropoda, clams are in the Bivalvia, and octopuses are in the Cephalopoda.

13. **a.** The segmentation of the annelids allows for more complex, coordinated movement.

14. **d.** Anthozoans include the corals that live symbiotically with photosynthetic dinoflagellates.

Application Answers

1. The mollusk body plan has three basic morphological components shared by all classes. The first trait is a muscular foot used for locomotion. In cephalopods the foot has been modified into arms and tentacles. The second trait is a mantle, which covers the third trait, the visceral mass. The mantle secretes a calcareous shell in many species and often houses the gills. The visceral mass consists of the internal organs.

2. The first of these characteristics to evolve was the tight junction. This type of cell junction occurs throughout the animal kingdom and was probably present in the common ancestor of all animals. Figure 32.1 shows that the evolution of a complete gut came next, and then bilateral symmetry. After the evolution of bilateral symmetry, the phylogenetic tree in Figure 32.1 splits into the deuterostomes and protostomes; as protostomes, the annelids have a mouth that develops from their blastopore. According to Figure 32.14, the next characteristics to evolve were spiral cleavage and a segmented body plan, respectively. The move from aquatic to terrestrial lifestyle is not covered in these figures, but given that the annelids are overwhelmingly aquatic, and even terrestrial annelids are confined to moist habitats, it seems likely that the terrestrial lifestyle was the most recent of the listed characteristics to evolve.

3. The vast majority of species that have lived on Earth have gone extinct without leaving fossils (see Chapter 22), so it is very unlikely that scientists will ever discover a fossilized body of the common ancestor of all bilaterally symmetrical animals. Some fossilized animal tracks from the Precambrian appear to have been made by a bilaterally symmetrical animal, but these offer limited information. The best source of information about the common ancestor of any group of animals is the set of characteristics that are common to all members of that group. All bilaterally symmetrical animals share an array of intercellular signaling systems and certain homeobox genes, and it is probable that these were also present in the common ancestor of this group.

33 Ecdysozoans: The Molting Animals

Important Concepts

Ecdysozoans are distinguished by an external skeleton that does not grow.

- Ecdysozoans have evolved an exoskeleton that ranges from thin and flexible to hard and rigid.

- Exoskeletons provide protection and support, but present obstacles to growth, locomotion, and gas exchange. Ecdysozoans have overcome these obstacles to become the most numerous and diverse animal lineage.

Marine ecdysozoans have a thin exoskeleton.

- A large number of marine ecdysozoans have a cuticle, a thin exoskeleton that is molted periodically. The cuticle allows for exchanges with the environment, but offers little protection.

- The priapulids (phylum Priapulida) are 16 species of unsegmented, tiny, wormlike animals.

- The kinorhynchs (phylum Kinorhyncha) are 150 species of marine worms, all less than 1 millimeter in length, that live in mud and feed on decaying organic matter. All kinorhynchs have bodies divided into 13 segments.

- The arrow worms (phylum Chaetognatha) have no circulatory system and no larval stage. They are major predators of small organisms in the open ocean (see Figure 33.3). All respiratory gas exchange occurs across the body surface.

A tough exoskeleton has evolved in some worms.

- Species of both the Nematomorpha and Nematoda phyla have tough cuticles.

- The horsehair worms (phylum Nematomorpha) are composed of 230 species and are long and very thin, as their name suggests. They live near the edges of ponds and streams and feed in the larval stage as parasites of terrestrial and aquatic insects and crabs (see Figure 33.4).

- The nematodes, or roundworms (phylum Nematoda), can reach up to 9 meters in length. About 25,000 species have been described, but this may represent less than one-quarter of all nematode species. Nematodes use their gut for both gas exchange and nutrient uptake (see Figure 33.5). They live in soil, on the bottoms of lakes and streams, in marine sediments, and as parasites of plants and animals.

- Parasitic roundworms have a life cycle that helps to transfer them from host to host. The larvae of *Trichinella*

spiralis infect mammals when they eat flesh containing the roundworm. The gut of the host activates the larvae, which attach to the intestinal wall and feed. They bore through the intestinal wall and make cysts in the muscles.

Arthropods have segmented external skeletons.

- Arthropods have a hard exoskeleton of chitin that provides protection, waterproofing, and muscle attachment sites. They also have jointed appendages that allow for a greater range of movement.

- The Onychophora and Tardigrada are close relatives of the arthropods that have unjointed legs and thin cuticles composed of chitin. They use their body cavities, which are fluid-filled, as hydrostatic skeletons (see Figure 33.6).

- Onychophorans have probably changed relatively little from their common ancestor with the arthropods.

- Trilobites, which lived in the Cambrian and Ordovician time period but are now extinct, had heavy armor and jointed appendages (see Figure 33.7).

- Four major arthropod lineages have survived to the present: the Crustacea (crabs, barnacles, etc.), the Hexapoda (insects), the Myriapoda (centipedes and millipedes), and the Chelicerata (spiders, mites, etc.).

Most marine arthropods are crustaceans.

- Crustaceans include many familiar animals (shrimp, lobsters, crabs, barnacles) as well as many less-familiar animals (isopods, copepods, and others). About 40,000 species have been described so far.

- Small crustaceans known as copepods may be the most abundant animals on Earth.

- The crustacean body is divided into the head, thorax, and abdomen, with the thorax and abdomen having one pair of appendages each. In many species, a carapace extends dorsally from the head to protect and cover the body.

- The typical crustacean larva is called a nauplius. It has three pairs of appendages and one eye (see Figures 33.10 and 21.1).

Most terrestrial arthropods are insects.

- About 1.4 million species of insects have been described so far. The insect body has three parts: the head with a pair of antennae, the thorax with three pairs of legs, and the abdomen. Gas exchange occurs in a system composed of a series of air sacs and tubular channels called tracheae.

- There are two classes of insects: the wingless Apterygota and the winged Pterygota.

- The Apterygota include springtails and silverfish. The common ancestor of all insects was probably more like an apterygote than any other modern insect group.

- The Pterygota typically have two pairs of wings that are attached to the thorax. They are divided into 29 orders with three major lineages: those that cannot fold their wings over their bodies, those that can fold their wings and undergo incomplete metamorphosis, and those that can fold their wings and undergo complete metamorphosis.

  - Insects that cannot fold their wings over their bodies are dragonflies, damselflies, and mayflies.

  - Insects that undergo incomplete metamorphosis include grasshoppers and crickets, termites, stone flies, earwigs, thrips, true bugs, aphids, cicadas, and many others. In these groups, the hatchlings resemble small, usually wingless adults; as they molt from one stage to the next, they gradually acquire more adult characteristics.

  - Complete metamorphosis occurs in lacewings, beetles, caddisflies, butterflies, moths, flies, wasps, bees, and ants, among others. In these groups, the larvae and adult forms are substantially different; the young pass through at least two stages (larva and pupa) before becoming adults.

- The evolution of wings probably occurred only once during insect evolution. Studies of homologous genes in insects and crustaceans suggest that insect wings are modified from a respiratory structure on the leg of an ancestral crustacean (see Figure 33.14).

Myriapods and chelicerates have only two body regions.

- The myriapods (phylum Myriapoda) include 3,000 described species of centipedes and 10,000 described species of millipedes. Members of both groups have similar body plans: a well-formed head and a long, segmented trunk. Centipedes have one pair of legs on each trunk segment; millipedes have two pairs on each segment (see Figure 33.15).

- The chelicerates (phylum Chelicerata) are broken into three classes: Pycnogonida, Merostomata, and Arachnida. All chelicerates have two body parts, and most have eight legs.

  - The pycnogonids are a rarely seen marine group (see Figure 33.16).

  - The class Merostomata contains only the five living species of horseshoe crabs and an extinct group known as giant water scorpions.

  - The class Arachnida is made up of the predatory spiders, scorpions, mites, and ticks. They have a simple life cycle, with young resembling small adults. Spiders, the most familiar arachnids, build webs of protein strands that they use to capture prey.

Protostome evolution occurred mostly in the oceans.

- Dissolved organic matter and very small organisms were the only food during much of animal evolution.

- A sessile lifestyle has evolved repeatedly with the advantages of access to local resources and sharing of resources.

The major disadvantages are the necessity to rely on external fertilization and the competition for space on which to grow.

- Specialized body coverings probably evolved as protection against predation. Eventually, these body coverings also led to new forms of locomotion.

The Big Picture

- The ecdysozoans include more species than all other lineages combined. The earliest groups had a flexible, unsegmented exoskeleton. Wormlike animals with a thin cuticle are the Priapulida, Kinorhyncha, and Chaetognatha. Unsegmented worms with a tough cuticle belong to the Nematomorpha and Nematoda phyla.

- A segmented body with a hard exoskeleton and jointed appendages characterize the Arthropods. A related phylum, the Onychophora, have soft bodies and unjointed appendages. The phylum Trilobita is extinct. The Chelicerata, Myriapoda, and Hexapoda have invaded the land; the Crustacea are the dominant arthropods in marine environments.

Common Problem Areas

- Learning systematics of any large animal group can be challenging, particularly when they are organisms with which you are unfamiliar. This will likely be the case with the ecdysozoans, represented by a large number of animal groups. Review Table 33.1 for an overview of ecdysozoan and lophotrochozoan characteristics.

Study Strategies

- In this chapter there are many facts about a large number of phyla. Try to use the phylogeny presented in Figure 33.1 as a base for learning about different groups. Understanding that certain characteristics are shared between related groups can take you a long way toward remembering which characteristics define each group.

- Sometimes learning the Latin root of a group you are studying can help you organize your thinking. Although the names are unfamiliar at first, they all "make sense" if you understand what they mean.

- Review the following activities on the Student Website/CD:
  Activity 33.1 Animal Classification
  Activity 33.2 Animal Features

Test Yourself

Knowledge and Synthesis Questions

1. Lobsters, millipedes, and butterflies all share which of the following traits?
   a. Parapodia
   b. Setae
   c. Gas exchange across the skin
   d. Jointed appendages

2. Which of the following phyla or subphyla has a specialized internal gas exchange system?
   a. Priapulida
   b. Chaetognatha
   c. Hexapoda
   d. Nematomorpha

3. Which of the following insect groups consists of species that undergo complete metamorphosis and can fold their wings back over their bodies?
   a. Coleoptera (beetles)
   b. Apterygota (silverfish and springtails)
   c. Orthoptera (grasshoppers and their kin)
   d. Odonata (dragonflies and damselflies)

4. Which of the following insect groups consists of species that undergo incomplete metamorphosis and can fold their wings back over their bodies?
   a. Coleoptera (beetles)
   b. Apterygota (silverfish and springtails)
   c. Orthoptera (grasshoppers and their kin)
   d. Odonata (dragonflies and damselflies)

5. Which of the following insect groups consists of winged species that *cannot* fold their wings back over their bodies?
   a. Coleoptera (beetles)
   b. Apterygota (silverfish and springtails)
   c. Orthoptera (grasshoppers and their kin)
   d. Odonata (dragonflies and damselflies)

6. The majority of terrestrial arthropod species belong to the
   a. Myriapoda.
   b. Hexapoda.
   c. Cirripedia.
   d. Tardigrada.

7. The typical crustacean larva is called a
   a. nauplius.
   b. pupa.
   c. instar.
   d. cyst.

8. Which of the following chelicerate groups has a long fossil history showing little morphological change?
   a. Pycnogonida (sea spiders)
   b. Arachnida (mites, ticks, and spiders)
   c. Merostomata (horseshoe crabs)
   d. None of the above

9. Which of the following groups of ecdysozoans is extinct?
   a. Chelicerata
   b. Crustacea
   c. Tardigrada
   d. Trilobita

10. Which of the following characteristics is *not* part of the arthropod body plan?
    a. Segmentation
    b. Jointed appendages
    c. A closed circulatory system
    d. A hard exoskeleton

Application Questions

1. A thick cuticle provides an animal with protection, but it also can seal it off from the world, obstructing the exchange of gases. How would you design a simple, hypothetical animal with a covering that provided for both protection and gas exchange?

2. The root *pter* occurs in the names of many insect groups. What does this root mean? Most good dictionaries give the meanings of word prefixes and roots; using such a dictionary, determine the meaning of the following names of insect groups:
   Apterygota
   Ephemeroptera
   Orthoptera
   Isoptera
   Hemiptera
   Homoptera
   Coleoptera
   Trichoptera
   Lepidoptera
   Diptera
   Hymenoptera
   Siphonaptera

Answers

Knowledge and Synthesis Answers

1. **d.** All of these animals have jointed appendages.

2. **c.** The Hexapoda have an internal system of tracheae for gas exchange.

3. **a.** The Coleoptera, or beetles, undergo complete metamorphosis and are able to fold their wings over their bodies.

4. **c.** The Orthoptera (grasshoppers, crickets, roaches, etc.) undergo incomplete metamorphosis and are able to fold their wings over their bodies.

5. **d.** The Odonata (dragonflies and damselflies), along with the Ephemeroptera (mayflies), have wings that they cannot fold over their bodies. Apterygotes (silverfish and springtails) are wingless.

6. **b.** The 1.4 million species of insects that have been described so far are only a small fraction of all insect species.

7. **a.** The larvae of crustaceans are known as nauplius larvae.

8. **c.** Horseshoe crabs have changed little during their fossil history; they are sometimes referred to as living fossils.

9. **d.** The Chelicerata include spiders, sea spiders, and horseshoe crabs; the Crustacea include decapods (crabs, shrimp, and others), barnacles, isopods, and copepods; the Tardigrada are a small (but not extinct) group also known as water bears. Of the groups listed, only the Trilobita are extinct.

10. **c.** Arthropods have a hard exoskeleton, a segmented body, jointed appendages, and an open circulatory system. (see Table 33.1).

Application Answers

1. The animal would need special mechanisms for gas exchange with the environment. The insects have overcome this problem with their system of tracheae. This network of tubes is found throughout the body and opens to the environment. The network is in close contact with all the cells of the animal. In an aquatic environment, the animal could have an internal chamber with gills used for respiration.

2. As a root or prefix, *pter* refers to wings (think of pterodactyls).
   Apterygota: The prefix *a-* in this case means *without*. The Apterygota are insects without wings.
   Ephemeroptera: *Ephemeral* means *short-lived*. Adult mayflies typically live only a few hours or days.
   Orthoptera: *Orth-* means *straight*. Orthopterans tend to have very straight wings (see Figure 33.13*b*).

Isoptera: *Iso-* means *equal* or *uniform*. All four wings on termites tend to be very similar.

Hemiptera: *Hemi-* means *half*. In many true bugs, the front half of the forewing is hardened, and they appear to have half a shell (like a beetle's) and half wings.

Homoptera: *Homo-* means *same*. Unlike their close relatives the hemipterans, homopterans have forewings and hind wings that are similar to each other.

Coleoptera: *Coleo-* means *sheath*. The forewings of beetles are modified into a protective sheath that usually covers the hind wings and abdomen.

Trichoptera: *Trich-* means *hair*. The bodies and wings of adult caddisflies are covered in small hairs.

Lepidoptera: *Lepid-* means *scale*. Butterfly and moth wings are covered with tiny scales.

Diptera: *Di-* means *two*. Unlike most winged insects, flies have only two wings.

Hymenoptera: *Hymen-* means *membrane*. Bees, wasps, and the reproductive castes of ants have membranous wings.

Siphonaptera: *Siphon-* means *tube* or *pipe*. This is a reference to the flea's feeding method. Note that this name ends in *-aptera*, not *-optera*: Fleas do not have wings (*a-* once again meaning *without*), though as pterygotes they are descended from winged ancestors.

34 *Deuterostomate Animals*

Important Concepts

The deuterostomes are echinoderms, hemichordates, and chordates (including humans).

- Yunnanozoans are likely ancestors of all deuterostomes. They had mouths, external gills, and segmented posterior bodies.
- Living deuterostomes fall into two clades: One clade includes the echinoderms and hemichordates, the other is the chordates (see Figure 34.2).

The Echinodermata evolved an internal skeleton.

- Two features distinguished the evolution of the Echinodermata: They have evolved a calcified internal skeleton, and they have a water vascular system used for gas exchange, locomotion, and feeding (see Figure 34.3).
- Echinoderm larvae are bilaterally symmetric planktonic organisms. Adults show pentaradial symmetry, with five units (or multiples of five units) radiating from a central point.
- Of the 20,000 species of Echinodermata known to have lived, about 7,000 are living today. They are found in two lineages: the Pelmatozoa and the Eleutherozoa.
- The only class of Pelmatozoa surviving today is the Crinoidea, with 80 species that include the sea lilies and feather stars. The body is formed of a cup-shaped structure and has anywhere from five to several hundred arms. The arms have tube feet that catch food particles. Cilia on the tube feet move the food to a groove that leads to the mouth. The tube feet also provide a surface for respiratory gas exchange and nitrogenous waste excretion.
- The eleutherozoans consist of the classes Echinoidea, Holothuroidea, Asteroidea, Ophiuroidea, and Concentricycloidea (see Figure 34.4).
  - The Echinoidea are the sand dollars and sea urchins, which both feed on algae and organic debris. Neither has arms, but the sea urchin has spines that are attached to the skeleton.
  - The class Holothuroidea are the sea cucumbers. The tube feet of the sea cucumber are used for anchoring the animal to the substrate rather than for moving. The anterior tube feet have evolved into tentacles that catch food particles.
  - The sea stars, class Asteroidea, are predators of many animals. Some species use their tube feet to pry open

the shells of mollusks. Once the mollusks are open, the sea star inserts its stomach into the shell of the mollusk and digests it. The tube feet are also sites of gas exchange.
  - The 2,000 species of the class Ophiuroidea (brittle stars) have five flexible arms and feed on the surface sediments and the accompanying organic material through their single opening to the digestive tract.
  - Little is known about the sea daisies (class Concentricycloidea) other than that they feed on decaying oceanic driftwood. Data on molecular evolution suggest a close relationship with the Asteroidea.

Hemicordata have pharyngeal slits.

- The animals of the phylum Hemichordata are composed of a trunk, a collar, and a proboscis (see Figure 34.5). The sticky proboscis is used for catching prey and digging. Cilia move the food from the proboscis to the mouth.
- A pharynx that opens to the outside through pharyngeal slits is found behind the mouth. The pharyngeal slits are used as respiratory surfaces; water is moved through the slits, and oxygen is extracted.

Chordates have a notochord.

- The chordates are evolved from a sister lineage of the acorn worm that had enlarged pharyngeal slits used to catch prey.
- All chordates share five anatomical features: They all have pharyngeal slits at some point in their life, a hollow dorsal nerve cord, a ventral heart, a tail beyond the anus, and a notochord. In tunicates the notochord is lost during metamorphosis from larval to adult forms, whereas in vertebrates it is replaced by skeletal structures.
- The subphylum Urochordata contains the 2,500 species of marine tunicates. Tunicate larvae have pharyngeal slits, a nerve cord, and a notochord. The adults lose the notochord and nerve cord but have an enlarged pharynx, the pharyngeal basket, for catching food (see Figure 34.6).
- The common ancestor of the chordates may have resembled tunicates.
- Most urochordates fall into three major groups: ascidians, thaliacians, and larvaceans.

Lancelets are small, fishlike animals.

- The lancelets, subphylum Cephalochordata, retain the notochord for their entire life and use their pharyngeal baskets to catch prey items (see Figure 34.7).

Vertebrates evolved an internal skeleton.

- The lineage from which the vertebrates evolved had enlarged pharyngeal baskets used to collect prey from mud.

- Vertebrates can be categorized by the following traits: a vertebral column replacing the notochord, an internal skeleton with two pairs of appendages, an anterior head with a large brain and protective skull, a closed circulatory system, and a large coelom containing the internal organs.

- The animals that gave rise to the fishes lacked jaws. The ostracoderms were a class of early jawless fish that were able to swim above the substrate and exploit the environment. The hagfish and lamprey are the only jawless fishes surviving today.

- The jawed fishes evolved in the Devonian period. The evolution of jaws provided access to new food sources. The early jawed fishes, such as the placoderms, were the major predators of the Devonian.

- The Chondrichthyes fishes, including sharks, skates, rays, and chimaeras, have a skeleton composed of cartilage and pairs of unjointed appendages known as fins. The fins allow for better control of movement in the water.

- Actinopterygii are fishes with a skeleton of bone and a protective covering of scales. A gas-filled sac to aid in respiration evolved in early ray-finned fishes. In most ray-finned fishes, the sac has become a swim bladder and aids in the control of buoyancy. With the aid of the swim bladder and fins, many fish can control their position in the water column with minimal energy expenditure.

Lobe-finned fishes and lungfishes have structures that may have led to the colonization of land by vertebrates.

- The lunglike sacs of ray-finned fishes evolved to aid oxygen uptake in low-oxygen water and may have allowed some fishes to survive out of water. The evolution of jointed fins may have helped the colonization of land.

- Fishes in the class Actinistia (lobe-finned fishes) have jointed fins and were thought to be extinct until 1938. One species, *Latimeria chalumnae*, composed mainly of cartilage, has been found in the waters of the Indian Ocean (see Figure 34.14).

- Three species of lungfishes (class Dipnoi) have survived to the present, one each in South America, Africa, and Australia. In addition to gills, these fishes have lungs, and they can burrow into mud to survive periods when their shallow ponds and swamps dry up.

Amphibians ventured onto land.

- The class Amphibia evolved from a common ancestor shared with the lungfishes. Amphibians have jointed walking legs, like reptiles, birds, and mammals.

- Amphibians live in moist environments and spend at least part of their life cycle in the water (see Figure 34.16). Many species come back to the water to lay their eggs, producing aquatic larvae.

- The skin surface is a common site of respiratory gas exchange in amphibians. They also have lungs used in gas exchange.

- Extant amphibians are found in the orders Gymnophiona (caecilians), Anura (frogs and toads), and Urodela (salamanders).

Reptiles invaded dry environments.

- Amphibians are confined to moist environments. Two major morphological adaptations occurred to allow reptiles and birds to colonize dry land: a shelled egg and waterproof skin.

- The animals that evolved a protective calcium-based shell are known as amniotes. The shell inhibits dehydration of the developing egg and allows eggs to be laid in dry environments.

- The class Reptilia branched off during the Carboniferous period about 250 mya. The Reptilia is not a monophyletic group because it does not contain the birds. This classification has been used in the classical system and is the one used in your textbook.

- The skin of reptiles is impermeable to water, decreasing water loss and at the same time decreasing gas exchange. Gas exchange occurs in ventilated lungs.

- The subclass Squamata contains the lizards, snakes, and amphisbaenians. Most lizards and snakes are carnivores, but some lizards are herbivores.

- Tuataras are a subclass that contains only two extant species.

- The Chelonia (turtles and tortoises) are characterized by dorsal and ventral bony plates that form a shell for protection. Their relationship to other reptiles is not clear.

- The Crocodylia (crocodiles and alligators) are all carnivores; they live only in tropical and warm temperate regions.

- Dinosaurs evolved from the reptiles around 215 mya and survived for about 150 million years.

Birds evolved from a dinosaur ancestor.

- Recent fossil discoveries in China have demonstrated that some dinosaurs had feathers.

- *Archaeopteryx* lived 150 mya and is the oldest known fossil from the subclass Aves. It was covered in feathers and had well-developed wings and a wishbone like present-day birds (see Figure 34.20).

- One theory about the evolution of birds is that they evolved from bipedal dinosaurs with feathers evolving to aid in insulation or courtship displays.

- Both the lightweight bones and feathers of birds have been modified for flight. Along with the ability to fly came a high metabolic rate needed to fuel flight.

- Birds typically lay eggs in nests and incubate the eggs until they hatch. Development of the embryos occurs rapidly, producing either chicks that are well developed with feathers or featherless chicks that require more parental care.

- The 9,600 species of birds are the largest subclass of terrestrial vertebrates (see Figure 34.21).

The Mammalia evolved from mammal-like reptiles.

- The mammals evolved from a group of mammal-like reptiles early in the Mesozoic era. Following the dinosaur extinction, mammals flourished in size and number.

- A number of unique traits have evolved in the mammal line. The jawbone is comprised of a single bone, they have mammary glands that provide nutrients for their young, and external body hair.

- The subclass Prototheria is composed of the Monotremata, an egg-laying order of mammals (see Figure 34.22).

- The subclass Theria contains the Marsupialia and eutherian mammals. Marsupial development occurs mainly in a pouch on the front of the mother (see Figure 34.23). The eutherian mammals develop in the mother's uterus.

Humans belong to the primates.

- The primates have relatively recently undergone an extensive evolutionary radiation.

- The earliest split in the primates occurred in the late Cretaceous, when two lineages originated: the prosimians (comprised of lemurs, bush babies, and lorises) and the anthropoids (comprised of the monkeys, apes, and humans).

- The anthropoids further split into the arboreal New World monkeys and the arboreal or terrestrial Old World primates. The apes split off about 20 mya.

- The ancestor of modern African apes (and thus of humans), *Dryopithecus*, migrated to Africa around 9 mya.

Early human ancestors evolved bipedalism.

- Ardipithecines (the earliest protohominids) and their descendants, the australopithecines, were adapted for bipedalism, freeing up their hands for other tasks and elevating the eyes.

- *Australopithecus afarensis* appears to be the ancestor of modern humans. From this species a number of *Australopithecus* species evolved, including the tool-using *A. garhi*, which may have given rise to the genus *Homo* (see Figure 34.29).

- The oldest member of the genus *Homo*, *Homo habilis*, lived 2 mya. Tools have been found with the bones of *H. habilis*. *Homo erectus* lived from 1.6 mya to 200,000 years ago and used fire and stone tools. *H. erectus* may have played a role in the extinction of *H. habilis*.

- In tropical regions, *Homo erectus* was replaced by *Homo sapiens* (our species) about 200,000 years ago.

- Brains of *H. sapiens* reached modern size by about 160,000 years ago. The rapid evolution of this trait may have been driven by an increasingly complex social life that required more sophisticated communication. Increased consumption of fish and mollusks may have provided the fatty acids necessary to support growth of larger brains.

- A number of types of *H. sapiens* lived from 1.5 mya to about 30,000 years ago. Neanderthals, a type of *H. sapiens*, lived between 75,000 and 30,000 years ago. The Cro-Magnons were most likely responsible for the demise of the Neanderthals. The Cro-Magnons used sophisticated tools, spread across Asia, and were in North America by about 20,000 years ago.

Humans have developed culture.

- With the evolution of larger brain sizes in our *Homo sapiens* ancestors, the capacity for language increased.

- Humans have developed culture and pass knowledge and traditions from generation to generation.

- Human societies were transformed from communities of hunters and gatherers to those of farmers.

Deuterostomes and protostomes evolved in the same environments.

- The deuterostomes and protostomes evolved similar solutions to similar environmental conditions. Whereas the protostomes invaded land from the oceans, the deuterostomes invaded from fresh water.

- The bodies of some members of both groups are divided into compartments, and some members of each group evolved ways to control buoyancy in water.

The Big Picture

- The deuterostomes are not as diverse or numerous as the protostomes. Most fit into two groups, the echinoderms and the vertebrates. The echinoderms are joined by the hemichordates in one clade; the urochordates and cephalochordates join the vertebrates in a second clade (the Chordata).

- Echinoderms have an internal skeleton and a special water vascular system used in feeding. They include the sea stars, sea lilies, sea cucumbers, and sea daisies, which all exhibit pentaradial symmetry as adults.

- Members of the phylum Chordata have several shared traits: pharyngeal slits, a nerve cord, a ventral heart, a tail beyond the anus, and a notochord. The Chordata are composed of the tunicates, lancelets, and vertebrates. The vertebrates have evolved a rigid endoskeleton for support and a dorsal vertebral column that replaces the notochord during development.

- The Vertebrates include the cartilaginous fishes (Chondrichthyes), ray-finned fishes (Actinopterygii), Amphibians (Amphibia), reptiles (Reptilia), birds (Aves), and mammals (Mammalia).

- Humans evolved from bipedal ancestors in Africa. *Australopithecus afarensis*, from 3.5 mya, appears to be an ancestor of the modern human. The most recent ancestor of humans was Cro-Magnon. During the evolution of the humans, brain size has increased, allowing humans to develop culture and learn language.

Common Problems

- Learning systematics is a difficult task. Luckily, in this chapter you should run across a number of phyla that you are familiar with. As you work your way through the phylogenies presented in the book, try to recall the representative organisms that you know from each group and what their special traits are. If you base your learning on information you already know, you will have an easier time learning the facts.

Study Strategies

- Many students become confused with the evolution of the chordates and vertebrates and the traits that separate them. Create a time line with all the major chordate classes. Insert the specific traits that distinguish one class from another. This will help you organize the evolution of the chordates so that memorizing it becomes a manageable task.

- Review the following tutorial and activities on the Student Website/CD:
  Tutorial 34.1 Life Cycle of a Frog
  Activity 34.1 Deuterostome Phylogeny
  Activity 34.2 Amniote Egg
  Activity 34.3 Deuterostome Classification

Test Yourself

Knowledge and Synthesis Questions

1. Which of the following is *not* a characteristic of species in the phylum Echinodermata?
   a. They have a water vascular system.
   b. They have an internal skeleton.
   c. They are protostomes.
   d. They have bilateral symmetry as larvae.

2. All but one of the following classes of echinoderm belong to the subphylum Eleutherozoa. Which is the exception?
   a. Asteroidea
   b. Crinoidea
   c. Echinoidea
   d. Holothuroidea

3. Which of the following groups of chordates evolved prior to the appearance of cartilaginous fishes?
   a. Ray-finned fishes and sea squirts
   b. Tunicates, lancelets, and agnathans
   c. Tunicates, lancelets, and ray-finned fishes
   d. Agnathans and ray-finned fishes

4. The vertebrate body plan includes all of the following traits except
   a. a ventral spinal cord.
   b. an internal skeleton.
   c. a well-developed circulatory system.
   d. organs suspended in the coelom.

5. The swim bladder of many fishes evolved from a lung-like sac. What important function does the swim bladder provide?
   a. It aids in prey capture.
   b. It controls swimming speed.
   c. It controls buoyancy.
   d. It aids in reproduction.

6. Which one of the following traits do the Chondrichthyes and Actinopterygii share?
   a. The gills are the major site of gas exchange.
   b. The skeleton is composed of cartilage.
   c. Their outer surface is covered with scales.
   d. They have a swim bladder.

7. The transition from aquatic to terrestrial lifestyles required many adaptations in the vertebrate lineage. Which of the following is *not* one of those adaptations?
   a. Switch from gill respiration to air-breathing lungs
   b. Improvements in water resistance of skin
   c. Alteration in mode of locomotion
   d. Development of feathers for insulation

8. The amniotes evolved the ability to reproduce by laying eggs with shells. What is the major advantage to laying shelled eggs?
   a. The embryo needs only a small amount of yolk for development.
   b. The shelled egg does not have to be laid in a moist environment.
   c. Evaporation from the egg is increased.
   d. Nitrogenous wastes can be excreted across the shell.

9. How do birds differ from reptiles?
   a. Birds have a lower metabolic rate.
   b. Reptiles lay eggs and birds give birth to live young.
   c. Birds can breathe and run at the same time, whereas reptiles cannot.
   d. They do not share a common ancestor.

10. Which of the following is *not* a trait that you could use to identify an animal as a mammal?
    a. Mammary glands
    b. Hair
    c. Sweat glands
    d. Kidneys

11. Which of the following vertebrate groups is arguably a living representative of the dinosaur lineage?
    a. Amphibians
    b. Birds
    c. Lobe-finned fishes
    d. Snakes

12. Which of the following statements about human evolution is *false*?
    a. Bipedalism was a hominid adaptation for life on the ground.
    b. Increases in the size of hominid brains preceded the appearance of language and culture.
    c. The extinction of the Neanderthal people was caused by the emergence of *Homo habilis*.
    d. Humans are not the direct descendants of modern-day chimpanzees.

Application Questions

1. Tunicates are armless and legless organisms that spend their adult lives attached to a substrate under water. They feed by pulling water into one tube, filtering out planktonic organisms, and pushing the water out another tube. Grasshoppers have legs, move around freely on land, and feed by ingesting food through a mouth. What evidence has led biologists to believe that humans are more closely related to tunicates than to grasshoppers?

2. The evolution of a hinged jaw resulted in an extensive radiation of the fishes into the many modern jawed forms. What was the main advantage and significance of hinged jaws in this radiation?

Answers

Knowledge and Synthesis Answers

1. **c.** Species from the phylum Echinodermata are deuterostomes.

2. **b.** The class Crinoidea belongs to the Pelmatozoa.

3. **b.** The tunicates, lancelets, and agnathans all evolved before the cartilaginous fishes.

4. **a.** Along with an internal skeleton, well-developed circulatory system, and organs suspended in a coelom, the vertebrates have a dorsal spinal cord.

5. **c.** The swim bladder of modern-day fishes are involved in controlling buoyancy.

6. **a.** In both the ray-finned fishes and cartilaginous fishes the major site of gas exchange is the gills. The ray-finned fishes have a skeleton of bone, are covered with scales, and have a swim bladder. These traits are not shared with the cartilaginous fishes.

7. **d.** In the move onto land, the development of feathers for insulation was not required. Amphibians and reptiles do not have an insulation layer, and many mammals have hair for insulation.

8. **b.** The shelled egg of the birds and reptiles allowed them to occupy dry terrestrial habitats because the shell decreases water loss from the egg.

9. **c.** Birds and reptiles differ in the morphology of their muscles that control movement and breathing. Birds can run and breathe at the same time, whereas reptiles must stop running to take a breath.

10. **d.** Kidneys are present in reptiles, birds, fishes, and mammals. The presence of kidneys is not a trait that could be used to identify a mammal.

11. **b.** The birds are now thought to be direct descendants of the dinosaurs.

12. **c.** The extinction of Neanderthals is thought to have been due in some part to the presence of Cro-Magnon. *Homo habilis* was extinct for more than a million years by the time the Neanderthals were present on Earth.

Application Answers

1. Although adult tunicates and adult humans seem like very different animals, our embryonic stages share certain important characteristics, including the presence of a notochord and (like all deuterostomes) a blastopore that develops into the anus (see Chapter 32). A look at embryonic grasshoppers, however, shows them to be fundamentally different: The embryonic grasshopper's blastopore develops into the mouth, putting grasshoppers squarely among the protostomes.

2. The hinged jaw of the fishes evolved during the Devonian period. The evolution of the jaw opened up a new untapped food source for these animals. With a jaw, fish could now grasp and kill larger living prey and allowed them to chew and tear the body parts of prey.

35 *The Plant Body*

Important Concepts

Vegetative structures serve a variety of roles, including nutrient procurement, support, water uptake, and transport.

- In flowering plants, vegetative organs are involved in all plant functions except reproduction.

- Angiosperms consist of two lines of plants with distinct characteristics, the monocots and the eudicots. Monocots, such as the grasses, lilies, orchids, and palms, differ from the eudicots, broad-leafed flowering plants, such as soybeans, roses, and maples, in a number of key anatomical characteristics.

  - Monocots have one cotyledon, parallel leaf veins, flower parts in multiples of three, and scattered primary vascular bundles.

  - Eudicots have two cotyledons, a netlike leaf venation, flower parts in fours or fives, and primary vascular bundles in a ring.

- Plants are composed of root systems and shoot systems (Figure 35.2). Roots are responsible for mineral and water uptake and support. Shoot systems consist of leaves (and leaf derivatives) involved in photosynthesis and stems for support.

- Root systems are generally one of two types.

  - Taproots are large single roots that grow deep within the ground and serve as storage tissue. Smaller lateral roots absorb water as the taproot enlarges.

  - Fibrous roots are numerous thin roots that spread far out from the plant, hold soil in place and rarely grow very deep.

- In some plants, adventitious roots occur along the stem and assist with rerooting or support.

- Stem systems produce buds, leaves, and flowers. Buds are embryonic shoots. At each node where leaves meet stem, a lateral bud is produced that may produce a new branch. At each stem tip is an apical bud responsible for elongation of that stem.

- Stems may be modified into tubers (such as potatoes) or runners. Stems may remain photosynthetic or become nonphotosynthetic woody material.

- Leaves are the site of photosynthesis but may have modifications to allow for specific jobs. The blade of a leaf is attached to the stem by a petiole. The leaf blade, with help from the petiole, is able to maintain a constant orientation toward sunlight for maximum photosynthesis. Leaves may be simple and consist of a single blade or may be compound with leaflets arranged along a central axis (see Figure 35.5). Vein patterns may be netted as in eudicots or parallel as in monocots.

- Some leaves are modified for storage of nutrients or water. Many have color modifications to attract pollinators. Some are modified into structures like tendrils and suckers to help the plant hold onto a support. Humans have taken advantage of these characteristics by breeding for these modifications.

Plant cells are specifically suited for their functions.

- Plant cells have the characteristics common to all eukaryotic cells but differ in that some have plastids, many have vacuoles, and every plant cell is bounded by cellulose-containing cell walls. Some plant cells are only functional after they die.

- Cell walls play important roles in plants. They provide turgor pressure, are chemically active, allow for transport, and become modified for specific functions. Cell walls are formed after cytokinesis. The daughter cells secrete the middle lamella and then begin laying down the primary cell wall. Once the cell is full size, it may begin secreting other substances to form a secondary wall.

- Cell walls do not block communication between cells. Small cytoplasmic strands called plasmodesmata extend through the walls allowing movement of substances from cell to cell. Pits in secondary cell walls (tiny holes) also allow materials to pass from cell to cell.

- Plant cells are often specialized to carry out specific functions.

  - Parenchyma cells are frequently photosynthetic or used for storage. They are thin-walled and have large central vacuoles. They may also continue dividing and serve as "parent cells" in a wounded area.

  - Collenchyma cells are support cells with special thickenings at the cell wall corners. They are elongated and allow for support without rigidity, which is an important characteristic for plants in windy areas.

  - Sclerenchyma cells with highly thickened cell walls provide rigid support after they die. They may be elongated fibers or variously shaped sclereids. Fibers strengthen bark and woody stems. Sclereids produce the gritty texture of pears.

- Xylem conducts water from the roots to the rest of the plant. Tracheary elements are functional after they die. In the gymnosperms, tracheids make up the xylem transport. Vessel elements are found in angiosperms. They secrete a specialized waterproofing substance that keeps them from "leaking" as water is moved.

- Living phloem moves carbohydrates and nutrients. Individual phloem cells are called sieve tube elements. Plasmodesmata enlarge where sieve tube elements join, making sieve plates. Sieve tube elements may lose nuclei and other organelles to prevent clogging of the sieve plates. These cells are filled with sieve tube sap that is high in carbohydrates. Adjacent companion cells regulate the function of the sieve tube elements.

Plant cells are organized into tissues.

- Simple tissues are composed of only one cell type; complex tissues are composed of multiple cell types. Xylem and phloem are examples of complex tissues that perform multiple functions by means of multiple cell types.

- Three tissue systems extend through the plant: vascular, dermal, and ground.

  - The vascular tissue system is made of xylem and phloem and is the conductive tissue of the plant.

  - The dermal tissue system provides the outer covering of the plant. This includes the epidermis and the layer of cuticle it secretes.

  - The ground tissue system composes the rest of the plant. This tissue functions in storage, support, and photosynthesis.

Plants develop in specific patterns from embryonic tissue.

- From the beginning, the plant is laid out along a main axis from root to shoot and in a concentric pattern that establishes the layout of the tissue systems.

- Plants frequently lose parts during growth. To counter this loss, plants are laid out in modules or units. A unit consists of a node and its attached leaf, the internode (a section of stem) below the node, and the lateral buds at the base of the internode. New units are formed as the plant grows. Leaves in some respects can be thought of as units themselves.

- Plants differ from animals in having indeterminate growth. They continue to grow throughout their life span. This is possible because plants have regions of continual cell division called meristems.

- Apical meristems give rise to the primary plant body and are located at the tips of roots and stems and in buds. They are responsible for elongating the plant body. Apical meristems give rise to primary meristems that produce the primary plant body. These are called the protoderm (dermal tissue system), the ground meristem (ground tissue system), and the procambium (vascular tissue system). All plant parts arise from division of the apical meristems.

- In some plants, lateral meristems lay down wood and bark. The vascular cambium arises from the lateral meristem and forms new secondary xylem and secondary phloem. The cork cambium produces new dermal tissues to accommodate increasing diameter and inhibits water loss with the production of periderm cells. The action of the vascular and cork cambiums is called secondary growth. Wood results from secondary xylem, and bark is everything outside the cork cambium.

- Root primary meristems (protoderm, ground meristem, and procambium) produce root tissues. At the tip of a root, the root apical meristem forms a root cap and quiescent center. The protoderm gives rise to the epidermis and root hairs. Ground meristem gives rise to the cortex and endodermis. The endodermis is specialized with a waxy coating of suberin to assist with water movement. The procambium gives rise to the stele, which houses three tissues: the pericycle, the xylem, and the phloem. In eudicots, the very center of the root is xylem, but in monocots the center is pith tissue (see Figures 35.16 and 35.17).

- The shoot primary meristem also gives rise to three primary meristems that produce the three tissue systems. Leaves arise from leaf primordia with bud primordia forming at each base. Shoot vascular tissues are arranged in vascular bundles. In eudicots, the vascular bundles are arranged in a cylinder allowing for woody growth. In monocots the vascular bundles are scattered.

- Stems and roots may undergo secondary growth from the lateral meristems. The orientation of the vascular cambium gives rise to cylinders of secondary xylem (wood) and secondary phloem. Connections of living cells to secondary phloem occur via vascular rays. Vascular rays are perpendicular to the vessel elements for conduction. Only eudicots have secondary growth.

- Annual rings seen in wood are a result of climate shifts in temperate zones, particularly water availability. You do not see rings in tropical trees. Heartwood in trees appears darker because the xylem has been clogged with minerals. Sapwood is still conducting at the time of cutting and appears lighter and less dense. Knots are the result of branches becoming buried in the expanding trunk.

- Stretching, breaking, and flaking off of epidermis and cortex produce bark. This leaves the secondary phloem at risk. Cells at the surface of the phloem produce cork that is thickened and reinforced with suberin. New cork is produced as secondary growth proceeds. The areas that allow gas exchange through the bark are known as lenticels.

Leaves are designed to maximize photosynthesis.

- Leaves have two zones of photosynthetic cells called mesophyll. The upper level of cylindrical mesophyll is called palisade mesophyll. The lower level is called spongy mesophyll. Air space around mesophyll cells is necessary for carbon dioxide to reach the photosynthesizing cells.

- Vascular tissue extends throughout leaves in veins. Veins carry water to cells and transport carbohydrates to sink tissues.

- The entire leaf is covered by a protective epidermis and is waterproofed by a waxy cuticle. Gas exchange occurs through guarded stomata. Guard cells open and close stomata to limit water loss.

The Big Picture

- A plant can be thought of as having vegetative structures that carry out the major functions of day-to-day life and reproductive structures that are responsible for reproducing the plant. You will see that there are cases in which the vegetative portions of the plant are quite adept at reproducing asexually. The vegetative plant body consists of the root system, which anchors the plant and absorbs water and nutrients, and the shoot system, which photosynthesizes and supports the plant against gravity. Modifications of the root and shoot systems lead to specialization of the plant.

- Plant cells are uniquely suited for support, transport, and the carrying out of cellular functions. Groups of cells form tissues that have specific roles within a plant. Most plants cells are totipotent, and under appropriate conditions they can become different cell types. Patterns of development are under the control of hormones and regulatory genes. Indeterminate growth and the modular organization of plants allow for regeneration of parts lost to damage and disease.

- Plant growth occurs from meristems. Apical meristems allow for elongation of the plant, whereas lateral meristems allow for secondary or woody growth. Not all plants exhibit secondary growth. All tissue types arise from the meristems and go through a process of elongation, then differentiation. Review stem, root, and leaf anatomy using the figures in your book.

Common Problem Areas

- This chapter covers plant anatomy. Basic plant anatomy is not difficult, but it is new to most students. The more time you spend looking at diagrams and live specimens, the easier it will be to understand the anatomy. As you study, think about the function of each structure. You will find that "form follows function."

- This chapter exposes you to many new vocabulary words. Resist the temptation to merely memorize. By looking at the words and understanding their roots, you will be better able to understand the terms.

Study Strategies

- The best study strategy for this material is to use the figures, pictures, and live material. Plant anatomy is structure. Structures are three dimensional and best understood visually.

- Make a vocabulary list. Though you don't want to memorize the terms, a list will help you organize your study.

- If your school has three-dimensional models, make use of them. There are 3-D models available now on the Internet. Using these will make it easier to understand the orientation of the structures.

- Review the following tutorial and activities on the Student Website/CD:
  Tutorial 35.1 Secondary Growth: The Vascular Cambium
  Activity 35.1 Eudicot Root
  Activity 35.2 Monocot Root
  Activity 35.3 Eudicot Stem
  Activity 35.4 Monocot Stem
  Activity 35.5 Eudicot Leaf

Test Yourself

Knowledge and Synthesis Questions

1. You are studying tropical plants in a Costa Rican cloud forest. You have identified a plant that has not been studied before. It has compound leaves with netted veins, vascular bundles arranged in a cylinder, and five petals on its very large and showy flower. This plant is most likely a
   a. monocot.
   b. eudicot.
   c. gymnosperm.
   d. None of the above

2. The plant described in Question 1 grows to approximately 40 m at maturity and is in a forest ecosystem. This plant most likely has which of the following root types?
   a. Fibrous
   b. Taproot
   c. Adventitious
   d. Rhizoids

3. You planted sweet peas along your back garden fence. You note that they are very effective at climbing the fence. Upon closer inspection, you notice that the plants are attached to the fence by tendrils. These tendrils are modifications of
   a. stems.
   b. roots.
   c. branches.
   d. leaves.

4. Plant cells are easily distinguished from animal cells because they have
   a. rigid cell walls.
   b. plastids.
   c. large vacuoles.
   d. All of the above

5. Plant cells that are photosynthetically active are found in the _____ layer of the leaf and are _____ cells.
   a. mesophyll; parenchyma
   b epidermis; parenchyma
   c. mesophyll; sclerenchyma
   d. epidermis; sclerenchyma

6. Water is conducted in _____ tissue, and carbohydrates and nutrients are transported in _____ tissue.
   a. xylem; phloem
   b. phloem; xylem
   c. parenchyma; phloem
   d. parenchyma; xylem

7. Plants are capable of indeterminate growth because of
   a. meristem tissues.
   b. regions of continually dividing cells.
   c. their modular nature.
   d. All of the above

8. Which of the following best describes the origin of wood?
   a. Xylem cells enlarge and deposit large amounts of lignin.
   b. Primary meristems increase the amount of xylem deposited.
   c. Lateral meristems contribute to continuous increases in vascular tissue.
   d. None of the above

9. Which of the following statements concerning wood is true?
   a. All woody plants show annual growth rings.
   b. Patterns of both cylindrical secondary growth and lateral vascular rays are visible in wood.
   c. Sapwood and hardwood result from patterns in primary xylem.
   d. All of the above

10. Which of the following best describes the function of the cork cambium?
    a. It lays down a protective cork covering over exposed phloem tissue.
    b. It allows for the sloughing of epidermal tissue.
    c. It allows for diameter expansion in stems and roots.
    d. All of the above

11. Sieve tube members have sieve plates where they join other sieve tube members. Which of the following best describes the sieve plates?
    a. Sieve plate pores are enlargements of plasmodesmata.
    b. Sieve plates are necessary to allow conduction between sieve tube cells.
    c. Sieve plates allow joining of cytoplasm between adjacent sieve tube cells.
    d. All of the above

12. Plants regulate gas exchange and water loss via
    a. the cuticle.
    b. guarded stomata.
    c. coated pits.
    d. sieve plates.

Application Questions

1. Draw a typical eudicot plant. Label the root system and shoot systems. Indicate on your drawing where you would find lateral buds and where you would find apical buds. Label the following structures: internode, petiole, leaf blade, taproot, lateral roots.

2. Differentiate between apical and lateral meristems. Do all plants have apical meristems? Do all plants have lateral meristems?

3. Describe the development of a xylem cell in a monocot from its origin in the apical meristem.

4. Draw a growing root. Label the primary meristems, root cap, cortex, stele, and epidermis. Discuss the function of each of these structures.

5. Compare and contrast monocots and eudicots. Identify the following plants as monocot or eudicot: corn, oak tree, day lily, rose, dandelion.

6. Look at the leaf in Figure 35.23 in your book. Which surface of that leaf faces the sun? How do you know? Why are the stomata opposite the palisade layer?

Answers

Knowledge and Synthesis Answers

1. **b.** See Figure 35.1 for a comparison of eudicots and monocots.

2. **b.** A taproot would be necessary to anchor a plant of that size. Most forest trees have taproots.

3. **d.** Tendrils are modified leaves. Some climbing plants produce tendrils and others produce suckers.

4. **d.** Plant cell shape is maintained by their cell walls. No animal cells have cell walls. Plants have plastids and vacuoles as well. Though these characteristics are seen in some protists, no animal cells have them.

5. **a.** Palisade and spongy mesophyll cells are photosynthetically active and derived from parenchyma cells.

6. **a.** Xylem tissue transports water from the roots throughout the plant. Phloem tissue transports carbohydrates and nutrients from source tissue to sink tissue.

7. **d.** The apical and (when present) lateral meristems are regions of continually dividing cells that can contribute to the growth of a plant throughout its life. A modular body plan also contributes to the ability of plants to grow continually.

8. **c.** Lateral meristems are responsible for the growth of new xylem and phloem. The secondary xylem gives rise to wood.

9. **b.** Wood grain is a result of patterns in the vascular cylinders. Wood porosity results from the lateral phloem rays.

10. **d.** The cork cambium is a meristematic region outside the secondary phloem. As girth increases and splits the epidermis causing loss of those protective layers, the cork cambium produces new cells to cover the expanding vascular tissue.

11. **d.** The sieve plates allow for the conduction of sap from sieve tube cell to sieve tube cell. The pores are due to enlargements of the plasmodesmata. These cells may also lose nuclei and organelles so that carbohydrates can more easily pass through the sieve tubes.

12. **b.** Guard cells at the edges of stomata respond to changes in osmotic pressure. These changes lead to the opening and closing of the stomata to regulate gas exchange and water loss.

Application Answers

1. See Figure 35.2 for a labeled diagram.

2. Apical meristems are responsible for elongation of the plant body. Lateral meristems are responsible for an increase in girth. Lateral meristems are found only in woody eudicots and are responsible for creating wood. All plants have apical meristems.

3. The apical meristem gives rise to the protoderm, the ground meristem, and the procambium. The procambium gives rise to the vascular tissue system, including xylem cells. Monocots do not have secondary xylem, so only primary xylem is possible in this plant.

4. See Figure 35.15 for a diagram of root growth. The protoderm gives rise to the epidermis for protection. The ground meristem gives rise to the cortex for storage. The procambium gives rise to the stele for transport. The root cap protects the meristem as it pushes through the soil.

5. See Figure 35.1 for a comparison of monocots and eudicots. Corn is a monocot. The easiest way to be sure of this is to look at the parallel veins in its leaves. An oak tree is a eudicot because only eudicots have woody secondary growth. Day lilies are monocots. They have the characteristic parallel veins and classically three-petaled flowers. Roses are eudicots. Many roses have woody secondary growth, petals in multiples of five, and netted veins. Dandelions have netted veins, but the clearest indication that they are eudicots is their well-developed taproot (see Figure 35.15).

6. The "top" of the diagram in your book would be the surface that faces the sun. This is the surface where photosynthetic cells, which need maximum sun exposure, are located. The stomata are on the opposite side to reduce water loss due to evaporation during photosynthesis.

36 *Transport in Plants*

Important Concepts

Water and nutrients are taken up in the roots of plants.

- The movement of water across a membrane is a special type of diffusion known as osmosis (see Figure 36.2).

  - For osmosis to occur across a semipermeable membrane, there must be a solute potential (or difference) great enough to initiate movement and a pressure potential small enough to allow for movement.

  - The overall tendency of a solution to take up water across a membrane is called water potential and is the sum of the negative solute potential and the positive pressure potential. All three parameters can be measured in megapascals (MPa). Water always moves to a region of more negative water potential.

- The structure of plants is maintained by osmotic phenomena. If a plant loses turgor by a decrease in pressure potential, it wilts. Movement of water from cell to cell depends on water potential. Long-distance movement depends on pressure potential and is referred to as bulk flow.

- Specialized membrane channel proteins in plant cells called aquaporins can increase the rate of water movement by allowing water to cross the plasma membrane without interacting with the hydrophobic bilayer that slows water flow. Though aquaporins can increase the rate of osmosis, they cannot influence the direction of flow.

- Mineral uptake requires active transport via proteins. When mineral concentrations are greater in the soil solution than in the plant, they are taken up by facilitated diffusion. If minerals are in smaller concentrations outside the plant than inside the plant, or if they must be moved against an electrical gradient, then the plant must rely on active transport.

- Plants rely on a proton pump for active transport of minerals into cells. Plants actively pump protons out of cells, causing the area outside the cell to become more positive. This assists facilitated diffusion of positive ions through protein channels. It also drives the movement of negatively charged ions into the cell by active transport (see Figure 36.3). The result of this pumping action is to make the internal environment of the plant cell highly negative compared to its environment. This difference in charge is called membrane potential.

- Water moves into a root because the root has a more negative water potential than the soil solution it is bathed in. Movement inside the root takes place because the stele (vascular tissue) has a more negative water potential than the cortex. Minerals dissolved in water are moved via bulk flow of water once it is in the vascular system.

- Minerals follow two paths for reaching the vascular tissue: the apoplast or the symplast.

  - The apoplast is formed by cell walls and intercellular spaces. Water and minerals may move unregulated through this space without ever having to cross a membrane.

  - The symplast is the living portion of the plant and is enclosed in plasma membranes. Movement of water and minerals in the symplast is highly regulated.

  - Water and minerals can travel through the apoplast as far as the endodermis. At the endodermis, water and minerals are stopped by the Casparian strips, which are waxy structures surrounding the endodermal cells.

  - Because of this, water can reach the stele only via the symplast. The transport proteins in the endodermal cells determine which minerals enter the stele.

  - Once past the endodermal barrier, water and minerals can again leave the symplast and move back to the apoplast. Some cells, known as transfer cells, are specialized for doing this.

- Transfer cells are parenchyma cells specialized with knobby projections allowing for increased surface area and increased transport of proteins. Transfer cells also have large numbers of mitochondria to manufacture ATP needed to drive active transport.

- Ultimately, water and minerals from the soil solution end up in xylem cells and are referred to as xylem sap.

Water and nutrients are moved through the plant in the xylem.

- It was originally thought that a pumping mechanism for the movement of fluids might be active in plants. This was shown to be false by experiments in 1893. Tree trunks immersed in poison showed progressive death of all living cells as the poison progressed through the plant. This experiment led to three important conclusions:

  - Because movement continued even as cells were killed, no "pumping" cells were active.

- Leaves were critical to transport because transport continued until the leaves died.

  - Transport did not depend on the roots because it occurred in the absence of roots.

- Root pressure, as shown by guttation (the forcing of water out of openings in leaves), was another theory for water movement. It was thought that the pressure exerted by root tissue might be sufficient to force water up the xylem. In actuality, xylem sap is under negative pressure as it is ascending.

- Pulling forces due to transpiration at the leaf surface are responsible for the movement of water through the xylem. Water evaporates from mesophyll cells during transpiration, creating tension on the water associated with the mesophyll cell wall. Transpiration generates tension on the water molecules in the xylem water column, which pulls them up from the roots and through the apoplast of the leaves. Water molecules are cohesive enough to resist the tension and pull other water molecules along because of their hydrogen bonding. This results in bulk flow (see Figure 36.8).

- The transpiration–cohesion–tension mechanism pulls water from the roots up through the plant. This is a passive process requiring no energy input by the plant.

- Minerals are drawn passively along with the water column. They may be redistributed via the phloem, but initial movement is through the xylem.

- Transpiration also assists with temperature regulation through evaporative cooling.

- Tension in the xylem sap can be measured with a pressure bomb (see Figure 36.9). By measuring tension in stems, the mechanism of water transport was tested and understood. If xylem sap is under tension and a stem is removed from a plant, the xylem sap should be drawn up in the xylem away from the cut. If the cut portion is placed under pressure, the sap should rise back to the surface of the cut end. The pressure that it takes to raise this is equal to the tension of the xylem sap. Using this mechanism, Per Scholander showed that all xylem is under tension.

- The rate of flow of xylem sap depends on a number of environmental factors such as temperature, light, and wind. Changes in K^+ concentration also influence xylem flow.

Water loss in a plant must be controlled.

- Leaf surfaces are covered with a waxy cuticle to prevent excessive water loss. However, gas exchange must also occur at the leaf. Any time a plant surface is open enough to allow gas exchange, water is also lost.

- Stomata with guard cells regulate gas exchange and water loss from a leaf. Guard cells, in response to osmotic differences, shrink and swell to open and close the stomata. Guard cells open when light is sufficient to maintain photosynthesis and carbon dioxide levels are low. Guard cells are also regulated by water potential. If the water potential in mesophyll cells is low, mesophyll cells release abscisic acid, which causes the guard cells to close.

- Blue light stimulates a proton pump that helps regulate guard cell activity. Guard cells open when potassium ions are actively transported into them as a result of the action of the proton pump. High potassium levels cause water to move in by osmosis. Turgor builds in the guard cells, and they are pulled apart to reveal the stoma. For guard cells to shut, the proton pump stops and potassium ions move back across the membrane. Water follows, and the cells go limp and seal off the stoma. See Figure 36.11.

- Scientists are looking for ways to reduce water loss by transpiration without reducing carbon dioxide availability to plants. This has been attempted through foliar sprays of antitranspirants like abscisic acid and more recently through genetic modification of plants to make them more sensitive to abscisic acid.

Phloem moves materials from sources to sinks by translocation.

- Sources are organs that produce more sugars than are used by metabolism, storage, and growth. Sinks are organs that do not make enough sugar for their own growth or storage needs. Sugars, amino acids, minerals, and other substances are translocated between sources and sinks in the phloem.

- Translocation proceeds in both directions along the stem. Translocation stops if phloem tissue is killed and is inhibited whenever respiration and the availability of ATP are limited.

- Scientists can sample the sieve tube sap of a single sieve tube member by using aphids. An aphid drills into a single cell, and sap is forced out. Once the aphid begins eating, it is frozen and its feeding organ is used as a tap to collect the sap from a single sieve tube.

- The pressure flow model explains how materials move through the phloem. Once in the sieve tubes, sieve tube sap moves via bulk flow, which requires no energy input by the plant. Energy is required for the loading of the sieve tubes at the sources and the unloading of solutes when the sink is reached. Sucrose is actively transported into the sieve tube members at the sources. This causes water to move into surrounding sieve tube members by osmosis, thereby increasing the pressure potential at the source end and pushing the contents toward the sink end of the tube, resulting in bulk flow.

- For the pressure flow model to work, the sieve plates must be open to allow uninterrupted flow of sieve tube sap from one sieve tube member to another. Microscopic analysis indicates that the sieve tube plates are open in undamaged cells.

- There must also be a mechanism for loading and unloading sucrose and amino acids. Neighboring cells assist with the loading and unloading of sucrose at sources and sinks.

- Plasmodesmata assist in moving substances from cell to cell and help with the loading and unloading of sieve tube members. Plasmodesmata in source areas differ from those in sink areas. They are larger and more abundant in sink tissues for unloading. Scientists are exploring the regulation of plasmodesmata in crops to enhance production.

The Big Picture

- In terrestrial plants water must be acquired from the soil, transported through the plant, and used in the leaves for photosynthesis. At the same time, the nutritional products of photosynthesis must be transported throughout the plant to nonphotosynthetic organs. This two-way transport is achieved through specialized cells that make up the vascular tissue of the plant. Water with dissolved mineral nutrients is absorbed through the roots and transported to cells in the xylem of the plant, where it is pulled up the stem to the leaves via the transpiration–cohesion–tension mechanism. Sugars and solutes are moved out of the leaves and to the rest of the plant through cells in the phloem.

- Water uptake is regulated by osmotic and water potentials in the root cells, and the rate of transport is controlled by the rate of evaporation at the leaf surface. Guard cell activity in the leaves regulates the opening and closing of stomata to match water availability, light, and drying conditions. Sucrose movement is regulated by active transport and facilitated diffusion in the phloem tissue. The rate of sucrose transport depends on the rate of loading and unloading at source and sink tissues.

Common Problem Areas

- Students often fail to look at the anatomy of the structures that are used in transport. Each structure is uniquely suited to its function. If you couple form and function, the process is much easier to understand.

Study Strategies

- Water and sucrose transport are pathways that can be understood by visually tracing a molecule of water or sucrose through the plant. Use the figures in your book to follow these pathways.

- In order to understand water transport, you need to understand osmosis and the properties of water molecules. If you are struggling with this chapter, review these concepts from earlier chapters.

- The basis of nutritional transport in plants is cell-to-cell transport. Review the sections in your book on diffusion, osmosis, and active transport.

- Review the following tutorial and activity on the Student Website/CD:
  Tutorial 36.1 The Pressure Flow Model
  Activity 36.1 Apoplast and Symplast of the Root

Test Yourself

Knowledge and Synthesis Questions

1. The function of the Casparian strips is to
   a. divert water and minerals through the membranes of endodermal cells.
   b. prevent water and minerals from entering the stele through the apoplast.
   c. provide regulation for water and mineral movement in the plant.
   d. All of the above

2. The primary difference between the apoplast and the symplast is that the
   a. apoplast consists of nonliving spaces and cell walls.
   b. apoplast relies on active transport.
   c. symplast consists of nonliving spaces and cell walls.
   d. apoplast prevents passive diffusion.

3. Which of the following regarding water transport is true?
   a. Root pressure is sufficient to drive xylem sap movement.
   b. Bulk flow is not a mechanism by which water and minerals are transported.
   c. The cohesive nature of water is central to water movement in a plant.
   d. None of the above

4. Tension is a result of which of the following?
   a. Transpiration at the leaf surface
   b. The cohesive nature of water
   c. The narrowness of the xylem tube
   d. All of the above

5. The fact that water transport continues as long as leaves are alive and active indicates that
   a. leaves pump water.
   b. leaves are necessary for transport of water.
   c. roots are active.
   d. water is not needed for leaves to remain alive.

6. Which of the following is true regarding transport in phloem?
   a. Transport in phloem is always in the direction of leaves to roots.
   b. Transport in phloem is from source tissue to sink tissue.
   c. Transport in phloem cells requires no energy inputs from the plant.
   d. None of the above

7. If the pressure potential is 0.16 megapascals (MPa) and the solute potential is –0.24 MPa, then the water potential would be
   a. 0.04 MPa.
   b. 0.08 MPa.
   c. –0.08 MPa.
   d. –0.24 MPa.

8. If you were to order the water potential of the following root cells or regions from least to most negative, which cell or region would be third?
   a. Xylem
   b. Soil next to root
   c. Cortex apoplast
   d. Stele apoplast

9. The movement of water up the stems of tall plants is least dependent on which of the following factors?
   a. Guttation
   b. Transpiration
   c. Cohesion of water molecules
   d. Tension within water columns

10. Which of the following is true of both xylem transport and phloem transport?
    a. Both are passive processes that do not require energy from the plant.
    b. Both involve only living cells.
    c. Both rely on a water potential gradient.
    d. The direction of flow can reverse in both.

11. Which of the following regulates stomatal opening and closing?
    a. Abscisic acid levels
    b. Light levels
    c. Carbon dioxide concentrations
    d. All of the above

12. Regulators of stomatal opening and closing work by activating the
    a. proton pump in guard cells.
    b. proton pump in stomata.
    c. sodium–potassium pump.
    d. All of the above

Application Questions

1. Differentiate between source and sink tissues. What happens relative to phloem in each?

2. Under what conditions does transpiration occur most rapidly? What effect will increased transpiration have on water flow in a plant? What happens if adequate water for the plant is not available?

3. Trace a water molecule as it moves from the soil solution to the stele of a plant. Identify where the molecule is traveling through the apoplast and where the molecule is traveling through the symplast.

4. Describe the role of the proton pump in moving minerals into the root.

5. Explain how transpiration, cohesion, and tension work together to move water in a large plant.

6. Describe the pressure flow model of phloem transport.

Answers

Knowledge and Synthesis Answers

1. **d.** Not all minerals that enter the apoplast of a plant's root are beneficial to the plant. The Casparian strips prevent water and minerals from reaching the stele through the apoplast, diverting them instead through the plasma membranes of the endodermal cells. Channel proteins in these plasma membranes determine which minerals can enter the symplast, and from there, the stele. The Casparian strips thus contribute to the regulation of water and mineral movement in the plant.

2. **a.** The intercellular spaces and cell walls of the plant constitute the apoplast.

3. **c.** Water movement depends on the cohesive nature of water to withstand the tension placed on the water column by transpiration.

4. **a.** Transpiration causes tension.

5. **b.** Leaves are necessary for transpiration to take place.

6. **b.** Transport in phloem does not always go from leaf to root, but it is always from source tissue to sink tissue. The plant must contribute energy to create the water pressure gradient by pumping solutes into the phloem at the source and out of the phloem at the sink.

7. **c.** Water potential is equal to pressure potential plus solute potential.

8. **c.** The xylem would be the most negative, followed by the stele, then cortex, then the area outside the root.

9. **a.** Guttation occurs under extremely humid conditions when water is plentiful.

10. **c.** Both xylem transport and phloem transport depend on water potential.

11. **d.** Abscisic acid, light, and carbon dioxide levels all regulate stomatal opening and closing.

12. **a.** Stomatal regulators work by activating and deactivating the proton pump in guard cells.

Application Answers

1. Source tissues produce sugars in excess of what can be used and stored. Phloem loading occurs in source tissues and creates a pressure potential that results in bulk flow of sieve tube sap toward sink tissues. Sink tissues produce fewer sugars than can be stored or used and unload phloem through active transport.

2. Transpiration will occur most rapidly in high light conditions when stomata are open, along with high wind conditions and low humidity when evaporation is greatest. This will result in faster bulk flow through the xylem and increased water demands by the plant. If water is not available, plant cells will lose turgor and the plant will wilt.

3. The water molecule enters the epidermis due to water potential. If it crosses the plasma membrane into an epidermis or cortex cell, it enters the symplast and can travel via plasmodesmata to the stele. If it does not enter the symplast, it can travel through the apoplast until it reaches the endodermis and a Casparian strip. There it crosses the plasma membrane of the endodermis and enters the symplast. Once past the endodermis, it may again enter the apoplast (see Figure 36.4).

4. Plants rely on a proton pump for active transport of minerals into cells. Plants pump protons out of cells, which causes the area outside the cell to be more positive. This assists facilitated diffusion of positive ions through protein channels. It also drives the movement of negatively charged ions into the cell by active transport (see Figure 36.3).

5. The transpiration–cohesion–tension mechanism pulls water from the roots up through the plant. Water evaporates from mesophyll cells during transpiration. This puts tension on the film of water associated with the mesophyll cell wall. The tension at the mesophyll cell draws water from the xylem of the nearest vein. This creates tension in the entire xylem column, and the column is drawn upward from the roots.

6. The difference in solute concentration between sources and sinks creates a pressure potential along sieve tubes, resulting in bulk flow. For this to occur, sugars must be loaded at the source tissue and unloaded at the sink tissue through active transport, and the sieve plates must remain open and unclogged along the phloem column.

37

Plant Nutrition

Important Concepts

All organisms must make nutrients through the cycling of other compounds or acquire nutrients from their environment.

- The basic nutrient requirements for all living things are carbon, hydrogen, oxygen, and nitrogen. These elements are the fundamental building blocks of all macromolecules.
- Carbon, oxygen, and hydrogen enter biological organisms through photosynthesis.
- Nitrogen's entry into biological organisms is dependent on nitrogen-fixing bacteria in the soil.
- Mineral nutrients are essential to life. Sulfur, phosphorous, magnesium, and iron are all essential components of many macromolecules.
- Mineral nutrients enter biological organisms through soil solutions, which plants take up through their roots.
- Plants form the basis of the terrestrial food chain because they make organic compounds that are available to other organisms. The ability to make organic compounds from inorganic compounds is called autotrophism. Plants rely on photosynthesis to produce the organic molecules that allow them to be autotrophs. Heterotrophs require preformed organic compounds.
- Plants are sessile, meaning they cannot move around and therefore cannot "search" for nutrients. They overcome this problem by growing toward new resources. A plant grows taller to procure more sunlight for itself and to shade other plants, and its roots spread to acquire mineral nutrients in the soil.

Mineral nutrients are vital to proper plant growth.

- Every plant requires specific essential nutrients necessary for growth and development. These nutrients cannot be replaced by another element and must be directly obtained for the day-to-day functioning of the plant. A deficiency in any essential nutrient leads to an unhealthy plant.
- Macronutrients are required at a rate of 1 g per 1 kg of dry plant matter, and micronutrients are required at a rate of 100 mg per 1 kg of dry plant matter.
- See Table 37.1 for a list of macro- and micronutrients, their sources, and their functions.
- Though many deficiencies ultimately lead to plant death, specific deficiency symptoms are displayed in a plant

before it dies. Nitrogen deficiency is the most common deficiency. Many farmers use chemical fertilizers to supplement nitrogen in soils so that crop plants have adequate nitrogen sources.
- Plant growth will slow to match the availability of nutrients until the nutrient loss becomes severe.
- Chlorosis, or yellowing of leaves, is the most common symptom of nutrient deficiency. This occurs when chlorophyll cannot be synthesized because of inadequate nitrogen. Chlorosis is also evident in iron deficiencies for the same reason. Chlorosis of young leaves is characteristic of iron deficiency because iron cannot be moved through a plant easily.
- Many essential elements have multiple roles in plants. Depriving plants of specific nutrients and observing them identified essential nutrients. Many of the experiments to determine the essential micronutrients have been conducted in special controlled laboratories. Missing nutrients that prevented a plant from completing its life cycle were deemed essential.

Soils are nature's nutrient sink.

- Plants and soils interact in a complex fashion. Plants change soils, and soils influence the growth of plants.
- Soils are composed of living and nonliving matter. The living portion of soil contains roots, protists, bacteria, fungi, and many small animals. The nonliving portion consists of rock fragments, clay, water, air spaces, and dead organic matter.
- All soils have a soil profile consisting of two or more horizons (layers). Water-soluble nutrients are leached to deeper horizons through rainfall. Three major horizons (also called zones) can be identified in soils. Zone A is topsoil that is organically rich and very biologically active, and is the most agriculturally important layer. Loam soil with sand, silt, and clay, high nutrient content, plentiful water and air spaces is ideal topsoil for agriculture. Sands typically do not hold nutrients or water well, and clays are too dense for root growth. Zone B is the subsoil, and holds many leached nutrients. Zone C is parent rock, which roots cannot penetrate.
- Soils form from weathering rocks. Mechanical and chemical weathering break down rocks. Mechanical weathering occurs through freeze/thaw cycles, rain, and drying. Chemical weathering occurs by oxidation, hydrolysis, or breakdown by acids, and can change the composition of rock. Chemical weathering is very important in clay formation.

- Mineral nutrients are tied to clay particles in the soil. Because many nutrients are cations, clays with a negative charge can hold these nutrients and make them available to plants. Roots release protons into the soil that bind to clay particles, and the cation minerals are released. This process is called ion exchange. Nutrients that are negatively charged and therefore do not participate in ion exchange are rapidly leached from soil. Thus, nitrate and sulfate are often not available to plants.

- Agricultural fertilizers add nitrogen, phosphorous, and potassium to soils and are rated by their "N-P-K" percentages. A 10-10-10 fertilizer contains 10 percent nitrogen, 10 percent phosphate, and 10 percent potash (potassium source). These nutrients need to be replenished in soils because they are negatively charged and leach from soils rapidly.

- Fertilizers may be organic in nature (manures, compost) or inorganic (chemical fertilizer). Organics release nutrients much more slowly and do not leach as quickly as inorganics. Inorganics provide a much more rapid release of nutrients.

- The pH of soils affects nutrient availability. Specific plants have specific pH needs. Soils tend to become slightly acidic from leaching and rain. The practice of liming raises soil pH. This has the secondary effect of making calcium available to plants. Adding sulfur reduces soil pH.

- Copper, iron, and manganese can be added by foliar spraying of nutrients.

- Plants affect the pH of soil, adding decaying matter in the form of humus, and altering soil temperature.

Nitrogen is vital to plant growth.

- Nitrogen gas is readily available in the atmosphere but plants cannot use it directly. Only a few bacteria species can convert nitrogen gas into biologically usable ammonia through nitrogen fixation.

- The most biologically important nitrogen fixers are associated with plant roots and release up to 90 percent of the nitrogen they fix to the soil. *Rhizobium* bacteria have a close association or mutualistic symbiosis with the roots of plants in the legume family; both species benefit from the association. These bacteria and the associated legumes are important agriculturally and are the basis of crop rotation.

- The fixation of nitrogen requires the enzyme nitrogenase to catalyze the reaction, lots of energy (ATP), and a strong reducing agent. It is inhibited by oxygen and therefore is active only under anaerobic conditions. The protein leghemoglobin binds with oxygen to keep its levels low.

- The anaerobic *Rhizobium* bacterium resides in a root nodule of the legume where it is protected from oxygen. The bacteria are provided with a low-oxygen growing environment, and the plant benefits from the released ammonia. To establish the symbiosis, the plant releases flavonoids to attract the bacteria. In response to the flavonoids, the bacterial *nod* genes are turned on. The resulting gene products cause formation of the nodule by the plant. Once housed in the nodule, the bacteria form swollen bacteroids capable of nitrogen fixation. The plant surrounds the bacteroids with leghemoglobin to support respiration (see Figure 37.7).

- Bacterial nitrogen fixation is not adequate to supply all the nitrogen needed for agriculture. Currently, nitrogen fertilizers are produced via industrial fixation through the Haber process, an energy-expensive process. Much research is looking into the possibility of engineering plants that have their own nitrogenase.

- Ammonia itself is toxic to plants. Therefore, soil bacteria that convert ammonia to nitrate are necessary. These bacteria are called nitrifiers.

- Plants, through their metabolism, reduce nitrate to ammonia. This is accomplished by the plant's own enzymes.

- The entire nitrogen cycle is completed by denitrifiers, which convert the nitrogen from waste and dead organic matter back to nitrogen gas. See Figure 37.8 for a review of the nitrogen cycle.

Some plants are heterotrophic.

- Some plants that live in soils that are nitrogen or phosphorous deficient are carnivorous. Carnivorous plants acquire nitrogen from trapped decaying animals. Examples of carnivorous plants are Venus flytraps, pitcher plants, and sundews. Although these plants can survive and grow with out consuming insects, they do best when they have a continuous supply of them.

- Some plants have lost the ability to photosynthesize. These plants must acquire their nutrients from other sources. Some are parasitic and acquire some or all of their nutrients from a host plant at the host plant's expense.

The Big Picture

- Plants require specific macro- and micronutrients. Deficiencies in any of these produce consequences for the health of the plant. Essential nutrients must be present, and no substitutions will sustain the plant. Nutrients are procured from the soil solution that bathes the roots of a plant. The availability of nutrients depends on quantity, solubility, and structure of soil. Many agricultural practices deplete soils of nutrients, which must be added back through fertilization.

- Nitrogen availability is essential to plant growth. Bacteria and plants are intrinsically linked in the nitrogen cycle (see Figure. 37.8). Biological nitrogen fixation is utilized in agriculture by rotating crops with those that harbor nitrogen-fixing bacteria in their root nodules. Commercial nitrogen fixation for chemical fertilizers is extremely energy demanding, and alternatives through genetic engineering are being sought.

- A small number of plants do not photosynthesize. These heterotrophic plants are often parasites of other plants and acquire their nutrients solely through their host plant.

Study Strategies

- The most difficult concept of this chapter is the nitrogen cycle. Use the figures in your book to visualize this process and understand how the organisms involved interact with the environment.

- Be sure you understand the consequences of nutrient shortages in plants.

• Review the following tutorial and activity on the Student Website/CD:
Tutorial 37.1 Nitrogen and Iron Deficiencies
Activity 37.1 Nitrogen Cycle

Test Yourself

Knowledge and Synthesis Questions

1. Which of the following nutrients is *not* considered essential for plant growth?
   a. Cadmium
   b. Nitrogen
   c. Manganese
   d. Potassium

2. Macronutrients are _____ than micronutrients.
   a. larger molecules
   b. needed in greater quantities
   c. more essential
   d. more important for growth

3. Carbon and oxygen are acquired through _____ and nitrogen and potassium are acquired from _____ .
   a. consumption; soil solution
   b. photosynthesis; nitrification
   c. photosynthesis; soil solution
   d. photosynthesis; fertilizer

4. You notice that the young leaves of your tomato plants are very yellow. What type of deficiency does this suggest?
   a. Nitrogen
   b. Carbon
   c. Water
   d. Iron

5. Years of cotton farming in the South has stripped away much of the A horizon of the soils. Subsequent agriculture has been difficult for farmers because the
   a. A horizon contains the most available nutrients.
   b. B horizon contains significantly less available nutrients.
   c. A horizon is most conducive to root growth.
   d. All of the above

6. Clay particles in soils are important for
   a. holding soil together.
   b. ion exchange.
   c. holding water.
   d. All of the above

7. Most clays form from the _____ of rock.
   a. mechanical weathering
   b. chemical weathering
   c. heaving
   d. All of the above

8. You purchase a commercial fertilizer at your local garden center. The label says that it is 10-20-10. This label refers to the
   a. percentages of nitrogen, phosphate, and potassium.
   b. percentages of nitrogen, carbon, and oxygen.
   c. rate at which nitrogen is released from the fertilizer.
   d. ratio of organic to inorganic matter in the fertilizer.

9. The relationship between *Rhizobium* and the roots of legumes can best be described by which of the following terms?
   a. Parasitic
   b. One-sided
   c. Mutualistic
   d. Carnivorous

10. Nitrogen gas is reduced to ammonia by which of the following enzymes?
    a. *Rhizobium*
    b. Nitrogenase
    c. Nitrification
    d. Denitrification

11. Plants are able to take up and use nitrogen in the form of _____ and _____.
    a. ammonia; nitrate
    b. ammonia; nitrogen gas
    c. nitrogen gas; nitrate
    d. All of the above

12. Carnivorous plants are often found in acidic and nutrient-poor environments. The main selective pressure for carnivory is
    a. lack of nitrogen and phosphorous sources.
    b. lack of iron and calcium sources.
    c. incomplete ion exchange.
    d. All of the above

Application Questions

1. Explain how it was determined whether a nutrient was essential. What were the drawbacks of early experimentation?

2. Sketch the nitrogen cycle. Which components of the cycle are impacted by plants and how?

3. Describe how plants and bacteria interact to form nitrogen-fixing root nodules. Why do many farmers plant crops such as alfalfa and soybeans without harvesting them?

4. Most carnivorous plants are found in boggy, wet, acidic environments. What effect would an environment such as this have on nutrient availability?

5. Explain how plants "grow into their nutrients."

6. Differentiate between micro- and macronutrients. Where are most of these nutrients acquired?

7. Discuss the impact agriculture has on plant nutrition. Which activities increase nutrition and which procedures rob nutrients from the soil?

Answers

Knowledge and Synthesis Answers

1. **a.** Cadmium is not one of the 14 essential micro- and macronutrients.

2. **b.** The main difference between micronutrients and macronutrients is in how much of them a plant needs to survive.

3. **c.** Carbon, oxygen, and hydrogen are all acquired through photosynthesis. Nitrogen and mineral nutrients are acquired in soil solution (dissolved in water).

4. **d.** If the older leaves are yellow, a nitrogen deficiency would be suspected because nitrogen is easily relocated in the plant whereas iron is not.

5. **d.** Topsoil, or the A horizon, is most conducive to root growth. Ample nutrients are available, as are air spaces and water for ease of root growth.

6. **d.** Clay particles are critical for ion exchange. They are also important for retaining water and for the integrity of soil.

7. **b.** Chemical weathering leads to clay formation.

8. **a.** Your fertilizer is 10 percent nitrogen, 20 percent phosphorous, and 10 percent potash (a potassium source).

9. **c.** The relationship is mutualistic in that both the plant and the bacteria benefit from the association.

10. **b.** Nitrogenase catalyzes the reduction of nitrogen gas to ammonia. This is an energy-expensive process.

11. **a.** Plants can take up and use nitrogen that is in the form of ammonia or nitrate.

12. **a.** Carnivory supplements insufficient nitrogen and phosphorous availability.

Application Answers

1. Plants were grown in cultures that lacked specific nutrients. If a plant could not complete its life cycle, the missing nutrient was said to be essential. Early experiments were unreliable because of impurities in the chemicals used to grow the plants. Experiments were ultimately improved through well-controlled hydroponic studies.

2. See Figure 37.8. Plants are actively involved in nitrogen fixation through the symbiosis with *Rhizobium*. *Rhizobium* fix nitrogen only in close association with the roots of plants in the legume family.

3. See Figure 37.7. By rotating crops (and even more by rotating and plowing legume crops under) farmers can organically add nitrogen to depleted soils. This results in significantly larger yields to offset the years when the land is not utilized for harvested crops.

4. Acidic environments limit decomposition and thus nitrogen sources. They also limit ion exchange. Both conditions result in reduced nutrient availability that can be offset by carnivory.

5. Plants cannot physically move to areas with greater nutrients. They are limited to extending their roots into soils that contain a larger nutrient reserve. When nutrients are limited, plant growth matches the availability of resources.

6. Macronutrients are needed at a rate of 1 g/1 kg of dry plant tissue. Micronutrients are needed at a rate of 100 mg/1 kg of dry plant tissue. The majority of these nutrients are in soil solution and are taken up as water is drawn into roots.

7. Agriculture has increased plant yields-per-unit land through the use of fertilizers and advanced farming techniques, but at a cost. Much of the nation's topsoil has been depleted through poor farming practices, thus limiting the nutrient resources available and forcing dependence on fertilizer. The industry is attempting to find new methods of farming that are more sustainable and less harsh to the environment.

$\mathscr{38}$ Regulation of Plant Growth

Important Concepts

Plant development is influenced by multiple regulatory factors.

- Plant development depends on the interplay of environmental cues, receptors that sense these cues, and hormones that mediate the effects of the environment. Enzymes influence development at all stages.

- Hormones are regulatory compounds that are produced in one area of a plant and translocated throughout it. The effects of the hormones result from their relative concentrations, and they play multiple regulatory roles in plants.

- Photoreceptors are proteins that sense light and are altered by light quality to induce changes within a plant.

- A plant's genome ultimately regulates its growth and development, under the influence of hormones and photoreceptors.

- Signal transduction mediates hormone and photoreceptor action. A receptor receives an environmental cue, a signal transduction pathway is initiated, and ultimately a cellular response is generated through a protein kinase.

Plant development depends on cell division, cell expansion, and cell differentiation.

- Plant seeds are dormant and remain so until seed germination. Germination is triggered by one or more mechanical or environmental cues. To germinate, a plant must imbibe water and draw nutrients from the endosperm or cotyledons. A plant is considered a seedling when its radicle breaks through the seed coat to end germination.

- Because most seeds germinate underground, there must be rapid elongation of the stem to reach light. This rapid growth is halted by signals from a photoreceptor. Subsequent plant growth is under hormonal control.

- Flowering is initiated by plant age, size, or season. Seasonal flowering is controlled by light duration and is sensed by photoreceptors. Flower formation is most likely induced by hormones, as is subsequent flower development.

- Perennials, which live year after year, have buds that enter dormancy during winter. Hormones such as abscisic acid maintain dormancy. In plants that lose their leaves, senescence is controlled by hormone interactions of ethylene and auxin. Plant death may follow senescent activity.

Defined events regulate the breaking of seed dormancy and germination.

- Dormancy, lasting for different amounts of time depending on the plant, involves exclusion of water or oxygen from the embryo by the seed coat, mechanical restraint of the embryo by the seed coat, and chemical inhibition of the embryo.

- Dormancy can be broken by mechanical abrasion or fire. Prolonged exposure to water may leach chemical inhibitors away from the seed and induce germination.

- Dormancy ensures survival through adverse conditions. Some seeds rely on other environmental cues before germination, such as cold temperatures, light, or the passage of a specific amount of time. These cues help ensure that the seed will germinate in the correct location and time.

- Germination begins with a seed's imbibing water. Water results in metabolic changes. DNA synthesis is halted until the radicle emerges from the seed coat. Breakdown of starch and protein reserves in the cotyledons and endosperm begins to provide nutrients for the developing embryo. In cereal seeds, the embryo secretes gibberellins, which promote enzymes that break down starch and proteins.

Gibberellins regulate growth from germination to fruiting.

- Gibberellins are produced in both plants and fungi. The first gibberellin was isolated from a fungus that infects rice plants and causes them to grow tall and spindly. Initial experiments indicated that the medium that the fungus was grown in was sufficient to cause rice plants to be spindly; therefore, the agent that caused the phenomenon was a chemical produced by the fungus.

- Experiments using mutant and normal corn plants showed that plants have innate gibberellins. Normal corn plants exposed to gibberellins showed no alteration in appearance, but dwarf plants exposed to them showed shoot elongation to near normal lengths. These experiments proved that the chemical was present in plants and that its addition did nothing to alter the plant when it existed in sufficient quantities. In plants lacking the chemical, however, application of gibberellin induced stem elongation.

- Gibberellins also have roles in fruit development. Developing seeds produce gibberellins that enhance development of fruit tissue. It is common agricultural practice to spray seedless fruits (especially grapes) with gibberellins to enhance fruit growth. Gibberellins also induce bolting and seed set in biennials under appropriate environmental cues.

- Additional roles of gibberellins include fruit development from unfertilized flowers, germination of some seeds, and breaking of winter dormancy.

Auxin affects plant growth and form.

- The discovery of auxin (indoleacetic acid) was the result of work of Charles and Francis Darwin in the 1880s. They were interested in phototropism and which part of an emerging plant coleoptile was responsible for sensing light. They found that the tip was the light-receptive portion, but that the growing region (responsible for the bending to or away from light) was some distance below the receptor region. From this they reasoned that some chemical must be transmitted from the tip to the growing region. Additional experiments with removed tips and tips on gelatin blocks indicated that a chemical was indeed moving from the tip to the growing region. Subsequent experiments by other researchers showed that gelatin exposed to the tips was sufficient to cause altered growth. The chemical was later isolated and determined to be auxin.

- Auxin movement in plant tissues is unidirectional and polar from apex to shoot. Auxin anion efflux carriers are carrier proteins responsible for export of auxin anions from cells and contribute to the unidirectional movement of auxin. Auxin enters the cell by passive diffusion and by active transport along with H^+. Because a gradient is established via transport, auxin acts as a morphogen and directs how the cells differentiate along the gradient.

- Redistribution of auxin laterally is responsible for phototropism and gravitropism. Figure 38.10 illustrates how the lateral distribution of auxin affects the curvature of coleoptiles in response to light and gravity. Higher auxin concentrations correspond to increased rates of growth.

- Auxin affects vegetative growth by initiating root growth in cuttings, causing abscission of leaves, maintaining apical dominance, and promoting stem elongation and inhibiting root elongation. Auxin also can stimulate parthenocarpy in fruits. Synthetic auxins are used as selective herbicides.

- The effects of auxin on growth are mediated at the cell walls. The cell wall determines the rate and direction of cell growth. Cells grow by taking up more water. The amount of water that can be taken up is restricted by a rigid cell wall. Cell walls must loosen, stretch, and add polysaccharides and cellulose to maintain structure. Cell walls exhibit elasticity when their shape is altered but later returns to normal. Plasticity means the shape is altered and does not return to normal. Plasticity is necessary for cell growth. It is thought that auxin affects plasticity and "loosens" cell walls so that a plant can grow. Auxin stimulates a cell-wall-loosening factor that results in a pH change. The pH change allows proteins called expansins to alter polysaccharide bonding so that they slide past one another during expansion.

- Because auxin has multiple functions, it stands to reason that there would be multiple auxin receptors and trans-

duction pathways. Research supports the notion of multiple receptors and pathways.

- Auxins and cytokinins regulate organ development based on the relative concentrations of the two. High auxin levels favor root formation, and high cytokinin levels favor bud formation.

Cytokinins have multiple roles within plants.

- Cytokinins are powerful cell division stimulators and initiate bud formation, aid in seed germination, inhibit stem elongation, and delay leaf senescence. They are synthesized primarily in the roots and are translocated throughout the plant.

- Each of these activities is regulated by the relative ratio of cytokinin to auxin.

Ethylene promotes senescence.

- Ethylene is a gaseous hormone that promotes senescence of leaves and ripening of fruit. Commercial use of ethylene spray to promote ripening is common for a variety of fruits. Ethylene scrubbers are used in fruit storage to prevent ethylene from accumulating and causing fruit to spoil. Other chemicals are used in the flower industry to inhibit ethylene's affects on flower senescence.

- Ethylene plays a role in maintaining the apical hook on emerging eudicots by inhibiting cells on the inner portion of the hook. Ethylene also inhibits stem elongation and promotes lateral swelling of stems. These three responses are known as the triple response.

- Figure 38.17 illustrates the signal transduction pathway initiated by ethylene. The process begins when ethylene binds to its receptor. This binding initiates a cascade that leads to the expression of genes that result in the physiological changes stimulated by ethylene.

Abscisic acid is responsible for maintaining dormancy.

- Abscisic acid inhibits seed germination and promotes storage during embryo formation. It accumulates during times of stress and is responsible for maintaining winter dormancy. It also inhibits stem elongation.

- Plants deficient in abscisic acid often show vivipary, in which seeds sprout while still attached to the plant.

- Abscisic acid regulates guard cell activity by opening calcium channels. Increased calcium levels eventually lead to the opening of potassium channels and the closing of the stoma as guard cells sag together.

Brassinosteroids are not completely understood, but have many effects.

- Brassinosteroids were originally isolated from a member of the Brassicaceae. They have been shown to stimulate cell elongation, pollen tube elongation, and vascular tissue differentiation and to inhibit root elongation, all of which are similar to the effects of auxin. It took the isolation of a mutant for scientists to determine definitively that brassinosteroids were themselves a group of hormones.

Light and photoreceptors interact to stimulate a variety of plant events.

- Photoreceptors interpret light, its duration, and its wavelength. Light regulates a wide variety of plant processes, including germination and flower production.

- The effects of red and dim-blue light are mediated by five phytochromes. Blue and red light promote germination. Far-red light reverses the effect of prior exposure to red light. Far-red light is 730 nm, whereas red light is 660 nm. Exposure to red and far-red light has different effects on plants. The "switching" effects occur because phytochrome can be shifted from one form (red P_r) to the other (far-red P_{fr}) upon absorption of light. When exposed to red light, P_r is converted to P_{fr}. When exposed to far-red light, P_{fr} is converted to P_r.

- Phytochromes regulate etiolation, which functions to conserve resources until plants can begin photosynthesizing. In seedlings that have not been exposed to light, all the phytochrome is the P_r type. Once exposed to light, it shifts to P_{fr} and begins a series of events that reverses etiolation.

- There are multiple phytochromes with multiple regulatory functions. They may be responsible for changes in the plant throughout the day in response to shading. It is thought that they interact with G proteins in a variety of cascades.

Cryptochromes and phototropin are blue-light receptors.

- Cryptochromes absorb blue and ultraviolet light. These receptors are found both in plants and animals. In plants, they carry out many of the same regulatory functions as phytochrome, but they are active in the nucleus rather than in the cytoplasm.

- Phototropin, under blue light, initiates the signal responsible for phototropism. Zeaxanthin plays a role in light-stimulated opening of the stomata.

The Big Picture

- Plant growth is controlled by interactions between a plant's environment, hormones, and its genetic makeup. Changes in plant growth are dependent on receptors to receive information, cascades of events involving enzymes and mediated by hormones, and alteration of gene expression. Receptors are often highly specific and respond to select signals. Hormones are produced in a specific region of the plant body and translocated throughout the plant; therefore, their effects are often concentration dependent. All growth in a plant is a result of changes in cell division, cell expansion, and cell differentiation.

- Seed germination, growth of the vegetative structures, reproduction, and senescence all proceed in defined patterns. Seed dormancy is broken and development begins when the seed coat is abraded, inhibitory chemicals are diluted, and the seed imbibes water. This begins a series of events that mobilize nutrients and induce growth. Once a seedling emerges from the soil, light begins to influence subsequent development under the control of hormones.

- Hormones exist in several classes, each with its own effects on growth and development. Some of these effects are antagonistic in nature, and therefore control is maintained via relative concentration of several hormones. Hormones influence every step of development, from breaking of seed dormancy to senescence.

- Light regulates plant processes through photoreceptors. Photoreceptors respond to very specific wavelengths of light and induce cascades that lead to changes in a plant.

Common Problem Areas

- Understanding how phytochrome shifts from the red to far-red forms frequently gives students difficulty. This is understandable, because phytochromes have given researchers a difficult time for many years.

- The tendency when studying hormones is simply to memorize functions. You need to understand what relative ratios of hormones will do in a plant, not what one individual hormone does.

- It is easy to get overwhelmed with information in this chapter. Chances are the material is completely new to you. Take your time in learning how growth is regulated and the effects of the various hormones.

Study Strategies

- Avoid focusing too much on details. Try to assimilate the big picture of how the environment, receptors, hormones, and the genome interact. From there, begin to work toward the details. A big mistake is to attempt to jump in and memorize functions of hormones or sequences of development without understanding the broad picture.

- Review the following tutorials and activities on the Student Website/CD:
  Tutorial 38.1 Tropisms
  Tutorial 38.2 Went's Experiment
  Tutorial 38.3 Auxin Affects Cell Walls
  Activity 38.1 Monocot Shoot Development
  Activity 38.2 Eudicot Shoot Development
  Activity 38.3 Events of Seed Germination

Test Yourself

Knowledge and Synthesis Questions

1. Plant growth is ultimately regulated by which of the following?
   a. Environmental cues
   b. Hormones
   c. Signal transduction pathways
   d. A plant's genome

2. Which of the following schemes best represents how environmental cues are transduced to changes in a plant?
   a. Receptors receive environmental cues, a signal transduction pathway is initiated, there is an alteration in the particular genes that are transcribed and translated, and a cellular response is generated.
   b. Receptors are triggered, hormones are released, signal transduction pathways are initiated, there is an alteration in expression of genes, and a cellular response is generated.
   c. Neither a nor b
   d. Both a and b

3. Which of the following triggers germination of a seed?
   a. It imbibes water.
   b. It is released from fruit.
   c. It undergos chemical changes.
   d. All of the above

4. Which of the following may function to break dormancy in seeds?
   a. Penetration of the seed coat
   b. Leaching of inhibitory compounds by water
   c. Exposure to fire
   d. All of the above

5. Which of the following hormones is responsible for bud break in the spring in deciduous trees?
   a. Auxins
   b. Cytokinins
   c. Gibberellins
   d. Ethylene

6. Auxin regulates cell growth by which of the following mechanisms?
   a. Altering the elasticity of cell walls
   b. Altering the plasticity of cell walls
   c. Synthesizing new cell walls
   d. Breaking down cell walls in growing cells

7. Which of the following hormones are responsible for maintaining bud dormancy in deciduous trees?
   a. Auxins
   b. Cytokinins
   c. Gibberellins
   d. Abscisic acid

8. You have installed an outdoor gas-burning grill on your back patio next to your favorite camellia bush. After the first few chilly nights of using your grill you notice that your camellia, which does not normally lose its leaves, is beginning to do so. Which of the following is the best explanation for what is happening?
   a. The bush is getting too warm next to your grill.
   b. Ethylene is a by-product of the gas you are burning and is causing senescence in your plant.
   c. Abscisic acid is a by-product of the gas you are burning and is causing senescence in your plant.
   d. The plant is a biennial and is bolting.

9. You are slicing a green pepper for a pizza you are making at home. As you slice into it, you notice lots of tiny pepper plants emerging from the seeds of the pepper. This pepper is exhibiting _____ and may be lacking in _____.
   a. parthenocarpy; gibberellins
   b. parthenocarpy; abscisic acid
   c. vivipary; gibberellins
   d. vivipary; abscisic acid

10. Which of the following light receptors is responsible for absorbing blue and ultraviolet light?
    a. Phytochrome P_r
    b. Phytochrome P_{fr}
    c. Cryptochrome
    d. Phototropin

11. Etiolated seedlings are produced by germinating seeds and keeping them in total darkness. Under which of the following conditions will plants kept in the dark begin to synthesize chlorophyll?
    a. After being given a pulse of blue light
    b. After being given a pulse of red light
    c. After being given a pulse of red light followed by a pulse of far-red light
    d. After being given a pulse of far-red light followed by a pulse of red light

12. Match the major site of production on the right with the plant growth substance produced at that site on the left. A substance may be produced at multiple sites.
    _____ Abscisic acid   a. Embryo
    _____ Auxin   b. Buds
    _____ Cytokinin   c. Leaves
    _____ Ethylene   d. Fruit
    _____ Gibberellin   e. Root tips
               f. Shoot tips

Application Questions

1. You have planted pea seeds in your garden. Trace the steps that occur between your planting of a seed and the emergence of the pea plant from your garden.

2. In the pea plant in Question 1, what hormonal influences are affecting it from the moment you put it in the ground until its emergence?

3. Discuss how gibberellins were discovered. How did researchers determine they were chemical in nature?

4. Explain how auxin distribution regulates phototropism and gravitropism.

5. Auxins and cytokinins appear to cancel out the effects of each other. Why would a hormone that affects bud growth be produced in the roots and one that affects root growth be produced in the shoots? How does this relate to polar distribution of hormones?

6. Genetic engineering has produced fruits that are deficient in the ability to produce ethylene. How is this deficiency useful in the storage and marketing of fruits?

7. You are experimenting in the laboratory with tissue culture methods. You have isolated pith tissue and grown an undifferentiated mass of cells (called a callus). You divide up the tissue and place it in the following culture media.
   a. Necessary nutrients plus indoleacetic acid
   b. Necessary nutrients plus zeatin (a cytokinin)
   c. Necessary nutrients plus equivalent concentration of indoleacetic acid and zeatin
   d. Necessary nutrients plus an excess of zeatin and minimal indoleacetic acid

   Explain what happens to the mass of tissue under each condition.

8. Commercial bananas are triploid; therefore, their seeds abort without development. Under normal conditions the fruits would not develop. Explain how commercial banana growers use hormones to produce sweet fruit with no seeds. Which hormones could they be using?

Answers

Knowledge and Synthesis Answers

1. **d.** Though the environment influences development, final regulation is at the genomic level.

2. **d.** Receptors receive environmental cues and trigger hormone release or a signal transduction pathway that involves hormones. The signal alters gene expression by altering which genes are transcribed and translated. Cellular response depends on expression of those genes.

3. **a.** The uptake of water by a seed begins the processes that lead to seed germination.

4. **d.** Mechanical abrasion, leaching of inhibitors by water, and exposure to fire may all trigger germination. Actual germination cannot begin until a seed imbibes water.

5. **c.** Gibberellins are responsible for bud break in deciduous trees.

6. **b.** Altering plasticity allows for permanent changes in cell wall shape. The cell wall must increase in size in order for cell growth to occur.

7. **d.** Abscisic acid is responsible for maintaining bud dormancy in winter in deciduous plants.

8. **b.** Ethylene gas promotes senescence and is one of the by-products of burning your gas grill. You should move your grill or your camellia bush.

9. **d.** Vivipary is the germination of seeds before they leave the parent plant and is caused by a deficit of abscisic acid.

10. **c.** Cryptochromes respond to blue and ultraviolet light wavelengths.

11. **c.** The pulse of red light converts P_r to P_{fr}. Subsequent pulses of far-red light stimulate changes that lead to chlorophyll synthesis.

12. **b.** Abscisic acid

    c., f. Auxin

    e. Cytokinin

    d. Ethylene

    a. Gibberellin

Application Answers

1. Upon planting a pea seed you watered it, promoting the leaching of germination inhibitors. At this point the seed also began to imbibe water, resulting in metabolic changes. DNA synthesis is halted until the radicle emerges from the seed coat. Breakdown of starch and protein reserves in the cotyledons and the endosperm begins to provide nutrients for the developing embryo. The apical hook begins to push through the soil. Upon its emergence and exposure to light, chlorophyll synthesis begins.

2. Dormancy of the seed is maintained by abscisic acid until it is leached from the seed by water. Once water is imbibed, cytokinins begin to influence germination. Gibberellins assist with the mobilization of storage products to the growing embryo. Ratios of auxins and cytokinins balance root and bud formation. The apical hook is maintained by ethylene.

3. The first gibberellin was isolated from fungus that infects rice plants and causes them to grow tall and spindly. Initial experiments indicated that the medium that the fungus was grown in was sufficient to cause rice plants to be spindly; therefore, the agent that caused the phenomenon was a chemical produced by the fungus.

4. Lateral distribution of auxin controls phototropism and gravitropism. Auxin accumulates in the shaded portions of a stem and stimulates cell growth. This uneven cell growth results in a bending of the stem toward the light. The same mechanism works in response to gravity. An accumulation of auxin occurs where the gravitational pull is the strongest. Cells grow in response to auxin, and stems bend upward, away from the gravitational force.

5. Roots need shoots and vice versa. Increases in roots require increases in shoots for photosynthesis. Regulation of the development of one by the other keeps growth in tandem. Because the distribution of the molecules is polar, a concentration gradient can be established. It is this relative gradient that controls development.

6. Fruits can be kept from ripening until they reach their destination. Once in markets they can be sprayed with ethylene to stimulate ripening. The result is fruit that can be shipped more easily yet can be ripe at any time in the market.

7. a. Roots will develop from the callus.
   b. Buds will develop from the callus.
   c. Both roots and buds will develop from the callus.
   d. Buds will develop from the callus.

8. Commercial banana and grape growers frequently use gibberellins to assist with fruit formation. Gibberellins, and to some extent auxins, can be used to stimulate parthenocarpy or fruit development without seeds.

39

Reproduction in Flowering Plants

Important Concepts

Plants may reproduce asexually or sexually.

- Asexual reproduction results in plants with the same genetic makeup as the parent plant. In agriculture, this may be desirable.

- Sexual reproduction is necessary for genetic recombination. This promotes genetic variability and is used in agriculture to reap the benefits of "hybrid vigor."

Sexual reproduction provides for genetic variability.

- Sexual reproduction is advantageous because it provides for greater adaptability within a population; however, it is energy consuming and can break up useful alleles.

- The flower is the basis of sexual reproduction in angiosperms. The basic flower structures include carpels, pistil, ovule, ovary, stigma, style, stamen, anther, filament, petals, sepals, tepals, calyx, corolla, and receptacle. Review the description of the flower in Chapter 29.

- Plant reproduction involves the concept of alternation of diploid and haploid generations. The flower is produced in the diploid sporophyte generation. The flower produces spores that develop into highly reduced gametophytes and produce gametes. The female gametophytes, called embryo sacs, develop in the megasporangia. Male gametophytes, called pollen grains, develop in microsporangia.

- Within the ovule, a megasporocyte produces four haploid megaspores through meiosis. Only one of these megaspores survives. It divides mitotically to produce eight nuclei within a single large cell. The nuclei migrate to either end of the cell, but two remain in the middle of it. Cell walls form, isolating the three nuclei at either end into individual cells, whereas the two nuclei in the middle remain together in one cell. At one end, the three cells become two synergid cells and one egg cell; at the other are three antipodal cells, which degenerate. In the large central cell are the two polar nuclei. This seven-celled embryo sac is the gametophyte (see Figure 39.1).

- Pollen grains develop when a microsporocyte undergoes meiosis. All products of meiosis are retained and undergo mitosis to form a two-celled pollen grain composed of the tube cell and the generative cell. Further development is halted until after pollination.

- Pollination allows for fertilization without water. Pollen grains are carried to female flowers via wind, animals, or other vectors. Some plants are able to self-fertilize within the same flower. Pollen grains must attach to the stigma for fertilization to occur.

- The stigma allows plants to practice "mate selection." Pollen of the right species sticks readily to the stigma via cell-to-cell signaling; however, pollen of other species does not adhere, and germination cannot occur. A single gene, the *S* gene, regulates self-incompatibility and prevents self-fertilization in many plants. Other plants self-fertilize without problem.

- When a pollen grain germinates, a pollen tube grows down the style toward the embryo sac. Chemical signals from synergids within the ovule direct the growth of the pollen tube.

- The pollen grain consists of two cells at the time of pollination; the tube cell and the generative cell. The tube cell controls the growth of the pollen tube. As the tube is growing, the generative cell undergoes meiosis and produces two haploid sperm cells. Once the pollen tube enters the embryo sac, the two sperm cells are released into a synergid that disintegrates and releases the sperm nuclei. One nucleus fuses with the egg cell, the other, with the polar nuclei. This results in the zygote ($2n$) and the endosperm ($3n$). All other cells disintegrate.

- Double fertilization resulting in a zygote and endosperm is a characteristic feature of angiosperms (see Figure 39.5).

Embryos develop within seeds.

- The success of the embryo depends on its own development, the development of the endosperm, the integuments, and the carpel. Ultimately a seed is produced that protects the dormant embryo.

- The zygote divides mitotically to produce the embryo itself and the suspensor. The suspensor pushes the embryo to the endosperm. The suspensor continues to divide and produces a filament, whereas the embryo produces a globular structure. This establishes polarity and symmetry in the embryo.

- In eudicots, the embryo progresses through the heart stage and torpedo stage as cotyledons and other organs are developed. The hypocotyl, with its apical meristems is also established. See Figure 39.6.

- The endosperm also develops and holds nutrient reserves. The cotyledons may absorb the nutrients from the endosperm and increase in size.

- Once development is nearly complete, the seed loses water, and the embryo ceases development. It remains in this state until germination.

- As the embryo and seed are developing, the ovary begins to form the fruit. Other parts of the flower and plant may be included in the fruit, but to be considered a fruit, only the ovary wall and the seed need be involved. The fruit disperses by various means including attaching to the coats of animals or being consumed and later deposited.

Specific environmental factors signal the onset of reproduction through flower development.

- Flowering may halt, interrupt, or have little effect on plant growth. Seed production is ultimately the purpose of the plant.

- The first transition from vegetative growth to floral production is the alteration of the apical meristem to a floral meristem. The floral meristem differs from the apical meristem in that growth is determined. An inflorescence is produced when flowers are orderly clustered. The floral meristem is programmed to produce four consecutive whorls of flower organs and to be active for a defined amount of time.

- A gene cascade leads to flower formation. Floral formation begins with the activation of a set of meristem identity genes. The products of the transcription and translation of these genes influence pattern formation genes known as cadastral genes, which direct the spatial organization of the floral organs. The expressions of floral organization genes specify successive whorls.

- Plants with different life cycle patterns fall into three categories. Annual plants go from seed to seed set and die within one growing season. Biennial plants require one vegetative growing season before reproducing. Perennial plants repeatedly flower and live for many years.

- The gene cascade leading to flower development is controlled by environmental cues. Seasonal flowering is a result of changes in photoperiod, specifically the duration of continuous darkness. Experiments have shown that plants actually respond to the length of the night, rather than amount of daylight. In experiments in which daylight was interrupted, there was little effect on flowering, but in experiments in which dark was interrupted, there were significant effects on flowering. The mechanism controlling this phenomenon is far from understood, however. Each plant type has a critical day length corresponding to light availability that induces flowering.

- Short-day plants flower when the amount of light available is shorter than the critical maximum. Long-day plants flower only when the day is longer (more light available) than the critical minimum. Some plants require complex combinations of day lengths. Some are day-neutral and depend on other factors, such as temperature, to flower.

- Plants, like all other organisms, have an internal mechanism for measuring the length of continuous dark periods. The duration of dark periods appears to be detected by special phytochromes and blue-light receptors.

- Manifestations of an internal biological clock are referred to as circadian rhythms. Circadian rhythms have a duration referred to as a period and an amplitude, or magnitude of change over time. Circadian rhythms have specific characteristics in all organisms. They are not influenced by temperature, they are highly persistent, they can be entrained to coincide with different light–dark cycles, and light exposure causes a phase shift in the rhythm. Plants provide numerous examples of 24-hour circadian rhythms.

- Plants use phytochromes and blue-light receptors to control the biological clock. The biological clock in *Arabidopsis* controls the expression of the *CONSTANS* gene and the build up of the gene product: CO protein. When CO protein levels are high, light absorbed by phytochrome A and the blue-light receptor cryptochrome B, leads to flowering (see Figure 39.14).

- Florigen is thought to be the hormone responsible for flowering even though it has yet to be isolated. A series of experiments provided evidence that once a leaf is light induced, it sends a hormonal signal to the rest of the plant to initiate flowering.

- Vernalization is the induction of flowering by low temperatures. The discovery that plants could be induced to flower by exposure to cold has been enormously useful to agriculture.

In certain conditions, asexual reproduction is advantageous to plants' success.

- Asexual reproduction occurs without genetic recombination. During the process of vegetative reproduction, asexual reproduction occurs through the modification of a vegetative organ. Stolons, tip layers, tubers, rhizomes, bulbs, corms, and suckers are all modifications of stems or roots that allow vegetative reproduction.

- Some plants such as dandelions reproduce asexually through seeds in a process called apomixis. Apomictic plants skip over meiosis and fertilization and produce seeds with the identical genetic makeup of the maternal plant. Sometimes apomixis does require pollination and the union of sperm nuclei with polar nuclei to form endosperm, yet no genetic recombination occurs.

- Asexual reproduction is used in agriculture. Cuttings are commonly used in horticulture. Grafting is widely used in fruit crops where rootstock is grafted to a scion (shoot system). This allows for a hardy rootstock to be combined with a good but often-fragile fruit producer. Tissue culture (production of entire new plants from a small tissue sample) and recombinant DNA technology are leading to daily advances in agriculture.

The Big Picture

- Though plants can reproduce both asexually and sexually, maintenance of genetic variability depends on sexual reproduction. The flower is the basis of sexual reproduction in plants. The flower not only produces the necessary gametes, but it is also integrally involved in ensuring pollination. Eggs are produced through megasporogenesis, and pollen is produced through microsporogenesis. Angiosperms exhibit double fertilization resulting in a diploid embryo and a triploid endosperm.

- Embryo development takes place in the seeds of angiosperms and progresses until cotyledon(s) and the hypocotyl with its apical meristems are developed. At that point, the embryo halts development, the seed desiccates, and dormancy begins until conditions are right for germination. The seed protects the embryo, and in many cases a fruit is formed for protection until maturation and to aid in dispersal of the mature seed.

- Flowering and seed set is triggered by the length of continuous dark exposure. Some plants flower in response to short days, others to long days, and yet others to complex combinations of the two. Day-to-day functions of plants are triggered by photoperiod and circadian rhythms. Plants may flower, set seed, and die in one growing season or two growing seasons, or they may continue to do so for even longer periods of time.

Common Problem Areas

- Megasporogenesis is probably the most difficult concept in this chapter, but it is relatively simple if you look at the source of each cell.
- Recall that flowering, like all other plant processes, is a result of environmental signals, receptors, enzyme cascades mediated by hormones, and alterations in gene expression.

Study Strategies

- Refer to the figures in the textbook to understand flower structure, megasporogenesis, microsporogenesis, and fertilization.
- Spend some time thinking about the advantages and disadvantages of sexual and asexual reproduction. Related questions often show up on exams.
- Be sure you understand the particular features of double fertilization.
- Refer to the following tutorials and activity on the Student Website/CD:
Tutorial 39.1 Double Fertilization
Tutorial 39.2 The Effect of Interrupted Days and Nights
Activity 39.1 Early Development of a Eudicot

Test Yourself

Knowledge and Synthesis Questions

1. You manage a greenhouse that produces roses for Valentine's Day. Roses normally bloom in June. Which of the following will most likely be the best lighting schedule for your roses?
   a. 16 hours of light, 8 hours of interrupted dark
   b. 16 hours of light, 8 hours of uninterrupted dark
   c. 10 hours of light, 14 hours of dark
   d. None of the above
2. After setting the correct photoperiod, you still don't have roses. Which of the following has most likely contributed to the problem?
   a. The heating system is allowing fluctuations in temperature of between 20°C and 25°C.
   b. The furnace mechanic accidentally turned off the lights for an hour two days in a row.
   c. The cleaning crew turned the lights on for an hour three nights in a row.
   d. None of the above

3. You have moved into a new house. During the first summer you notice lots of plants that do not bloom. During the second summer your yard is a sea of blooms. It is now spring of the third year, and there are no plants. This can best be explained by which of the following?
   a. Your plants are annuals.
   b. Your plants are biennials.
   c. Your plants are perennials.
   d. Your plants are being affected by drought.
4. You mother gives you a houseplant. You notice it sending out long stems with what look like "little plants" attached. You allow one of these to rest in a cup of water and note that roots form. This is an example of
   a. asexual reproduction.
   b. apomixis.
   c. heterospory.
   d. parthenogenesis.
5. If plants produce pollen and mature eggs at the same time, how are self-pollination and fertilization prevented?
   a. Some plants have self-incompatibility genes.
   b. Some plants rely on physical barriers.
   c. Both a and b
   d. None of the above
6. Most wine grape vines are grafted onto rootstock of another species. How does this help grape yields?
   a. A hardy rootstock can replace a weak rootstock.
   b. A high-producing vine stock can replace a low-producing vine stock.
   c. It allows vintners to select for pest resistance without losing grape quality.
   d. All of the above
7. The induction of flowering by means of exposure to low temperature is called
   a. vernalization.
   b. frigidation.
   c. apomixis.
   d. None of the above
8. The cyclic variation in an organism that corresponds to a set light–dark cycle is referred to as an organism's
   a. biorhythm.
   b. circadian rhythm.
   c. period.
   d. photoperiod.
9. The production of seeds without fertilization is called
   a. apomixis.
   b. parthenogenesis.
   c. conception.
   d. circadian rhythm.
10. Identify the following structures as sporophyte or gametophyte.
    a. Embryo sac _____
    b. Antipodal cells _____
    c. Polar nuclei _____
    d. Integument _____
    e. Receptacle _____
    f. Ovary _____
    g. Anther _____
    h. Pollen grain _____

11. In the transition from vegetative growth to floral growth the _____ must be transformed to the _____. This involves a shift from _____ growth to _____ growth.
    a. apical meristem; floral meristem; indeterminate; determinate
    b. lateral meristem; floral meristem; indeterminate; determinate
    c. apical meristem; floral meristem; determinate; indeterminate
    d. apical cambium; floral cambium; determinate; indeterminate

12. Which of the following best describes the fate of the generative cell of the pollen grain?
    a. It coordinates growth of the pollen tube.
    b. It divides by meiosis to produce two sperm nuclei.
    c. It divides by mitosis to produce two sperm nuclei.
    d. None of the above

Application Questions

1. Compare and contrast asexual and sexual reproduction in plants. In which category does self-fertilization belong?

2. Describe egg formation in angiosperms. Begin with the sporophyte.

3. Much effort is spent detasseling corn (removing male flowers). Explain why corn plants cannot be sprayed with a meiosis inhibitor to halt pollen production. (Hint: Male and female flowers occur on the same corn plant.)

4. Flowering is stimulated when light sets off a gene cascade. Explain how this may be hormonally controlled.

5. Describe double fertilization. What is the ploidy level of the products of double fertilization?

6. When is asexual reproduction beneficial to plants? Under what conditions is sexual reproduction beneficial?

7. Draw and label an angiosperm flower. Identify the minimum parts necessary for reproduction.

8. Define a fruit. Which of the following are fruits: tomato, pear, potato, banana, cucumber, snow pea, peanut, sunflower seed?

Answers

Knowledge and Synthesis Answers

1. **b.** Flowering is regulated by darkness. Interruptions in darkness can prevent flowering.

2. **c.** The interruptions in the dark cycle by the cleaning crew have affected the signals that tell the plant to begin flowering.

3. **b.** The plants are most likely biennials that spend one season growing vegetatively and one season producing flowers before dying.

4. **a.** The runner (stolon) is asexually propagating this plant.

5. **c.** Some plants have self-incompatibility genes that prevent the growth of the pollen tube through the style. Others have mechanical barriers to their own pollen.

6. **d.** Grafting allows selection of hardy rootstock, well-producing vine stock, and disease resistance.

7. **a.** Vernalization is noted in winter wheat and is used agriculturally to grow high-yielding wheat.

8. **b.** Circadian rhythms roughly match a 24-hour cycle and can be reset by altering light and dark cycles.

9. **a.** Dandelions and other plants produce seeds without meiosis and fertilization (called apomixis). These seeds are genetically identical to the parent plant.

10. **a.** Embryo sac   gametophyte
    b. Antipodal cells   gametophyte
    c. Polar nuclei   gametophyte
    d. Integument   sporophyte
    e. Receptacle   sporophyte
    f. Ovary   sporophyte
    g. Anther   sporophyte
    h. Pollen grain   gametophyte

11. **a.** Apical meristems and floral meristems differ because growth from the apical meristem is indeterminate and growth from the floral meristem is determinate, leading to four whorls of floral structures.

12. **b.** The two sperm nuclei used in double fertilization are derived from the generative cell of the pollen grain.

Application Answers

1. Asexual reproduction does not involve meiosis, fertilization, or genetic recombination. The offspring from asexual reproduction are genetically identical to the parent plant. Sexual reproduction requires a meiotic event and fertilization. Self-fertilization is sexual reproduction, even though genetic recombination is limited. This is because a meiotic event occurs and fertilization is necessary for reproduction to take place.

2. See Figure 39.1 in the textbook.

3. Meiosis occurs in the megasporocyte as well as the microsporocyte. Inhibition of pollen formation via a meiosis inhibitor will also inhibit egg formation.

4. Light induction begins with the leaves. A single isolated leaf may stimulate floral production throughout the plant. In experiments in which leaves from one plant were grafted on other plants, an "induced" leaf caused a plant to flower even though the plant had not been exposed to the required amount of darkness. Therefore, the leaf must send a signal (most likely a hormone) that begins the gene cascade leading to flower formation.

5. Double fertilization occurs when the two sperm nuclei produced from the generative cell of the pollen grain unite with the egg nucleus and two polar nuclei respectively. The resulting zygote is diploid and the endosperm is triploid.

6. Asexual reproduction is beneficial in stable environments in which many genetically identical plants are sustainable. This process can help to colonize a habitat or a population of plants to spread. The disadvantage is lack of genetic diversity. Lack of diversity can be detrimental if the environment changes rapidly, requiring adaptability of the population.

7. See Figure 29.6. Only the ovule or the anther is required for reproduction, provided that the corresponding structure (anther or ovule) is found on another plant of the same species.

8. A fruit is the seed, the ovary wall, and may include other structures of a flowering plant. With the exception of the potato, all of the listed structures are fruits.

40 Plant Responses to Environmental Challenges

Important Concepts

Plants and their pathogens interact, each giving clues to the other.

- The action of a pathogen signals a plant to mount a defense. The plant's defense signals the pathogen. These signals go back and forth until one or the other "wins."

- The first system of plant defense is nonspecific; an attempt to prevent entry by pathogens. Cutin, suberin, and waxes cover the cork and epidermis to exclude pathogens. Plants cannot repair damaged and invaded tissues, so they seal them off to prevent systemic infection. The damaged area is replaced.

- Upon recognition of a pathogen, polysaccharides are deposited in the cell wall and seal off the plasmodesmata to form a barrier between cells. Subsequently, lignin is deposited, which reinforces the barrier and acts as a pathogen toxin.

- Once invaded, plants produce chemical defenses against pathogens. Phytoalexins are nonspecific antifungal and antimicrobial compounds. Pathogenesis-related proteins (PR proteins) may function to break down the walls of pathogens or serve as signaling molecules.

- Salicylic acid, a relative of aspirin, is produced in plants due to a hypersensitive response. Pathogen-containing tissue and surrounding tissues die upon infection. As these cells are dying, they release phytoalexins and salicylic acid. These compounds "arm" the plant against subsequent invaders.

- Systemic acquired resistance protects a plant against further infection. This may be due to the stimulation of PR proteins after an infection event. Plants also release methyl salicylate, which can become airborne and serve as a signal to other plant parts or to neighboring plants to mount a defense via PR proteins.

- Gene-for-gene resistance is a highly specific type of resistance and depends on a plant's having an allele for a gene that matches an allele in the pathogen. Dominant *R* genes code for resistance in plants, and dominant *Avr* genes result in less-virulent strains of the pathogen by coding for a substance that elicits a defensive response in the plant. Though this mechanism is not completely understood, the gene product interactions lead to production of PR proteins.

- Plants can use interference RNA (RNAi) to respond to attack by RNA viruses. A plant produces small pieces of short interfering RNA (siRNA) from interactions with the viral RNA. These siRNA degrade the viral mRNA.

Herbivory confers advantages and disadvantages to plants.

- Some plants depend on the grazing of herbivores to increase productivity due to coevolution of the plant and the herbivore. Removal of some leaves triggers increased photosynthesis in remaining leaves and a better partitioning of resources. Grazing also allows light to reach young leaves in plants such as grasses. Plants that are grazed have basal rather than apical growth and continue growing after being grazed. Some produce leaves longer and flower more abundantly.

- For the majority of plants, grazing is a problem; therefore, defenses against grazing exist. Plants produce specific secondary metabolites in defense. The effects of these chemicals are diverse, ranging from neurotoxins to hormone mimics. See Table 40.1 for classes of defensive secondary metabolites.

- Canavanine is a defensive secondary metabolite that plays multiple roles in a plant. It assists with nitrogen storage in the seed, and because it is similar to arginine, it is incorporated into herbivore proteins after consumption. The failure to distinguish between arginine and canavanine during protein synthesis alters the tertiary structure of proteins and can be lethal to the organism consuming the plant. Some organisms have the ability to distinguish between the two and are not harmed by canavanine.

- Defense is the result of a series of chemical signaling events. Figure 40.5 details the events leading from wounding to production of an insecticide. Wounding causes the release of systemin, which travels to other parts of a plant and induces membrane breakdown. By-products of this breakdown are jasmonates, which enter the nucleus and activate production of a protease inhibitor that inhibits digestion in the predator.

- Some plants produce products that make them resistant to attack. These chemicals can be isolated and the genes producing them identified. Resistance can be bred into plants or transferred through biotechnology.

Plants use a variety of strategies to prevent harming themselves with their own defensive mechanisms.

- Plants may isolate toxic secondary metabolites into compartments. Water-soluble toxins are stored in vacuoles, hydrophobic poisons are stored in lactifers along with latex, while others are dissolved in the waxes covering the epidermis.

- Some toxic substances are produced only in already-damaged or dying tissue. The precursors of toxic material are stored in different compartments than the enzymes that convert them to active poison, and these combine only once the cell is damaged.

- In some plants, the plant's enzyme mechanisms do not recognize the toxin.

- Insects sometimes find ways to get around the mechanisms used by plants to deter them. One example is milkweed and a beetle that feeds on its leaves.

Plants must be able to cope with water availability.

- Some plants, called xerophytes, are specifically adapted to dry areas. They survive in these areas by completing their entire life cycle in the short period that water is available. Others may have special modifications such as thickened cuticles, epidermal hairs, and sunken stomata to prevent water loss. Succulents have water-storing leaves, others drop leaves during drought, and yet others have water-storing stems and highly reduced leaves.

- Roots can also be modified for drought conditions. Long taproots can reach water supplies far underground. Shallow, fibrous root systems that grow only during rainy seasons are also adapted to desert growth. Xerophytic plants are able to extract more water when it is available by changing the osmotic and water potential of their root cells. They do this by storing proline in their vacuoles.

- Too much water can be as dangerous as too little to plant survival. Roots submerged in water do not get adequate oxygen to sustain respiration. Some plants that grow in standing water overcome this by producing pneumatophores, which are root extensions that grow above water and provide oxygen to the entire root system. Others rely on fermentation and thus grow very slowly. Yet others have leaf modifications called aerenchyma for buoyancy and oxygen storage.

Some plants are adapted to survive in saline environments.

- Saline environments have negative water potential and pose an osmotic stress to plants. For a plant to survive, it must have greater negative water potential to be able to extract water from the environment. Halophytes (salt-loving plants) are adapted to do this.

- Halophytes are the only plants that accumulate sodium and chloride ions, which they store in vacuoles in leaves. The increased salt concentration of the halophyte means it has a more negative water potential.

- Some plants are able to excrete salt so that it does not reach toxic proportions. Salt glands move salt to the leaf surface where it can be lost to wind or rain. Salt glands on the leaf assist with water procurement from the roots and with reduction of water loss to evaporation.

- Many modifications, such as succulence, allow for survival in both drought and saline conditions.

Heavy metal accumulation in soils is detrimental to most plants.

- Aluminum, mercury, lead, and cadmium are poisonous to most plants. Some geographic areas are rich in these metals as a result of normal geological processes, and others have been contaminated by human activity.

- Plants living in these areas have adapted to these conditions through selection. These plants are important in bioremediation and for cleanup of these areas. The exact mechanism of survival of plants in these areas is not known.

Temperature extremes pose yet another stress to plant survival.

- High temperatures denature proteins and destabilize membranes. Cold temperatures decrease membrane fluidity, and freezing causes rupturing of membranes if ice crystals form.

- Transpiration provides evaporative cooling to plants in high temperatures, but at the expense of water loss. Adaptations to high temperatures are similar to adaptations to xerophytic conditions. Plants produce heat shock proteins that help stabilize protein structure against denaturation.

- Plants can adjust to cold through a process of cold-hardening, which involves repeated exposure to cool non-damaging temperatures. This alters the composition of membranes and makes them more resistant to rupture. Heat shock proteins are also produced in response to cold temperatures. Some plants have antifreeze compounds that prevent ice crystal formation.

The Big Picture

- Plants must respond to invasion by pathogens, physical damage by natural events and herbivory, variable water supplies, temperature fluctuations, and variable soil conditions. Natural selection has made plants uniquely suited to areas they occupy. Some thrive in very harsh environments, whereas others cannot.

- Interactions between plants and pathogens stimulate a series of chemical changes that ward off further infection. Plant strategies to isolate pathogens include sealing plasmodesmata, releasing proteins that interact with pathogens, and signaling other parts of the plant.

- Herbivory can be advantageous for plants adapted to grazing or detrimental to those that are not. In the latter case, a plant produces chemicals to prevent herbivory and mounts defenses around the damaged tissues.

- Specific adaptations allow plants to withstand adverse conditions. Saline environments, water loss, and high temperatures can be survived if water uptake is maximized and water loss is minimized. Sunken stomata, reduction of surface area, and increased water potential of cells are adaptations to these conditions. Extremes in temperature are survived with heat shock proteins that stabilize metabolic proteins during temperature extremes.

Common Problem Areas

- Gene-for-gene resistance can be difficult to understand, but remember that in any pathogen situation, the pathogen and the host are continually interacting; therefore, over time, interplay among the genomes has been selected for.

- There are many examples of the ways plants deal with pathogens and herbivory. This chapter provides examples, but these are not exhaustive.

Study Strategies

- The best strategy is to break this chapter up into components and study them one at a time so that all of the defenses/modifications do not run together. Focus first on pathogen interaction, then herbivory resistance, and so on.
- Think about how form follows function as you look at how plants are adapted to specific harsh environments.
- Review the following tutorial and activity on the Student Website/CD:
  Tutorial 40.1 Signaling Between Plants and Pathogens
  Activity 40.1 Concept Matching

Test Yourself

Knowledge and Synthesis Questions

1. Primary nonspecific defensive strategies in plants include which of the following?
   a. PR proteins
   b. Suberin
   c. Cutin
   d. Both b and c

2. Plants exhibit systemic acquired resistance much the same way people acquire resistance to pathogens; however, the mechanism is quite different. Which of the following has (have) a role in acquired resistance in plants?
   a. Salicylic acid
   b. PR proteins
   c. Methyl salicylate
   d. All of the above

3. Gene-for-gene resistances depend on which of the following?
   a. Compatible alleles in plant and pathogen
   b. Incompatible alleles in plant and pathogen
   c. Recessive *Avr* genes
   d. Recessive *R* genes

4. Upon infection by a pathogen, plant cells increase synthesis of polysaccharides. The function of these is to
   a. reinforce cell walls and seal off plasmodesmata.
   b. synthesize antibodies to the pathogen.
   c. isolate the pathogen in the invaded tissue.
   d. Both a and c

5. Canavanine is toxic to many herbivores but not to plants. Which of the following can best explain this differential toxicity?
   a. Canavanine is confused with arginine in predators.
   b. Canavanine is incorporated into proteins and causes them to fold improperly.
   c. Plants are able to differentiate between arginine and canavanine.
   d. All of the above

6. Which of the following best describes how plants produce their own insecticide?
   a. Wounded cells release systemin, systemin directly induces synthesis of protease inhibitors, protease inhibitors act as insecticides.
   b. Wounded cells release jasmonates, jasmonates stimulate systemin synthesis, systemin causes the production of protease inhibitors, protease inhibitors act as insecticides.
   c. Wounded cells release systemin, systemin causes membrane breakdown, membrane breakdown releases jasmonates, jasmonates induce synthesis of protease inhibitors, protease inhibitors act as insecticides.
   d. Wounded cells release systemin, systemin causes membrane breakdown, membrane breakdown releases jasmonates, jasmonates act as insecticides.

7. Which of the following is (are) strategies for coping with drought conditions?
   a. Water-storing tissues
   b. Leaf loss during drought
   c. Sunken stomata
   d. All of the above

8. Halophytes are different from all other types of plants in that they
   a. can accumulate sodium and chloride ions.
   b. have a positive water potential.
   c. are unaffected by saline conditions.
   d. All of the above

9. Which of the following conditions stimulates the production of heat shock proteins?
   a. Abnormally high temperatures only
   b. Abnormally low temperatures only
   c. Both abnormally high or abnormally low temperatures
   d. Heat shock proteins are continually available in a plant.

10. The main function of heat shock protein is to
    a. stabilize proteins necessary to a cell's survival.
    b. reinforce membranes that lose fluidity.
    c. cause the plant to enter dormancy.
    d. act as an antifreeze compound.

Application Questions

1. Describe strategies used by plants to prevent herbivory.

2. Explain what is meant by coevolution of plants and herbivores. How can grazing increase the productivity of some plants?

3. It has long been a practice in rural areas to nail fencing to trees. Describe how the act of nailing a fence to a tree trunk can introduce pathogens into the tree. What measures does the tree take to ward off disease?

4. Explain how altering the osmotic and water potential of a cell can increase its ability to take up water. Under what conditions is this capability necessary?

5. At a T-intersection in a small town in the snowbelt region of Pennsylvania, a local business plants several ornamental shrubs. During the winter, snowplows push snow from the intersection around the shrubs. In an attempt to save the shrubs, the business covers the shrubs with burlap. Will this save the shrubs? What conditions in the intersection are likely to cause the greatest damage to the shrubs?

6. Bioremediation of mine and ore refinery sites is an area of intense study. How can plants help with the cleanup of sites contaminated with heavy metals?

7. Many plants produce toxins that damage eukaryotic predators. Considering that the cell structure of both the predator and the plant is similar, why is the plant not affected by its own toxins?

8. Pansies are planted in the winter in the southeastern United States. To ensure that the plants are ready to enter cool ground, nurseries cold-harden the plants for several days. Describe this process and explain why it is done.

Answers

Knowledge and Synthesis Answers

1. **d.** Cutin and suberin, along with waxes, aid in preventing entry of pathogens into a plant by forming a barrier.

2. **d.** Plants that have been exposed to pathogens wall off invaded tissue. This tissue releases salicylic acid to the rest of the plant, which stimulates PR proteins. The tissue also releases airborne methyl salicylate, which causes release of PR proteins in distant regions of the plant and even in neighboring plants.

3. **a.** The plant must have an *R* allele, and the pathogen must have an *Avr* allele. Compatibility between these alleles leads to resistance.

4. **d.** Polysaccharides seal plasmodesmata and reinforce cell walls. The purpose of this is to contain the pathogen in a minimal number of cells.

5. **d.** Canavanine's mechanism of action is that it is taken up by the predator and incorporated into its protein structures. Because the structure of canavanine is different from that of the arginine it replaces, the predator's proteins take up a structure that is lethal to the predator. The plant itself and some predators have enzyme mechanisms to distinguish canavanine from arginine and prevent its incorporation into proteins.

6. **c.** See Figure 40.5 in the textbook.

7. **d.** Water storage tissues allow the plant to take advantage of any water that is available and hold on to it for lean times. Leaf loss, when photosynthesis cannot be sustained, prevents loss of resources. Sunken stomata prevent excessive water loss to evaporation.

8. **a.** They can accumulate sodium and chloride ions. This makes their water potential more negative, and they are able to extract more water from their surroundings than plants that do not store the ions.

9. **c.** Heat shock proteins are produced in response to extreme high and low temperatures.

10. **a.** Heat shock proteins are chaperonin proteins. They function to stabilize and prevent denaturation of essential proteins.

Application Answers

1. Plants prevent herbivory by producing secondary compounds that affect potential herbivores. These compounds may act as neurotoxins, inhibit digestion, or have a whole host of other consequences (see Table 40.1).

2. Coevolution of herbivores and plants occurs when the herbivore has resistance to toxins the plant may produce, and the plant develops strategies to cope with herbivory. In some cases, regular grazing increases photosynthetic productivity of the plant by reducing shading and allowing better partitioning of resources.

3. Nailed fencing on trees penetrates the nonspecific primary defenses of a tree. In response to pathogens introduced by the nail, the tree begins to isolate the affected tissues. Polysaccharides are synthesized to plug plasmodesmata, and lignin reinforces the cell walls and acts as a toxin to the invading pathogens. Phytoalexins and PR proteins are released, and salicylic acid also may be released. The tissue around the nail begins to die, and pathogens are isolated.

4. Altering the osmotic and water potential of a cell will affect the rate of osmosis. Water will flow to an area of more negative water potential. Anything that lowers the water potential of a cell will enhance water flow into that cell.

5. The burlap will most likely do little for the shrubs. The roots are in most danger. Because streets are heavily salted in the winter, the snow around them creates a saline environment. If the shrubs are not adapted to a saline environment, little will help them survive the salt accumulation around their roots.

6. Only plants that have evolved a strategy for taking up and coping with heavy metals can grow in these sites. By isolating and specifically growing these types of plants, the slow process of removing the metals from the soil can begin.

7. Plants may isolate the toxins into vacuoles within their cells so that toxins do not come into contact with the active machinery of the cells. Plants may restrict toxin production to already-damaged tissue. Plants may also develop enzymes that differentiate toxins and render them harmless.

8. Cold-hardening is gradual exposure to adverse temperatures. This allows for structural changes in cell membranes to help them cope with extreme temperatures. It also allows for the production of heat shock proteins and other compounds necessary for survival at temperature extremes.

41 Physiology, Homeostasis, and Temperature Regulation

Important Concepts

Control of the internal environment is necessary for proper cell function.

- Due to their small size, single-celled organisms can receive all their required nutrients from the surrounding media. Small multicellular organisms, in which all the cells are only a few layers from the environment, can receive nutrients from the environment without the help of specialized cells. Larger multicellular organisms can maintain an internal environment that differs from the external environment. Specialized cells and groups of cells help to control the makeup of the internal environment and maintain the differences between the internal and external environments (see Figure 41.1).

- Homeostasis is the state of having a constant, regulated internal environment, and homeostatic mechanisms help maintain this state.

Groups of cells are organized as tissues, organs, and organ systems.

- A group of similar cells with the same form and function constitutes a tissue. There are four types of tissues: epithelial, connective, muscle, and nervous tissues.

- Epithelial tissues make up the lining of the inner and outer body surfaces, such as the skin or the lining of the various organs. Some epithelial cells have secretory functions, whereas others have cilia to help move substances over surfaces. Still other epithelial cells function in protection, absorption, or the provision of information to the nervous system.

- Connective tissue is made up of an extracellular matrix that holds together a dispersed group of cells. The protein collagen makes up the bulk of the extracellular matrix. Cartilage, bone, adipose tissue, and blood are all connective tissues.

- Muscle tissue is either skeletal, smooth, or cardiac muscle.

- Neurons and glial cells are the two types of cells found in nervous tissue.

- Organs are groups of tissues that carry out a specific function in the body. Most organs are composed of all four of the major tissue types. The integration of a number of organs to carry out a specific function is known as an organ system, such as the digestive system (see Table 41.1).

Control systems help to regulate physiology and maintain homeostasis.

- Physiological regulatory systems require both a set point at which the parameter is to be held and feedback that provides information about the state of the system (see Figure 41.4). Differences between the set point and the feedback information indicate to the regulatory mechanism which corrective measures are required to restore a parameter to its set point.

- Physiological control systems must take information from the regulator systems and effect changes. Negative feedback results in a slowdown or reversal of a process, returning the variable controlled by that process back toward its set point or regulated level.

- In contrast to negative feedback controls, positive feedback results in amplification of a response.

- The set point maintained by an organism for a given physiological process can also be changed by feedforward information.

Temperature plays an important role in life.

- Most living cells function only between a temperature of 0°C and 45°C, with most organisms having much narrower limits in their range of survival temperatures.

- All physiological processes in organisms are sensitive to temperature. Increases in temperature usually increase the rate of a given process. The Q_{10} is a measure of change in a physiological process as the temperature increases or decreases by 10°C (see Figure 41.5). Most biological Q_{10} values range from 2 to 3.

- The metabolic rate of an organism is the total energy used by the animal and is usually measured as oxygen consumption.

- Acclimatization is the change in a physiological process that occurs over time in response to a change in the environment. Acclimatization typically results in an enhanced probability of thriving in the new environmental conditions. Metabolic compensation—acclimatization of metabolic rate—occurs in fish with changes in the seasons (see Figure 41.6). During the cold winter, fish may have a higher metabolic rate at any given body temperature than they had during the warmer summer.

Animals can be classified as ectotherms or endotherms.

- Animals can be classified into two main groups based on their source of heat. Endotherms regulate their body temperature with internal heat production by means of a high metabolic rate and effective insulation. Ectotherms have a low metabolism and little insulation and must rely on the environment to provide body heat.

- Ectotherms are poikilotherms, with body temperatures that track the environmental temperature. An endotherm maintains a constant body temperature even as the environmental temperature changes (see Figure 41.7). Endotherms are either homeotherms, which always maintain a constant temperature, or heterotherms, which do so only part of the time (such as hibernating mammals). But both have the ability to maintain a high body temperature.

- Both endotherms and ectotherms use behavioral mechanisms to regulate body temperature. Ectotherms shuttle between warm and cool environments during the day to maintain body temperature within a given range (see Figure 41.8). Endotherms also use behavioral thermoregulation as a first line of thermoregulation (see Figure 41.9).

- There are a number of routes through which heat can be gained or lost to the environment (see Figure 41.10). Evaporation from the skin or respiratory tract cools the body. Solar radiation from the sun heats the body. Thermal radiation, convection, and conduction can all either warm or cool the body, depending on the thermal environment.

- Both ectotherms and endotherms use changes in blood flow to control body temperature. Increases in blood flow result in increased heat exchange with the environment, whereas decreases in blood flow result in decreased heat exchange (see Figure 41.11).

- A number of ectothermic insects are able to generate heat to raise their body temperature. This increases their physiological performance, particularly for flight.

- Some large fishes have the ability to maintain a body temperature higher than the surrounding water. They use special vascular countercurrent heat exchangers that retain metabolically produced heat in blood that remains in the core (see Figure 41.12).

Endotherms have special adaptations to help in controlling body temperature.

- For endotherms, the basal metabolic rate per gram of tissue decreases as animals get larger (see Figure 41.13). The reason for this pattern is unclear.

- The range of environmental temperatures at which the metabolic rate of endotherms does not change is known as the thermoneutral zone (see Figure 41.14). The upper and lower critical temperatures are the environmental temperatures at which metabolic rate begins to increase. Increases in heat production occur by nonshivering and shivering thermogenesis when the environmental temperature is below the lower critical temperature. Nonshivering thermogenesis involves a specialized adipose tissue, brown fat (Figure 41.15). Metabolism increases when environmental temperature is above the upper critical temperature because the animal must expend energy to cool down.

- Decreases in surface area are seen in animals that live in cold climates. This decrease in surface area decreases the

area over which heat can be lost. Animals that live in hot areas display increases in surface area to promote heat exchange with the environment (see Figure 41.16a).

- Fur and feathers provide insulation in mammals and birds experiencing cold environments (see Figure 41.16b). These structures help to retain heat produced in the body and help maintain a constant body temperature.

- Evaporative cooling provides an avenue for heat loss to the environment by sweating or panting.

The hypothalamus is the vertebrate thermostat.

- The hypothalamus at the bottom of the brain is involved in the control of homeostasis for many regulatory systems, including thermoregulation. The hypothalamus has a set point temperature, and changes in hypothalamus temperature result in thermoregulatory responses to offset any temperature change (see Figure 41.17). The hypothalamus also integrates thermal information received from the skin.

- In mammals, hypothalamic set points are different at different environmental temperatures (see Figure 41.18). Other factors, including the sleep-wake cycle and level of activity, also influence set points.

- The body has the ability to increase the set-point temperature in response to infections, resulting in fever. Some animals are also able to decrease body temperature either on a daily scale (daily torpor) or over longer time periods (hibernation) (see Figure 41.19).

The Big Picture

- One of the most important roles of tissues, organs, and organ systems is to help maintain homeostasis within the body's cells. Only by maintaining relatively constant intracellular conditions can basic metabolic reactions continue.

- Animals can either generate their own body heat (endotherms) or rely on the environment to determine body temperature (ectotherms). Endotherms typically have much greater insulation than ectotherms as well as a higher metabolic rate. The metabolic rates of endotherms and ectotherms respond differently to changes in environmental temperature. Endotherms increase metabolism as environmental temperature decreases. The metabolic rate of ectotherms is temperature-dependent and rises and falls with the rise and fall of environmental temperature.

- Control of body temperature can occur by either behavioral or physiological mechanisms. Both endothermic and ectothermic vertebrates use the brain's hypothalamus as the thermostat for temperature regulation.

Common Problem Areas

- Homeostasis does not necessarily mean holding every body variable absolutely constant, but rather making sure that body variables are held at the correct level for the circumstances. Even body temperature in humans, which we think of as being a constant 37°C, drops slightly in the early morning hours and rises considerably during extended vigorous exercise.

- Understanding the difference between an ectotherm and an endotherm can be difficult. Think of temperature regulation in ectotherms and endotherms as the temperature regulation of a house. A house with good insulation, furnace, and thermostat is a kind of endotherm, whereas a shed with poor insulation, a small wood stove, and no thermostat is an ectotherm.

Study Strategies

- Use Table 41.1 to acquaint yourself with the various organ systems and their functions. This will help you when you study each of the organ systems in later chapters.
- Refer to the following tutorial and activities on the Student Website/CD:
  Tutorial 41.1 Control, Regulation, and Feedback
  Activity 41.1 Thermoregulation in Endotherm
  Activity 41.2 The Major Organ Systems

Test Yourself

Knowledge and Synthesis Questions

1. Which of the following contributes to the proper maintenance of homeostasis in the bodies of animals?
   a. pH buffering of the blood
   b. Control of blood flow to the skin
   c. Production of hormones
   d. All of the above

2. In an environment with an ambient temperature lower than an animal's core body temperature, which of the following would be an *inappropriate* physiological or behavioral response if the animal needed to eliminate excess body heat?
   a. Wallowing in a pool of water
   b. Preventing blood flow to peripheral vessels
   c. Sweating
   d. All of the above

3. A lizard lives in a desert environment where the temperature at night is low and during the day is high. What might this animal do to maintain the highest stable body temperature?
   a. Stay in a burrow during the night and shuttle between the sun and shade on the surface during the day
   b. Stay on the surface during the night and move to a burrow during the day
   c. Increase metabolism and heat production during the night and decrease it during the day
   d. None of the above

4. On a warm day your dog has been running around fetching his Frisbee and returning it to you. You notice that he is panting heavily. Which of the following does *not* describe the condition of your dog?
   a. Your dog is using evaporative cooling to lower core body temperature.
   b. Your dog is operating at a basal metabolic rate.
   c. Your dog is at the upper end of its thermoneutral zone.
   d. All of the above

5. Evaporative cooling is an effective way to increase heat loss, but carries with it the physiological drawback of
   a. increased use of ATP and substantial water loss.
   b. lowering the hypothalamic thermal set point.
   c. decreased use of ATP and substantial water gain.
   d. exhaustion of the supply of brown fat.

6. Fever
   a. is a higher than normal body temperature that is always dangerous.
   b. decreases the metabolic rate of the body to conserve energy.
   c. is regulated by chemicals that reset the thermostat of the body to a higher set point.
   d. causes the liver to release large amounts of calcium, which seems to inhibit bacterial growth.

7. Which of the following is a characteristic unique to connective tissue?
   a. Highly modified cells that show the special property of contractility
   b. Diverse anatomy with distinct specializations for information transfer
   c. Lining of the inner surfaces of the intestines and lungs in mammals
   d. Loose array of cells embedded in an extracellular matrix

8. Elephants use their ears to release heat to the environment. What mechanisms might they employ to increase heat loss from the ears?
   a. Increased convection due to flapping of the ears
   b. Moving into the sun
   c. Increased blood flow to the ears
   d. Both a and c

9. Which organ system is involved in the maintenance of homeostasis in humans?
   a. Endocrine
   b. Digestive
   c. Muscle
   d. All of the above

10. What is the difference between a negative feedback mechanism and a positive feedback mechanism?
    a. Negative feedback mechanisms only exist in the circulatory system.
    b. Negative feedback mechanisms return a system to a set point, and positive feedback mechanisms amplify a response.
    c. Negative feedback mechanisms move away from a set point, and positive feedback mechanisms stabilize toward a set point.
    d. Negative feedback mechanisms stabilize toward a set point, and positive feedback mechanisms reset the set point.

11. A number of physiological processes can undergo acclimatization. Which of the following statements about acclimatization is true?
    a. It occurs in response to seasonal temperature changes.
    b. Acclimitization of metabolic rate occurs because enzyme expression changes.
    c. It involves the changing of a set point.
    d. All of the above

12. In fast-swimming cool-water fishes such as sharks and tunas, which of the following contribute(s) to the generation and maintenance of core body temperatures in excess of ambient water temperatures?
    a. Specializations in the size and arrangement of blood vessels
    b. Low rates of activity in the swimming muscles
    c. An ability to acclimatize rapidly to the surrounding water
    d. Metabolic rates that are insensitive to temperature change

13. Which of the following statements about hypothalamic function is *false*?
    a. Circulating blood temperature is monitored by the hypothalamus.
    b. Hypothalamic thermal set points never change.
    c. The hypothalamus is a part of the central nervous system.
    d. Different thermoregulatory responses have different hypothalamic set points.

14. Some animals use brown fat as a source of heat generation. Which of the following statements about brown fat is true?
    a. Brown fat is involved in nonshivering thermogenesis.
    b. The protein thermogenin is involved in uncoupling proton movement from ATP production.
    c. Brown fat is found in endotherms.
    d. All of the above

15. During childbirth, the pressure exerted on the mother's cervix by the emerging infant leads to increased contraction of the uterus. This interaction can be explained by
    a. negative feedback.
    b. feedforward feedback.
    c. positive feedback.
    d. metabolic compensation.

Application Questions

1. Having measured the metabolism of two similar sized animals at 15°C and 25°C, you find that the metabolic rate of the first animal is 115 and 55 ml of oxygen per hour at 15°C and 25°C, respectively. The metabolic rate of the second animal is 5.5 and 11.5 ml of oxygen per hour at 15°C and 25°C, respectively. Calculate the Q_{10} for the metabolic rate in both animals and determine if they are endotherms or ectotherms.

2. You have measured the metabolism of the first species in Question 1 at both low and high environmental temperatures. You find that the metabolic rate at 40°C is higher than at 30°C. What physiological phenomenon explains why the metabolic rate of this animal rises when environmental temperature is above the upper critical limit?

3. What are the different mechanisms that animals use as insulation, and what physiological controls exist to increase insulation?

4. Animals have the ability to undergo acclimatization and metabolic compensation. Describe each of these processes and how they relate to temperature regulation.

5. Why is it important for an organism to maintain homeostasis?

Answers

Knowledge and Synthesis Answers

1. **d.** All of the choices are mechanisms to help to maintain the internal environment of an animal.

2. **b.** If the environment were cooler than the animal, the animal would want to lose heat to the environment across the skin. Increasing peripheral blood flow would move heat from the core of the animal to the surface, where it could be lost to the cooler environment.

3. **a.** In the desert, if an ectothermic lizard wants to maintain the most stable body temperature, it should be in a burrow during the night, where temperatures do not drop very much, and then shuttle between the sun and shade during the day.

4. **b.** After exercise, your dog will have a high metabolic rate and will pant in an attempt to cool down. Its body temperature will most likely be above the upper thermoneutral zone.

5. **a.** Evaporative cooling is a costly means to lower body temperature. ATP must be used either during sweating or panting. Water loss can also be high during evaporative cooling.

6. **c.** Fevers are brought about by a resetting of the thermoregulatory set point in the hypothalamus in response to pyrogens.

7. **d.** Connective tissues are blood, cartilage, and bone, all of which contain a loose population of cells embedded in an extracellular matrix.

8. **d.** Elephants use both physiological mechanisms (increased blood flow to the ears) and behavioral mechanisms (flapping of ears) to release heat into the environment.

9. **d.** All of these organ systems are involved in maintaining homeostasis.

10. **b.** Negative feedback loops use information about the state of a system to bring the internal environment back toward the set point values. Positive feedback loops amplify the response of a system, moving the system away from the set point.

11. **d.** Physiological processes acclimate in response to a seasonal change in the environment. Acclimatization may involve changes in enzyme expression and resetting of a physiological set point.

12. **a.** Large fishes such as the tuna are able to keep core body temperature above ambient temperature by using a number of mechanisms. They have a special countercurrent heat exchanger built into their blood vessels. They also have high metabolic rates and thus produce heat in their swimming muscles.

13. **b.** The hypothalamic thermal set points change daily, monthly, and seasonally.

14. **d.** Brown fat is a mechanism by which thermogenin protein uncouples proton movement from ATP production. This occurs in the endotherms, and the process of heat production is termed nonshivering thermogenesis.

15. **c.** Increased uterine contraction in response to pressure on the cervix is an example of a positive feedback.

Application Answers

1. The metabolic rate of the first animal increases with a decrease in environmental temperature, making this animal an endotherm. The metabolic rate of the second animal decreases with a decrease in environmental temperature, making this animal an ectotherm. Using the equation $Q_{10} = (R_T / R_{T-10})$, the Q_{10} for metabolic rate can be calculated. The Q_{10} of the endotherm's metabolic rate is 2.1, whereas the Q_{10} of the ectotherm metabolic rate is 0.48 using the equation provided.

2. An endotherm in a hot environment will attempt to keep core body temperature from increasing. The heat-loss mechanisms available to an endotherm, sweating and panting, require the use of ATP and result in an increased

metabolic rate. Endotherms will expend energy in order to protect body temperature and lose heat to the environment.

3. Mammals use fur and fat as effective insulators, whereas birds use feathers. Fur and feathers are good insulators because they trap still, warm air from the body. Increases in the length of the fur or feathers will increase their effectiveness as insulators. Still air is one of the best insulators. Fat works in a similar manner because it has a low thermal conductance. Humans do not have sufficient hair on their bodies for it to provide good insulation; therefore, they wear clothes as an insulation layer. Physiologically, animals can change their blood flow patterns to increase or decrease the amount of warm blood that is received by the skin.

4. Animals undergo acclimatization in response to changes in seasonal conditions such as temperature. Often these changes are brought about by the production of enzymes that function better at the new temperature. One example of acclimatization is metabolic compensation. As seasonal temperatures change, the enzymes controlling resting metabolism adjust to compensate for the change in temperature (see Figure 41.6).

5. Homeostasis is the maintenance of a constant internal environment. The body functions with the help of proteins and enzymes. Changes in the internal environment such as pH, temperature, glucose level, and oxygen and carbon dioxide levels can all effect cellular function. Loss of homeostasis can lead to improper function of proteins and cell membranes and ultimately lead to cell death.

42 Animal Hormones

Important Concepts

Hormones are chemical signals used in physiological regulation.

- Hormones are secreted by endocrine cells, present as either individual cells or as groups of cells known as glands.

- Most hormones are secreted into the blood, where they are carried to target cells and bind with a receptor to trigger its function. Not all hormones act on a global scale; some are secreted and taken up by local cells. These latter hormones act locally and are known as paracrine hormones.

- Histamine, growth factors, and neurotransmitters are examples of paracrine hormones. Histamine acts locally in response to damage to tissue by causing dilation of blood vessels. Growth factors stimulate the growth and differentiation of cells. Neurotransmitters are paracrine hormones that carry a message from one neuron to the target cells over a very small distance.

- Hormones that act on the cells that secrete them are known as autocrine hormones.

Hormones are involved in the control of development in insects.

- Hormones control molting and metamorphosis in insects.

- In an elegant set of experiments, Wigglesworth used the blood-sucking bug *Rhodnius* to examine the control of molting. Decapitation of the bugs at various times after feeding showed that there was a substance secreted in the heads of these bugs that stimulates molting.

- The brains of *Rhodnius* and other insects produce brain hormone that is stored in the corpora cardiaca attached to the brain. After a meal, brain hormone moves to the prothoracic gland, where it stimulates the release of ecdysone, a hormone that stimulates molting.

- In insects with incomplete metamorphosis, juvenile hormone results in further molts as a juvenile. In the absence of juvenile hormone, an insect molts into the adult form. In insects with complete metamorphosis, juvenile hormone results in molt into a larva, declining levels result in molt into a pupa, and in the absence of juvenile hormone a pupa molts into an adult.

The pituitary gland is the master endocrine gland in vertebrates.

- The pituitary gland is called the "master gland" because it produces hormones that control many of the other endocrine glands in the body. The pituitary is made up of the anterior pituitary and posterior pituitary, each utilizing different control mechanisms and releasing different hormones.

- The posterior pituitary controls the release of antidiuretic hormone and oxytocin. Both of these hormones are produced by neurons in the hypothalamus and move down the neurons into the posterior pituitary. Antidiuretic hormone (ADH) acts on the kidneys to stimulate reabsorption of water to help maintain blood pressure and solute concentration of the blood. During childbirth, oxytocin stimulates uterine contractions in the mother. After birth, oxytocin starts the flow of milk from mammary glands.

- The anterior pituitary gland releases four tropic hormones that control the function of the thyroid, adrenal cortex, testes, and ovaries. The anterior pituitary also produces growth hormone, prolactin, endorphins, and enkephalins. Growth hormone stimulates the liver to release hormones that stimulate bone and cartilage growth. In female mammals, prolactin stimulates milk production.

- The anterior pituitary gland is controlled by hormones produced in the hypothalamus and transported to the anterior pituitary gland by portal blood vessels. Two of the controlling hormones are thyrotropin-releasing hormone, causing the release of thyrotropin that stimulates the thyroid, and gonadotropin-releasing hormone, stimulating the release of hormones that control the gonads.

- The anterior pituitary gland is also controlled by a negative feedback loop. Hormones released from target glands can have a negative effect on the release of further hormones from the anterior pituitary gland.

The thyroid gland regulates metabolic rate and blood calcium levels.

- The thyroid gland helps regulate metabolic rate through the action of several hormones that it produces; thyroxine (T_4) and triiodothyronine (T_3) increase metabolic rate in mammals. Thyrotropin from the anterior pituitary gland stimulates the thyroid gland to produce thyroxine. Poor regulation of thyroxine production can result in an enlarged thyroid gland known as goiter.

- Another thyroid hormone, calcitonin, regulates calcium levels in the blood by stimulating osteoblasts, which deposit new bone, and inhibiting osteoclasts, which break down bone. The parathyroid gland produces parathyroid hormone, which increases blood concentrations of calcium by stimulating osteoclasts to dissolve bone. These two hormones act antagonistically, countering the effect of each other.

- Vitamin D is a hormone produced by skin cells and activated when it passes through the liver and kidneys. In the kidneys, vitamin D works with parathyroid hormone to decrease the loss of calcium in the urine.

The pancreas regulates blood sugar levels.

- Endocrine cells in the pancreas, called the islets of Langerhans, produce the hormones insulin and glucagon, which regulate glucose levels in the blood. Insulin binds to receptors on cells, allowing glucose to enter the cell from the bloodstream. When insulin is absent or the receptors become insensitive to it, glucose accumulates in the blood, resulting in the disease diabetes mellitus.

- Glucagon stimulates the liver to convert glycogen into glucose to be used as an energy source when blood glucose levels are low.

- A rapid rise in blood glucose levels stimulates the pancreas to release somatostatin, which inhibits the release of insulin and glucagon.

The adrenal gland produces hormones with numerous regulatory functions.

- The adrenal gland has two regions, the adrenal medulla, making up the core of the gland, and the adrenal cortex, which surrounds the medulla.

- Epinephrine and norepinephrine are produced in the adrenal medulla and are involved in the fight-or-flight response.

- The adrenal cortex produces glucocorticoids, mineralocorticoids, and sex steroids from cholesterol. The main glucocorticoid is cortisol, which mediates our response to stress by stimulating cells that are not necessary for the fight-or-flight response to decrease their use of glucose. The main mineralocorticoid is aldosterone, which is involved in the reabsorption of sodium and excretion of potassium at the kidney.

The gonads produce sex steroids that act as hormones.

- In males, testes produce androgens (testosterone), whereas in females, ovaries produce estrogens and progesterone. Early in development, the sex organs of human embryos are similar. At about week 7, presence of the Y chromosome in male embryos causes the undifferentiated gonads to begin producing androgens. In response to androgens, the reproductive system develops into a male system. In the absence of androgens, the reproductive system develops into a female system.

- Sex steroids increase in concentration at the time of puberty. Both luteinizing hormone (LH) and follicle-stimulating hormone (FSH) from the anterior pituitary gland control the production of sex steroids. Increased levels of LH and FSH in females and LH in males stimulate the gonads to produce more sex hormones. These hormones in turn have profound physiological effects on the development of the secondary sexual characteristics.

The pineal gland produces melatonin.

- Melatonin is involved in regulating the daily cycles in the physiology of an organism. The release of melatonin by the pineal gland occurs in the dark.

Hormones act on the receptors of the target cells.

- Most hormones are released in very small quantities. A hormone works only on cells that have specific receptors for it.

- Hormones are either lipid-soluble or water-soluble. Lipid-soluble hormones are able to cross the plasma membrane of cells and act on receptors inside the cell. Water-soluble hormones cannot pass through the plasma membrane and thus act on receptors spanning the plasma membrane. Such binding usually triggers the target cell's response by activating enzymes within the cell.

- The level of hormone receptors can be adjusted to help control the action of hormones. Downregulation involves a decrease in the number of receptors in response to high levels of a hormone. Upregulation occurs when a hormone stimulates a target cell to increase the number of hormone receptors.

The Big Picture

- Hormones are regulatory substances that circulate broadly throughout an animal's body, regulating functions ranging from growth to sexual maturity. In numerous ways, the endocrine system complements the nervous system as the second major regulator of body function and homeostasis.

- Every organ system in the vertebrate body is a target for hormones. Hormones are actively involved in the maintenance of homeostasis.

- Hormone action is controlled by a number of different mechanisms. In many instances there is a "hormone cascade" in which a hormone controls the release of a hormone, which controls the release of a hormone, and so on. Negative feedback occurs when a released hormone inhibits (negatively feeds back on) the tissue that may have stimulated its release in the first place.

Common Problem Areas

- Appropriate for its title "master gland," the pituitary produces numerous regulatory hormones. The functions of the anterior and posterior pituitary are quite distinct, and keeping them separate in your mind is important to your understanding of the endocrine system.

- Remember that most hormones are broadly distributed throughout the body and so all cells are exposed to the same level of hormone concentration. Hormones have specific actions because different cells have different levels of sensitivity to them, not because the hormone concentration differs throughout the body.

Study Strategies

- The many hormones may seem to constitute a long laundry list. Try to learn the function along with the name of each hormone as this will help you remember the target tissues.

- Some hormones have extremely specific actions (e.g., follicle-stimulating hormone), whereas others have broad effects on many target tissues (e.g., epinephrine, thyroid hormones). Learning which hormones are "generalists" and which are "specialists" will help you better understand the endocrine system.

- Biologists studying the endocrine system frequently use abbreviations for the names of hormones. Learn these abbreviations along with the full name of a hormone.

- Refer to the following tutorial and activities on the Student Website/CD:
  Tutorial 42.1 Complete Metamorphosis
  Activity 42.1 Endocrine Glands in Humans
  Activity 42.2 Concept Matching: Vertebrate Hormones

Test Yourself

Knowledge and Synthesis Questions

1. A paracrine hormone
   a. is a local hormone that acts on the cell that released it.
   b. always acts on a wide variety of target tissues.
   c. is a local hormone produced at one site but active at a different site in the body.
   d. None of the above

2. A target cell's response to a hormone depends on
   a. the amount of hormone released.
   b. the number of receptors on that target cell.
   c. how well the hormone binds with the receptor.
   d. All of the above

3. "Upregulation" of hormone receptors refers to
   a. increase in abundance of hormone receptors with low hormone levels.
   b. increase in abundance of hormone receptors with high neurotransmitter levels.
   c. increase in hormone levels produced by increase in abundance of hormone receptors.
   d. decrease in hormone levels produced by decrease in abundance of hormone receptors.

4. Hormones have various regulatory functions. Which of the following statements does *not* describe how hormones function?
   a. They act in very low concentration
   b. They act at sites distant from where they are produced
   c. They are transported in blood
   d. None of the above

5. In insects, juvenile hormone
   a. is produced by the corpora cardiaca.
   b. is also known as brain hormone.
   c. prevents maturation.
   d. All of the above

6. Vitamin D
   a. is a hormone secreted by skin cells.
   b. is activated on passing through the liver and kidneys.
   c. decreases loss of calcium in the urine.
   d. All of the above

7. Which of the following glands is considered the "master" endocrine gland in vertebrates?
   a. Adrenal gland
   b. Thyroid gland
   c. Hypothalamus
   d. Pituitary gland

8. Which of the following glands is involved in the function of our "biological clock"?
   a. Pineal gland
   b. Thyroid gland
   c. Adrenal glands
   d. Ovaries

9. The half-life of hormones in the plasma ranges from
   a. minutes to hours.
   b. minutes to days or weeks.
   c. days to weeks.
   d. hours to months.

10. The target tissues of hormones are those tissues that
    a. the particular hormone can actually penetrate.
    b. have specific enzymes with which hormones directly interact.
    c. have high concentrations of the "second messenger."
    d. have receptors for the particular hormone.

11. Which of the following sets of vertebrate hormones are produced in the anterior pituitary gland?
    a. Somatostatin, vasopressin, insulin
    b. Prolactin, growth hormone, enkephalins
    c. Oxytocin, prolactin, adrenocorticotropin
    d. Estrogen, progesterone, testosterone

12. Hormones that are secreted by one endocrine gland and control the activities of another endocrine gland are called _____ hormones.
    a. growth
    b. obstructive
    c. tropic
    d. selective

13. Which of the following secretes hormones but is not considered traditional glandular tissue?
    a. Pancreas
    b. Heart
    c. Testes
    d. Adrenal glands

Application Questions

1. Describe the modes of action of lipid-soluble and water-soluble hormones.

2. The pituitary gland is made up of the anterior and posterior pituitary. What is the relationship between these two parts of the pituitary gland and the hypothalamus?

3. Hormones are regulated in numerous ways. One common method of hormonal regulation is by negative feedback. Discuss the negative feedback loop involved in the response of the thyroid to cold.

4. Wigglesworth conducted a number of studies on the insect *Rhodnius* to examine the control of molting. Describe what would happen to a fourth-instar *Rhodnius* if Wigglesworth were to either partially or fully decapitate the insect one week after a blood meal.

Answers

Knowledge and Synthesis Answers

1. **c.** Paracrine hormones act on local cells that are different from those that secrete them. Autocrine hormones act on the very same cells that secrete them.

2. **d.** The response of target cells to a hormone depends on how much of the hormone is present and acting on the target, how many receptors are present on the surface of the cell that the hormone can act on, and how well the hormone binds with the receptor.

3. **a.** "Upregulation" of hormone receptors on a cell is the production of more receptors when a hormone is present over time in the blood or other fluids surrounding the cell.

4. **d.** All of the statements are correct. Hormones are found and act at low concentrations in the body. They are transported by the blood or diffuse through interstitial space to reach target cells that may be some distance from the secreting gland.

5. **c.** Juvenile hormone is secreted by the corpora allata of insects and prevents maturation. Brain hormone is secreted by the corpora cardiaca and is involved with ecdysone in molting.

6. **d.** Vitamin D is a hormone secreted by skin cells that becomes more active once it passes through the liver and kidneys. Vitamin D is involved in raising levels of calcium in the blood by reducing loss of calcium in the urine.

7. **d.** The pituitary gland is considered the master gland because it produces hormones that control endocrine glands throughout the body.

8. **a.** The pineal gland secretes melatonin, a hormone that controls the daily cycles and acts as a "biological clock."

9. **b.** Hormones tend to act over time periods of minutes to days or weeks. Epinephrine has a half-life of up to three minutes, whereas cortisol has a half-life of many days.

10. **d.** For a hormone to act on a target cell, the target cell must have the receptors for the specific hormone to bind and trigger the hormonal action.

11. **b.** Prolactin, growth hormone, and enkephalins are produced in the anterior pituitary.

12. **c.** Hormones that control endocrine gland function are known as tropic hormones.

13. **b.** The heart is not typically considered a gland, but it secrets atrial natriuretic hormone that acts on the kidneys to promote sodium excretion.

Application Answers

1. The plasma membrane of cells is hydrophobic. Lipid-soluble hormones, which as their name suggests dissolve in lipids (fats), cross the plasma membrane and act on receptors inside the target cells. Water-soluble hormones cannot cross the lipid-based plasma membrane and act on receptors on the outside of target cells. Such binding by water-soluble hormones initiates changes in the cell by activating enzymes. Because lipid-soluble hormones cross the plasma membrane and enter the cell, they tend to take longer to act, but they also remain active for a longer time period than water-soluble hormones.

2. The hormones secreted by the posterior pituitary gland are produced by neurons in the hypothalamus. The neurohormones move down neurons into the posterior pituitary gland, where they are stored and secreted. The anterior pituitary produces the hormones it secretes. However, hormones from the hypothalamus control the release of these hormones.

3. In response to cold the body tends to increase metabolic rate mediated by increased levels of thyroxine. The hypothalamus is stimulated by the cold to release thyrotropin-releasing hormone. Thyrotropin-releasing hormone stimulates the anterior pituitary to secrete thyrotropin. Thyrotropin stimulates the thyroid gland to release thyroxine, resulting in increased metabolic rate. The increase of thyroxine in the system acts as a negative feedback to inhibit the secretion of thyrotropin by the anterior pituitary gland.

4. The insect *Rhodnius* can live for long periods of time after it has been decapitated. If the insect is partially decapitated, leaving the corpora allata, the insect will molt into a fifth-instar juvenile. This is because the corpora allata produces juvenile hormone, which prevents maturation into an adult. The animal that was fully decapitated would molt into an adult rather than another juvenile instar. The fully decapitated animal had enough brain hormone circulating one week after the meal that it would molt; however, there would be no juvenile hormone present, and thus the animal would molt into an adult.

43 *Animal Reproduction*

Important Concepts

Some animals reproduce asexually.

- Many invertebrates employ asexual reproduction, producing genetically identical offspring from the parent.

- New offspring can be produced either by budding off outgrowths to produce a new individual or by regenerating a new individual from pieces of an animal (see Figure 43.1).

- Animals can produce new individuals asexually through parthenogenesis—the development of unfertilized eggs into new offspring.

- Sometimes sexual behavior is required for asexual reproduction (see Figure 43.2).

Sexual reproduction produces genetic diversity.

- Sexual reproduction in animals produces genetic diversity when two haploid gametes join to form a diploid cell.

- Sexual reproduction has three stages: gametogenesis, mating, and fertilization.

- Gametogenesis produces haploid gametes using meiotic cell division.

  - Gametogenesis occurs in an animal's gonads. Male gametes—sperm—are produced in the testes, and female gametes—eggs or ova—are produced in the ovaries. Gametes are derived from germ cells to produce diploid oogonia and spermatogonia. They then turn into primary oocytes and primary spermatocytes.

Males produce sperm by spermatogenesis.

- Spermatogenesis begins with mitosis of a male germ cell ($2n$) into a spermatogonium (see Figure 43.3*a*). The first meiotic division produces secondary spermatocytes (n), and a second meiotic division produces spermatids (n).

- Mammalian spermatids share gene products of the X chromosomes at this stage.

- The spermatids differentiate and mature into the sperm cell gametes.

Females produce eggs by oogenesis.

- Oogenesis begins with differentiation of a female germ cell ($2n$) into an oogonium ($2n$) and then into primary oocytes ($2n$) (see Figure 43.3*b*).

- The primary oocytes enter prophase and in human females remain in prophase for up to 50 years. No new primary oocytes will be produced during the life of the female.

- Upon exiting prophase, the primary oocyte goes through a meiotic division to produce a secondary oocyte and a polar body. The second meiotic division produces the ootid and another polar body. This second meiotic division may not be completed until fertilization in some species.

Fertilization is the joining of a sperm with an egg to form a diploid zygote.

- For successful fertilization to occur, the sperm and egg must recognize each other.

- During the acrosomal reaction, the acrosomal membrane of the sperm breaks down, releasing enzymes that digest a path for the sperm through the protective layers surrounding the egg (see Figure 43.4).

- Fusion of the plasma membranes of the sperm and egg triggers blocks to polyspermy that prevent more than one sperm from entering the egg.

- Soon after entry by a sperm, the egg is metabolically activated and stimulated to begin development. The diploid nucleus of the zygote is created when the egg and sperm nuclei fuse.

Fertilization occurs differently in the water and on land.

- Many aquatic animals have external fertilization of eggs. The eggs and sperm are released into the water, leaving possible contact and fertilization to chance.

- Many fish and amphibians employ special behaviors to bring gametes closer to each other and ensure fertilization.

- Some mite and scorpion males produce a sperm-filled spermatophore that they deposit into the environment. When a female comes across it she straddles it and takes the sperm into her reproductive tract.

- Internal fertilization requires accessory sex organs that pass the sperm on to the female during copulation. The penis is a male accessory sex organ and the vagina, in some species, is the female accessory sex organ.

Some animals have both male and female reproductive systems.

- Dioecious species have separate male and female individuals.

- In monoecious species, a single hermaphroditic individual may produce both egg and sperm. There can be either simultaneous hermaphrodites (both male and female at the same time), or sequential hermaphrodites (changing sexes at some point during their life span). Sequential hermaphrodites have less of a chance of inbreeding than simultaneous hermaphrodites.

The move to land required a reproductive system that functions in a dry environment.

- Amphibians were the first vertebrates to live on land, but still require water to reproduce.

- Many reptiles and all birds have solved the problem of a desiccating environment by being oviparous; they lay protective shelled eggs termed amniotic eggs. Internal fertilization must occur prior to shell formation. Some reptiles are ovoviviparous, with the embryo developing in a shelled egg that is retained in the female.

- All mammals except the monotremes are viviparous, retaining the embryo within the uterus.

In humans, males produce sperm and deliver it to the female.

- Males produce semen containing sperm (*n*) and fluids that facilitate fertilization. In the testis, sperm cells are produced by spermatogenesis in the seminiferous tubules. Sertoli cells surround developing sperm cells and provide protection and nourishment (see Figure 43.9).

- Sperm develop a flagellum at the back and at the front, an acrosome cap that contains digestive enzymes. Mitochondria provide energy for the flagellum (see Figure 43.9). Sperm are stored in the epididymis, which connects to the vas deferens and then to the urethra, which opens to the outside of the body through the penis.

- The other fluids in semen come from various glands. The bulbourethral glands produce an alkaline and mucoid secretion that helps control pH and lubricates the tip of the penis. The seminal vesicles produce seminal fluid containing energy for sperm in the form of fructose, mucus, and proteins. The prostate gland produces the milky prostate fluid and a clotting enzyme that works on the protein in seminal fluid to turn semen into a gelatinous mass.

- During sexual arousal, the penis becomes engorged with blood, and contraction of smooth muscle in the ducts and at the base of the penis results in emission and ejaculation of semen. After sexual stimulation, the autonomic nervous system causes the vessels at the base of the penis to constrict, leading to loss of erection.

- Testosterone, produced in the Leydig cells of the testes, is the male sex hormone that controls sexual function and sperm production (see Figure 43.10).

Females produce eggs, receive sperm, and incubate the embryo.

- The ovaries release eggs into the body, where they enter the fallopian tube or oviduct and move toward the uterus (see Figure 43.11). Sperm swim up the vagina, through the opening to the uterus (cervix), through the uterus, and into the oviduct. Embryos develop in the uterus after fertilization of the eggs by sperm in the oviducts.

- The initial division of the zygote produces a blastocyst that moves into the uterus, where it embeds into the uterus wall (the endometrium) to form the placenta. The placenta provides nutrients and removes wastes from the developing embryo.

- Mature eggs are produced during a 28-day ovarian cycle (see Figure 43.12). Initially, six to twelve primary oocytes begin to mature and are surrounded by follicle cells to produce individual follicles. After one week, one primary oocyte continues to grow while the others shrink. Prior to ovulation, the primary oocyte matures into a secondary oocyte and is expelled from the ovary. The remaining follicle cells become the corpus luteum, which produces progesterone and estrogen.

- The uterine cycle, in concert with the ovarian cycle, prepares the endometrium to receive the blastocyst. If a blastocyst does not implant by around day 14 of the uterine cycle, the endometrium is broken down and expelled in the process of menstruation. The uterine cycles of most mammals other than humans do not include menstruation. Other mammals resorb the endometrium at the end of the uterine cycle.

- Luteinizing hormone (LH) and follicle-stimulating hormone (FSH) control the timing of the ovarian and uterine cycle (see Figure 43.13). The first step of the uterine and ovarian cycle is menstruation. Prior to day 10 of the uterine cycle, FSH and LH levels are kept low by gonadotropins, estrogen, and progesterone. Around days 12–14, estrogen stimulates a rise in FSH levels, which in turn stimulates the follicles to begin to mature. LH levels also increase, triggering the release of the egg from the mature follicle and the growth of the corpus luteum. Estrogen and progesterone produced by the corpus luteum stimulate the growth and maintenance of the endometrium. In the absence of fertilization, the corpus luteum degenerates, estrogen and progesterone levels decrease, and menstruation occurs to start the cycle again.

- When the egg is fertilized and implants in the uterus, human chorionic gonadotropin is produced, which keeps the corpus luteum functional. Estrogen and progesterone stimulate development of the endometrium, preparing the uterus for pregnancy.

Humans use several methods to prevent pregnancy.

- Methods of contraception differ in their mode of action and failure rate (see Table 43.1).

- The rhythm method involves avoiding sex during the time period that the female is most likely to be fertile and has a 15 to 35 percent failure rate.

- Condoms, diaphragms, and cervical caps work by creating a barrier to sperm. They have a failure rate of from 3 to about 25 percent.

- A number of methods employ hormones to prevent ovulation by the female. Oral contraceptives contain synthetic estrogen and progesterone that interfere with the development of the ova and suspend the ovarian cycle. The failure rate of oral contraceptives is less than 3 percent.

- Intrauterine devices work by blocking the implantation of a fertilized egg in the uterus and have a failure rate less than 6 percent.

- Both males and females can be sterilized. Males can undergo a vasectomy in which the vas deferens is cut, whereas females can have their oviducts cut in tubal ligation procedure (see Figure 43.16).

- The termination of pregnancy is known as abortion. A spontaneous abortion occurring early in pregnancy is commonly called a miscarriage. RU-486 blocks progesterone receptors in the uterine lining and when used early in pregnancy causes shedding of the endometrium and embryo.

- Controlling female fertility is much easier than controlling male fertility. Males continuously produce millions of sperm. Continuous chemical intervention would be needed to block spermatogenesis completely.

Medical science has found ways to help overcome infertility.

- A number of problems may prevent couples from conceiving. A male may not be producing adequate amounts of sperm, or the environment of the uterus may be poor for hosting sperm or the fertilized egg.

- Artificial insemination places sperm in the female reproductive tract where fertilization can occur to increase the chance of fertilization.

- Assisted reproductive technologies take eggs from a female and then either fertilize them outside the body or place a mixture of sperm and egg in the female oviduct.

Some diseases are sexually transmitted.

- Sexually transmitted diseases have been present since ancient times.

- Over 10 million new cases of STDs occur in the United States each year.

- STDs can be caused by viruses, bacteria, yeasts, or protozoans.

- At present, latex condoms are the best protection against STDs.

The Big Picture

- Animals reproduce both asexually and sexually. Asexual reproduction includes budding, regeneration, and parthenogenesis. Sexual reproduction has an advantage over asexual reproduction in that it helps to increase genetic diversity. Sexual reproduction has three main stages: gametogenesis, mating, and fertilization.

- Gametogenesis is the process by which the sex cells are produced. Haploid sperm develop from diploid spermatogonia by the process of spermatogenesis. The haploid egg of the female develops from diploid oogonia by the process of oogenesis.

- The female has two cycles, the uterine and the ovarian cycles, during which an egg is produced and the uterus is prepared for pregnancy. These two cycles occur in parallel and are coordinated by changes in hormone levels. Luteinizing hormone, follicle-stimulating hormone, gonadotropins, estrogen, and progesterone all play a role in the control of these cycles. The uterine cycle prepares the uterus for implantation, and the ovarian cycle produces and releases an egg.

- Mating in aquatic animals involves external fertilization of the egg by sperm. On land, animals use internal fertilization with the help of secondary sex organs to ensure fertilization. The eggs and sperm of a given species recognize each other.

- A number of methods are available both for preventing pregnancy and for overcoming infertility.

Common Problem Areas

- Understanding the ploidy level of the developing egg and sperm can be very confusing, but it is essential in determining how genetic inheritance works.

- Following the hormone changes and hormone cues associated with the human menstrual cycle can be confusing. Create a chronological sequence of hormones and the events they trigger during the 28-day uterine cycle.

Study Strategies

- Recognize that although sperm and egg formation are completely separate processes, each with its own characteristics, the same general events are happening in both. This will help you to recognize and remember the stages of spermatogenesis and oogenesis.

- To better understand the changes in ploidy in gametes and the products of their combination, create a flowchart that shows how the haploid state changes over one or two cycles.

- Review the following tutorial and activities on the Student Website/CD:
  Tutorial 43.1 Fertilization in the Sea Urchin
  Activity 43.1 Human Female Reproductive Tract
  Activity 43.2 Spermatogenesis
  Activity 43.3 Human Male Reproductive Tract

Test Yourself

Knowledge and Synthesis Questions

1. Asexual reproduction is an effective strategy in stable environments because
   a. gametogenesis is most efficient under these conditions.
   b. the resulting offspring, genetically identical to their parents, are preadapted to their environment.
   c. asexual parthenogenesis produces a large amount of genetic diversity.
   d. animal cells tend to be more totipotent under stable conditions.

2. An important difference between a sperm and an egg is
   a. their size.
   b. the amount of cytoplasm they contain.
   c. whether or not they are motile.
   d. All of the above

3. Which of the following would *not* be present in the body of or lifestyle of a terrestrial male animal that reproduces sexually?
   a. Internal fertilization
   b. A secondary sexual organ as part of the genitalia
   c. Synchrony of reproductive physiology with a female
   d. Oogenesis

4. Which of the following best represents the normal path of a sperm cell as it makes its way from the point of entry into a female's reproductive tract to the location where fertilization typically occurs?
   a. Cervix, vagina, ovary, oviduct
   b. Vagina, cervix, uterus, oviduct
   c. Uterus, cervix, vagina, oviduct
   d. Vagina, uterus, cervix, oviduct

5. The function of the seminal vesicle is to
   a. produce a solution of fructose to provide energy for the mitochondria of the sperm.
   b. secrete alkaline fluids that neutralize the acidity of a female's reproductive tract.
   c. initiate the muscular contractions that lead to emission.
   d. produce lubrication for the tip of the penis.

6. Which of the following is an example of positive feedback control in the reproductive cycle of males or females?
   a. The increased response of the hypothalamus and anterior pituitary gland in response to estrogen
   b. The decreased response of the hypothalamus and anterior pituitary gland in response to estrogen
   c. The inhibition of luteinizing hormone by high levels of testosterone
   d. The stimulation of luteinizing hormone by low levels of testosterone

7. Which of the following statements about oogenesis is *false*?
   a. The polar bodies degenerate after the second meiotic division.
   b. The ovum produced is haploid.
   c. The major growth phase of the primary oocyte occurs in prophase I.
   d. The primary oocyte is haploid.

8. Which of the following is *not* an accessory sex organ?
   a. Penis
   b. Prostate gland
   c. Gonad
   d. Breast

9. For approximately how long during the human female's menstrual cycle are progesterone concentrations high enough to maintain the uterus in a proper condition for pregnancy?
   a. All of the cycle
   b. None of the cycle
   c. During the first half of the cycle
   d. During the second half of the cycle

10. Which of the following statements about the birth control pill is *false*?
    a. It works by preventing ovulation.
    b. It works by preventing implantation.
    c. The ovarian cycle is suspended by the birth control pill.
    d. It contains low doses of estrogen and progesterone.

11. Which of the following is *false*?
    a. Sex hormones control negative feedback of the hypothalamus.
    b. Contraception through complete chemical blockage of sperm production is possible.
    c. Surgical transfer of gametes from one area of the reproductive tract to another is possible.
    d. Reproductive organs can be susceptible to cancer.

12. If you compared the genetic makeup of a female animal produced by parthenogenesis with that of its mother, which of the following would you expect?
    a. About 100 percent genetic similarity
    b. About 50 percent genetic similarity
    c. No genetic similarity
    d. Parthenogenetic animals do not have mothers.

13. Which of the following statements about oviparity is *false*?
    a. Only birds and reptiles are oviparous.
    b. The large amount of yolk in the egg provides the nutrients for the developing embryo.
    c. The shell protects the egg from dehydrating.
    d. Both oxygen and carbon dioxide can diffuse through the shell.

14. During spermatogenesis a single male germ cell produces _____ sperm cell(s).
    a. 1
    b. 2
    c. 3
    d. 4

15. Which of the following statements about fertilization is *false*?
    a. The egg permits several sperm to enter it.
    b. The plasma membranes of sperm and egg fuse.
    c. The egg is activated and stimulated to begin development.
    d. Species-specific recognition occurs between egg and sperm.

Application Questions

1. *Daphnia* (water fleas) can reproduce sexually or asexually. When and why might these animals switch between sexual and asexual reproduction?

2. While visiting the shore of a small pond, you notice two frogs mating in the water. Based on what you know about external fertilization, what will ensure that the female's eggs will be fertilized?

3. How is it possible for a sea star to reproduce both sexually and by regeneration? How will the genetic makeup of offspring from each method differ?

4. The following graph depicts the average blood concentrations of three circulating hormones collected from 100 healthy adult women. Each hormone plays a role in regulating some aspect of the menstrual cycle. Identify the hormone that each curve most likely represents, and explain the criteria used in making your choice.

Answers

Knowledge and Synthesis Answers

1. **b.** The parents that have survived to reproduce asexually are able to survive in the current stable environment. Therefore, the offspring should be preadapted for this stable environment.

2. **d.** There are many differences between an egg and sperm. They are produced by different sexes, sperm is motile and an egg is not, and an egg is much larger than sperm.

3. **d.** In males, sperm is produced during spermatogenesis. An egg is produced during oogenesis. Males have a secondary sexual organ such as the penis that aids in internal fertilization. External fertilization is not practical on land.

4. **b.** A sperm is ejected by the male into the vagina. From the vagina the sperm move through the cervix into the uterus and finally the oviduct, where fertilization occurs.

5. **a.** The seminal vesicles are involved in producing the seminal fluid. One of the components of the seminal fluid is an energy source for the sperm in the form of fructose.

6. **a.** During days 12 through 14, a positive feedback occurs in response to estrogen. The hypothalamus and anterior pituitary gland are stimulated to release LH and FSH.

7. **d.** During oogenesis, the primary oocyte is diploid; after the first meiotic division into the secondary oocyte the cell becomes haploid.

8. **c.** The gonads are not secondary sex organs; they are the primary sex organs producing the sperm and the egg.

9. **d.** High levels of progesterone are needed to maintain the uterus in the proper condition for pregnancy. The levels of progesterone are high only during the second half of the uterine cycle.

10. **b.** The birth control pill interferes with the maturation of the follicles and the ova, inhibiting the release of an egg.

11. **b.** Males constantly make sperm, so any chemical block of sperm production would have to be constant. Chemical blocks of sperm would also have serious medical side effects.

12. **a.** Parthenogenesis is a form of asexual reproduction in which offspring develop from unfertilized eggs; there is no sexual recombination of genes, so a female animal born parthenogenetically will be nearly identical to its mother genetically. (This kind of reproduction occurs in aphids, for example.) In many hymenopterans (bees, wasps, and ants), only males are born from unfertilized eggs. Such males are haploid, so they are not genetically identical to their diploid mothers.

13. **a.** Birds and reptiles are not the only oviparous species— a group of mammals, the monotremes, are also egg-layers.

14. **d.** During spermatogenesis a single male germ cell will under go two meiotic divisions, resulting in the production of 4 haploid sperm cells.

15. **a.** During fertilization an egg uses blocks to polyspermy to prevent more than one sperm from entering it.

Application Answers

1. Female *Daphnia* tend to produce new females by parthenogenesis when conditions are favorable. When conditions turn unfavorable, they produce males instead of females. After males are produced, the females reproduce sexually with the them, producing a resting egg. Mating with the males allows for genetic mixing of the population and the production of a resting egg that will produce a new parthenogenetic female when conditions turn favorable again.

2. For successful external fertilization in water, the two frogs must engage in a behavior that will bring the male's sperm in close proximity to the female's eggs. The second important thing that must occur is that the male and female must both produce large numbers of gametes to ensure that eggs are fertilized. Finally, the male and female must be at the same reproductive stage for successful mating to occur.

3. If a sea star loses one of its arms, it can regenerate a new one to replace it. If the detached arm remains alive, it can regenerate an entire new sea star. The new sea star will be genetically identical to the original one. Sea stars also produce eggs and sperm that join during sexual reproduction to produce new individuals. The new sea stars will share a genetic makeup with both parents, increasing genetic diversity.

4. The corpus luteum produces progesterone during the second half of the uterine cycle to maintain the endometrial lining of the uterus. The levels drop off by day 28 if implantation has not occurred. As the follicle grows during the first half of the ovarian cycle, follicle-stimulating hormone (FSH) slowly increases. When the follicle ruptures and releases the egg, FSH is inhibited, and the levels decrease. Luteinizing hormone peaks between days 12 and 14 to stimulate ovulation.

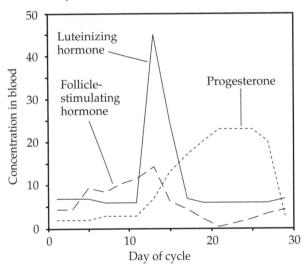

44 *Neurons and Nervous Systems*

Important Concepts

The nervous system consists of neurons and glial cells.

- The nervous system is made up of two major types of cells: neurons and glial cells.
- Neurons transmit information as electric signals from sensory cells to effectors such as muscles.
- Simple animals such as the sea anemone possess a nerve net (see Figure 44.1).
- The nervous system of invertebrates is made up of groups of neurons distributed throughout the body. These are known as ganglia.
- Vertebrates have a central nervous system—the brain and spinal cord—and a peripheral nervous system—the neurons in the rest of the body.
- A neuron is composed of four parts: the cell body, dendrites, axons, and axon terminals (see Figure 44.2). The cell body contains the nucleus, with dendrites coming off it and receiving information from other neurons. The plasma membrane of neurons has an electrical charge or action potential that helps transmit information away from the cell body along an axon. Axon terminals interact with the neuron's target cells to form a synapse.
- Neurons work together in the nervous system. The simplest neuronal network is a sensory neuron connected to a motor neuron connected to a muscle cell.
- Glial cells serve many functions in the nervous system, such as supplying nutrients to neurons, maintaining proper ionic balance, and removing waste particles. Glial cells that insulate axons are Schwann cells and oligodendrocytes. Astrocytes found around blood vessels in the brain form a barrier to some toxins as well as to certain drugs.

Neurons have an electrically charged cell membrane at rest.

- Resting neurons have a negative charge inside and a positive charge outside, resulting in a difference in electrical charge across the membrane known as the membrane potential. In an unstimulated neuron, this voltage difference is called a resting potential (see Figure 44.4).
- The electrical charge across the membrane at rest is due to differences in concentrations of the charged ions' sodium (Na^+), chloride (Cl^-), potassium (K^+), and calcium (Ca^{2+}).

- The lipid bilayer of the plasma membrane is impermeable to ions. Ions move across the plasma membrane through channels or by ion pumps.
- The sodium–potassium pump maintains higher concentrations of Na^+ ions outside the cell and higher concentrations of K^+ ions inside it. This pump transports Na^+ out of the cell and K^+ into it (see Figure 44.5).
- Ion channels are selective pores in the plasma membrane that allow specific ions to diffuse through the membrane down their concentration gradient. Ion channels can be voltage-gated, responding to changes in the voltage across the membrane, or chemically gated, responding to the presence of a specific chemical.
- At rest, neurons have a specific charge due to K^+ movement to the outside of the cell, resulting in a resting potential. K^+ channels are the most common open channels, allowing K^+ to diffuse out of the cell down the concentration gradient that has been set up by the Na^+–K^+ pump (see Figure 44.6).

Neurons generate action potentials to conduct nerve impulses.

- Changes in resting membrane potential ("depolarization") produce action potentials or nerve impulses.
- Depolarization occurs when the inside of a neuron becomes less negative compared to the resting potential (see Figure 44.8a). Hyperpolarization occurs when the inside of a neuron becomes more negative compared to the resting potential (see Figure 44.8b).
- Changes in polarity of the plasma membrane due to opening and closing of ion channels are passed down along an axon to transmit a signal as an action potential.
- Action potentials are very short-lived changes in membrane potential (see Figure 44.9). Action potentials are generated when the membrane reaches a threshold potential, Na^+ voltage-gated channels open, and Na^+ enters the cell to make the inside of the axon positive. Voltage-gated K^+ channels then open, allowing positive-charged K^+ to leave the axons to help return the membrane potential back to the resting level. As the K^+ channels open, the Na^+ channels close and cannot be opened for about 1 to 2 milliseconds, which is the refractory period. The Na^+–K^+ pump helps return the concentration of ions back to the resting levels.
- Action potentials travel down an axon by a positive feedback mechanism that stimulates adjacent regions of an

axon to generate the action potential. The Na$^+$ ions that enter during an action potential flow to adjoining regions of the axon, stimulating depolarization and the movement of the action potential along the axon (see Figure 44.10).

- The refractory period during which the Na$^+$ channels cannot act keeps an action potential moving in one direction.

- An action potential is an all-or-nothing response. The depolarization must reach a threshold level for an action potential to occur.

Ion channels can be studied directly using patch clamping.

- Patch clamping is a technique that allows the recording of voltage differences due to the movements of ions through channels in an isolated patch of plasma membrane (see Figure 44.11).

- Sakmann and Neher developed patch clamping in the 1980s and received a Nobel Prize in 1991.

Action potentials can jump down an axon.

- Some axons are myelinated. The protein myelin wrapped around some axons insulates portions of them from depolarization (see Figure 44.12).

- Nodes of Ranvier are gaps in the myelin wrapping at which depolarization can occur. Depolarization jumps from node to node, increasing the speed of transmission of an action potential along the axon in a process known as saltatory conduction.

Neurons communicate across synapses.

- Transfer of information from one cell to another by either a chemical or electrical message occurs in a small junction called a synapse. Presynaptic nerve cells send a message across a synapse to postsynaptic cells.

- A chemical synapse uses a chemical messenger, or neurotransmitter, to communicate between the presynaptic and postsynaptic cells. The neurotransmitter is produced in the axon terminal and packaged into vesicles. The neurotransmitter is released into the synaptic cleft when the vesicle binds with the presynaptic membrane. The neurotransmitter crosses the synaptic cleft and binds with receptors on the surface of the postsynaptic cell.

Acetylcholine is a common synaptic neurotransmitter.

- The neurotransmitter acetylcholine is the chemical messenger carrying information between motor neurons and muscle cells at neuromuscular junctions (see Figure 44.13). Acetylcholine is enclosed in a vesicle at the presynaptic synapse, which fuses with the membrane to release the acetylcholine into the 20–40-nm-wide gap between the two cells, the synaptic cleft.

- Ca^{2+} channels open when the action potential reaches the axon terminal, causing Ca^{2+} to rush in and regulate the fusing of the acetylcholine-containing vesicles to the presynaptic membrane.

- Acetylcholine released into the synaptic cleft binds with receptors in the postsynaptic membrane, the motor end plate, which opens Na$^+$ channels, resulting in depolarization of the motor end plate (see Figure 44.14).

- Acetylcholinesterase breaks down acetylcholine in the synapse junction to halt the action of the released acetylcholine.

Synapses can either depolarize or hyperpolarize postsynaptic membranes.

- Synapses in vertebrates can be excitatory and depolarize the postsynaptic membrane or inhibitory and hyperpolarize the postsynaptic membrane. Hyperpolarization occurs by the neurotransmitters gamma-aminobutyric acid (GABA) or glycine, which activate Cl$^-$ channels.

- Neurons may receive synaptic inputs from many neurons. Excitatory and inhibitory postsynaptic potentials are summed by adding simultaneous potentials (spatial summation) or by summing rapid firing of one postsynaptic potential (temporal summation) (see Figure 44.15).

- The region of the cell body at the base of the axon, called the axon hillock, is the "decision-making" area of a neuron. If the axon hillock is depolarized, the axon will fire an action potential.

Two main types of neurotransmitter receptors exist.

- Ionotropic receptors found on the postsynaptic membrane are ion channels that are activated by the binding of the neurotransmitter.

- Metabotropic receptors found on the postsynaptic membrane act by initiating a second messenger, such as G proteins, to activate ion channels (see Figure 44.16).

- Metabotropic receptors are categorized depending on the neurotransmitter that activates them. For example, muscarinic receptors are stimulated by acetylcholine.

- There are over 25 neurotransmitters, including amino acids, dopamine, norepinephrine, serotonin, peptides, and nitric oxide (see Table 44.1). One neurotransmitter can act on a number of different receptors. The action of a neurotransmitter depends on the receptor to which it binds.

- In the CNS, the neurotransmitters are glutamate, glycine, and GABA.

- The action of neurotransmitters is stopped by enzymes, diffusion away from the synaptic cleft, or active uptake.

NMDA and AMPA receptors may have a role in memory.

- Two types of receptors—NMDA and AMPA—may play a role in learning. NMDA receptors allow for a slow, long-lasting influx of Na$^+$, whereas the AMPA receptors allow for rapid influx. The NMDA receptors also let Ca^{2+} into the postsynaptic cell, resulting in long-term cellular changes.

- Both of these receptors are excited by glutamate.

- Potentiation is an enhanced response in the postsynaptic neuron with repeated stimulation (see Figure 44.18). Long-term potentiation may be involved in memory.

- Mice with NMDA receptors that stayed open longer than usual learned and remembered tasks better than normal mice.

Electrical synapses connect some neurons.

- Electrical synapses, or gap junctions, are formed by direct contact between adjacent neurons.

- Two neurons forming an electrical synapse are joined by connexons, which are tunnels between the two neurons allowing ions to pass between the two cells.

- Electrical synapses (gap junctions) are very fast connections and can transmit an action potential in either direction, making them good for rapid communication.

- Gap junctions are less common than chemical synapses. With electrical synapses, there can be no temporal summation of synaptic inputs. Electrical synapses also require large areas of contact and cannot be inhibitory.

The Big Picture

- The neuron, with support from surrounding glial cells, is the functional unit of the nervous system. The neuron is composed of a cell body, dendrites, and an axon. The membrane of a neuron has a difference in voltage across it. Nerve impulses are passed down the axon of a neuron as action potentials. An action potential is a temporary disruption of the "battery-like" state of the resting neuron membrane due to the opening and closing of sodium and potassium voltage-gated channels. The action potential is an all-or-nothing response that occurs when the depolarization of an axon reaches a threshold level.

- At the end of the axon synapse, information flow is controlled through excitation or inhibition of synapses. There are many different neurotransmitters that transmit a nerve impulse from the presynaptic cell to the postsynaptic cell. The action of a specific neurotransmitter will depend on the receptor with which it binds.

Common Problem Areas

- One of the most difficult challenges when learning about the nervous system is understanding how the resting membrane potential and action potential are produced. Remember that the neuron at rest is like a battery and that the charge of the membrane is dependent on the ions that are on either side and moving across the membrane.

- Students often have difficulty with the function of a synapse and how the action potential is transmitted across the synapse. Remember that in many cases the signal goes from an electrical signal, to a chemical signal, and back to an electrical signal.

Study Strategies

- Students should understand the concept of "pre-" and "postsynaptic" neurons. Try to remember the sequence of events and relate them to the function of the neuron transmitting a stimulus from one cell to another.

- One of the major problems that students have is in understanding how the membrane potential is set up. Students should understand what each important anion and cation does at rest and during the action potential. Then, piece by piece, put together a sense of how the neuron membrane is charged at rest, and what happens during an action potential.

- Review the following tutorials and activity on the Student Website/CD:
  Tutorial 44.1 The Resting Membrane Potential
  Tutorial 44.2 The Action Potential
  Tutorial 44.3 Synaptic Transmission
  Activity 44.1 Concept Matching: Neurotransmitters

Test Yourself

Knowledge and Synthesis Questions

1. The extensions of postsynaptic neurons that provide the main receptive surface for presynaptic neurons are
   a. nuclei.
   b. somas.
   c. axons.
   d. dendrites.

2. The substance that wraps the axon of many neurons and provides for increased conduction speed is
   a. dendrase.
   b. histamine.
   c. acetylcholine.
   d. myelin.

3. The long extensions from the neurons that provide the pathway for action potentials to synapse are
   a. dendrites.
   b. cell bodies.
   c. axons.
   d. presynaptic membranes.

4. The threshold of a neuron is the
   a. amount of inhibitory neurotransmitter required to inhibit an action potential.
   b. membrane voltage at which an axon potential will be suppressed.
   c. amount of excitatory neurotransmitter required to elicit an action potential.
   d. membrane voltage at which the membrane potential develops into an action potential.

5. When a membrane is at the resting potential, the concentration of
   a. sodium and potassium ions is higher on the inside of its membrane.
   b. sodium and potassium ions is higher on the outside of its membrane.
   c. sodium ions is higher on the inside of its membrane and of potassium ions is higher on the outside.
   d. sodium ions is higher on the outside of its membrane and of potassium ions is higher on the inside.

6. Glial cells are specialized to do all of the following *except*
   a. receive neural impulses.
   b. insulate axons.
   c. supply neurons with nutrients.
   d. help maintain a proper ionic environment for the neuron.

7. The cells that create the blood–brain barrier, keeping toxic substances from entering the brain are _____ and belong to a type of neural tissue called _____.
   a. endothelial cells; Schwann cells
   b. astrocytes; glial cells
   c. glial fibers; axons
   d. dendrites; synapses

8. A particular disease of the nervous system specifically involves the Ca^{2+} ion channels at the chemical synapses of motor neurons where neurotransmitter is stored and released. In other words, this disease affects the
   a. axon terminals of the presynaptic cell and the release of acetylcholine.
   b. axon terminals of the postsynaptic cell and the release of K^+ ions.
   c. electrical synapses.
   d. axon terminals of the presynaptic cell and the release of K^+ ions.

9. Which of the following statements about gap junctions or electrical synapses is *false*?
   a. Connexons form molecular tunnels between two cells.
   b. Electrical synapses cannot be inhibitory.
   c. Electrical synapses do not allow for temporal summation.
   d. Electrical transmission is very slow.

10. One group of neurotransmitter receptors is the metabotropic receptors. Which of the following statements about metabotropic receptors is *false*?
    a. Muscarinic receptors are metabotropic receptors.
    b. Metabotropic receptors are coupled with A proteins.
    c. Metabotropic receptors are not ion channels.
    d. Muscarinic receptors bind with acetylcholine.

11. The rapid depolarization of a neuron during the first half of an action potential is due to the
    a. exit of K^+ ions from the cell through gated potassium channels.
    b. rapid reversal of ion concentration caused by the action of the sodium–potassium pump.
    c. entry of Na^+ ions into the cell through gated sodium channels.
    d. movement of both Na^+ and K^+ ions through appropriate open channels.

12. The refractory period of a neuron results from
    a. the period when the sodium–potassium pump is nonfunctional.
    b. activation of voltage-gated chloride channels.
    c. closing of inactivated voltage-gated sodium channels.
    d. the action potential reaching the synapse.

13. Which of the following statements about the process of summation in a neuron is *false*?
    a. Slight perturbations of the membrane potential spread across the postsynaptic cell body.
    b. Axons that terminate closer to the axon hillock have more influence on the summation process than those that don't.
    c. Summation essentially consists of comparing the total of excitatory and inhibitory postsynaptic inputs.
    d. The concentration of voltage-gated sodium channels is highest in the dendrites of the postsynaptic cell.

14. The amount of neurotransmitter released from a presynaptic cell is a function of
    a. the voltage of the individual action potential in the presynaptic neuron.
    b. the frequency of arriving action potentials in the presynaptic neuron.
    c. whether the action potential is excitatory or inhibitory.
    d. the receptors in the postsynaptic cell.

15. The electrical events labeled as EPSPs are the result of
    a. hyperpolarization of the postsynaptic membrane.
    b. depolarization of the postsynaptic membrane.
    c. hyperpolarization of the presynaptic membrane.
    d. depolarization of the presynaptic membrane.

Application Questions

1. You are trying to determine the role of the potassium channels in a neuron. To do this you have knocked out all the functional potassium channels, and you depolarize the membrane potential. What will happen to the membrane potential after depolarization?

2. The active ingredients in many nerve gases belong to a class of chemicals called anticholinesterases. Suggest a possible synaptic mechanism to explain how these chemicals can damage an animal's nervous system.

3. The following diagram represents a monosynaptic reflex loop. Label the sensory neuron, motor neuron, and effector in the simple neural circuit. Also indicate which part of the circuit is the sensory side and which is the motor side.

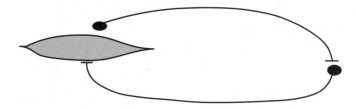

4. Explain how an action potential travels faster down an axon wrapped in myelin than one without myelin.

Answers

Knowledge and Synthesis Answers

1. **d.** The neuron is composed of a cell body, an axon, and dendrites. The dendrites form synapses with presynaptic cells to create the junction where information from one neuron is transferred to another neuron.

2. **d.** The cells that coat the axon of some neurons are myelin. Myelin is a specific type of protein found only in glial cells.

3. **c.** The neuron is composed of the cell body, the dendrite, and the axon. The axon carries action potentials away from the cell body to the synapses.

4. **d.** For an action potential to occur in an axon, the membrane must be depolarized above a certain level. This level is known as the threshold.

5. **d.** The resting potential of a neuron membrane occurs when the sodium ion concentration is higher on the outside and the potassium ion concentration is higher on the inside.

6. **a.** Glial cells perform many functions in the nervous system, but they do not receive neural impulses.

7. **b.** The blood–brain barrier is formed by astrocytes that wrap around the blood vessels traveling through the brain. Astrocytes are a special kind of glial cell.

8. **a.** If the disease acts on a chemical synapse where the neurotransmitter is stored and released, it is affecting the axon terminals of the presynaptic cell. Ca^{2+} channels are involved in regulating the release of acetylcholine by allowing Ca^{2+} to enter the presynaptic cell and promoting the fusing of acetylcholine-containing vesicles to the membrane.

9. **d.** Electrical synapses join two cells with protein tunnels known as connexons. These junctions provide for very fast transmission between cells.

10. **b.** Metabotropic receptors are coupled with G proteins that bind GDP. When the receptor binds a neurotransmitter, the GDP is turned into GTP, and a subunit of the G protein separates and activates effector proteins.

11. **c.** The first step in an action potential is the influx of
Na$^+$ ions leading to a depolarization of the axon membrane. Na$^+$ ions rush into the cell due to the higher concentration outside of the cell and the negative membrane
potential.

12. **c.** After the spike of the depolarization, the sodium voltage-gated channels close. One of the properties of these
channels is that they will open again only after a short
delay. This short delay is known as the refractory period
when the sodium voltage-gated channels are inactive.

13. **d.** Dendrites, and most of the cell body, have few gated
sodium channels. These channels mediate the action
potentials that travel down the axon, where their levels
are high.

14. **b.** Neurotransmitter is released from the presynaptic cell
when an action potential reaches the synapse. Increases
in the frequency of arriving action potentials in the presynaptic neuron will increase the release of neurotransmitter into the synapse.

15. **b.** Excitatory postsynaptic potentials (EPSPs) make it
easier for an action potential to occur, so they depolarize
the postsynaptic membrane.

Application Answers

1. The potassium voltage-gated channels are responsible for
setting up the resting potential of a membrane. Potassium
ions have a tendency to diffuse out of the cell, leaving a
negative charge inside. Knocking out the function of the
potassium voltage-gated channels would result in the
cell's being unable to maintain resting potential. If the cell
was depolarized by the opening of sodium voltage-gated
channels, it might not repolarize because the potassium
channels that help to repolarize the membrane would not
be functioning.

2. The neurotransmitter at many synapses is acetylcholine.
Acetylcholine transmits the action potential from a presynaptic cell to a postsynaptic cell. Acetylcholinesterase is
found in the synaptic cleft and it cleaves acetylcholine to
help remove it from the synaptic cleft after an action
potential. A nerve gas composed of anticholinesterases
would block the action of acetylcholinesterase, causing
acetylcholine to build up in the synaptic cleft. This buildup would mean that the receptors on the postsynaptic
cell would remain bound with acetylcholine.

3.

4. The conduction of an action potential down a myelinated
axon is known as saltatory conduction. The myelin acts
to insulate areas of the axon, preventing depolarization.
The areas of the axon between the myelin are known as
nodes of Ranvier. Depolarization can only occur at these
nodes. As the action potential moves down a myelinated
axon, the influx of sodium ions at one node diffuses
down the axon. This results in the depolarization of the
next node of Ranvier. Depolarization can occur only in
the downstream nodes because the upstream nodes are in
a refractory period. As a result, the action potential
moves quickly down the axon to the synapse.

45 *Sensory Systems*

Important Concepts

Sensory cells detect stimuli and transmit the information to the CNS.

- Sensory cells are modified neurons specialized to detect internal and external stimuli and to transmit information concerning these stimuli as action potentials to different sites in the central nervous system (CNS).
  - Sensory cells are involved in maintaining homeostasis by transmitting information about the status of the internal environment to the CNS.
- Sensory organs are groups of sensory cells that, along with other cells, collect, filter, and amplify stimuli.
- Sensory transduction is the process by which a mechanical, thermal, or chemical stimulus to a sensory cell is transformed ("transduced") into an action potential. Sensory transduction can occur by direct activation of receptor proteins that open or close ion channels to change membrane potential. Alternatively, the stimulus can activate a second messenger within the cell that couples with a G protein to eventually open ion channels.
- Receptor potentials are transmitted into action potentials by the release of a neurotransmitter or by generation of action potentials within the sensory cell though the opening of Na^+ voltage-gated channels.
- Many sensory cells have diminished responses to a stimulus over time in a process known as adaptation. Adaptation allows organisms to ignore background conditions and focus on new information.

Chemoreceptors produce action potentials in response to a chemical stimulus.

- Smell, taste, and chemicals circulating within the body stimulate chemoreceptors.
- Insects use chemical signals called pheromones to attract mates by remotely stimulating their target's chemoreceptors (see Figure 45.3). The concentration of the pheromone released by a female creates a gradient that provides information about her specific location.
- Olfaction is the sense of smell. The olfactory sensors of vertebrates are neurons with the axon extending to the olfactory bulb of the brain and the dendrites exposed as hairs to the environment (see Figure 45.4). Olfactory receptor proteins are found on the hairs, and each receptor binds with specific odorants. The ability to discriminate many different odors is due to the high number of

specific receptors. The strength of a smell is related to the number of odorant molecules that bind to receptors.

- Binding of an odorant results in depolarization of the cell through a G protein that activates a second messenger that then opens sodium channels.
- Gustation is the sense of taste. Taste buds contain sensory cells that are not neurons (see Figure 45.5). Binding of the stimulus to receptor proteins causes a change in membrane potential and the release of neurotransmitters that stimulate the underlying sensory neurons.

Mechanoreceptors detect mechanical force.

- Mechanical force causes distortion of the membranes of mechanoreceptors, causing ion channels to open and an action potential to be generated.
- The skin has a number of different types of mechanoreceptors (see Figure 45.6). Meissner's corpuscles are nerve endings located in the outer layer of skin and are sensitive to light touches of the skin. Merkel's discs adapt slowly and provide information about objects touching the skin. Pacinian corpuscles and Ruffini endings are deep in the skin and respond to vibrations.
- Stretch receptors are mechanoreceptors that sense position and stress on limbs. Muscle spindles are stretch receptors in the connective tissue of muscles that perceive muscle stretch. The Golgi tendon organs are found in the tendons and ligaments and provide information about forces in contracting muscles. Collectively, these mechanoreceptors provide information on body posture and limb position, as well as muscle and tendon tension.
- Hair cells are mechanoreceptors that release neurotransmitters in response to the bending of stereocilia on the hairs. In mammals, the inner ear uses hair cells in the semicircular canals and the vestibular apparatus to maintain equilibrium. The lateral line system of fish is composed of hair cells located under the scales that detect movements around the fish (see Figure 45.8).
- The inner ear of vertebrates has hair cells that detect a body's position relative to gravity (see Figure 45.9). In the semicircular canal of the ear, flow of semicircular fluid shifts the cupulae containing the hair cells when the head changes position. In the vestibular apparatus, layers of calcium carbonate resting on hair cells are moved by gravity when the head changes position. Both of these stimuli to the hair cells inform the organism of its position in relation to gravity.

The auditory system contains mechanoreceptors specialized for sound reception.

- The auditory system takes in sound as pressure waves and transforms them into action potentials (see Figure 45.10).

- The ear pinna is the outer portion of the mammalian ear that directs sound waves into the auditory canal. The end of the auditory canal is covered with a tympanic membrane, which vibrates and transmits sound waves to ear bones (ossicles) in the middle ear. The ossicles are the malleus (hammer), incus (anvil), and stapes (stirrup), and together they transmit sound waves to the membrane called the oval window.

- Sound travels through the oval window into the fluid-filled cochlea, where pressure waves are turned into action potentials. The cochlea, or inner ear, of a mammal is a three-canal chamber with two membranes: Reissner's membrane and the basilar membrane divide the chamber. The organ of Corti contains hair cells with stereocilia that are in contact with a tectorial membrane that transduces movement of the basilar membrane into action potentials to the auditory nerve (see Figure 45.10).

- The upper and lower chambers of the cochlea are separated at the distal end by the basilar membrane. Sound waves enter through the oval window as pressure waves. The waves move down the cochlea, reach the end, and move back down, where they are absorbed by the flexible membrane of the round window.

- If the oval window vibrates at faster frequencies, the wave cannot make it all the way down the cochlea and it crosses the basilar membrane, causing it to flex. The flexing of the membrane stimulates hairs and the organ of Corti. Different pitches of sound cause the basilar membrane to flex at different locations, stimulating different hairs (see Figure 45.11).

Light-sensitive pigments are used in photoreceptors.

- Rhodopsin is a family of pigments made up of the protein opsin and the light-absorbing prosthetic group 11-*cis*-retinal (see Figure 45.12). The 11-*cis*-retinal absorbs photons of light and changes conformation to all-*trans*-retinal, causing opsin to change conformation and become photoexcited rhodopsin.

- Photoexcited rhodopsin triggers a cascade that ultimately leads to changes in membrane potential and the photoreceptor's response to light. Excited rhodopsin activates the G protein transducin. Transducin converts cyclic GMP (which opens Na^+ channels) to 5'-GMP, allowing Na^+ channels to close and the cell to hyperpolarize.

- In vertebrate eyes, rod cells are photoreceptor cells that contain an inner segment, a synaptic terminal, and an outer segment with many rhodopsin molecules (see Figure 45.13*a*). Rods are found in the retina, along with a layer that transduces visual information into action potentials.

- Rod cells become hyperpolarized in response to light and transmit this by decreasing the levels of neurotransmitter released (see Figure 45.13*b*).

Invertebrates use rhodopsin in their visual systems.

- Flatworms have photoreceptor cells organized into eye cups which are used to orient away from light sources.

- Arthropods have compound eyes with 800 to 10,000 ommatidia per eye. The ommatidia contain light-sensitive photoreceptors called retinula cells (see Figure 45.15). The inner border of the retinula cells contains rhodopsin-containing microvilli.

Components of the image-forming eyes of vertebrates

- The vertebrate eye is surrounded by the sclera, formed from connective tissue (see Figure 45.16). The cornea is the transparent sclera through which light passes.

- The iris controls the amount of light entering the eye through the pupil.

- The lens focuses images onto the retina at the back of the eye by changing from a spherical to a flattened shape. We focus on near and far objects by the contraction and relaxation of ciliary muscles that round up or flatten the lens to bring objects into focus (see Figure 45.17).

- The center of the retina is the fovea, an area with the highest density of photoreceptor cells. A blind spot with no photoreceptors in the retina is located where the blood vessels and optic nerve exit the back of the eye.

- The human retina contains both rods (which are more light-sensitive) and cones (which absorb light of various wavelengths, allowing for color vision). Cones have different opsin molecules that absorb blue, green, yellow, or red and are the only cells found in the fovea (see Figure 45.19).

The vertebrate eye focuses light onto photoreceptors.

- There are five layers of neurons in the retina, with the photoreceptive rods and cones in the last layer (see Figure 45.20).

- The first layer of cells consists of ganglion cells (which create the action potential) and the axons of the optic nerve. Bipolar cells are stimulated by neurotransmitters from the photoreceptors to transmit the signal from the photoreceptor to the ganglion cells by the release of a neurotransmitter.

- The two other layers are the horizontal cells (which connect adjoining groups of photoreceptors and bipolar cells) and amacrine cells (which connect adjoining groups of bipolar cells and ganglion cells).

- One ganglion cell receives and integrates the information from many groups of photoreceptors that make up a circular receptive field with a center and a surround (see Figure 45.21). On-center receptive fields excite ganglion cells and are connected by the bipolar cells. The off-center receptive fields inhibit ganglion cells and connect though horizontal cells. Receptive fields of ganglion cells can overlap.

Some animals can sense another world.

- Humans can see light from only a narrow region of the electromagnetic radiation of the environment.

- Insects are sensitive to ultraviolet light, which allows them to see ultraviolet light patterns on many flowers.

- Some snakes are sensitive to infrared radiation, detecting heat differences in the environment.

- Bats and marine mammals use echoes of sound to visualize the environment.

- The lateral lines of some fishes are able to pick up electrical fields that help locate prey.

- Elephants can hear and communicate with very low frequency sounds called infrasound.

The Big Picture

- Sensory structures work by converting some form of stimulus—mechanical, chemical, light, sound—into action potentials in the nervous system, which are then interpreted by the central nervous system as a perceived sense.

- Receptors are named on the basis of their sensitivity. Chemoreceptors respond to chemical stimulation, mechanoreceptors respond to mechanical stimulation, auditory receptors respond to sound, and visual receptors respond to light.

- Different animals have different sensitivities of senses, as well as different types of senses.

Common Problem Areas

- The action potentials produced in the ear, eye, knee, or stomach are identical. The action potentials coming from the eye, for example, are interpreted as light because of the region of the brain that receives and analyzes them.

- Many of the more complexly structured receptors have both neural and non-neural components. The non-neural components (e.g., the ear pinnae) help channel or otherwise alter or filter the stimulus that will arrive at the neural component of the receptor.

Study Strategies

- The travel of sound in the ear can be very confusing. View the cochlea in the uncoiled form as in Figure 45.11. This will help you to visualize how pressure waves of different wavelengths produce different sounds.

- All of the senses have what appear to be very different mechanisms for the transmission of information to the brain. To help sort all of this out, remember that there are only a few types of receptors that respond to stimuli and that they all generate action potentials. The steps in sensory transduction are also similar between all the different sensory systems.

- Review the following tutorial and activities on the Student Website/CD:
  Tutorial 45.1 Sound Transduction in the Human Ear
  Activity 45.1 Structures of the Human Ear
  Activity 45.2 Structure of the Eye
  Activity 45.3 Structure of the Retina

Test Yourself

Knowledge and Synthesis Questions

1. An electrode is inserted into a chemosensory nerve leading away from a taste bud in the mouth of a dog. A mild acid solution is then flushed continuously over the sensors associated with this nerve. Initially the nerve responds to this stimulation but over time ceases to carry action potentials. This observation would best be explained by
   a. translocation.
   b. adaptation of the sensory cells.
   c. depletion of neurotransmitter in the sensory nerve.
   d. second messenger influences that increase cell membrane potentials.

2. Which of the following statements about sensory cells is *false*?
   a. Mechanoreceptors detect stimuli that distort membranes.
   b. Electroreceptors enable some fish to detect weak electric fields generated by prey.
   c. Chemoreceptors are involved in smell, taste, and hearing.
   d. Photoreceptors exhibit a conformational change when stimulated by light.

3. Silkworms use chemosensory molecules for mate attraction. These are known as
   a. rhodopsin.
   b. hormones.
   c. pheromones.
   d. G proteins.

4. Stretch receptors in the aorta and carotid artery sense changes in arterial pressure. These receptors are
   a. chemoreceptors.
   b. thermoreceptors.
   c. electroreceptors.
   d. mechanoreceptors.

5. All but which of the following systems employs hair cells as its transducer?
   a. Meissner's corpuscle
   b. Lateral line
   c. Organ of Corti
   d. Vestibular apparatus

6. Which of the following statements about human gustation is *false*?
   a. Taste bud cells are relatively short lived because of the high degree of abrasion they encounter.
   b. Taste buds are confined to the oral cavity.
   c. Changes in the membrane potential of the taste bud sensory cells cause them to release neurotransmitter onto the dendrites of sensory neurons.
   d. Humans perceive only three categories of tastes: sweet, sour, and bitter.

7. Which of the following statements about the photosensitive molecule rhodopsin is *false*?
   a. Opsin is converted from the 11-*cis* to the all-*trans* form upon absorbing a photon of light.
   b. The retinal is the light-absorbing group.
   c. Photoexcited rhodopsin triggers a cascade of reactions that ultimately alters the membrane potential of a photoreceptor cell.
   d. None of the above

8. Which of the following cells in the human visual system send information directly to the brain?
   a. Amacrine cells
   b. Bipolar cells
   c. Ganglion cells
   d. Rods and cones

9. Through which of the following cell layers must a photon of light pass before striking a cone cell in the eye of a hawk?
   a. Amacrine
   b. Bipolar
   c. Ganglion
   d. All of the above

10. Which of the following statements about animal vision is *false*?
    a. Photoreceptors in simple invertebrates produce image vision.
    b. There are several types of cones in vertebrate retinas, each of which detects different wavelengths.
    c. Ommatidia make up the compound eye of an arthropod.
    d. Although retinal actually absorbs the photon, the wavelength it absorbs best is determined by opsin.

11. Which of the following about stimulation of an off-center receptive field is true?
    a. Stimulating the surround inhibits it.
    b. Stimulating the surround excites it.
    c. Stimulating the center excites it.
    d. Both a and c

12. Which of the following is *not* a membrane found in the cochlea?
    a. Reissner's membrane
    b. Tectorial membrane
    c. Tympanic membrane
    d. Basilar membrane

13. Which of the following about receptor potentials is *false*?
    a. They are changes in the resting membrane potential of a sensory cell in response to a stimulus.
    b. The receptor potential spreads from the cell body of a sensory cell to the axon hillock, where action potentials are generated.
    c. They must be converted to action potentials to travel long distances.
    d. A receptor potential always releases a neurotransmitter to induce an action potential.

Application Questions

1. This past winter was an awful year for colds. You had a cold for a number of weeks, during which time your sense of smell was diminished. What was the cause of your loss of smell?

2. After a loud rock concert your hearing appears to be dampened. What portion of the ear has been altered to dampen your hearing?

3. You have just given a presentation in your biology class that had many elaborate red- and green-colored slides. Afterward, a male friend tells you that he could not see any of the differences you were reporting. Why could your friend not see the differences?

4. Label each of the following structures on the diagram of the human eye shown below: sclera, cornea, iris, pupil, lens, retina, fovea, and blind spot.

Answers
Knowledge and Synthesis Answers

1. **b.** When a sensor cell is stimulated by an unchanging, steady-state stimulus, it will become adapted. This allows the sensory system to ignore the unchanging stimulus while still being able to respond to new information.

2. **c.** Chemoreceptors are involved in smell and taste but not hearing.

3. **c.** Pheromones are chemical molecules that a female releases into the environment. The male uses chemoreceptors to follow the pheromone to the source.

4. **d.** The stretch receptors of the aorta, which detect changes in pressure, are examples of mechanoreceptors.

5. **a.** The Meissner's corpuscle of the skin does not use hair cells to sense a stimulus. The cell membranes of the Meissner's corpuscle deform in response to light touching of the skin.

6. **d.** Humans can perceive five tastes: sweet, salty, sour, bitter, and umami (a meaty taste). The combination of taste and smell provides the complex subtle flavors of the food we eat.

7. **a.** Rhodopsin contains two groups: the protein opsin and the light-sensitive group retinal. Retinal, not opsin, is converted from the 11-*cis* to the all-*trans* form upon absorbing a photon of light. Opsin does change conformation in response to a change in the rhodopsin to signal the detection of light.

8. **c.** The ganglion cells transmit information from the bipolar cells to the brain. The axons of the ganglion cells connect with the optic nerve.

9. **d.** The photoreceptive cells are located at the back of the retina. Light must pass through a layer of ganglion cells, a layer of amacrine and bipolar cells, and a horizontal cell layer.

10. **a.** The photoreceptors in simple invertebrates such as the flatworm only detect light–dark contrast.

11. **b.** The stimulation of the surround of an off-center receptive field excites it, whereas the stimulation of the center inhibits it. The reverse is true of an on-center receptive field.

12. **c.** Although the tympanic membrane is found in the human ear, it is not found in the cochlea. The tympanic membrane is the membrane that transmits sounds from the auditory canal and the middle ear.

13. **d.** The receptor potential does not always release a neurotransmitter to induce an associated neuron to generate an action potential. Sometimes the receptor potential generates action potentials within the sensory cell itself.

Application Answers

1. Your sense of smell depends on olfactory cilia receptors binding with odorant molecules. This triggers an action potential that is sent to the olfactory bulb. The olfactory cilia line the surface of the nasal epithelium. Usually there is a thin layer of protective mucus covering this epithelium. However, when you have a cold the production of mucus increases. This increased mucus covers the epithelium and the olfactory cilia, making it more difficult for odorant molecules to penetrate the mucus and

reach the cilia. Thus you are not receiving as much stimulation, and your sense of smell is decreased.

2. Sound is transmitted into the ear and into the fluid-filled cochlea. Sounds that are too loud will eventually damage the hair cells of the organ of Corti. Because these hair cells transmit the sound to the organ of Corti, any damage will decrease the response of the organ of Corti to sounds. This damage to the hair cells is cumulative and permanent.

3. Your friend is red-green color blind. The cones in our eyes allow us to see color. We have cones for red, green, and blue. Your friend either has red and green cones that do not function properly, or lacks them altogether.

4.

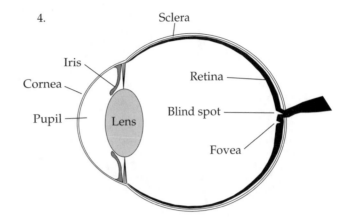

46 The Mammalian Nervous System: Structure and Higher Functions

Important Concepts

The mammalian nervous system consists of the central nervous system and the peripheral nervous system.

- The brain and the spinal cord together constitute the central nervous system (CNS).

- The nerves of the peripheral nervous system transport information between every tissue in the body and the CNS.

- A nerve is a bundle of axons that carries information in both directions between the organs of the body and the CNS.

- The peripheral nervous system is composed of afferent nerves that carry information such as hearing and vision to the CNS, and efferent nerves that carry information from the CNS to muscles and glands (see Figure 46.1).

The CNS develops from the embryonic neural tube.

- The neural tube is a tube of neural tissue running down the length of a vertebrate embryo in its early developmental stages (see Figure 46.2). The anterior end of the neural tube will develop into the main parts of the brain—the hindbrain, the midbrain, and the forebrain—with the spinal cord developing from the rest of the neural tube.

- The hindbrain is made up of the medulla and pons, which are involved in control of physiological functions, and the cerebellum, which is involved in coordinating commands leaving the brain with the actual actions of the muscles.

- The midbrain receives and processes auditory and visual information.

- The forebrain consists of a central diencephalon region and a surrounding telencephalon region. The diencephalon contains the thalamus, a relay station for sensory information, and the hypothalamus, which is involved in maintenance of homeostasis. The cerebral hemispheres (cerebrum) make up the telencephalon.

The spinal cord is the main neural highway between the body and the brain.

- The spinal cord is a bidirectional neural pathway for information flow between the peripheral nervous system and the brain.

- The spinal cord also converts some afferent information from the peripheral nervous system into efferent information sent back to the peripheral nervous system in a process known as a spinal reflex (see Figure 46.3).

- A monosynaptic reflex, such as the knee-jerk reflex, involves only an afferent neuron, an efferent neuron, and one synapse. Leg muscle stretch receptors trigger the sensory neuron to conduct action potentials to the spinal cord upon stretching. The action potential is passed through the synapse of the sensory and motor neuron located in the central gray matter of the spinal cord.

- Flexor muscles, which flex limbs, and extensor muscles, which straighten limbs, control limb movement. The motor neurons that stimulate these muscles are antagonistic and polysynaptic, with each sensory neuron stimulating one motor neuron while inhibiting the other motor neuron at two distinct synapses.

The reticular system and limbic system interact with the forebrain.

- The reticular system contains groups of neurons forming a nucleus. Such nuclei are located through the medulla, pons, and midbrain.

- Many of the connections involved in functional control of the body occur in the reticular system, along with the control of sleep and waking.

- The limbic system is involved in instincts and emotions (see Figure 46.4). A portion of the limbic system known as the hippocampus helps to transfer short-term memory to long-term memory.

The cerebrum is the largest portion of the mammalian brain.

- The cerebral hemispheres are covered by a convoluted cerebral cortex, a sheet of gray matter that processes sensory information and higher-order information in the association areas (see Figure 46.5). The convolutions are ridges known as gyri and valleys known as sulci. Underneath the gray matter is white matter containing axons that connect cell bodies in the cortex and other areas of the brain. Each cerebral hemisphere is broken down into four main regions: the temporal lobe, the frontal lobe, the occipital lobe, and the parietal lobe. Each region has an association area that is involved in various functions.

- The size of the cerebral cortex is larger and its and complexity greater in animals that exhibit complex behaviors.

- The temporal lobe processes auditory information, and the association areas are involved with recognition, identification, and naming. Damage to the temporal lobe results in a condition in which individuals are unable to identify a stimulus even though they can perceive it. Integration of spoken language can also be diminished with damage to the temporal lobe.

- The frontal lobe contains the primary motor cortex made up of axons that project to muscles in the body. The primary motor cortex can be mapped according to the locations that control movements of various body parts (see Figure 46.7). The association areas of the frontal lobe are involved in planning.

- The central sulcus separates the parietal lobe from the frontal lobe. The primary somatosensory cortex is located in the parietal lobe. As with the motor cortex, the location of the body that is sensed by the somatosensory cortex can be mapped on the brain (see Figure 46.7). Contralateral neglect syndrome results from damage to the right parietal lobe and is characterized by the inability to detect stimuli from the left side of the body (see Figure 46.9).

- Visual information is processed by the occipital lobe of the cerebrum. The association areas for the occipital lobe are involved in translation of visual stimuli into language.

The sympathetic and parasympathetic systems make up the autonomic nervous system.

- The sympathetic and parasympathetic divisions of the autonomic nervous system have antagonistic effects on the organ systems they innervate and play a large role in the maintenance of homeostasis.

- The sympathetic division is involved in the fight-or-flight response of increased heart rate, blood pressure, and cardiac output. The parasympathetic division slows down heart rate and decreases blood pressure and cardiac output; however, it accelerates digestive activities.

- Preganglionic neurons of the autonomic efferent pathway have cell bodies in the brain stem or spinal cord. They use the neurotransmitter acetylcholine at synapses where the ganglion lies outside the CNS. The axon that runs from the ganglion to the target cells is known as the postganglionic neuron.

- Norepinephrine is the neurotransmitter in postganglionic neurons of the sympathetic system, whereas acetylcholine is the neurotransmitter in postganglionic neurons of the parasympathetic system.

- Both acetylcholine and norepinephrine stimulate the pacemaker cells in the heart to control heart rate.

- The parasympathetic system has preganglionic neurons that come from the brain stem and last segment of the spinal cord. In the sympathetic system they come from the middle and upper regions of the spinal cord (see Figure 46.10).

Visual images are constructed in the occipital cortex.

- Information from the retina is transferred by the optical nerve to the thalamus and then to the visual cortex located in the occipital cortex.

- The visual cortex is composed of cells with specific receptive fields associated with areas of the retina that respond to specific light patterns. Simple cells are stimulated by static bars of light with specific orientations (see Figure 46.11). Complex cells receive input from several simple cells that have receptive fields in different locations on the retina. Mental images of the world are determined by analyzing the edges of the patterns of light falling on the retina.

- The optic nerve from the two eyes joins at the optic chiasm (see Figure 46.12). At the optic chiasm, half of the axons from one eye cross over to the opposite side of the brain. Binocular vision occurs because the overlap in the field of view for each eye is transmitted to the same location in the visual cortex, the binocular cells. Binocular cells receive overlapping visual information from both eyes and interpret the disparity between the overlapping information from the two eyes to produce a three-dimensional image.

Sleep involves complex cerebral function.

- An electroencephalogram (EEG) measures the electrical activity of neurons in the cerebral cortex. An electromyogram (EMG) records electrical activity of muscles.

- There are two main states of sleep in humans: rapid-eye movement (REM) sleep and non-REM sleep (see Figure 46.13).

- Non-REM sleep has four states progressing from stage 1 to the restorative stages 3 and 4. During non-REM sleep the neurons in the thalamus and cerebral cortex become hyperpolarized due to the opening of K^+ channels. During hyperpolarization, the cells fire off in synchronized bursts due to Ca^{2+} channels deactivating, which produces a slow-wave pattern.

- During REM sleep, dreams and nightmares occur along with complete paralysis of skeletal muscles. Neurons that were hyperpolarized during non-REM sleep return to waking levels, allowing information to be processed. Afferent and efferent pathways are inhibited during REM sleep.

- During a night of sleep, the brain cycles between REM sleep and non-REM sleep, with 80 percent of sleep being non-REM sleep.

Learning and memory involve long-lasting synaptic changes.

- Long-lasting synaptic changes must occur for learning to take place. Long-term potentiation (LTP) occurs when high-frequency electrical stimulation makes circuits of the hippocampus more sensitive to subsequent stimulation. Ion channels controlled by glutamate allow both Na^+ and Ca^{2+} ions to enter neurons. This acts to enhance their responsiveness.

- Long-term depression (LTD) results when hippocampus circuits become less sensitive to low-level stimulation.

- Associative learning in animals involves linking two unrelated stimuli, as in the conditioned reflex of Pavlov's dogs. The dogs were conditioned to associate eating with the ringing of a bell, so that even in the absence of food the mere ringing of a bell stimulated salivation. Another example of a conditioned reflex involves the eye-blink reflex of rabbits. By repeatedly pairing a tone and a puff of air, rabbits are conditioned to blink in response to the tone alone. Further experiments showed that a specific part of the cerebellum was necessary for rabbits to learn the conditioned reflex (see Figure 46.14).

- Immediate memories are very short-term vivid memories of what has just happened. Short-term memories last between 10 and 15 minutes and contain less information than long-term memories. Repetition enhances the transfer of short-term memory to long-term memory.

- Declarative memory is memory of people, places, events, and things. Procedural memory is memory of how to perform motor tasks.

The language center is located in the left cerebral hemisphere.

- In most people the ability to produce and interpret language occurs in the left hemisphere of the cerebrum.

- The two cerebral hemispheres are connected by the corpus callosum. This connection allows the two hemispheres to communicate. Severing the corpus callosum results in an individual being unable to express in language the knowledge that is in the right hemisphere.

- A number of areas have been located that are important for language (see Figure 46.15). The frontal lobe contains Broca's area, which is involved in speech. Wernicke's area is located in the temporal lobe and involves sensory aspects of language. The angular gyrus is thought to be involved in integrating spoken and written language.

- Language involves the flow of information between the various regions mentioned earlier. Damage to any of the areas involved in language will break the chain of information and result in aphasia, the condition of not being able to use or understand written or spoken words.

The Big Picture

- The mammalian nervous system can be divided both anatomically and functionally. Anatomically, it consists grossly of the central and peripheral nervous systems. Functionally, it consists of the sympathetic and parasympathetic nervous systems. Afferent nerves carry information toward the central nervous system from the peripheral nervous system, and efferent nerves carry information from the central to the peripheral nervous system.

- The brain can be divided anatomically and functionally into many different regions, each responsible for its many vital actions. The "higher" brain centers of the cerebrum are responsible for conscious thought and deliberate (voluntary) movements. The "lower" brain centers such as the cerebellum, pons, and medulla regulate involuntary movements and are involved primarily in maintaining homeostasis throughout the body.

Common Problem Areas

- It is easy to confuse "afferent" and "efferent," and it is important in your understanding of the nervous system to differentiate them. Think of *a*fferent as *a*rriving, and *e*fferent as *e*xiting the reference point—usually the CNS.

- It is important to recognize that although the neurotransmitters acetylcholine and norepinephrine always have antagonistic effects on each other, there is no universal pattern as to which stimulates tissue and which inhibits tissue. For example, acetylcholine causes the smooth muscle in blood vessels in many regions of the body to relax, but causes the smooth muscle in the stomach and intestines to contract.

- It is often difficult for students to appreciate that, when dealing with sensory input to the brain, specific regions of the body "map onto" specific regions of the cerebrum. Similarly, with regard to control of movements, specific regions of the brain "map onto" specific regions of the body. Moreover, the amount of brain matter devoted to a particular body region depends on the amount of muscle control and sensors contained in that body region. Thus, a relatively small area of the cerebral hemispheres is devoted to the upper leg (which has a relatively limited range of movement and sensation), whereas the tongue, with its many sensory receptors and high degree of mobility, commands more of the tissue of the cerebrum.

Study Strategies

- There are several key terms in this chapter that occur repeatedly: parasympathetic, sympathetic, afferent, efferent, agonist, antagonist, preganglionic, postganglionic. Until you master this vocabulary, it will be difficult to put together a comprehensive picture of the nervous system. Create your own list of the crucial terms you see repeatedly, and make sure that you understand their meanings.

- Brain anatomy is complex. Start by dividing up the structures into forebrain, midbrain, and hindbrain regions. Then identify each of the subcomponents. Your understanding will be more complete if you learn the general functions of each brain section as you go along. That is, rather than learning the anatomy of the brain and then starting over to learn the functions, learn the two at the same time.

- Review the following tutorial and activities on the Student Website/CD:
  Tutorial 46.1 Information Processing in the Spinal Cord
  Activity 46.1 The Human Cerebrum
  Activity 46.2 Brain Language Areas
  Activity 46.3 Brain Structures

Test Yourself

Knowledge and Synthesis Questions

1. A man has damage to his brain that affects his ability to recognize the faces of people he knows. Where did the damage occur?
   a. Hypothalamus
   b. Temporal lobe
   c. Parietal lobe
   d. Frontal lobe

2. The secretion of hormones from an endocrine gland is most likely under the control of which of the following components of the nervous system?
   a. Autonomic
   b. Voluntary
   c. Dendritic
   d. Limbic

3. The motor cortex of the cerebrum is
   a. mapped from the head region on the lower side of the cortex to the lower part of the body on the upper side of the cortex.
   b. areas of the brain where the most sensitive body areas are represented by the largest areas of the cortex.
   c. located in the parietal lobe.
   d. None of the above

4. Which of the following statements about the sympathetic division of the autonomic nervous system is *false*?
   a. It increases heart rate.
   b. It relaxes the urinary bladder.
   c. It stimulates digestion.
   d. It increases blood pressure.

5. Which of the following is *not* part of the central nervous system?
   a. Brain stem
   b. Spinal gray matter
   c. Cerebellum
   d. Neuronal cell body of a sensory afferent

6. Observations of patients with aphasia indicate that
   a. only Broca's area is essential for normal language skills.
   b. language skills depend on proper neural flow between the temporal lobes and motor cortex.
   c. the right hemisphere dominates the production and use of language in humans.
   d. most people can control language functions from either cerebral hemisphere.

7. A friend wakes you from sleep, and you have the sensation of having just experienced a vivid dream. Which of the following would be *false*?
   a. Your hands and feet have been twitching slightly.
   b. Your eyes have been twitching.
   c. Most of your voluntary body muscles have been inactive.
   d. Your cerebral cortex was not as active as when you were awake.

8. If you want to find a complex cell in a cat's visual system, you would look in
   a. Broca's area.
   b. the spinal cord.
   c. the occipital cortex.
   d. the reticular system.

9. Which of the following statements about the peripheral nervous system is *false*?
   a. Afferent portions of the peripheral nervous system carry information to the CNS.
   b. Efferent portions of the peripheral nervous system carry information from the CNS.
   c. The peripheral nervous system communicates only with the circulatory and digestive systems.
   d. The efferent portion is broken down into voluntary and involuntary divisions.

10. Which of the following statements about the developing CNS is *false*?
    a. The CNS develops from a solid neural cylinder.
    b. The midbrain becomes part of the brain stem.
    c. The forebrain develops into both the diencephalon and the telencephalon.
    d. The hindbrain develops into the medulla, the pons, and the cerebellum.

11. In which of the following cortical lobes can association areas and long-term memory be found?
    a. Temporal
    b. Parietal
    c. Occipital
    d. All of the above

12. Which of the following animals would likely have the smallest ratio of telencephalon size to body size?
    a. Fishes
    b. Amphibians
    c. Mammals
    d. Reptiles

13. Which of the following statement about the knee-jerk reflex is *false*?
    a. It is a monosynaptic reflex.
    b. The leg extends in response to a hammer tap.
    c. Chemoreceptors sense the hammer tap.
    d. The afferent nerve travels from the receptor to the spinal cord.

14. The ridges of the cerebral cortex are called _____ and the valleys of the cerebral cortex are the _____.
    a. gyri; interneuron
    b. gyri; sulci
    c. sulci; gyri
    d. interneuron; gyri

Application Questions

1. What are the differences between the brain functions involved in reading a written sentence out loud and repeating a sentence one has just heard?

2. You have been sitting up all night watching movies in the living room of your apartment, and your roommate enters from a bedroom, apparently sleepwalking. What type of sleep is your roommate in at the moment, and how can you tell?

3. You have been eyeing hard candy in a dish and finally decide to unwrap a piece and eat it. As you begin to suck on the candy, your salivary glands begin to secrete saliva. What parts of the nervous system are involved in the sequence of events?

4. Humans have the ability to see things in three dimensions. What allows us to have binocular vision?

Answers
Knowledge and Synthesis Answers

1. **b.** The temporal lobe is involved in recognition of people and objects. A person who has had damage to the temporal lobe will not be able to identify someone by face and must use other cues.

2. **a.** Secretion of hormones by endocrine glands is involuntary. The autonomic nervous system consists of efferent pathways that link the CNS with many physiological functions.

3. **a.** The motor neurons in the head region control specific parts of the body. Parts of the body can be mapped on the motor cortex, from the head region on the lower side to the lower part of the body at the top.

4. **c.** The sympathetic division of the autonomic nervous system inhibits digestion rather than stimulating it.

5. **d.** The cell bodies of the sensory neurons are located in the periphery and send their axons to the CNS.

6. **b.** Speaking a written or heard word requires neural flow from Wernicke's area in the temporal lobe to Broca's area in the frontal lobe. Therefore, language skills depend on proper neural flow between the temporal lobes and the motor cortex.

7. **d.** During REM sleep the body is paralyzed except for twitching of muscles. During this stage of sleep the brain is as active as when you are awake.

8. **c.** The occipital cortex is the higher brain region controlling vision.

9. **c.** The peripheral nervous system is in contact with every tissue in the body sending and receiving information to the CNS.

10. **a.** The CNS develops from a hollow tube composed of neural tissues.

11. **d.** Association areas and long-term memory are present in all these regions of the cerebral cortex.

12. **a.** The fishes have the most undeveloped telencephalon of the group listed, resulting in the smallest ratio.

13. **c.** The knee-jerk reflex is an example of a monosynaptic reflex. Stretch receptors sense the hammer tap on the tendon.

14. **b.** The ridges on the surface of the cerebral cortex are the gyri and the valleys are the sulci.

Application Answers

1. Both speaking written language and repeating heard language follow similar paths in the brain. The main difference has to do with the initial region of the brain perceiving the word. In reading a word, the area at the back of the cerebrum is used to visualize it. The spoken word uses an area of the cerebrum just behind the area used for speech. Once the word has been processed by the initial centers, the path used for speaking the word is the same. Wernicke's area is stimulated, followed by Broca's area, and then the motor area.

2. Your roommate is sleepwalking during non-REM sleep. During REM sleep the skeletal muscles of the body become paralyzed, and the sleeper is not able to move. You can tell that your roommate is not in REM sleep because you will not see the eye movement typical of this stage of sleep.

3. The peripheral system contributed to both seeing the candy and the movements of the arms and legs that you used to pick it up. The parasympathetic branch of the autonomic nervous system stimulated salivation. The central nervous system was involved in the recognition and decision to unwrap and eat the candy.

4. The right side of your brain receives visual information from the left visual field, and the left side receives information from the right visual field. In the visual cortex, the cells are organized in columns that alternate between receiving information from the right and left eyes. At the borders of the columns, the inputs from the right and left eyes overlap. The cells that receive the overlap are called binocular cells, and they interpret the disparity between what the two eyes sense. This provides a three-dimensional image.

47 *Effectors: Making Animals Move*

Important Concepts

Microtubules and microfilaments generate movement at the cellular level.

- Microtubules are involved in movement of cilia and flagella and in helping to change the shape of cells.
- Small multicellular animals use cilia to move through liquids. Cilia also move fluids over the bodies of animals, helping to deliver nutrients and oxygen.
- Flagella provide movement of a body or a fluid over a body's surface.
- Actin is a microfilament that is involved, along with the protein myosin, in movement of muscles in vertebrates. Amoeboid movements are produced by the interaction of actin and myosin.

Smooth, skeletal, and cardiac muscle types are found in the human body.

- Actin and myosin are arranged in filaments in muscle cells.
- Smooth muscles are involuntary muscles comprised of long, simple cells, each with a single nucleus (see Figure 47.1). Smooth muscles are controlled by acetylcholine and norepinephrine of the autonomic nervous system (see Figure 47.2). When smooth muscle is stretched it contracts with strength proportional to the stretch of the muscle.
- Cardiac muscle is found in the heart and is composed of branched muscle cells that form a strong meshwork (see Figure 47.1). Intercalated discs add additional strength by holding the cells together. Heartbeats originate at the pacemaker cardiac muscle cells and spread rapidly through gap junctions in the muscle. The heartbeat is said to be myogenic because it is generated by the heart muscle itself.
- Skeletal muscles are striated voluntary muscles made of large muscle fibers (see Figure 47.1). Muscle fibers have many nuclei and are composed of actin and myosin arranged into myofibrils.

Muscle contraction is described by the sliding filament theory.

- The contracting unit of myofibrils is the sarcomere, which contains actin and myosin filaments and has very distinct repeating patterns (see Figure 47.3).

- In a sarcomere, the actin filaments are anchored by the Z lines and myosin filaments are found at the center in the A band.
- In relaxed muscle, the H zone and I band are the regions in which there is no overlap of actin and myosin. During muscle contraction, the Z lines move toward each other and the H zone and I band shrink in size due to the sliding of actin filaments along the myosin filaments.

The proteins myosin and actin are the key to muscle contraction.

- Myosin molecules are made of two polypeptide chains wrapped around each other, each with a globular head at one end, much like two twisted golf clubs. A myosin filament is composed of many myosin molecules (see Figure 47.4).
- Actin filaments are composed of two monomer chains in a helical arrangement, appearing as two linear strings of pearls wrapped around each other. The proteins tropomyosin and troponin are associated with actin (see Figure 47.4).
- Myosin heads change conformation when they bind to actin filaments at myosin binding sites, forming a cross-bridge connection. The conformational change in the myosin pulls the actin in toward the middle of the sarcomere. ATP then binds to an ATP binding site on myosin, resulting in the bound actin's release from the myosin and the myosin's return to the original conformation. Many myosin molecules cycle through binding with actin to shorten a sarcomere.

Calcium ions regulate myosin and actin filament movements.

- In relaxed muscle, tropomyosin and troponin cover the myosin binding sites on actin filaments, inhibiting muscle contraction. Calcium regulates contraction by binding with troponin, causing the tropomyosin to change conformation and expose the actin–myosin binding sites on the actin filaments (see Figure 47.6).
- Action potentials form at motor neuron synapses with hundreds of muscle fibers all contracting at the same time. All the fibers activated by a single motor neuron constitute a motor unit. Action potentials spread deep into the cytoplasm or sarcoplasm of the muscle through transverse tubules or T tubules that are in contact with sarcoplasmic reticulum throughout the sarcoplasm (see Figure 47.5). The sarcoplasmic reticulum takes up and

releases Ca^{2+} ions into the sarcoplasm, thereby controlling relaxation and contraction of the myofibrils.

- In smooth muscle, contraction is controlled by a calmodulin–Ca^{2+} complex. This complex activates myosin kinase, an enzyme that phosphorylates the myosin head to cause contraction. Myosin phosphatase works in the opposite direction by dephosphorylating myosin.

Skeletal muscles show graded contractions, ranging from twitches to tetanus.

- An action potential in a skeletal muscle fiber causes twitch or contraction of the muscle. Twitches can occur as discrete contractions, or if they occur frequently enough, the twitches are summed together (see Figure 47.7).

- Maximum muscle tension, or tetanus, occurs when there is a high rate of stimulation by action potentials. Temporal summation occurs when there is fast twitching of individual fibers, and spatial summation occurs when more than one motor unit stimulates contraction.

- During tetanic contraction, actin and myosin bonds cycle to help keep a muscle fiber from stretching. ATP levels control the length of a tetanic contraction because it is the molecule that provides the energy for myosin to break the bond with actin.

- Slow-twitch muscle fibers are highly resistant to fatigue because they have a lot of myoglobin, mitochondria, and blood vessels. They are also called "red muscle" (see Figure 47.8).

- Fast-twitch muscle fibers fatigue quickly but rapidly develop maximum tension. They have lower levels of mitochondria, myoglobin, and blood vessels and are called "white muscle" (see Figure 47.8).

The strength of a muscle fiber is related to its length.

- The amount of force a sarcomere can generate depends on its resting length (see Figure 47.9).

- Stretching a muscle causes the sarcomeres to shorten, resulting in less overlap between actin and myosin filaments and less force.

Exercise enhances the strength and endurance of muscle.

- Anaerobic exercise increases strength. Such exercise induces the formation of new actin and myosin filaments in existing muscle fibers and thus produces bigger fibers and bigger muscles.

- Aerobic exercise increases endurance. Such exercise increases myoglobin, number of mitochondria, density of capillaries, and enzymes involved in energy utilization.

Muscles have three systems for obtaining ATP.

- Muscles use the immediate, glycolytic, and oxidative systems to obtain ATP needed for contraction (see Figure 47.10).

- The immediate system utilizes preformed ATP and creatine phosphate.

- The glycolytic system follows the immediate system within seconds and metabolizes carbohydrates to lactate and pyruvate.

- The oxidative system is fully activated within about one minute and completely metabolizes carbohydrates or fats to water and carbon dioxide.

Invertebrates use hydrostatic skeletons and exoskeletons for support and movement.

- Many soft-bodied invertebrates have a fluid-filled body cavity that acts as a hydrostatic skeleton. Earthworms have circular muscles and longitudinal muscles that oppose each other to act on the hydrostatic skeleton and control elongation and shortening of body segments (see Figure 47.11).

- Arthropods have an exoskeleton, or hard outer shell, composed of chitin that offers protection and provides sites for muscle attachment.

- As an animal with an exoskeleton grows, it undergoes the process of molting. With each molt the old exoskeleton is shed, revealing a new one that has developed underneath.

The vertebrate endoskeleton provides a frame for support and movement.

- Endoskeletons are growing, living tissue that provide sites for muscle attachment and support for the body.

- The human skeleton is composed of a central axial skeleton and an appendicular skeleton (see Figure 47.12).

- The pliable portions of the endoskeleton, such as the framework of the nose, and the surface of the joints of the endoskeleton are composed of cartilage containing the protein collagen.

- The bone in the endoskeleton is strong and solid and is composed mainly of collagen and calcium phosphate. Bone is constantly being reworked by two types of cells, osteoblasts, which lay down new bone, and osteoclasts, which break down bone (see Figure 47.13).

- Developing bone is created as membranous bone growing on a scaffolding of connective tissue or as cartilage bone hardening from an initial cartilage state. Compact bone is solid, whereas cancellous bone is lightweight, with many cavities. Compact bone is known as Haversian bone; it is composed of concentric rings with blood vessels running through the middle known as Haversian systems (see Figure 47.15).

Movable bones come together at joints.

- There are six types of joints where bones meet, allowing for a variety of bone movement (see Figure 47.17).

- The muscles attached to bones at joints work antagonistically (see Figure 47.16). The flexor muscles bend the joints, whereas the extensor muscles straighten them.

- Two types of connective tissues hold joints and bones together. Ligaments hold bone to bone, and tendons hold muscle to bone.

Effectors can be cells other than muscle cells.

- Nematocysts are specialized cells found in hydras, jellyfish, and cnidarians that are fired at prey animals either to entangle or poison them.

- Chromatophores are specialized cells that contain pigment and allow an animal to change color.

- Glands also act as effector organs.

- Electric organs in some animals are able to produce large electrical currents.

The Big Picture

- Actin and myosin are the "universal" proteins for motion. Whether they are located in a unicellular animal or the leg muscle of a human, the molecular interactions of actin and myosin produce movement in the structures in which they reside.

- The sliding filament theory provides a model for how actin and myosin filaments slide past each other, resulting in the shortening (contraction) or lengthening (relaxation) of muscle cells. The whole process of actin and myosin interaction is tightly regulated by the movement of calcium ions into and out of the intracellular spaces of muscle cells, which in turn is activated by the arrival of action potentials in motor neurons. Muscle contraction requires the use of energy in the form of ATP.

- Muscles act in concert with skeletons that are either external (invertebrates) or internal (vertebrates). In vertebrates, bones are articulated, forming joints that provide for specialized directional movements of the tissues supported by the bones (e.g., limbs). Muscles controlling joint movement are often located in pairs acting antagonistically, with one set of muscles causing bending, or flexion, of the joint and the other causing straightening, or extension, of the joint.

Common Problem Areas

- The sarcomere is the functional unit of the muscle cell, and until you understand its fine structure—Z lines, H lines, etc.—it will be difficult for you to appreciate how the sarcomere shortens through the actions of actin and myosin.

- The way that movements of multiple muscle groups cause both flexion and extension of a joint can be confusing. Thinking of joints in terms of levers and pulleys may help you understand their actions.

Study Strategies

- The interactions of myosin, actin, troponin, tropomyosin, and calcium and the role of action potentials in stimulating muscle contraction make up a complex, multistep process.

  - First, break the process down into its constituents and learn their location and general structures.

  - Second, determine how actin and myosin move relative to each other through a series of power strokes.

  - Finally, ascertain how calcium ions released from the sarcomeres by action potentials initiate and maintain the whole process of muscle contraction.

- Review the following tutorial and activities on the Student Website/CD:
  Tutorial 47.1 Molecular Mechanism of Muscle Contraction
  Activity 47.1 The Structure of a Sarcomere
  Activity 47.2 The Neuromuscular Junction
  Activity 47.3 Joints

Test Yourself

Knowledge and Synthesis Questions

1. The small-scale movements of cilia and flagella are generated by
   a. microfilaments.
   b. microtubules.
   c. sarcomeres.
   d. None of the above

2. A motor unit is best described as
   a. all the nerve fibers and muscle fibers in a single muscle bundle.
   b. one muscle fiber and its single nerve fiber.
   c. a single motor neuron and all the muscle fibers that it innervates.
   d. the neuron that provides the central nervous system with information on the state of contraction of the muscle.

3. Which of the following statements is *false*?
   a. Cardiac muscle is striated.
   b. Smooth muscle does not contain actin.
   c. Skeletal muscle is considered voluntary.
   d. Smooth muscle is found in the digestive tract and the walls of the bladder.

4. The oxygen-binding molecule in skeletal muscle is
   a. myoglobin.
   b. hemoglobin.
   c. ATP.
   d. myokinase.

5. The action potential that triggers a muscle contraction travels deep within the muscle cell by means of
   a. sarcoplasmic reticulum.
   b. transverse (or T) tubules.
   c. synapses.
   d. motor end plates.

6. A sarcomere is best described as a
   a. moveable structural unit within a myofibril bounded by H zones.
   b. fixed structural unit within a myofibril bounded by Z lines.
   c. fixed structural unit within a myofibril bounded by A bands.
   d. moveable structural unit within a myofibril bounded by Z lines.

7. ATP provides the energy for muscle contraction by allowing for
   a. an action potential formation in the muscle cell.
   b. the breaking of actin-myosin bonds.
   c. the formation of actin-myosin bonds.
   d. release of calcium by sarcoplasmic reticulum.

8. Ca^{2+} binds to _____ in skeletal muscle and leads to exposure of the binding site for _____ on the _____ filament.
   a. troponin; myosin; actin
   b. troponin; actin; myosin
   c. actin; myosin; troponin
   d. tropomyosin; myosin; actin

9. Tropomyosin is moved by which of the following proteins?
   a. Calmodulin
   b. Acetylcholine
   c. Actin
   d. Troponin

10. Summation of frequent muscle twitches to give maximum contraction is called
    a. motor unit summation.
    b. twitch.
    c. facilitation.
    d. tetanus.

11. _____ are responsible for the dynamic remodeling of bone that is always underway.
    a. Osteoblasts
    b. Osteoblasts and osteoclasts
    c. Osteoclasts and osteocytes
    d. Osteoblasts, osteoclasts, and osteocytes

12. The primary difference between an endoskeleton and an exoskeleton is
    a. the presence of both circular and longitudinal muscles.
    b. whether or not the skeleton is on the inside of the body.
    c. the presence or absence of joints.
    d. the amount of fluid in the body.

13. A soccer player suffers a knee injury that damages the tissue holding his upper and lower leg bones together. The damaged tissue is probably a
    a. muscle.
    b. tendon.
    c. ligament.
    d. cartilage.

Application Questions

1. Smooth muscle contracts involuntarily, whereas skeletal muscle contraction is under voluntary control. What is the difference between the mechanisms that control smooth muscle and skeletal muscle contraction?

2. White muscle and red muscle are found in different parts of the body and are used for different types of movement. What are the physiological and morphological characteristics that distinguish the two types of muscle?

3. Smooth muscle contracts involuntarily in the digestive tract, blood vessels, and urinary bladder. Describe the two main ways that smooth muscle contraction and the membrane potential of smooth muscle are controlled.

4. Earthworms have a hydrostatic skeleton. Describe how a hydrostatic skeleton is used to move an earthworm through the soil.

Answers

Knowledge and Synthesis Answers

1. **b.** Microtubules generate the movements of cilia and flagella.

2. **c.** A motor unit is a single motor neuron and all the muscle fibers that it innervates.

3. **b.** Although contraction of smooth muscle is controlled differently from that of skeletal muscle, smooth muscle does contain actin and myosin.

4. **a.** Myoglobin is the main oxygen-carrying molecule in skeletal muscle.

5. **b.** The action potential arriving to the muscle travels into the muscle through the transverse (or T) tubules.

6. **d.** A sarcomere is a structural unit within a myofibril bounded by Z lines that contain actin and myosin.

7. **b.** ATP provides energy that is used to break actin–myosin bonds.

8. **a.** Calcium is released from the sarcoplasmic reticulum and binds with troponin, resulting in the myosin binding site on actin being exposed.

9. **d.** The binding of calcium with troponin causes a conformational change in tropomyosin.

10. **d.** Tetanus is the maximum level of muscle contraction.

11. **d.** Osteoblasts, osteoclasts, and osteocytes are all involved in remodeling bone.

12. **b.** Endoskeletons (such as those of mammals) are found inside the body, and exoskeletons (such as those of insects) are found outside the body.

13. **c.** Ligaments hold bones together.

Application Answers

1. Smooth muscle contraction and skeletal muscle contraction are regulated by the presence of Ca^{2+}. In smooth muscle, Ca^{2+} joins with the protein calmodulin to activate myosin kinase in the sarcoplasm. Myosin kinase then phosphorylates the myosin head, allowing the myosin head to bind with actin. In skeletal muscle, Ca^{2+} binds with troponin on the actin filaments. This binding of Ca^{2+} causes a conformational change in tropomyosin and uncovers the myosin binding sites on the actin, allowing myosin to bind with actin.

2. Fast-twitch fibers are known as white muscle and have few mitochondria, small amounts of myoglobin, and few blood vessels. Slow-twitch fibers are known as red muscle. Red muscle has many mitochondria, large amounts of myoglobin, and many blood vessels.

3. Smooth muscle is sensitive to stretching. It will depolarize in response to any stretching and contract, with the strength of the contraction proportional to the amount of stretch in the muscle. Parasympathetic inputs of acetylcholine also cause depolarization of the muscle membrane resulting in contraction, whereas norepinephrine hyperpolarizes the muscle membranes.

4. The hydrostatic skeleton of the earthworm is an incompressible fluid-filled cavity surrounded by longitudinal and circular muscles. Contraction of the longitudinal muscles causes the segments to contract and the body to shorten. Contraction of the circular muscles causes the body segments to elongate and the body to lengthen. Alternating contractions between the longitudinal and the circular muscles move the animal in a push and pull manner. Bristles on the body help hold the animal in place after elongation, and the body is pulled forward during shortening.

48 *Gas Exchange in Animals*

Important Concepts

Respiratory gas exchange occurs by diffusion.

- The respiratory gas oxygen (O_2) is required by cells to produce energy in the form of ATP.
- Carbon dioxide (CO_2) is one of the waste by-products of ATP production and must be eliminated from an animal's body.
- The transfer of these gases occurs by simple diffusion in the respiratory systems of animals.

Several physical parameters affect respiratory gas exchange.

- There are major differences in the O_2 capacity of air and water.
- Water, because of its density, is more expensive to move than air. Water also contains far less O_2 than does air, and O_2 also diffuses far more slowly in water than in air.
- Temperature affects respiration of aquatic animals because there is less O_2 in warm water than in cold, and an animal's need for oxygen increases with temperature (see Figure 48.2).
- The tendency for a gas to move by diffusion across gills, skin, or lungs depends on its partial pressure. The sum of all the gases' partial pressures in a gas mixture is the total partial pressure, which in the atmosphere equals the atmospheric pressure.
- Oxygen makes up 20.9 percent of the atmospheric pressure. As elevation increases and atmospheric pressure decreases, the total amount of oxygen in air decreases.
- CO_2 diffuses across the respiratory organs in both air and water. In air, diffusion is rapid because of the large partial pressure gradient between the blood and the atmosphere. In aerated water, CO_2 diffuses easily, but if the water is stagnant a large partial pressure gradient may not exist, and CO_2 diffusion will be slower.

Fick's law of diffusion describes the exchange of gas in the respiratory system.

- The partial pressure gradient is the difference between the partial pressure of a gas at two locations.
- The partial pressure gradients of O_2 and CO_2 drive the movement of O_2 into the body and CO_2 out of the body.
- Rate of diffusion, described by Fick's law of diffusion, depends in part on a diffusion coefficient that varies according to temperature, the medium, and the diffusing molecules. Rate of diffusion also depends on the cross-sectional area over which the gas is diffusing and the partial pressure gradient.

Animals have adaptations to maximize gas exchange.

- Animals maximize respiration by decreasing the distance that gases must diffuse between the blood and the external environment. They achieve this with very thin respiratory surface membranes.
- Animals also actively move the environmental respiratory fluid over the gills and through the lungs and perfuse the internal side of the respiratory organ to carry the respiratory gases.
- Some aquatic amphibians and insects have external gills with large surface areas for exchange of respiratory gases with water (see Figure 48.3a).
- Internal gills have a large surface area and are protected from the environment by an animal's body cavity. They must be actively ventilated with water to achieve gas exchange.

Diverse types of respiratory organs have evolved in animals.

- Insects have tracheae—air tubes that end at the tissue cells as air capillaries and open to the environment through spiracles. Gases diffuse through the tracheae into the air capillaries, but are also moved by movement of the animal's body parts (see Figure 48.4).
- Some aquatic insects carry a bubble of air with them underwater. As they use the O_2 in the air bubble, the P_{O_2} in the bubble decreases, resulting in a P_{O_2} gradient between the water and the air bubble. Oxygen diffuses into the air bubble from the water down the P_{O_2} gradient, restoring oxygen in the bubble.
- Fish have internal gills with a large surface area that they ventilate with unidirectional flowing water to maximize external P_{O_2} levels. Each gill has hundreds of gill filaments with folds, or lamellae, that act as the respiratory gas exchange surfaces. Countercurrent flow of blood in the lamellae and water over the lamellae maximize the P_{O_2} gradient between the water and the blood (see Figures 48.5 and 48.6). As blood flows through the lamellae it is always in contact with water that has a higher oxygen level, resulting in a continuous oxygen gradient to maximize the uptake of oxygen into the blood.

- Birds create a unidirectional flow of air through air sacs and lungs, the site of gas exchange, which make up the avian respiratory system. Air requires two cycles of inhalation and exhalation to move from the posterior air sac, to the lung, to the anterior air sac, and out of the respiratory tract (see Figure 48.8). Gas exchange occurs in the parabronchi and air capillaries of the lung (see Figure 48.7).

The lung is the respiratory organ in mammals.

- Mammalian lungs have a very large surface area and a very short gas diffusion pathway.

- Air enters at the oral cavity or nasal passage, travels down the pharynx through the larynx, and then through the trachea, which branches into two bronchi, one leading to each lung. In the lung, more branching occurs to produce bronchioles and finally the site of gas exchange—small air sacs, the alveoli (see Figure 48.10).

- Oxygen diffuses across the thin-walled alveoli (less than 2 μm) into many surrounding capillaries.

- The alveoli have surface tension due to an aqueous layer that makes it difficult to inflate them. This is overcome by a surfactant produced by the cells forming the alveolar walls. The aqueous layer is needed to ensure the diffusion of gases across the membrane.

- Mucus, produced along the lung's larger airways, catches and removes inhaled dust and microorganisms. Cilia along the airways move the mucus up the airways and into the throat, acting as a mucus escalator.

Mammalian lungs are ventilated by tidal ventilation.

- The tidal breathing of mammals can be described by a series of lung volumes (see Figure 48.9). The normal volume of air moved during one cycle is known as the tidal volume. The total lung capacity is made up of the resting tidal volume, plus the additional volume of air that can be taken in with a large breath and the additional volume of air that can be forced out with a large exhale. Not all of the air can be forced out of the lungs. Some air remains in the bronchi and trachea, making up the dead space.

- Lungs are inflated by negative pressure created by contraction of the diaphragm muscle in the thoracic cavity. Closed pleural membranes line the thoracic cavity and allow for the development of negative pressure. During inhalation, external intercostal muscles on the ribs pull the ribs up and help expand the pleural cavities. During exhalation, internal intercostal muscles pull the ribs down (see Figure 48.11).

Oxygen is transported in the blood by hemoglobin in vertebrates.

- Gases diffuse between the respiratory organ and the circulatory system, where they are transported in blood. The liquid blood plasma transports only a small amount of dissolved O_2, with the vast majority of O_2 transported by the oxygen-binding pigment hemoglobin found in the red blood cells of vertebrates.

- Hemoglobin contains four polypeptide subunits, each with a heme group that reversibly binds O_2. The amount of O_2 bound to hemoglobin depends on the partial pressure of O_2. At the high blood P_{O_2} that is characteristic of

blood leaving the lungs, hemoglobin is almost 100 percent saturated with O_2. At the tissues, the P_{O_2} is lower and hemoglobin releases approximately 25 percent of its bound O_2. The relationship between the P_{O_2} of the environment and the percentage of hemoglobin bound with oxygen is known as the S-shaped oxygen-binding curve (see Figure 48.12). At low levels of P_{O_2}, only one of the hemoglobin subunits can bind O_2. Once it binds a single O_2 molecule, the conformation of the hemoglobin changes, making it easier for the other three sites on the hemoglobin to bind oxygen in a process known as positive cooperativity.

- Fetal hemoglobin has a different structure from the adult form, with a higher affinity for O_2.

- The Bohr effect is the shifting of the hemoglobin binding curve to the right, in response to a decrease in pH of the blood, resulting in more oxygen being released at a given environmental P_{O_2} at the lower pH.

- Hemoglobin's affinity for O_2 is lowered by 2,3 bisphosphoglyceric acid (BPG), which binds with hemoglobin to cause a conformational change in the hemoglobin, resulting in a right shift in the oxygen-binding curve.

- In muscle, myoglobin is the O_2 carrying molecule and functions primarily to store O_2 intracellularly. The O_2 affinity is higher in myoglobin than in hemoglobin, but myoglobin has only one unit to bind one O_2 molecule.

Carbon dioxide is transported as dissolved CO_2 and as bicarbonate ions.

- Tissues produce CO_2 as a by-product of metabolism. CO_2 must be moved from the tissues and excreted into the environment.

- Significant amounts of CO_2 dissolve in the plasma and are carried in this form to the lungs.

- Most of the CO_2 is transported as bicarbonate ions. CO_2 reacts with water to make bicarbonate ions and a proton. Carbonic anhydrase speeds up the conversion of CO_2 to carbonic acid, which then dissociates into bicarbonate ions.

Breathing is controlled by the brain and blood pH levels.

- Breathing is controlled by the autonomic nervous system from neurons in the medulla.

- Chemoreceptors on the medulla are sensitive to CO_2 levels and the pH of cerebral spinal fluid. Breathing increases when CO_2 levels increase and pH decreases.

- Control of breathing is relatively insensitive to blood O_2 levels, but chemoreceptors in the arteries leaving the heart can signal the medulla in response to low levels of O_2.

The Big Picture

- Cellular metabolism requires O_2 and produces CO_2 as a waste product that must be eliminated. Animals have evolved a variety of structures and mechanisms for exchanging these gases with the environment. Tracheal systems in insects, gills in fishes, and lungs in terrestrial vertebrates are three examples of gas-exchange organs. All are characterized by very short diffusion distances and very large surface areas—adaptations designed to maximize gas exchange.

- Oxygen is transported by the protein hemoglobin in vertebrates. Hemoglobin–oxygen binding is determined by the P_{O_2} levels surrounding hemoglobin. If P_{O_2} is high, hemoglobin accepts O_2, but it readily gives up its O_2 when the P_{O_2} falls. These properties allow hemoglobin to bind O_2 in the gas-exchange organ, transport it to the metabolizing tissues, and then release its oxygen for consumption in cellular metabolism.

Common Problem Areas

- It is easy—though incorrect—to think that the exchange of O_2 and CO_2 is a "two-way street," with O_2 taking the immediate place of CO_2 in the lungs and CO_2 then taking the place of O_2 in the tissues. In fact, the mechanisms for exchange of these two gases are completely different.

- Understanding how O_2 is picked up, transported by, and released from hemoglobin can be quite confusing. Think of the oxygen dissociation curve as a "tool." Think first of the situation at the top of the curve where P_{O_2} is high (the lungs or gills, for example). If the P_{O_2} is high, hemoglobin will maximally bind oxygen. Think, then, of transporting that blood to a region with a given P_{O_2}, and read from the curve what must happen to O_2 saturation under the new P_{O_2} level. The O_2 given from hemoglobin is now available for tissue respiration.

Study Strategies

- Use Fick's law of diffusion as a guideline for understanding gas exchange. The various components of the law—distance for diffusion, partial pressure gradient, surface area over which gas exchange occurs—provide a good framework for understanding why gas-exchange organs have evolved with certain characteristics in common. For example, gas-exchange organs tend to have large surface areas, have very short diffusion distances, and experience large partial gradients for O_2 across their surfaces. All of this makes sense in the context of Fick's law of diffusion. Fick's law of diffusion also governs the movement of O_2 and CO_2 in the body, and it can help you remember where and why these gases are picked up and released in the body.

- Review the following tutorials and activities on the Student Website/CD:
Tutorial 48.1 Airflow in Birds
Tutorial 48.2 Airflow in Mammals
Activity 48.1 Human Respiratory System
Activity 48.2 Oxygen-Binding Curves
Activity 48.3 Concept Matching

Test Yourself

Knowledge and Synthesis Questions

1. Which of the following is *not* a reason that breathing air is easier than breathing water?
   a. A given volume of air is easier to move across the respiratory organs than the same volume of water.
   b. Air holds more oxygen per unit volume than does water.
   c. Water breathers have a difficult time ridding themselves of CO_2 because CO_2 does not dissolve well in water.
   d. Temperature increases affect the O_2 content of water more than they do that of air.

2. External gills, tracheae, and lungs share which of the following sets of characteristics?
   a. Part of gas-exchange system; exchange both CO_2 and O_2; increase surface area for diffusion
   b. Used by water breathers; based on countercurrent exchange; use negative pressure breathing
   c. Exchange only O_2; are associated with a circulatory system; found in vertebrates
   d. Found in insects; employ positive-pressure pumping; based on crosscurrent flow

3. Which of the following represents a larger volume of air than is normally found in the resting tidal volume of a human lung?
   a. Residual volume
   b. Inspiratory reserve volume
   c. Expiratory reserve volume
   d. All of the above

4. Which of the following is shared by bird and mammal lungs?
   a. Both need two cycles of inhalation and exhalation for air to move through the lungs.
   b. Both contain alveoli as the terminal end of the lung.
   c. Both have an anatomical dead space.
   d. Both exchange O_2 and CO_2 with blood in capillaries.

5. According to Fick's law of diffusion, which of the following does not play a role in the diffusion of oxygen across a membrane?
   a. Surface area
   b. Volume
   c. Difference in concentration or partial pressure
   d. Diffusion distance

6. Because of the relatively high altitude of Antonito, Colorado, the town has a normal barometric pressure of about 600 mm Hg rather than 760 mm Hg as at sea level. The partial pressure of oxygen in Antonito's air is approximately
   a. 75 mm Hg.
   b. 126 mm Hg.
   c. 160 mm Hg.
   d. 760 mm Hg.

7. Gas flows into the lungs of mammals during inhalation because the
   a. pressure in the lungs falls below atmospheric pressure.
   b. volume of the lungs decreases.
   c. pressure in the lungs rises above atmospheric pressure.
   d. diaphragm moves upward toward the lungs.

8. The movement of O_2 and CO_2 between the blood in the tissue capillaries and the cells in tissues depends most directly upon
   a. active transport of O_2 and CO_2.
   b. total atmospheric (barometric) pressure differences across the cell membranes.
   c. diffusion of O_2 and CO_2 down a concentration gradient.
   d. diffusion of O_2 and CO_2 down a partial pressure gradient.

9. The alveoli of the lungs do not contain "air" because
   a. we normally do not ventilate our lungs at a high enough rate.
   b. the lungs have too many alveoli to ventilate.
   c. there is dead space in the trachea and bronchi.
   d. the trachea and bronchi are too small in volume.

10. Which of the statements about hemoglobin is *false*?
    a. Hemoglobin allows the blood to carry a large amount of O_2.
    b. Hemoglobin contains a single polypeptide chain with very high affinity for O_2.
    c. Hemoglobin is packaged inside red blood cells
    d. Fetal hemoglobin is structurally different from adult hemoglobin.

11. As blood becomes fully O_2-saturated, hemoglobin is combining with _____ molecule(s) of oxygen.
    a. 1
    b. 2
    c. 4
    d. 8

12. The Bohr shift describes the
    a. outward movement of Cl^- from the blood cell in exchange for HCO^-_3 moving into the cell.
    b. leftward shift of the entire oxygen equilibrium curve when temperature rises.
    c. rightward shift of the entire oxygen equilibrium curve when pH rises.
    d. rightward shift of the entire oxygen equilibrium curve when pH falls.

13. The presence of CO_2 in blood will lower pH because CO_2 combines with _____, with the rate of reaction increased by _____.
    a. H_2O to form H^+ and HCO^-_3; carbonic anhydrase
    b. H_2O to form only HCO^-_3; carbonic anhydrase
    c. H_2O to form only H^+; carbonic ions
    d. H^+ to form HCO^-_3; oxyhemoglobin

14. The largest proportion of CO_2 carried by the blood is in the form of
    a. bicarbonate ions (HCO^-_3) carried in the plasma.
    b. molecular CO_2 dissolved in the plasma.
    c. bicarbonate ions (HCO^-_3) carried within the red blood cells.
    d. molecular CO_2 chemically bound to hemoglobin.

15. Which of the following is involved in the control of breathing?
    a. Neurons in the medulla
    b. Chemoreceptors on the surface of the medulla
    c. Chemoreceptors in large blood vessels leaving the heart
    d. All of the above

Application Questions

1. In a dredge, you have just captured an aquatic animal that has no specialized respiratory surfaces. Describe how the organism can get enough oxygen to all of its cells to survive.

2. The gills of fish employ a countercurrent exchange mechanism to ensure that a large amount of oxygen is extracted from the water. Describe how the countercurrent exchange works.

3. What is the difference between the myoglobin oxygen-binding curve and the hemoglobin oxygen-binding curve?

Answers

Knowledge and Synthesis Answers

1. **c.** Water breathing is more difficult than air breathing because the higher density of water makes it more expensive to move across the respiratory surfaces than air, it has less oxygen than air, and the oxygen content is very dependent on the temperature of the water. CO_2, however, dissolves easily into water and is easy to get rid of.

2. **a.** The external gill, tracheae, and lungs are all examples of gas-exchange systems used by various animals to exchange both CO_2 and O_2. One of the special properties of all respiratory systems is that they increase surface area for diffusion.

3. **d.** All have a larger volume than the resting tidal volume.

4. **d.** The lungs of birds and mammals both are sites for the exchange of O_2 and CO_2 with blood in capillaries. The lungs of birds take two cycles for air to move through them, whereas the mammalian lung only takes one cycle. The mammalian lung is the only one that ends in alveoli and has an anatomical dead space.

5. **b.** The volume of the gas is not included in Fick's law of diffusion. All the other parameters are important in determining the rate of diffusion of gases in the respiratory system of animals.

6. **b.** Air at sea level with an atmospheric pressure of 700 mm Hg has a partial pressure of oxygen of 160 mm Hg ($700 \times 21\%$). Therefore, the partial pressure of O_2 at Antonito, Colorado, is $600 \times 21\% = 126$ mm Hg.

7. **a.** Inhalation in the mammalian lung occurs by means of negative pressure produced by contraction of the diaphragm. Thus the pressure in the lungs falls below atmospheric pressure.

8. **d.** Movement of O_2 and CO_2 from the blood to the tissues always occurs by diffusion of O_2 and CO_2 down their partial pressure gradients.

9. **c.** The alveoli do not contain "air" with 20.9 percent O_2 because incoming air is mixed with air left in the dead space of the trachea and bronchi that has had some of the O_2 removed by the lungs.

10. **b.** Hemoglobin is composed of four subunits, which act with positive cooperativity to bind O_2. Oxygen-binding myoglobin found in muscle has only one unit.

11. **c.** Hemoglobin has 4 subunits, each of which binds to 1 molecule of O_2 for a total of 4 molecules of O_2 bound to 1 hemoglobin molecule.

12. **d.** The Bohr effect describes the action of pH on the oxygen-binding curve. Decreases in pH result in a net right shift of the entire oxygen equilibrium curve.

13. **a.** Carbon dioxide combines with H_2O in the plasma to form H^+ and HCO^-_3. The enzyme carbonic anhydrase catalyzes the reaction.

14. **a.** The majority of CO_2 is carried as bicarbonate ions (HCO^-_3) in the plasma.

15. **d.** All are involved in the control of breathing.

Application Answers

1. Having no respiratory surfaces, the animal must rely on simple diffusion from the environment into its cells. In order for simple diffusion to be an effective means of oxygen transport, the animal must be very small, with all of its cells only a few cell layers from the environment. The metabolic rate of this animal is most likely very low in order to accommodate the lack of any special respiratory surfaces.

2. In the countercurrent exchange in the gills of fish, water and blood move in opposite directions. This results in blood with low oxygen content entering the gills and flowing past water that has a higher oxygen content. Oxygen diffuses down the partial pressure gradient into the blood. As blood moves along the lamella, it picks up oxygen, but it always has a lower oxygen content than the water moving in the opposite direction. In this way the blood is able to optimize oxygen uptake during the entire trip through the lamella.

3. The oxygen-binding curve of myoglobin is shifted to the left in relationship to the oxygen-binding curve of hemoglobin. This allows myoglobin to pick up O_2 and hold it at lower partial pressures of O_2.

49 *Circulatory Systems*

Important Concepts

Small organisms do not need circulatory systems.

- An animal's circulatory system transports nutrients and wastes to and from tissues and cells.

- In the smallest multicellular animals, every cell within the body is no more than one to two cells from the environment. Such small animals do not require a circulatory system because materials can move efficiently into and out of cells by diffusion alone.

- Some aquatic invertebrates have a gastrovascular cavity in which exchange of gases and nutrients occurs. Other organisms are flattened to increase the surface-to-volume ratio, ensuring that cells are close to the outside.

- Large animals need a circulatory system to carry wastes, gases, and nutrients to and from all the cells in the body.

Circulatory systems may be classified structurally as open or closed.

- In an open circulatory system, blood and other tissue fluids are not separated (see Figure 49.1). A pump may be present to help move body fluids through the various large compartments, assisted by movements of the animal.

- In a closed circulatory system, blood is pumped through a series of interconnected closed vessels (see Figure 49.2). There are numerous advantages to closed systems compared to open ones: rapid transport of fluid in the closed vessels, the ability to divert the flow of blood between different tissues, and direct hormone and nutrient transport to the tissues.

- The closed system typical of vertebrates is composed of arteries and arterioles, which carry blood from the heart to the body; veins and venules, which carry blood from the body to the heart; and an intervening capillary bed. In birds and mammals, the systemic circuit flows from the heart to the body tissues, and the pulmonary circuit flows from the heart to the gas exchange organ.

The gills and systemic tissues play an important role in the fish circulatory system.

- Fishes pump blood with a heart that has one atrium and one ventricle. Blood leaves the heart and enters the gills, where gas exchange occurs. Some arterial blood pressure is lost in passing through the gills.

- Blood leaving the gills collects in the aorta and travels to the systemic organs and tissues, eventually returning to the heart.

- African lungfish have a well-vascularized lung for gas exchange that is formed embryonically from an outpocketing of the gut. A heart with a partially divided atrium helps to separate oxygenated blood returning from the lung and deoxygenated blood returning from the tissues as they enter the single ventricle. A large proportion of oxygenated blood bypasses the gills and goes directly to the tissues.

Amphibians have partial separation of pulmonary and systemic blood.

- Amphibians have two atria, one that receives deoxygenated blood from the tissues and one that receives oxygenated blood from the lungs.

- Anatomical features of the ventricle help direct deoxygenated blood preferentially to the lungs and skin (which also acts as a gas exchange organ in amphibians) and oxygenated blood to the systemic body tissues. Even though the two circuits are not fully divided, blood tends to flow in parallel.

Reptiles control their intracardiac shunt to regulate the distribution of oxygenated and deoxygenated blood leaving the heart.

- Turtles, snakes, and lizards have two atria and a ventricle that is internally divided into subchambers that allow for control of the distribution of blood to the lungs and body.

- During air breathing, blood from the right side of the ventricle is preferentially pumped to the lungs because there is a higher resistance in the systemic circuit than in the pulmonary circuit. Also, there is an asynchronous ejection from the heart because pumping occurs first from the right and then the left side of the ventricle. At the stage in the cardiac cycle in which the blood from the left side leaves the heart, the resistance is higher in the pulmonary circuit. Thus the last blood leaving the heart tends to be deoxygenated blood and is preferentially directed to the pulmonary circulation.

- When the reptile stops breathing, vessel constriction in the lungs causes pulmonary circuit resistance to increase, causing blood from the right side to bypass the pulmonary circulation and flow to the systemic circuit.

- Alligators and crocodiles are unique in that they have a four-chambered heart, resembling that of a bird or mammal. However, alligators and crocodiles retain a left aorta that arises from the right side of the heart. There is a small passageway between the left and right aorta through which blood can bypass the pulmonary circulation. Thus, crocodiles and alligators can operate their heart like that of a mammal, distributing blood equally to the pulmonary and systemic circulations, or they can partially bypass the lungs by directing blood from the left aorta into the right aorta and on to the systemic circulation.

Birds and mammals have a completely parallel circulatory system.

- Birds and mammals have a four-chambered heart that allows for complete separation and no mixing of systemic and pulmonary blood.
- This cardiovascular arrangement maximizes O_2 transport by eliminating mixing between pulmonary and systemic circulations.
- The pulmonary circulation operates at low pressure, protecting the delicate lung membranes, whereas the systemic circuit operates at high pressures, allowing the perfusion of tissues and organs distant from the heart.

The human heart contains separate right and left sides pumping to the lungs and tissues, respectively.

- The human heart has two atria and two ventricles (see Figure 49.3). The atrioventricular valve between the atrium and ventricle stops backflow of blood from the ventricle into the atrium. The pulmonary and aortic valves reside between the ventricles and arteries and help maintain one-way flow through the heart.
- Deoxygenated blood flows from the body through the superior and inferior vena cava into the right atrium and into the right ventricle by passive filling during heart relaxation. During heart contraction, blood flows from the right ventricle to the pulmonary artery and the lungs, where blood becomes oxygenated.
- Oxygenated blood from the lungs flows back to the heart and into the left atrium, then into the left ventricle, and finally out through the aorta to the systemic tissues.
- Resistance in the blood vessels of the body's systemic tissues is higher than in the lungs, so the left ventricle must be larger and more muscular than the right ventricle in order to generate more force while pumping the same volume of blood.
- Systole is the contraction of the ventricle, and diastole is the relaxation of the ventricle. At the very end of diastole, the atria contract. Together these two phases make up the cardiac cycle.
- During the cardiac cycle the pressure is highest during ventricle contraction (systole) and lowest during relaxation (diastole; see Figure 49.4).

The human heartbeat is generated in the heart's pacemaker.

- The pacemaker, or sinoatrial node, is a small section of highly modified muscle tissue located at the junction of the superior vena cava and right atrium.

- The pacemaker rhythmically produces action potentials that pass on to the rest of the heart and stimulate a heartbeat (see Figure 49.7).
- Gap junctions connect cardiac muscle cells and thus allow the action potentials to spread rapidly from cell to cell.
- The action potential moves from the atrium to the ventricles through the atrioventricular node (AV node). The AV node slows down the transmission of the action potential from the atrium to the ventricle. The action potential passes rapidly down the bundle of His and to the Purkinje fibers fanning out in the ventricular tissue. From the Purkinje fibers the action potential spreads through both ventricles, causing contraction to occur slightly after the contraction of the atrium.
- The autonomic nervous system also plays a role in the control of the heartbeat. Sympathetic activity increases heart rate, and parasympathetic activity decreases heart rate (see Figure 49.6).
- The electrocardiogram (EKG) measures the electrical activity of the heart as a complex wave pattern (Figure 49.9). The wave has five parts: P (depolarization and contraction of the atrium), Q, R, and S (depolarization of the ventricles), and T (relaxation and repolarization of the ventricles).

The vascular system is composed of arteries, capillaries, and veins.

- Arteries carry blood away from the heart to the body under relatively high pressure. Veins carry blood toward the heart from the body under low pressure (see Figure 49.10). The very fine capillaries interconnect the arteries with the veins in the tissues, dissipate pressure, and are sites for all exchanges of gases, nutrients, and wastes.
- Arteries and the smaller arterioles are elastic, allowing them to stretch during systole and rebound during diastole. In vertebrates the walls of vessels contain smooth muscle that controls the vessel's resistance by causing changes in its diameter.
- Blood returns to the heart at low pressure in expandable veins. Because of the low pressure, veins that act against gravity have valves, and blood flowing toward the heart is helped by the movement of the surrounding skeletal muscle (see Figure 49.13). Gravity can cause pooling of blood in the veins if a person is inactive for a long period of time.

Capillaries are sites for exchange of gases, nutrients and wastes, and the locations of considerable fluid loss and recovery.

- Capillaries connect the arteries with the veins in the tissues. Capillaries have a large cross-sectional area compared with arteries and veins and extremely thin walls to allow the diffusion of gases, nutrients, and wastes (see Figure 49.10).
- As blood enters the capillaries it is moving slowly. Blood pressure from the arterial side forces fluids, ions, and small molecules out of the capillaries through small holes called fenestrations in a process known as filtration. Near the venule end of the capillaries, the difference in osmotic potential between the plasma inside the capillaries and the surrounding fluids creates an osmotic pressure result-

ing in fluid recovery back into the capillaries. Thus the net flow of water between the plasma and tissue fluid is determined by the difference in blood pressure and osmotic pressure (see Figure 49.12).

- Some capillaries are selective about which ions and molecules can pass through them, whereas others are less selective. The blood–brain barrier consists of highly selective brain capillaries.

The lymphatic system returns tissue fluid not reclaimed by the capillaries to the blood.

- In humans, lymph is moved through the lymphatic vessels up the body by surrounding skeletal muscle contractions. Lymph is returned to the blood through the thoracic ducts that empty into large veins at the base of the neck.
- Lymph nodes exist in mammals and birds and contribute to the defenses of the body. Foreign materials and microorganisms are removed by the phagocytic action of white blood cells in the lymph nodes.

Cardiovascular disease affects the arteries and veins.

- Hardening of the arteries is known as atherosclerosis.
- Plaque deposits begin to form at sites of endothelial damage, and lipids tend to accumulate on the deposits. Connective tissue invades the deposits and makes the wall of the artery less elastic and narrow. This can lead to the formation of a blood clot, or thrombus.
- The coronary arteries that supply the heart with blood are highly susceptible to blockage, resulting in a heart attack.
- An embolus is a piece of a thrombus that breaks off, travels in the vessels, and then lodges in a vessel. When this happens in the brain, it is known as a stroke. A stroke results in denial of blood flow to a particular region of the brain, causing paralysis, loss of cognitive abilities, or death.

Blood is a tissue composed of fluid plasma and many different cell types.

- The fluid matrix of blood is known as plasma and contains water, salts, and proteins. The hematocrit is the percentage of blood volume composed of cells.
- A number of different cell types in blood carry out a wide range of functions (see Figure 49.15). The cell types include the erythrocytes, basophils, eosinophils, neutrophils, lymphocytes, monocytes, and platelets.
- Erythrocytes, or red blood cells, are shaped like biconcave discs when mature. Their function is to transport respiratory gases.
- Erythrocytes are produced in stem cells located in the marrow of bone. Once they are released into the bloodstream they survive for about 120 days. The hormone erythropoietin is produced in the kidneys and regulates the production of red blood cells.
- The spleen serves as a reservoir for red blood cells.
- Platelets, which help to start blood clotting, are also found in blood and are produced by the breaking off of cell fragments from megakaryocytes in the bone marrow. Platelets bind to the edges of a break in a vessel wall, not only providing a mechanical plug but also starting a

chemical chain reaction that leads to the conversion of the circulating protein prothrombin into thrombin. Thrombin stimulates the formation of sticky fibrin threads that form a meshwork and cover the hole in the vessel (see Figure 49.16).

- Plasma contains many gases, ions, nutrients, proteins, and other molecules. Albumin in the plasma helps to maintain the osmotic pressure in the capillaries. The plasma also contains the hormones that are circulated by the blood to target cells.

The circulatory system is under hormonal, neural, and local control.

- Blood flow in the capillary beds of many tissues is controlled locally by autoregulatory mechanisms. Precapillary sphincters at the arterial end of the capillaries control the flow of blood, with contraction of these sphincters limiting or stopping blood flow through the capillary bed (see Figure 49.17). Low levels of O_2 and high levels of CO_2 can open the precapillary sphincters.
- Norepinephrine from sympathetic neurons and epinephrine released from the adrenal medulla cause arteries and arterioles to contract, reducing blood flow and increasing central blood pressure.
- Autonomic control of heart rate and constriction of blood vessels occurs from the medulla. Stretch receptors in the aorta and carotid arteries help regulate blood pressure. With rising blood pressure they inhibit sympathetic output and cause the heart to slow and arterioles to dilate. When pressure is low, sympathetic output is stimulated, increasing heart rate and constriction of arterioles.
- The hypothalamus also helps in control by releasing vasopressin, which stimulates arterioles to constrict, leading to higher blood pressure.

Diving mammals conserve blood oxygen by slowing their heart rate during dives.

- Marine mammals exhibit a diving reflex during which the heart rate slows and blood flow to all tissues except the brain drops dramatically. During dives, the reduced blood to tissues switches the tissues to anaerobic metabolism and suppresses the metabolism of the tissue.
- Humans also have a diving reflex that results in a mild slowing of the heart rate when the face is under water. The reflex is controlled by the vagus nerve and parasympathetic nervous system.

The Big Picture

- Cardiovascular systems are required when animals grow too big to acquire nutrients and eliminate waste by diffusion alone. Animals have evolved a wide variety of cardiovascular systems, which can be classified generally as open or closed. Open systems have relatively low pressures and have large sinuses that bathe tissues and organs in circulating fluid. Closed systems have an interconnected series of vessels that keep blood separated from tissue but allow gases, nutrients, and wastes to diffuse through their walls.
- During the evolution of vertebrates, the heart has changed from the simple two-chambered heart of fish, to the complex four-chambered heart of birds and mammals. However, the hearts of amphibians and

reptiles do provide for complex intracardiac shunting and bypass of blood flow to the lungs, which is an adaptation that allows them to stop breathing or to breathe intermittently.

- The human heart is really two pumps in one. The left pump provides oxygenated blood under high pressure to the systemic tissues, and the right pump sends deoxygenated blood under low pressure to the lungs. A cardiac pacemaker sets the cardiac rhythm.

- The flow of blood through capillaries is highly regulated at the local level by changes in their local surroundings; blood flow through arteries and veins is regulated by the combined action of neural activity and circulating hormones.

Common Problem Areas

- It is frequently thought that the hearts of amphibians and reptiles are primitive structures, awaiting "repair through evolution" to the more highly evolved bird and mammal heart. In fact, the hearts of amphibians and reptiles are highly adapted to suit their particular lifestyle, specifically their intermittent breathing and relatively low metabolic rate. By being able to shunt blood within the heart, these animals can save the energy that would otherwise be wasted in sending blood to the temporarily nonventilated lungs.

- Students often fail to appreciate that the human heart is actually two entirely distinct pumps—the right heart pumping blood to the lungs and the left heart pumping blood to the body. The two pumps just happen to be packaged in the same structure—the heart. In fact, cardiologists refer to the "left heart" and "right heart," emphasizing the separate nature of these two pumps. Figuring out the flow of blood through the human heart becomes easier when it is considered from this perspective.

Study Strategies

- Determining the pattern of blood flow through the human heart will be easier if you mentally "unfold" the circulation into a great circle, in which there are two pumps at opposite sides of the circle with a vascular bed intervening between each pump. Once you have mastered this concept, you will see that the only way to place the two pumps in proximity in the same structure (the heart) is to twist the circle into a folded figure eight—the typical diagram of the circulation in the textbook.

- Following from the preceding, learn the various chambers and valves in the context of a sequence of structures in each of the right and left pumps of the heart, rather than as a list of terms printed on a complex, anatomically accurate diagram of the heart.

- Refer to the following tutorial and activities on the Student Website/CD:
  Tutorial 49.1 The Cardiac Cycle
  Activity 49.1 Vertebrate Circulatory System
  Activity 49.2 The Human Heart
  Activity 49.3 Structure of a Blood Vessel

Test Yourself

Knowledge and Synthesis Questions

1. Which of the following is *not* one of the reasons that closed circulatory systems are more efficient than open circulatory systems?
   a. Closed systems rely exclusively on simple diffusion for transport, whereas open systems rely on pumping mechanisms.
   b. Transport within closed systems is more rapid than in open systems.
   c. Blood can easily be directed to specific areas in closed systems, but not in open systems.
   d. Closed systems operate better under higher pressure than do open systems.

2. How does the pattern of blood flow change when reptiles undergo periods of intermittent breathing?
   a. Blood is shunted from the systemic circuit to the pulmonary circuit.
   b. Blood is shunted specifically to the brain.
   c. Blood is shunted from the pulmonary circuit to the systemic circuit.
   d. The pattern does not change during intermittent breathing.

3. In which of the following would you record the highest blood pressure?
   a. The ventricle supplying blood to the gills of a fish
   b. The anterior dorsal artery of an ant
   c. The pulmonary vein of a frog
   d. The ventricle supplying blood to the systemic circuit of a bird

4. Blood consists of a fluid fraction consisting of _____ and a solid fraction consisting of _____.
   a. plasma; water, erythrocytes, and platelets
   b. erythrocytes; leukocytes, macrophages, and platelets
   c. plasma; erythrocytes, platelets, and leukocytes
   d. leukocytes; erythrocytes and platelets

5. Red blood cells are
   a. biconcave cells containing hemoglobin.
   b. spherical cells containing hemoglobin.
   c. spherical cells capable of ameboid motion and containing hemoglobin.
   d. biconcave cells that contain platelets.

6. Blood clotting pathways cause the conversion of
   a. vitamin K to prothrombin.
   b. fibrin to fibrinogen.
   c. thrombin to prothrombin.
   d. None of the above

7. Which of the following regions of the vascular bed is the actual site of gas exchange with surrounding tissue?
   a. Arteries
   b. Capillaries
   c. Lymphatic vessels
   d. Veins

8. Which is the correct sequence of parts through which cardiac action potentials pass?
   a. Purkinje fibers, AV node, SA node, bundle of His, atrial fibers
   b. AV node, atrial fibers, SA node, bundle of His, Purkinje fibers
   c. SA node, bundle of His, atrial fibers, AV node, Purkinje fibers
   d. SA node, atrial fibers, AV node, bundle of His, Purkinje fibers

9. The purpose of the AV node is to _____, and the purpose of the Purkinje fibers is to _____.
   a. create simultaneous atrial and ventricular depolarization; speed up transmission of the cardiac impulse into the ventricle

b. delay ventricular depolarization relative to atrial depolarization; insulate the cardiac impulse from the general ventricular fibers

c. delay ventricular depolarization relative to atrial depolarization; transmit the cardiac impulse to very small localized groups of ventricular fibers

d. delay atrial depolarization relative to ventricular depolarization; transmit the cardiac impulse to very small localized groups of ventricular fibers

10. The atrial walls are _____ than the ventricular wall, and pressure generated in the atrial chambers is _____ than in the ventricles.
    a. thinner; higher
    b. thinner; lower
    c. thicker; higher
    d. thicker; lower

11. The left ventricle exceeds the right ventricle in the
    a. amount of blood that enters during heart contraction.
    b. volume expelled during contraction.
    c. pressure developed during contraction.
    d. All of the above

12. Which of the following structures of the lymphatic system acts primarily as a filter for detecting and destroying microorganisms in lymph traveling through major lymph vessels?
    a. Lymph nodes
    b. Thymus
    c. Lymph capillaries
    d. Tonsils, but not the appendix

13. The net loss of fluid from blood capillaries will increase if
    a. plasma filtration decreases.
    b. the osmotic pressure of plasma increases.
    c. blood pressure increases in the capillaries.
    d. the colloid osmotic pressure of interstitial fluid decreases.

14. Which of the following statements about the control of circulation in humans is *false*?
    a. Sympathetic nerve input to skeletal muscle causes the blood vessels in the muscle to dilate.
    b. Blood flow can be regulated by autonomic nerve signals emanating from cardiovascular control centers in the medulla of the brain.
    c. Carotid artery chemoreceptors detect low O_2 levels in the blood and promote increased blood pressure.
    d. Hormones such as angiotensin and vasopressin cause venules to constrict.

Application Questions

1. What are the important differences between the circulatory system of a fish and that of a reptile?

2. The human heart pumps blood to the pulmonary and the systemic circuit. Describe the path that blood takes as it moves through the circulatory system, starting at the left ventricle.

3. You are about to take a final exam in your freshman biology course and are very nervous. What responses will your circulatory system have to your nervousness?

4. Blood moves through the arteries and veins at a faster velocity than the blood moving through the capillaries. What is the cause and effect of blood slowing down as it moves through the capillaries?

Answers

Knowledge and Synthesis Answers

1. **a.** Both closed systems and open systems rely on pumping mechanisms to distribute fluid throughout the body.

2. **c.** During intermittent breathing, the blood is shunted from the pulmonary circuit to the systemic circuit. The resistance in the pulmonary circuit increases, resulting in the shunt.

3. **d.** Of the vertebrate groups, mammals and birds tend to have the highest blood pressures. Therefore, the ventricle supplying blood to the systemic circuit of a bird has the highest pressure.

4. **c.** The fluid portion of blood is the plasma; three components of the solid fraction are erythrocytes, platelets, and leukocytes.

5. **a.** Red blood cells are biconcave cells that contain the oxygen-binding hemoglobin.

6. **d.** Blood clotting involves none of the answer option listed. Prothrombin is converted into thrombin and fibrinogen is converted into fibrin.

7. **b.** Gas exchange at the tissues occurs across the capillaries.

8. **d.** The cardiac action potential passes through the SA node, atrial fibers, AV node, bundle of His, and finally the Purkinje fibers.

9. **c.** The AV node delays the ventricular depolarization relative to atrial depolarization, so atrial contraction occurs before ventricular contraction. The Purkinje fibers transmit the cardiac impulse to very small localized groups of ventricular fibers.

10. **b.** The atrium has thinner walls and generates lower pressures than the ventricles.

11. **c.** The left ventricle generates a greater pressure in the blood flowing to the systemic circuit than the right ventricle with blood flowing to the pulmonary circuit.

12. **a.** The lymph nodes filter and destroy microorganisms that are traveling through the lymphatic system.

13. **c.** Changes in blood pressure will change the amount of filtration that occurs at the capillaries. Thus, increases in blood pressure will result in greater rates of filtration.

14. **d.** Angiotensin and vasopressin act to control constriction of the arterioles, not the venules.

Application Answers

1. Fish have a two-chambered heart, with blood flowing from the heart of the fish to the gills and then directly to the body. Reptiles have a more complex three-chambered heart, with a subdivided ventricle that directs outflow to the pulmonary circuit and to the systemic circuit. Thus the circulatory system of fish delivers blood to the respiratory organ and the body tissues in succession, whereas the circulatory system of reptiles delivers blood to the respiratory organ and the body tissues in parallel.

2. Blood from the right ventricle exits through the pulmonary valve into the pulmonary artery and then goes to the lungs. From the lungs, the blood returns to the left atrium and passes through the atrioventricular valve into the left ventricle. Blood from the left ventricle is pumped to the systemic circuit through the aortic valve and into

the aorta. From the aorta, blood moves to the tissues in arteries and arterioles, passes through tissues in the capillaries, and returns via the venules, veins, and ultimately the superior vena cava and inferior vena cava to the right atrium of the heart. From the right atrium blood is pumped back where we started—into the right ventricle through the atrioventricular valve.

3. Stress and nervousness elicit the fight-or-flight response. Signals from the medullary cardiovascular control center stimulate sympathetic inputs from the autonomic nervous system. In response to sympathetic inputs, the adrenal gland releases epinephrine into the bloodstream. Epinephrine stimulates an increased heart rate and arterial pressure. The blood flow to the smooth muscles decreases due to constriction of blood vessels. By constriction of smooth muscles, blood is diverted away from areas such as the digestive tract to the skeletal muscle needed for fight or flight.

4. Blood is pumped from the heart and into the arteries, where it has a relatively fast velocity. Upon entering the capillaries much of the velocity is lost. The capillaries have a much higher total cross sectional area than the arteries. Blood flow velocity is inversely proportional to the cross-sectional area; thus, the more area through which the blood flows, the slower it will go. This is similar to a stream that widens to become a river. Water in the upstream areas moves faster than the water in wider downstream areas. In the same way, the blood moves slowly through the capillaries and faster through the arteries and veins. The slow velocity of blood through the capillaries allows for greater exchange of materials with the tissues.

50 Nutrition, Digestion, and Absorption

Important Concepts

All animals must eat to acquire energy for survival.

- Animals are heterotrophs, meaning that they must get their energy and nutrients from the fats, carbohydrates, and proteins of other animals and plants. The energy that animals acquire from food is used to make ATP, a high-energy phosphate compound that provides energy for cellular work. The conversion of ATP during cellular work creates heat as a by-product.

- The calorie and kilocalorie (Calorie) are the measures of heat energy.

- Metabolic rate is a measure of the total energy used by an animal. Basal metabolic rate is the resting metabolic rate or the energy consumption needed for all essential physiological functions.

- Energy needed for metabolism can be obtained from food taken in or from stored food. Energy is stored as carbohydrates in liver and muscles. Fat has the highest energy content and is the most important form of energy stored in an animal's body.

Essential molecules must be obtained from a food source.

- Animals cannot produce acetyl groups through their own metabolism but instead must obtain them from another source (see Figure 50.4).

- Amino acids needed to make proteins are obtained by breaking down proteins in food. Some amino acids can be synthesized, but the essential amino acids can only be obtained from outside sources.

- Essential amino acids are needed for proper protein production. In humans these are isoleucine, leucine, lysine, methionine, phenylalanine, threonine, tryptophan, and valine.

- Ingested proteins are not used in their ingested form because, as macromolecules, they are not absorbed by cells in the gut. Also, their structures and functions are species-specific. Additionally, foreign proteins in the body would be recognized as invaders and attacked. Instead, proteins are broken down into peptides and amino acids.

- Essential fatty acids, such as linoleic acid, must also be acquired in food and are necessary for producing membrane phospholipids.

- The nutrients needed for survival are categorized as either macronutrients or micronutrients, depending on the quantity needed (see Table 50.1).

- Vitamins are used as coenzymes and are required in small amounts. Fat-soluble vitamins accumulate in the liver, whereas water-soluble vitamins are eliminated in urine. Vitamin D is needed to help in the uptake of calcium into the body.

Diseases can be caused by nutrient deficiency.

- Protein deficiency in humans can result in kwashiorkor, which leads to swelling, distension of the abdomen, degeneration of the liver, and other problems.

- Shortages in vitamins can lead to a large number of diseases (see Table 50.2). A deficiency in thiamin or vitamin B_1 results in beriberi, which causes loss of appetite and fatigue.

- Deficiencies may be due to the body's inability to absorb essential nutrients even when they are present in the diet. Anemia is caused by the inability to absorb vitamin B_{12} due to inadequate production of the intrinsic factor peptide needed for absorption.

There are a number of different ways for animals to feed.

- Two types of organisms feed on dead matter: decomposers, or saprotrophs, absorb nutrients from decaying organic matter, whereas detritivores actively feed on dead matter.

- Teeth are composed of a hard outer surface (enamel), a bony layer (dentine), and a pulp cavity that contains blood vessels and nerves (see Figure 50.7a).

- The shapes and organization of mammalian teeth reflect different diets (see Figure 50.7b). Herbivores eat plants. Since plant tissue is difficult to break down, herbivores have teeth that are good for shearing, crushing, tearing, and grinding, to break down vegetation.

- Carnivorous animals eat other animals that they capture. Their teeth are adapted for stabbing, ripping, and shredding their prey. Omnivorous animals eat both plants and animals and have teeth adapted for multiple purposes.

Digestion occurs in either a cavity or a tube in most animals.

- Food is generally digested extracellularly by enzymes that break it down. Most enzymes are produced in an inactive form called a zymogen and are converted to the active form at the site of use. Enzymes do not digest the gut because of a protective layer of mucus that lines it. There are a number of classes of digestive enzymes:

- Carbohydrases hydrolyze carbohydrates.
- Proteases hydrolyze proteins.
- Peptidases hydrolyze peptides.
- Lipases hydrolyze fats.
- Nucleases hydrolyze nucleic acids.

- In simple animals, a single opening that functions as both a mouth and an anus connects a gastrovascular cavity to the environment.

- The tubular gut includes the mouth where food enters and, in some animals, is broken up by grinding mechanisms that may include teeth or mandibles. Food then moves to a chamber for digestion and storage, such as the stomach, crop, or gizzard. It then enters the midgut or intestine where most of the nutrients are absorbed. The hindgut absorbs water and ions and stores waste until it is expelled from the anus during defecation.

- The parts of the gut that absorb nutrients have increased surface area in the form of fingerlike projections known as villi (see Figure 50.9).

The gut wall is composed of the same layers of tissue types throughout its length.

- Four layers of different cell types form the wall of the gut (see Figure 50.11).

- The innermost layer is a layer of mucosa that surrounds the cavity of the gut, or lumen. It secretes enzymes and hydrochloric acid and absorbs nutrients.

- The submucosa, lying under the mucosa, contains blood and lymph vessels that transport the nutrients.

- Smooth muscle surrounds the submucosa in both a circular and longitudinal layer that helps to move the contents of the digestive tract by peristalsis.

- The serosa and peritoneum make up the last two layers of the gut.

Food passes through the gut by smooth muscle contraction.

- Food enters the mouth and is swallowed, passing through the pharynx into the esophagus (see Figure 50.12a).

- Food is then moved along by peristalsis, which is a wave of muscle contraction that moves food through the gut (see Figure 50.12b).

- The esophageal sphincter helps keep food from reentering the esophagus from the stomach, and the pyloric sphincter controls the movement of chyme into the intestines.

A number of enzymes help break down food.

- The enzyme amylase begins the breakdown of starch in the mouth.

- In the stomach, the major enzyme is pepsin, which breaks down proteins into shorter peptides. Pepsin is initially secreted by gastric glands of the stomach in the zymogen form, pepsinogen (see Figure 50.13). Low pH activates the conversion of pepsinogen to pepsin through autocatalysis.

- The secretion of hydrochloric acid by the gastric glands in the wall of the stomach maintains a low pH.

- Hydrochloric acid and pepsin can produce an ulcer under certain conditions, particularly in the presence of the bacterium *Helicobacter pylori*.

- The acidic mixture of digesting food and gastric juices is known as chyme, which slowly passes into the intestines through the pyloric sphincter.

Most chemical digestion occurs in the small intestine.

- Carbohydrate and protein digestion continue in the small intestine.

- In humans, the small intestine can be more than six meters in length, with villi and microvilli contributing to a large surface area. The small intestine is divided into the duodenum, jejunum, and ileum.

- Fat digestion begins in the small intestine with the help of secretions and enzymes from the liver and pancreas.

- The liver secrets bile through the hepatic duct into the gallbladder, where it is stored. When lipids are present, bile is secreted by the gallbladder into the duodenum through the bile duct (see Figure 50.15). Bile molecules have hydrophobic ends, which attach to the tiny fat droplets to stabilize particles of fat in the form of micelles. This increases the exposed surface area of the lipids for digestion by fat-digesting enzymes, the lipases (see Figure 50.16).

- The zymogen trypsinogen is secreted by the pancreas and is converted into active trypsin by enterokinase. The pancreas produces a large number of zymogens that are released into the duodenum through the pancreatic duct. The zymogens are converted to the active form by trypsin. The pancreas also helps maintain a slightly alkaline pH in the duodenum by secreting bicarbonate ions.

The small intestine is the site of most nutrient absorption.

- Final digestion occurs among the microvilli of the small intestine. Proteins are broken down into amino acids by peptidases. The enzymes maltase, lactase, and sucrase break down disaccharides into monosaccharides. The mucosal cells secrete the peptidases and maltase, lactase, and sucrase.

- Absorption occurs via active transport for ions (sodium, calcium, and iron), amino acids, and sugars.

- Fats move through the plasma membranes as fatty acids and monoglycerides (see Figure 50.16). In the mucosal cells they are converted into triglycerides and combine with cholesterol and phospholipids to form chylomicrons that pass into the lymphatic system. Bile is not absorbed across the membrane of the duodenum, but is actively resorbed in the ileum and returned to the liver.

The large intestine is the site of water and ion absorption.

- Material entering the large intestine has had most of the nutrients removed, but still contains important water and ions. The water and ions are absorbed across the large intestine, leaving behind semisolid feces.

- Bacteria living in the large intestine produce vitamin K and biotin.

Herbivores employ fermenting microorganisms to help break down plant matter.

- Cellulase enzymes, which break down plant cellulose, cannot be produced by most herbivores.

- Microorganisms in the digestive tracts of herbivores produce the cellulase enzymes that break down cellulose into fatty acids.

- In place of a stomach, ruminants have a large four-chambered organ in which fermentation occurs (see Figure 50.17). The rumen and reticulum, the first two chambers, contain many microorganisms that break down cellulose.

- From the rumen, food passes through the omasum, where water absorption occurs, into the abomasum, or true stomach. In the abomasum, hydrochloric acid and proteases kill the microorganisms and digest them.

- Coprophagy is the reingesting of feces to increase fermentation. In coprophagous animals the fermentation vat is found in a branch off the large intestine called the cecum.

Digestion is controlled by hormones and autonomic reflexes.

- Salivation and swallowing are autonomic reflexes.

- The duodenum secretes secretin, a hormone that stimulates the pancreas to release bicarbonate ion into the small intestine to control pH (see Figure 50.18).

- Fats and proteins stimulate the mucosa of the small intestine to secrete cholecystokinin, which stimulates the release of bile and pancreatic digestive enzymes.

- Gastrin is secreted into the blood and is transported to the upper regions of the stomach, where it stimulates stomach secretions and movement.

Insulin and glucagon help control blood glucose levels.

- The absorptive period is the period after a meal when digestion occurs. The postabsorptive period is the period when the gut is empty and the body uses energy reserves.

- During the absorptive period, the liver takes up glucose from the blood and converts it to glycogen and fat, fat cells take up glucose, and the body preferentially uses glucose as the energy source.

- During the postabsorptive period, the liver breaks down glycogen into glucose, fatty acids are supplied to the blood by adipose tissues, and the body preferentially uses fatty acids as an energy source.

- The pancreatic hormones insulin and glucagon control metabolic fuel use (see Figure 50.19).

- Insulin is released in response to high blood glucose levels. Insulin stimulates cells to burn glucose, fat cells to use glucose to make fat, and liver cells to convert glucose to glycogen and fat.

- At postabsorptive blood glucose levels, insulin secretion is diminished and the liver and fat cells break down glycogen.

- At low levels of blood glucose, the pancreas releases glucagon, which stimulates the liver to break down glycogen and produce glucose.

Lipoproteins allow fats to be transported in the circulatory system.

- Lipoproteins are particles of fat and cholesterol covered with proteins that originate in the liver.

- Very-low-density lipoproteins (VLDL) are mostly triglyceride fats. Low-density lipoproteins (LDL) contain 50 to 60 percent cholesterol. High-density lipoproteins (HDL) are thought to carry cholesterol from tissues to the liver.

Food regulation may be controlled by leptin.

- Hunger and satiety are influenced by the region of the brain known as the hypothalamus. Both the ventromedial hypothalamus and lateral hypothalamus play a role in food intake.

- Leptin, a protein produced by fat, acts on the hypothalamus and is involved in control of hunger and satiety.

Toxins become concentrated in animals.

- Plants and animals may contain toxic compounds. Human activities also introduce toxic compounds into the environment.

- Hydrophilic water-soluble toxins can be metabolized quickly and filtered out of the body. Hydrophobic lipid-soluble toxins are metabolized slowly and tend to accumulate in adipose tissues.

- Bioaccumulation is the concentration of toxins in animals occupying higher and higher positions in the food chain. Long-lived species at the top of the food chain run the risk of heavy body burdens of toxins that are lipid soluble.

- The cytochrome P450 enzymes are involved in detoxification of natural chemicals. Some synthetic chemicals are not broken down by P450s and therefore bioaccumulate.

- Some synthetic chemicals may also act as hormone mimics.

The Big Picture

- Animals have evolved a wide variety of mechanisms for acquiring needed energy and raw materials for metabolism. All animals are heterotrophs; they can achieve energy and nutrient input by eating plants, animals, or both.

- The digestive system is specialized for efficient digestion of the type of food it must break down. The relatively short guts of carnivores have acid and protein-reducing enzymes. The longer guts of herbivores have numerous holding chambers for the laborious process of breaking down plant material.

- Food passing down the length of the gut is first broken down mechanically into small pieces and then broken down chemically into small molecules that can be easily absorbed across the gut. As nutrients and water are absorbed from plant material, the remaining material forms the feces, which are eventually eliminated from the body.

- The process of digestion is under both neural and hormonal control. Numerous hormones control the peristaltic movements of the gut and the secretion of enzymes and other chemicals involved in digestion.

Common Problem Areas

- The study of digestive function may seem uninteresting or, worse, unpleasant to students. Yet, some of the best and clearest examples of negative feedback and homeostasis are to be found in the study of digestion.

- Sometimes students think that there is a single digestive mechanism for food and are surprised to learn of the complexity of the process. Be sure to recognize that different food components—proteins, carbohydrates, fats—are digested, absorbed, and transported by very different mechanisms.

Study Strategies

- Although the guts of higher animals are structurally complex, they can be divided functionally into a foregut, midgut, and hindgut. If you break down the structure of the gut into these sections and determine which physiological process occurs in each section, you will have divided the task into smaller pieces.

- Because different types of food are digested and absorbed by different mechanisms, learn how a particular type of food is digested and match the appropriate enzymes with that process. For example, it will be much easier to remember how proteins are digested, rather than to recount a long list of digestive enzymes and then try to pick the one that might break down protein. In short, learn the process, not the list!

- Review the following tutorials and activities on the Student Website/CD:
  Tutorial 50.1 The Digestion and Absorption of Fats
  Tutorial 50.2 Insulin and Glucose Regulation
  Activity 50.1 Mineral Elements Required by Animals
  Activity 50.2 Vitamins in the Human Diet
  Activity 50.3 Mammalian Teeth
  Activity 50.4 Human Digestive System

Test Yourself

Knowledge and Synthesis Questions

1. Certain amino acids are essential to the diet of animals because
   a. they prevent overnourishment.
   b. they are cofactors and coenzymes that are required for normal physiological function.
   c. an animal cannot directly synthesize them through the transfer of an amino group to an appropriate carbon skeleton.
   d. animals need these substances in order to make stored fats that are used during hibernation and migration.

2. Which of the following descriptions describes the digestive characteristics of a sheep?
   a. It is a saprobe: It engulfs food and performs intracellular digestion.
   b. It is an autotroph: It synthesizes organic nutrients and performs extracellular digestion.
   c. It is an herbivore: It ingests food and performs extracellular digestion.
   d. It is a detritivore: It ingests food and performs intracellular digestion.

3. Protection of the walls of the stomach against the action of its own digestive juices
   a. results from the presence of an antienzyme chemical formed by the gastric glands.
   b. results from the nervous reactions of the lining of the stomach.
   c. is controlled by a center in the medulla of the brain.
   d. results from the mucus coating that covers its inner surface.

4. Chylomicrons are produced in the
   a. mouth.
   b. stomach.

   c. lumen of the small intestines.
   d. epithelial cells of the small intestines.

5. The gallbladder
   a. produces bile.
   b. is part of the liver.
   c. stores bile produced by the liver.
   d. produces cholecystokinin.

6. The pancreas
   a. produces exocrine products involved in chyme digestion.
   b. is exclusively an endocrine gland that produces salivary amylase.
   c. contains villi to increase surface area.
   d. produces urobiligen (a bile pigment).

7. Hydrochloric acid
   a. is secreted by the gastric glands of the liver.
   b. is secreted by the gastric glands of the stomach.
   c. produces a low pH in the small intestine.
   d. Both a and c

8. Bile produced in the liver is associated with which of the following?
   a. Emulsification of fats into tiny globules in the small intestine
   b. Digestive action of pancreatic amylase
   c. Emulsification of fats into tiny globules in the stomach
   d. Digestion of proteins into amino acids

9. Most of the chemical digestion of food in humans is completed in the
   a. small intestine.
   b. appendix.
   c. ascending colon.
   d. stomach.

10. Which of the following does *not* contribute to the large surface area available for nutrient absorption in the small intestines?
    a. Villi
    b. Intestinal length
    c. Microvilli
    d. Bile duct

11. Waves of muscle contractions that move the intestinal contents are
    a. caused by contraction of skeletal muscle.
    b. regulated by liver secretions.
    c. called peristalsis.
    d. voluntary.

12. What is the function of enterokinase?
    a. It converts pepsinogen to pepsin.
    b. It converts trypsinogen to trypsin.
    c. It digests proteins.
    d. It activates HCl.

13. Digestive enzymes responsible for breaking down disaccharides include
    a. pepsin, trypsin, and trypsinogen.
    b. amylase, pepsin, and lipase.
    c. sucrase, lactase, and maltase.
    d. pepsin, trypsin, and chymotrypsin.

14. Which of the following is characteristic of the large intestine?
    a. It has almost no bacterial populations.
    b. It contains chyme.

c. It absorbs much of the water remaining in waste materials.

d. It is the site of most of digestion.

15. The innermost layer of the digestive tract is the
    a. serosa membrane.
    b. mucosa membrane.
    c. submucosa membrane.
    d. lumen.

16. Which of the following hormones stimulates gluconeogenesis?
    a. Glucagon
    b. Insulin
    c. Estrogen
    d. All of the above

Application Questions

1. You have just eaten a small pizza with a lot of cheese. Because this food is high in lipids, your digestive system will have to perform specific tasks to digest and absorb the fats. What are the steps that your digestive system will take in order to digest and absorb the fats in this meal?

2. During the absorptive period your body is digesting and absorbing nutrients, whereas during the postabsorptive period your body is using up stored fuel. What are the main differences in fuel use and control mechanisms during these two periods?

3. Ingested food typically contains many useful proteins. However, these proteins are not used in the form in which they are ingested, but are broken down into smaller peptides and amino acids. Why does the digestive system break down these ingested proteins?

4. Humans have been polluting the environment for many years, dumping toxins into many of Earth's ecosystems. What effect does this have on toxin levels in the bodies of animals at the top of the food chain?

Answers

Knowledge and Synthesis Answers

1. **c.** Essential amino acids must be acquired through diet because an animal cannot directly synthesize all the amino acids needed for protein production.

2. **c.** A sheep is an herbivore that ingests plant material for nutrition. Digestion occurs extracellularly in the digestive tract of the animal.

3. **d.** The stomach is protected from digestive enzymes and low pH by the mucus secreted over its inner surface.

4. **d.** The epithelial cells of the small intestine produce chylomicrons by combining triglycerides with cholesterol and phospholipids, allowing lipids to pass into the lymphatic system.

5. **c.** Bile is produced by the liver, stored in the gallbladder, and released into the small intestine to aid in lipid digestion.

6. **a.** The pancreas produces exocrine products such as lipases, nucleases, amylases, and trypsin, all of which are involved in chyme digestion.

7. **b.** Hydrochloric acid is a strong acid secreted by gastric glands in the lining of the stomach. It lowers the pH of the stomach fluid.

8. **a.** Bile aids in the digestion of lipids in the small intestine by changing fat droplets into small fat particles.

9. **a.** The small intestine is the main site for chemical digestion in humans.

10. **d.** The length of the intestines and the presence of microvilli and villi increase the surface area of the small intestine.

11. **c.** Food is moved through the digestive tract by wavelike contractions of smooth muscle called peristalsis.

12. **b.** Enterokinase converts the zymogen trypsinogen to trypsin.

13. **c.** The digestive enzymes sucrase, lactase, and maltase are involved in the breakdown of disaccharides into the monosaccharides glucose, galactose, and fructose.

14. **c.** The large intestine is the site of water and ion absorption. The large populations of bacteria found in the large intestine contribute useful vitamins to their hosts.

15. **b.** The membranes of the digestive tract are, from the inside to the outside: mucosa, submucosa, circular and longitudinal muscles, serosa.

16. **a.** Glucagon increases the rate of gluconeogenesis, and insulin reduces the rate of gluconeogenesis.

Application Answers

1. The digestion of fats begins in the small intestine. Lipases are water soluble, but lipids tend to aggregate into large droplets in an aqueous environment. Bile helps to increase the exposed surface area of the lipids by forming small micelles with the lipids. This allows lipases to break down the fats into free fatty acids and monoglycerides. The fatty acids dissolve into the plasma membrane of the intestinal epithelial cells and will be absorbed across this surface. In mucosal cells, the free fatty acids and monoglycerides are turned into triglycerides, which are incorporated into chylomicrons. The chylomicrons can then pass out of the mucosal cells and into the lymphatic system (see Figure 50.16).

2. During the postabsorptive period, cells tend to metabolize fats, keeping the blood glucose reserves for the nervous system. Insulin levels are low during the postabsorptive period. During the absorptive period the body attempts to store the nutrients and fuel that are being absorbed. The preferred fuel source at this time is glucose, and insulin is released to help facilitate glucose uptake.

3. Proteins in digested food are not absorbed and used in their original form for a number of reasons. First, the proteins are too large to be easily absorbed by the cells of the digestive tract, whereas amino acids are readily absorbed. Second, proteins are species-specific, and thus a protein that is optimal for a prey species may not be optimal for the predator species. Lastly, the proteins of another species would be considered foreign and attacked by the immune system.

4. Toxins in the environment tend to bioaccumulate in top predators. Lipid-soluble toxins accumulate in body tissues of predators over their life spans. Therefore, as predators eat more contaminated food, concentrations of toxins in their bodies increase, resulting in increased body burdens.

51 *Salt and Water Balance and Nitrogen Excretion*

Important Concepts

Excretory organs control the solute concentrations of animals.

- Excretory organs control the level of solutes and the volume of tissue fluid and excrete urine. Urine composition depends on the environment surrounding an animal.

- Water movement in the body occurs either by pressure differences or differences in solute potential that produce water movement by osmosis. There is no active transport of water.

- Excretory systems use filtration across membranes, active secretion, and active resorption to control water and solute balance.

- Osmolarity is the concentration of salts in a fluid.

- Osmoconformers do not actively regulate the osmolarity of their tissues; rather, they allow them to come to equilibration with their environment. Marine invertebrates tend to be osmoconformers unless the concentrations in the environment are extreme.

- Osmoregulators actively regulate the osmolarity of their tissues, even as environmental osmolarity may change. The brine shrimp is an osmoregulator that maintains its tissue fluid osmolarity below that of the environment in high osmolarity waters and above that of the environment in low osmolarity waters (see Figure 51.1).

- Birds use a nasal salt gland to excrete salt from consumed seawater (see Figure 51.2).

Nitrogen is a potentially toxic metabolic waste.

- Ammonia (NH_3) is a highly toxic nitrogenous waste that must be either excreted or converted to the less toxic urea and uric acid (see Figure 51.3).

- Ammonotelic animals excrete ammonia to the aquatic environment, usually across gill membranes.

- Ureotelic animals excrete nitrogenous waste as water-soluble urea.

- To conserve water, uricotelic birds and reptiles excrete nitrogenous waste as insoluble uric acid.

Reflecting their diversity, invertebrates employ a number of excretory mechanisms.

- Marine invertebrates are typically osmoconformers.

- Freshwater and terrestrial invertebrates use active mechanisms to regulate their osmolarity.

- Flatworms have a protonephridium made up of tubules and flame cells that conserves ions and excretes water to produce a dilute urine (see Figure 51.4).

- Annelid worms filter the blood across the capillaries into the coelom, where the fluid enters a metanephridia in each segment through a funnel-like opening called a nephrostome. Tubules of the metanephridia actively secrete and absorb various ions and end in nephridiopores, which open to the environment to excrete dilute urine (see Figure 51.5).

- Insects use Malpighian tubules as their excretory organ. Malpighian tubules actively transport uric acid and sodium and potassium ions from the tissue into the tubules, with water following passively (see Figure 51.6). The tubules lead into the hindgut, where uric acid precipitates and water is reabsorbed.

The vertebrate excretory system is composed of nephrons in a kidney.

- The nephron of the vertebrate kidneys is composed of the glomerulus, renal tubules, and peritubular capillaries, which are further broken down into vascular and tubule components (see Figure 51.7).

- There are two sets of capillary beds in series in a nephron. The glomerulus is composed of the first set of capillary beds with blood entering at the afferent arteriole and exiting at the efferent arteriole. From the efferent arteriole the peritubular capillaries of the second capillary bed emerge and surround the tubule component of the nephron (see Figure 51.7).

- The renal tubule begins with Bowman's capsule, which encloses the glomerulus to form the renal corpuscle. Within Bowman's capsule, podocyte cells wrap around the capillaries of the glomerulus. Blood is filtered by the glomerulus into Bowman's capsule. The endothelial walls of the capillaries in the glomerulus have pores that allow water and small molecules to leave the blood. Only small molecules and water can enter Bowman's capsule.

- Arterial blood pressure is the driving force for filtration at the glomerulus. The porous capillary beds in the glomerulus also contribute to the high filtration rate.

- The filtrate entering the tubules is similar to blood plasma, but as the fluid moves through the tubules, ions and molecules are actively resorbed and secreted. This controls the dilution and composition of the urine.

Marine and freshwater fishes employ two opposing mechanisms for controlling water and ion balance.

- Freshwater fishes live in an environment hypotonic to their own body fluids, with water tending to move into the body by osmosis. To counter this, they excrete large amounts of dilute water and actively take up ions.

- Marine fishes live in an environment hypertonic to their body fluids, so water tends to be drawn out of the body by osmosis. They produce little concentrated urine, drink water, and actively secrete salts across the gills and the renal tubules.

- Cartilaginous fishes act as osmoconformers, allowing urea and trimethylamine oxide concentrations to increase, which increases the osmolarity of their tissues. The rectal gland excretes salts in these fishes.

Terrestrial and aquatic amphibians have different mechanisms for water management.

- Amphibians living in or near fresh water have a large water influx, and in response produce large amounts of dilute urine.

- Amphibians living in terrestrial environments have low metabolic rates and skin with reduced water permeability, resulting in low water turnover. Some also use their bladders as canteens to store water for later use.

Reptiles and birds conserve water in their terrestrial habitat.

- Reptiles lay shelled eggs, which allows them to live in a fully terrestrial environment. They have scaly, dry skin to decrease evaporative water loss, and they excrete uric acid with little water. They do not have the ability to produce concentrated urine.

- Birds have similar mechanisms as reptiles to control and conserve water, although some birds have the ability to produce concentrated urine.

The kidneys are the main excretory organs in mammals.

- In mammals, such as humans, the nephron is used to help conserve water when excreting nitrogenous wastes (see Figure 51.9).

- Humans have two kidneys that filter blood and produce urine. Urine exits the kidney through the ureter and travels to the bladder where it is stored until it is excreted through the urethra.

- The human kidney is shaped like a kidney bean, with a central medulla and a surrounding cortex. There are close to a million nephrons in the human kidney spanning the cortex and medulla regions.

- The glomerulus and adjoining proximal convoluted tubules are located in the cortex. The descending limb of the renal tubule runs down into the medulla, and the ascending limb returns to the cortex in what is known as the loop of Henle. From the ascending limb, the tubules become the distal convoluted tubule, which connects to the collecting ducts and runs down the medulla.

- Blood vessels also run in a similar pattern to the tubules. An afferent arteriole carries blood to the glomerulus where it is drained into the efferent arteriole, which gives rise to the peritubular capillaries. Capillaries run through the medulla, parallel to the loop of Henle, to form a vasa recta.

Filtration occurs at the glomerulus and resorption occurs largely at the proximal convoluted tubule.

- The typical glomerulus filters about 180 liters of blood per day in an adult human. However, 98 percent of this fluid is resorbed into the blood, so that only 2 percent of glomerular filtrate is excreted as urine.

- The majority of the filtrate is resorbed in the proximal convoluted tubules. Sodium ions and other solutes are actively transported out of the proximal convoluted tubule and water follows passively, all to be taken up by the peritubular capillaries.

The loop of Henle sets up a concentration gradient in the medulla.

- The loop of Henle produces a concentration gradient in the medulla by acting as a countercurrent multiplier system (see Figure 51.10). The purpose of this gradient is to move water across fluid compartments by osmosis because there is no mechanism for active transport.

- The concentration of the fluids surrounding the loop of Henle is raised by the active resorption of Cl^- and the passive flow of Na^+ into the tissue from the thick ascending limb. The thick ascending limb is impermeable to water, so only solutes leave the tubules.

- The thin descending limb is permeable to water but not sodium and chloride ions. Water moves out of the tubules into the surrounding tissues due to the higher concentration of Na^+ and Cl^- in the surrounding tissues.

- In the thin ascending limb, the fluid is more concentrated than the fluid in the surrounding tissues, so Na^+ and Cl^- move out of the tubule. Water cannot move out because of the low permeability of the thin ascending limb.

- The fluid reaching the distal convoluted tubule is less concentrated than blood plasma.

Concentration of urine occurs in the collecting duct.

- The collecting duct passes through the medulla, with its high solute concentration set up by the loop of Henle.

- Water moves out of the collecting duct and into the tissue, resulting in concentrated urine. The permeability of the collecting ducts can be adjusted hormonally to regulate the amount of water excreted.

- Longer collecting ducts result in a greater ability to concentrate urine.

Enzymes and hormones regulate glomerular filtration rate.

- Glomerular filtration rate is a function of the blood flow and pressure to the kidneys.

- The kidneys have autoregulatory mechanisms that help maintain their high filtration rate (see Figure 51.13).

- In response to decreasing arteriole resistance or decreased arterial blood pressure, the afferent renal arterioles open, or dilate, to increase flow to the glomerulus. If this does not increase glomerulus filtration rate sufficiently, the kidneys release the enzyme renin into the blood. Renin converts an angiotensin precursor into active angiotensin.

- Angiotensin constricts the efferent renal arterioles to increase pressure through the glomerulus. Angiotensin also constricts peripheral blood vessels, stimulates the release of aldosterone from the kidney, and stimulates thirst.

- Aldosterone stimulates the resorption of sodium by the kidney in an effort to maintain blood volume.

The kidneys help to regulate the pH of the blood.

- Bicarbonate ion are the major buffer in the blood. The kidneys control the level of H^+ and bicarbonate ions in the blood by varying their rate of secretion into the urine (see Figure 51.12).

Antidiuretic hormone regulates blood pressure and blood osmolarity.

- Antidiuretic hormone (ADH) is produced in the hypothalamus and released from the posterior pituitary gland (see Figure 51.14). ADH controls the permeability of the collecting ducts to water by stimulating production of water channels known as aquaporins over the long-term, and insertion of aquaporins into the membrane over the short-term.

- Stretch receptors in the aorta and carotid arteries inhibit the release of ADH in response to increased blood pressure, causing water to leave the body and blood volume to decrease.

- Osmosensors stimulate the release of ADH in response to increased osmolarity, resulting in increased water resorption and dilution of the blood.

- ADH also causes constriction of peripheral blood vessels.

- Atrial natriuretic peptide (ANP) is released by the heart in response to high blood pressure. At the kidneys, this hormone decreases sodium resorption, resulting in decreased water resorption and a lowered blood volume.

The Big Picture

- Different animals experience different types of stresses related to water and salt balance. Animals living in marine environments have to overcome problems with water loss due to osmosis, whereas animals living in freshwater environments have to overcome problems with water gain due to osmosis. Some invertebrates circumvent these problems by allowing their body fluid ion concentration to reflect that of their surroundings. Other invertebrates, and all vertebrates, retain their body fluids at more constant levels, expending energy to eliminate excess water and/or salts.

- All animals produce metabolic wastes (notably nitrogenous waste) as a by-product of metabolism. Invertebrates have evolved a variety of structures to eliminate wastes, whereas the vertebrates have evolved variations on the kidney. The nephron is the "functional unit" of the vertebrate kidney. By eliminating wastes without losing valuable nutrients and salts, the kidneys cleanse the blood of metabolic waste products.

- Kidneys work by differentially processing nitrogenous waste, ions, and water. In the kidney, large volumes of plasma are filtered from the blood into the interior of the nephron. Then, as the fluid passes through the various specialized regions of the nephron, the desirable components to be retained (most water and salts, nutrients) are pulled back into the tissues, leaving behind in the nephron the waste products in an ever more concentrated form of urine. As urine leaves the nephron, water content

is adjusted in the collecting ducts through hormonal alteration of their water permeability. If the collecting ducts are highly water permeable, then water will be drawn by osmosis out of the collecting ducts into the region of high salt concentration created by the action of the loop of Henle.

Common Problem Areas

- Depending on whether it is living in fresh water or saltwater, fishes may drink water, excrete copious amounts of dilute urine or small amounts of more concentrated urine, or take up ions at their gills. Given that living in either of these environments creates osmoregulatory problems, think about whether the tissues have higher or lower solute concentration relative to the surrounding water, and then predict water and ion movements on that basis.

- Understanding how the nephron functions, and in particular how the loop of Henle works as a "countercurrent multiplier system," is one of the greater challenges in this textbook.

Study Strategies

- Nephron function can be understood more easily if you place yourself in the position of a single Na^+ ion in the blood entering the nephron. Follow your pathway—and your potential pathways—as you traverse the nephron, eventually ending up in the urine. Note especially that it may take you some time to pass through the loop of Henle as you leave the ascending limb, only to recycle back into the descending limb—a process that could be repeated many times before finally escaping the countercurrent multiplier. Repeat this imaginary journey again as a water molecule.

- Review the following tutorial and activities on the Student Website/CD:
  Tutorial 51.1 The Mammalian Kidney
  Activity 51.1 Annelid Metanephridia
  Activity 51.2 The Vertebrate Nephron
  Activity 51.3 The Human Excretory System

Test Yourself

Knowledge and Synthesis Questions

1. What mechanisms does a marine fish use to maintain homeostasis?
   a. It drinks copious amounts of water, excretes only small amounts of water, and pumps sodium out of its body.
   b. It drinks little water, excretes large amounts of water, and pumps sodium into its body.
   c. It drinks little water, excretes only small amounts of water, and pumps sodium out of its body.
   d. None of the above

2. Which of the following statements about the excretory system of insects is *false*?
   a. Active transport moves materials from the coelomic fluid into the Malpighian tubules.
   b. The Malpighian tubules can produce a highly concentrated urea solution, allowing insects to inhabit some of Earth's driest habitats.
   c. Resorption of salts takes place mostly in the gut.
   d. Water resorption is by osmotic movement only.

3. What pathway is taken by water and solutes as they travel through a nephron?
   a. Glomerulus, to Bowman's capsule, to proximal tubule, to loop of Henle, to distal tubule, to collecting ducts
   b. Bowman's capsule, to glomerulus, to distal tubule, to loop of Henle, to proximal tubule, to collecting ducts
   c. Glomerulus, to Bowman's capsule, to distal tubule, to loop of Henle, to proximal tubule, to collecting ducts
   d. Glomerulus, to Bowman's capsule, to proximal tubule, to collecting ducts, to distal tubule, to loop of Henle

4. Na^+ and Cl^- are actively transported out of the tubules to help set up the countercurrent multiplier. Which of the following are sites of active Na^+ and Cl^- transport in the nephron?
   a. Proximal tubule, ascending limb of the loop of Henle
   b. Descending limb of the loop of Henle, ascending limb of the loop of Henle
   c. Ascending limb of the loop of Henle, proximal tubule
   d. Collecting duct, descending limb of the loop of Henle

5.–9. Use the following diagram to complete the statements about the mammalian nephron.

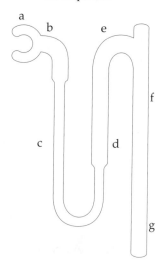

5. The composition of the filtrate would be most like plasma in the tubule next to letter _____.

6. The NaCl concentration in the extracellular fluid would be greatest in the area of letter _____.

7. The osmolarity of the filtrate next to letters _____ is similar to the osmolarity of blood plasma.

8. The urine would be most concentrated in the collecting duct next to letter _____.

9. Most of the glomerular filtrate is resorbed into the blood in peritubular capillaries next to letter _____.

10. The sole mechanism for water resorption by the renal tubules is
    a. active transport.
    b. osmosis.
    c. cotransport with sodium ions.
    d. cotransport with bicarbonate ions.

11. A number of hormones help to regulate water and solute uptake and release in the nephron. Antidiuretic hormone (ADH) promotes _____ in response to _____.
    a. active transport of Cl^-; increased solute concentration
    b. active transport of Na^+; increased blood pressure

    c. increased permeability of the collecting duct to water; decreased blood pressure
    d. decreased permeability of the collecting duct to water; increased solute concentration

12. Which of the following is *not* a normal constituent of the glomerular filtrate?
    a. Red blood cells
    b. Urea
    c. Sodium ion
    d. Glucose

13. If the afferent arteriole that supplies blood to the glomerulus becomes dilated,
    a. the protein concentration of the filtrate decreases.
    b. hydrostatic pressure in the glomerulus decreases.
    c. the glomerular filtration rate increases.
    d. All of the above

14. Osmoreceptors in the brain detect changes in blood ion concentration that can reflexively result in _____ by the kidneys.
    a. water retention
    b. water loss
    c. ion retention
    d. All of the above

15. If the human kidneys filter 150 liters of plasma in a 24-hour period, what is the typical amount of urine produced and eliminated in that time period?
    a. 0.15 liters
    b. 1.5 liters
    c. 15 liters
    d. 30 liters

Application Questions

1. Based on your knowledge of how the nephron works, explain why mammals produce concentrated urine whereas reptiles cannot.

2. What would happen to the composition of excreted urine if all active transport processes in the nephron came to a stop?

3. The drug urizadole inhibits antidiuretic hormone secretion from the pituitary. How will this change glomerular filtration rate, urine flow, and urine concentration in patients taking this medication?

4. You have just eaten a large number of very salty potato chips, and the osmolarity of your blood has increased. What response will your body have in order to bring your blood osmolarity back to homeostasis?

5. Earthworms and humans are very different, yet their excretory systems share many characteristics. List a few of the ways in which the excretory systems are similar in form and function.

Answers

Knowledge and Synthesis Answers

1. **a.** Marine fishes live in an environment in which salts will tend to be drawn into the body and water will tend to be drawn out of the body. To counter these effects, marine fishes drink lots of water, excrete small amounts of water, and actively pump sodium out of the body across the gills.

2. **b.** All are true of insects except that they excrete uric acid, not urea.

3. **a.** The route of water and solutes through the nephron is from the glomerulus, to Bowman's capsule, to proximal tubule, to loop of Henle, to distal tubule, to collecting ducts.

4. **a.** The proximal tubule is the site of active transport of Na⁺ out of the tubule. Na⁺ also moves out of the tubule at the ascending limb of the loop of Henle, but this is passive transport, with Cl⁻ being actively transported out.

5. **a.** At Bowman's capsule the filtrate is most similar to plasma.

6. **g.** The sodium concentration is the highest in the extracellular fluid near the middle of the medulla.

7. **a., b.,** and **e.** The filtrate has a similar osmolarity to plasma in the cortex of the kidney, including the proximal convoluted tubule, the glomerulus and Bowman's capsule, and the distal convoluted tubule.

8. **g.** The highest concentration of the filtrate in the collecting ducts will be near their ends, deep in the medulla.

9. **b.** The bulk of the water and solute that are resorbed from the filtrate are resorbed at the proximal convoluted tubule.

10. **b.** The sole mechanism for water resorption in the renal tubules is by osmosis.

11. **c.** Antidiuretic hormone acts on the collecting ducts by increasing permeability to water. Antidiuretic hormone secretion is stimulated by a decrease in blood pressure.

12. **a.** Red blood cells are too large to be filtered out of the blood at the glomerulus and thus will not be found in the filtrate.

13. **c.** Changes in the afferent arteriole pressure affect glomerular filtration rate. Dilation of the afferent arteriole increases pressure, which will increase filtration rate.

14. **d.** Osmoreceptors in the brain detect changes in blood ion concentration, resulting in reflexive increases or decreases in both ion and water retention, depending on the status of blood osmolarity.

15. **b.** About 1 percent of the filtrate is excreted as urine, so about 1.5 liters of the original 150 liters would be urinated.

Application Answers

1. The loop of Henle acts as a countercurrent ion multiplier in the medulla of mammals. This allows mammalian kidneys to produce concentrated urine. Reptiles lack the countercurrent multiplier of the loop of Henle.

2. Active transport of Na⁺ and Cl⁻ in the nephron provides the ions that set up the countercurrent multiplier, allowing for the production of concentrated urine in mammals. If active transport in the nephrons were to stop, the urine produced would eventually be isotonic with blood plasma because the countercurrent multiplier would disappear.

3. Inhibition of antidiuretic hormone by urizadole will affect the permeability of the collecting ducts to water. This will result in increased urine flow because water will not be able to be resorbed across the collecting ducts. Blockage of ADH will have no effect on the glomerular filtration rate.

4. In response to increased blood osmolarity, osmosensors in the hypothalamus stimulate the release of ADH from the pituitary into the blood. ADH acts to increase the permeability of the collecting ducts to water so that increased amounts of water can be resorbed to bring down the blood osmolarity. The osmosensors will also stimulate thirst, causing you to increase your water intake.

5. The metanephridia of the earthworm and the nephron of humans are both composed of tubules through which filtrate flows. In the tubules of both, certain molecules are actively transported out or secreted in. Both systems produce urine with an osmolarity different from that of the body fluid. The urine of the earthworm is hypotonic, whereas that of humans is usually hypertonic.

52 *Animal Behavior*

Important Concepts

Behaviors can be inherited or learned.

- The study of the function and evolution of behavior is known as ethology.

- Some behaviors are highly stereotypic and species-specific.

- Behaviors, such as web spinning in spiders, require no prior experience and are inherited (see Figure 52.1).

- Determination of inheritance of behaviors can be made by deprivation experiments, in which all relevant experience is denied, or by hybridization experiments, in which parents with known behaviors are crossed and the offspring are studied for their behavior.

- Hybridization experiments by Konrad Lorenz on ducks, in which the mating behavior was different for similar species, showed that hybrids performed mating behaviors that were a combination of those of the two parents. Nonhybrids were not attracted to hybrid displays. Thus, the species-specific displays may function to prevent hybridization (see Figure 52.2).

- Triggers or releasers may be necessary for animals to exhibit genetically-based behaviors. Releasers can be a simple subset of sensory information available to an animal. In herring gull chicks, a red dot on a parent's beak triggers a begging response (see Figure 52.3). The red dot on a parent's beak acts as a releaser for the genetically based begging behavior.

- Most behaviors involve the interaction between genes and learning. Spatial learning involves learning features in the environment and using them as cues. Wasps use spatial cues from the environment to help them locate their nests (see Figure 52.4).

- Behaviors with a genetic basis can be highly adaptive when making a mistake is costly or fatal. Examples of such behaviors are predator avoidance and mating rituals.

Bird songs involve both imprinting and genes.

- Imprinting of offspring on parents is a learned response during a critical period that helps in recognition. The critical period for imprinting is a short time during a developmental or hormonal state of the animal.

- Birds use songs in territorial displays and in courtship.

- In some species, imprinting of the species-specific song is required in the nestling in order for it to sing the song as an adult, even though the song is never sung by the juvenile bird. Imprinting forms a memory of the song that is recalled as the bird approaches adulthood. The adults need both the initial imprinting of the song as a juvenile and the ability to match the song with auditory feedback in order to sing the correct song (see Figure 52.6).

- Once a bird has perfected its song, the song is said to have "crystallized," and it will be sung in the correct form every spring.

Hormones can play an important role in controlling the behavior of animals.

- Males and females of a species frequently display different behaviors that are the result of sex steroids.

- Experiments on rats have revealed the influence of sex steroids on the development and activation of sexual behavior (see Figure 52.7).

- Sex steroids are necessary for the expression of sexual behavior, with testosterone present in males and estradiol present in females. Exposing male or female rats to testosterone during development results in the potential to display male sexual behavior in adults of both sexes; testosterone is needed to activate male sexual behavior in adulthood. Normal development of male sexual behavior requires the presence of testosterone, whereas normal development of female sexual behavior does not require estradiol.

- In male birds, testosterone levels control singing by inducing certain regions of the brain to grow during the breeding season (see Figure 52.8). During the nonbreeding season, these regions of the brain associated with singing are reduced in size.

- Female birds that are treated with testosterone during the spring will also develop the species-specific song. In these birds, testosterone has stimulated growth in those areas of the brain associated with singing.

There is a strong genetic component to animal behavior.

- Hybridization in crickets reveals the genetic basis for singing behavior. Crossing two species with distinct songs results in offspring song that contains features of both parental species. A female cricket's recognition of songs also has a genetic basis, with hybrid females choosing mates with the hybrid song.

- Artificial selection in dogs provides a good example of how different breeds have genetically-based behavioral tendencies.

- Nest-cleaning and non-nest-cleaning strains of honeybees also provide evidence for genetic control. Hygienic bees uncap cells and remove dead larvae from the hive, making them resistant to a bacterium, whereas another strain does not engage in nest-cleaning behavior. Crossing experiments have shown that two recessive genes control the cleaning behavior (see Figure 52.9). In crosses involving the second generation, nonhygienic bees showed a mix of behaviors.

- In the fruit fly, a hierarchy of genes has been found to influence sexual differentiation and courtship behavior. Courtship in male fruit flies is species-specific, requires no learning, and appears to be controlled by one gene, *fruitless* (*fru*).

- The *sex-lethal* (*sxl*) gene is expressed in female fruit flies and controls sexual differentiation and behavior. The *sxl* gene controls the *transformer* (*tra*) gene and its product, the Tra protein, which is required for development into a female. Tra protein in turn controls the *doublesex* (*dsx*) gene involved in male differentiation and the *fruitless* (*fru*) gene involved in male courtship behavior (see Figure 52.10).

Many animal behaviors involve some form of communication between individuals of a species.

- Communication is used by animals to transmit information and reduce uncertainty between individuals of a species.

- Pheromones are chemical signals used to communicate among individuals of a species. Pheromones can be long lasting, like the trail-marking and territory-marking pheromones of ants or cats, or they can be short lasting, like the sex-pheromones released by moths. The chemical properties of the pheromones dictate how long they last in the environment and how far they travel.

- Visual signals provide rapid, directional communication. One drawback to visual signals is the need for light.

- Sound has advantages over sight in that it can travel in complex environments, such as a forest, and over long distances, such as the sound produced by whales in the water. Sound also communicates directional information.

- Touch is commonly used in communication. Honeybees dance to communicate the location of a food source in the environment (see Figure 52.12). Round dances are used to communicate that food is less than about 80 meters from the hive. A waggle dance is used to communicate distance and direction to food that is more than 80 meters from the hive. Speed of the waggle indicates distance to the food source; direction of the straight run of the waggle dance indicates the direction of the food source relative to the sun.

- Some fish use electrical fields to sense their surroundings. In the glass knife fish, different individuals emit different frequencies that relate to their status in the population.

The timing of behaviors is controlled internally.

- Circadian rhythms are daily cycles in activity, sleep, foraging, and other physiological processes and behaviors that are controlled by an endogenous clock (see Figure 52.13).

- The length of one cycle in a rhythm is defined as one period. Any point in the cycle is known as a phase. Two cycles can be in phase if the rhythms match, or be phase-advanced or phase-delayed if they do not match.

- Circadian rhythms can be reset by environmental cues, such as the light–dark cycle, during entrainment. In constant conditions an animal will have a natural period that is different from the 24-hour period of the day. Such an individual is said to be free-running. Shifts of no more than 20–30 minutes can occur in humans per day, explaining why you get jet lag when crossing time zones.

- The circadian clock in mammals is located in the suprachiasmatic nuclei (SCN) just above the optic chiasm (see Figure 52.14). Removal of the SCN results in the loss of circadian rhythms. In some birds the pineal gland functions as the master clock.

- The genes that are involved in circadian rhythms are highly conserved across a wide range of organisms. In the fruit fly, the *per* gene is involved in the length of rhythms, and the absence of the *tim* gene results in the total loss of circadian rhythms (see Figure 52.15).

- The length of the day or photoperiod can trigger seasonal rhythms. An endogenous seasonal rhythm known as the circannual rhythm is found in hibernators. Seasonal rhythms are important in controlling the timing of breeding and hibernating. When the changing length of available daylight influences the physiology and behavior of a species, the species is said to be photoperiodic.

Behaviors control navigation.

- Recognition of landmarks is used to navigate during piloting.

- Homing is the ability to return to a specific location. Piloting is one simple method used in homing. However, many seabirds fly over the oceans where there are no environmental cues. Migrating birds are able to fly great distances and return to the same breeding ground each season.

- Distance-and-direction navigation involves knowledge of direction and distance to a destination (see Figure 52.17). Biological rhythms can help in distance determination, and the position of the sun and stars and can help in direction determination. The stars offer both moving and fixed points for navigation.

- Bicoordinate navigation is the process of using location and destination to plot a course. Bicoordinate navigation is also known as true navigation.

- There is considerable redundancy in animals' abilities to sense direction. For example, homing pigeons use, among other things, magnetic cues and a time-compensated sun compass (see Figure 52.18).

Human behavior

- Human behavior can be modified by experience over time.

- Learned behavior in humans is passed down from generation to generation as part of culture.

- Chimpanzees, like humans, exhibit different cultures in different physical locations.

- Motor patterns such as facial expressions appear to have a genetic component.

The Big Picture

- Animals exhibit a wide range of species-specific behaviors that can be either learned or instinctive. Instinctive, genetically based behaviors may require triggers or releasers to stimulate an animal. Nevertheless, patterns of behavior typically have genetic and environmental influences.

- Hormones are important in controlling the behavior of animals. Sex steroids are frequently involved in stimulating the different behaviors of males and females of a species. Daily, seasonal, or annual rhythms are often carried out through the actions of hormones.

- Communication is an integral part of animal behavior. Communications can be chemical (involving pheromones), visual (involving displays of fins, feathers, fur, etc.), auditory (sounds created by either general or specialized structures and received by auditory sensors), tactile, or electric. Such communication is an important component of territory-marking and reproductive behavior.

- Animals typically have more than one method that they can use to orient and navigate.

Common Problem Areas

- It can sometimes be difficult to distinguish between behaviors founded in genetics (instinctive behaviors) and those that are learned during the course of an animal's lifetime. Further muddying the waters, the ability to learn new behavior is, of course, a heritable trait!

- Hormones control many behaviors, and hormone production is often a function of the time of day or the time of year. Thus, a particular hormone that stimulates (or inhibits) behaviors under one condition at one point in time may not have the same effect at another time. Be sure to learn the conditions and caveats that accompany hormone actions.

Study Strategies

- As you learn about various animal behaviors, it may be useful to design "mind experiments" in your imagination to help you to distinguish between an instinctive and learned behavior. This also may help you to work out the nature of the behavior.

- Learn to distinguish the various behavior cycles associated with internal controls via "biological clocks" as opposed to behaviors that are directly stimulated by an immediate event in an animal's surrounding environment.

- Review the following tutorials and activity on the Student Website/CD:
  Tutorial 52.1 Hormonal Control of Sexual Behavior
  Tutorial 52.2 Circadian Rhythms
  Tutorial 52.3 Time-Compensated Solar Compass
  Activity 52.1 Honeybee Dance Communication

Test Yourself

Knowledge and Synthesis Questions

1. A newborn male rat is castrated and then allowed to mature. When he reaches adulthood, he is given an injection of male sex steroids and exposed to a normal adult female rat exhibiting lordosis. Which of the following will occur?
   a. The male will mount the female and attempt to copulate.
   b. The male will exhibit lordosis.
   c. The male will attempt to copulate and exhibit lordosis.
   d. The male will exhibit no sexual behavior at all toward the female.

2. Select the behavior that is least likely to be a fixed action pattern.
   a. Egg-laying in a turtle
   b. Courtship display
   c. A predator's prey-search behavior
   d. Web spinning in a spider

3. Which of the following statements about hormonal influences on bird song is *false*?
   a. Male songbirds castrated as nestlings do not sing proper songs as adults.
   b. Female songbirds can be induced to sing by treatment with testosterone.
   c. Females learn their species song, but do not normally express it under natural conditions.
   d. Testosterone is not necessary for singing in males once they have learned their song.

4. Studies of which of the following organisms have led most directly toward connecting a specific gene with a specific behavior?
   a. Honeybees
   b. Dabbling ducks
   c. *Drosophila*
   d. Domestic animals

5. Which of the following is not a valid conclusion resulting from studies done on hygienic behavior in honeybees?
   a. The inheritance of this behavior can be explained with simple Mendelian genetics.
   b. The uncapping and larval-removal components of this behavior seemed to be determined by a single gene.
   c. Because all F_1 bees resulting from a cross between hygienic and nonhygienic strains were nonhygienic, the trait must be determined by recessive alleles.
   d. A backcross of the F_1 to the nonhygienic strain should yield one-quarter hygienic to three-quarters nonhygienic individuals.

6. Which of the following statements about the action of releasers in herring gull chicks is *false*?
   a. The red dot on a parent's beak acts as a releaser.
   b. The shape of the parent's head is not important in eliciting a response from the chick.
   c. This behavior has a strong learning component.
   d. Begging is the behavior triggered by a releaser.

7. Which of the following statements about genetically determined behaviors is *false*?
   a. Inherited behaviors that involve mating and courtship are highly adaptive.
   b. In hybridization experiments, closely related species are bred to determine the behavior of the offspring.
   c. Web spinning in spiders is species-specific.
   d. Most behaviors have only a genetic component.

8. Animals exhibit daily rhythms in their behavior and physiology. Which of the following statements about circadian rhythms is *false*?
   a. Animals that are active at night do not display circadian rhythms.
   b. A circadian clock can be entrained by environmental cues.
   c. A circadian clock is located in the suprachiasmatic nuclei in mammals.
   d. In fruit flies, loss of the *tim* gene will result in the loss of circadian rhythms.

9. Which of the following is an example of piloting?
   a. Silkworms following a trail of pheromones
   b. Marine migration over a featureless ocean
   c. Bees learning the direction of a food source from a dance
   d. Gray whale migration along the pacific coast of America

10. There are a number of different navigational methods used by animals. Which of the following definitions of a navigational method is correct?
    a. Distance-and-direction navigation requires knowledge of latitude and longitude.
    b. Bicoordinate navigation requires knowledge of direction of and distance to a destination.
    c. The stars offer two sources of information about direction; fixed points and moving constellations.
    d. Piloting involves the magnetic mineral magnetite.

11. Which of the following animals would *not* be expected to have an endogenous circannual rhythm?
    a. A ground squirrel that hibernates
    b. An Arctic hare
    c. A cave salamander
    d. A migratory bird that summers in North America and winters in South America

12. In studies using fruit flies, it has been found that the _____ gene is involved in the development of male courtship behavior, and the _____ gene is involved in anatomical differentiation of males.
    a. *sex-lethal; transformer*
    b. *sex-lethal; doublesex*
    c. *fruitless; doublesex*
    d. *fruitless; transformer*

Application Questions

1. You go to a nature center that has a honeybee hive that you can see into. As you look at all the busy bees, you notice that one bee is doing a dance. What is the purpose of this dance, and what might it be telling the other bees?

2. A starling has been trained to feed at food bins at the eastern end of a circular cage from which it can see the sky. There are food bins at the N, NE, E, SE, S, SW, W, and NW ends of the cage. Knowing this, answer the following questions.

   a. If the cage is covered and a fixed light source is presented at the east end of the cage at the time of sunrise, where will the starling search for food at noon?

   b. If a mirror is used to shift the apparent position of the sun at noon from the south to the northwest, where will the starling search for food?

   c. The bird has been placed in a light-controlled environment for three weeks and phase-delayed by six hours. If the starling is returned to the cage under natural lighting conditions at sunset, where will it search for food?

Answers

Knowledge and Synthesis Answers

1. **d.** The development of normal male sexual behavior requires the presence of testosterone. In castrated newborn males, the testosterone is lacking. When given testosterone as an adult, this animal will not exhibit sexual behavior toward a female.

2. **c.** Fixed-action behaviors are acquired by inheritance, they are not learned. The behavior of a predator may be a learned response rather than an inherited one.

3. **d.** Young male songbirds must hear the species song and then later be able to hear themselves sing. In the adult, the presence of testosterone is needed to increase the size of regions in the brain associated with singing during the breeding season. Absence of testosterone will result in the inability to perform the correct song.

4. **c.** *Drosophila* has been used in many genetic studies on behavior.

5. **b.** In hygienic bees, uncapping and larval removal appear to be controlled by two separate genes.

6. **c.** The begging response that occurs when herring gull chicks observe a red dot is largely instinctive.

7. **d.** Behaviors tend to involve a complex interaction between learning and genetics.

8. **a.** All animals exposed to a daily cycle display circadian rhythms, regardless of when they are active.

9. **d.** Piloting is a means of navigation using visual cues and landmarks. Following the coast during migration is considered piloting.

10. **c.** Many animals use the stars for navigation. The stars present two types of information: Some are used as a fixed reference point, whereas others provide moving constellations.

11. **b.** Hibernators, cave-dwellers, and equatorial migratory species lack the necessary daily cues at some stage during the year, so they most likely have endogenous circannual rhythms. Thus the Arctic hare would be the only animal that could use circadian cues.

12. **c.** The *fruitless* gene regulates development of male courtship behavior, and the *doublesex* gene regulates male anatomical development.

Application Answers

1. Honeybees dance to inform other members of the hive about the distance and location of a food source. If food is less than about 80 meters from the hive, the bees will perform a round dance. If food is farther from the hive, the bees will perform a waggle dance. Only the waggle dance contains information about direction.

2. a. The starling will eat from the northern bin at noon. Normally at noon the sun is in the south, and the bird would eat from the eastern bin that is 90° to the left of the sun. With the light in the east, the bird would eat in the north, which is 90° to the left of the light.

   b. The bird will go to the bin that is 90° from the sun at noon, and will feed from the southwestern food bin.

   c. At sunset the eastern bin with food would normally be 180° from the sun in the west. After the bird has been phase-delayed six hours, it will think that the setting sun is the noon sun. At noon the bird usually eats from the eastern bin, which is located 90° to the left of the sun. Therefore, the bird will eat from the southern bin.

53 **Behavioral Ecology**

Important Concepts

Ecology is the study of the interactions of organisms with one another and with their nonliving environment.

- Ecology is a complex field of inquiry dealing with levels of organization ranging from the interactions of individual organisms with the environment to the structure of ecosystems.

- Behavioral ecology is the study of how animals decide when and where to carry out their activities; how to respond to environmental changes; how to select the resources they need; and how to respond to predators, competitors, and other members of their own species.

- Animals do not make conscious "decisions," but their behavioral choices do influence their chances of survival and reproduction and thus are molded by natural selection.

- The environment of an organism includes both abiotic (chemical and physical) and biotic factors. Organisms both influence and are influenced by their environments.

All organisms have the ability to respond adaptively to changes in their environment.

- Environmental changes to which organisms respond occur at different time scales. Some changes require immediate responses; others permit a more gradual response.

- All organisms have the ability to change their location, either actively or passively, at some stage of their lives.

- The habitat of an animal is the environment in which it normally lives. In choosing their habitat, animals use cues that are good predictors of general conditions suitable for future survival and reproduction.

- The success of already-settled individuals may indicate habitat quality.

- The territory of an animal is an area from which other individuals of its own species (and sometimes individuals of other species) are excluded. By establishing a territory, an animal may improve its fitness by gaining exclusive use of the resources of part of its habitat.

- Natural selection molds behavior (such as defending a territory) in accordance with costs and benefits. There are three aspects to the cost of behaving:

  1. *Energetic cost*—the amount of energy the animal expends during the behavior

  2. *Risk cost*—the amount of risk the behavior entails for the animal

  3. *Opportunity cost*—the benefits the animal forfeits by not engaging in other behaviors instead

- Foraging theory is an attempt to understand how animals select what kind of food to eat and where and when to search for it. These questions are investigated by optimality modeling, which assumes that natural selection has molded the behavior of animals so that they make the best choices available to them. To decide whether a model is realistic, scientists test its predictions using experiments that measure the actual foraging behavior of animals.

- Foraging animals frequently seek out sources of mineral nutrients. Human enjoyment of spices may have evolved because most spices used in cooking have antimicrobial activity.

- One important choice of associates an animal makes is mate selection. Because the quantity of sperm produced by males is typically far greater than the number of eggs produced by females, males and females approach courtship very differently. Males of most species can increase their reproductive success by mating with many females. In contrast, the reproductive success of females depends primarily on the quality of genes she receives from her mate, the resources he controls, and the amount of care he provides for their offspring. As a result, males usually initiate courtship and frequently fight for access to a female. Females seldom fight over males and often reject courting males.

- A male may attract females by courtship behavior that signals his good health and genetic quality. The behavior also may show that he is a good provider of parental care.

- By their ability to recognize the signals at which males cannot cheat, females have favored the evolution of "reliable" signals of male quality.

Social behavior evolves when individuals that cooperate with others of the same species have higher rates of survival and reproductive success than those achieved by solitary individuals.

- Social systems are best understood not by asking how they benefit the species as a whole, but by asking how the individuals that join together benefit by the association.

- Social systems are dynamic: Individuals constantly communicate with one another and adjust their relationships.

- Individuals in a social system experience different benefits and costs, depending on their age, sex, physiological condition, and status.

- Group living may benefit both predator and prey species. For predators, it may improve hunting efficiency or increase the size of the prey that can be captured. For prey, it may provide increased protection against predators.

- Living in a group imposes costs as well as benefits. Individuals in groups may compete for food, interfere with one another's foraging, injure one another's offspring, inhibit one another's reproduction, or transmit diseases to their associates.

- The most widespread social system is the family, an association of one or more adults and their dependent offspring. More complex social systems in animals have evolved via an extended family.

- Altruistic acts evolve most easily through kin selection, by which an individual can increase the representation of its own genes in the population by helping its relatives. Together, individual fitness and kin selection determine the inclusive fitness of an individual. An altruistic behavior pattern can evolve if it increases the inclusive fitness of the altruist—that is, if the benefit of increasing the reproductive success of relatives is greater than the cost of decreasing the altruist's own reproductive success.

- Eusocial species are those whose social groups include sterile individuals. The evolution of eusociality in the Hymenoptera (ants, bees, and wasps) has been facilitated by their sex determination system, in which males are haploid and females are diploid, with the result that sisters are genetically more similar to one another than to their own offspring. In eusocial species in which both sexes are diploid (termites, naked mole-rats), inbreeding and the difficulty of establishing independent colonies have favored the evolution of eusociality.

Decisions that organisms make about habitats, food, and associates have many implications for the structure and functioning of ecological systems.

- Social animals may achieve extraordinary abundance. Examples include ants, termites, and wildebeest. The increase in human abundance has occurred because of specialization, leading to advances in agriculture, medicine, and hygiene.

- Interspecific interactions, such as the behavioral dominance of one species over another, influence animal distributions.

The Big Picture

- Behavioral ecologists attempt to understand how the decisions animals make are related to their survival and reproductive success. In this effort, they not only draw on concepts from evolution, ecology, and animal behavior, but also make important contributions to those fields.

- As you learned from Chapter 23 (The Mechanisms of Evolution), the cost–benefit approach applied in this chapter to the analysis of behavior can be extended to any kind of adaptation of organisms.

- Because, according to standard Darwinian theory, only traits that contribute to individual fitness are favored by natural selection, explaining the evolution of altruistic behavior has presented a major challenge to evolutionary biologists. The concepts of kin selection and inclusive fitness have proved to be very fruitful in understanding altruism in animals and have also been applied to many aspects of human social behavior (though not without controversy).

Study Strategies

- In studying any area of biology, you should try to develop a sense of the kinds of questions that scientists in the field are trying to answer. Behavioral ecologists are especially interested in the adaptive value of behavior (i.e., how it contributes to fitness). As you know from reading the textbook, biologists in many other fields focus primarily on understanding the underlying mechanisms (e.g., physiological) that produce the properties (including behavior) of living organisms. These two approaches are complementary, and both are essential for a full understanding of life.

- Review the following tutorials and activity on the Student Website/CD:
  Tutorial 53.1 The Costs of Defending a Territory
  Tutorial 53.2 Foraging Behavior
  Activity 53.1 Concept Matching

Test Yourself

Knowledge and Synthesis Questions

1. Which of the following is not usually included within the domain of ecology?
   a. Interactions between conspecifics
   b. Modifications of the environment by organisms
   c. Evolution of different social organizations
   d. Modification of the environment by physical processes

2. Birds spend some of their time scanning the horizon for predators. While scanning, they cannot be foraging for food. Which of the following does this situation illustrate?
   a. Cooperative hunting
   b. Energetic cost
   c. Risk cost
   d. Opportunity cost

3. Experimental data suggest that human enjoyment of spices in food is probably
   a. of no adaptive significance, because a taste for spices appears to be entirely a learned (or culturally determined) behavior.
   b. adaptive, because most spices help to prevent the growth of microbes in food.
   c. adaptive, because spices mask the bad taste of spoiled food, thereby making it palatable.
   d. maladaptive, because spices often bind to mineral nutrients in the diet and thus prevent their absorption.

4. Two species of mice live in the same geographical region, but Species 1 prefers open fields, whereas Species 2 lives in forests. When presented in an experiment with simulated "fields" and "forests," individuals of each species born in a laboratory preferred the environment in which they normally lived. This experiment illustrates the concept of
   a. habitat selection.
   b. optimal foraging strategy.

c. territoriality.
d. Both a and b

5. Suppose a predator has two different prey: Species 1 and Species 2. You perform a series of experiments in which the density of Species 1 is varied while the density of Species 2 is kept constant. At each Species 1 density, you determine how many prey of Species 1 and 2 the predator takes. The two curves (*a* and *b*) plotted in the following graph show the outcome of this study.

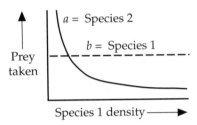

Select the statement that correctly interprets these curves.
a. Curve *a* indicates that Species 2 is the preferred prey.
b. Curve *b* indicates that Species 1 is the preferred prey.
c. Curve *b* indicates that Species 1 is not the preferred prey.
d. Neither of these curves provides insight into the prey preference of the predator.

6. You observe an example of an apparently altruistic act by an animal that seems to reduce its near-term likelihood of reproductive success. What would be the *least* plausible explanation for its behavior?
a. The act aids the reproductive success of individuals sharing a high proportion of genes with the altruistic individual.
b. The act is only apparently altruistic; over the long term, the behavior actually contributes to individual fitness.
c. The act increases the inclusive fitness of the animal performing it.
d. The act is advantageous because it helps the species to survive and reproduce, even if the altruist itself does not.

7. Some birds give a species-specific vocalization called an "alarm call" when they see a predator, although this call may direct the predator toward them. Other members of their species respond to these calls by taking cover. This would be an example of altruistic behavior that is beneficial to the calling bird if
a. the bird giving the vocalization survives the attack.
b. the inclusive fitness of the bird giving the vocalization is increased.
c. all birds survive the attack.
d. the birds benefiting are offspring of the bird giving the vocalization.

8. Which of the following characteristics is/are shared by all eusocial species?
a. A sex determination system in which males are haploid and females are diploid
b. Queens that mate with a single male
c. Presence of sterile classes
d. All of the above

9. Which of the following statements would explain why one or more social species have become very abundant?
a. Social living allows social species to exploit the services of other organisms in harvesting resources.
b. Social living enables organisms to exploit temporally and spatially patchy food resources.
c. Social living permits individuals to be specialized in their activities, thereby increasing the efficiency of the group in utilizing resources.
d. All of the above.

10. An individual of Species A displaces an individual of Species B from food that the latter has discovered. This is an example of
a. behavioral dominance.
b. territoriality.
c. habitat selection.
d. Both a and b

Application Questions

1. Describe the experiments done with bluegill sunfish preying on water fleas and the implications of these studies for foraging theory.

2. For each of the two *y*-axes in the following graph, draw a labeled curve that correctly summarizes observations made on goshawks attacking wood pigeons, as described in the textbook.

3. Based on the unusual sex determination system in Hymenoptera, describe W. D. Hamilton's hypothesis for the evolution of eusociality.

4. How has the social organization of humans contributed to the great abundance of the human population at present?

Answers

Knowledge and Synthesis Answers

1. **d.** All except the modification of the environment by physical factors are normally included within the domain of ecology.

2. **d.** The forfeited benefits of behaviors that could not be performed as a result of performing a different behavior, like scanning, comprise the opportunity cost of the performed behavior.

3. **b.** Experiments indicate that the majority of spices commonly used in cooking inhibit the growth of more than one kind of food-borne bacteria.

4. **a.** The environment in which a species normally lives is its habitat.

5. **b.** Because the number of Species 2 taken by the predator increases markedly at low Species 1 densities, you can conclude that the predator includes Species 2 in its diet only when insufficient Species 1 individuals are available. The number of Species 1 taken is independent of Species 1 density.

6. **d.** For altruistic behavior to evolve by natural selection, it must increase the fitness of the individual performing the behavior, usually through kin selection. Any explanation of an altruistic act based on its purported benefit to the species as a whole violates this principle.

7. **b.** Altruistic behavior is beneficial to the performer when the improvement in the reproductive success of relatives exceeds the reduced reproductive success of the individual performing the act. This would be true if the behavior improves the inclusive fitness (individual fitness + kin selection) of the performer.

8. **c.** By definition, eusocial species live in social groups with sterile castes. The sex determination mechanism in which males are haploid and females are diploid is found in the Hymenoptera (ants, bees, and wasps) but not in termites and naked mole-rats. Within the Hymenoptera, there are species in which queens mate with many males.

9. **d.** The social ants and termites frequently exploit other species in harvesting resources. Wildebeest are an example of an abundant social species that utilizes patchy foods. Human beings, as well as many eusocial species, exhibit specialization of behavior that improves the efficiency of resource utilization (e.g., food procurement).

10. **a.** The scenario presented in the question clearly illustrates behavioral dominance of Species A over Species B. Not enough information is given in the question to justify the assumption that either species was defending a territory.

Application Answers

1. In studies in which bluegill sunfish were provided with small, medium, and large water fleas at three different prey densities, the fish took equal proportions of the three prey sizes at low densities, but mostly large water fleas at high prey densities. This agrees with predictions from foraging theory on how a predator should behave to maximize its energy intake.

2. Your curves should show a positive relationship between the hawk's distance when spotted and pigeon flock size and a negative relationship between the hawk's attack success and pigeon flock size (see below).

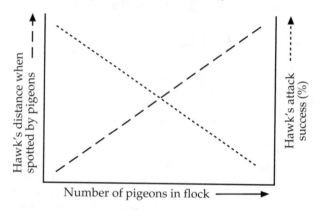

3. All species of the Hymenoptera (ants, bees, and wasps) have a sex determination system in which males are haploid and females are diploid; thus females share 75 percent of their genes with their sisters but only 50 percent of their genes with any offspring they could produce. Hamilton's hypothesis predicts that in species whose sexes are determined in this way, females can increase their inclusive fitness by foregoing reproduction and helping to raise sisters. This hypothesis cannot explain the evolution of eusociality in species without this mode of sex determination or in the many species of hymenopterans in which the queen mates with many males.

4. Social living allowed members of human groups to specialize in different activities. Occupational specialization led to advances in such areas as agriculture, hygiene, and medicine. By increasing food production and decreasing mortality from disease, these advances enabled our species to become enormously abundant.

54 *Population Ecology*

Important Concepts

A population consists of all the individuals of a species within a given area.

- Population structure includes such features as the population's size, density, spacing pattern, and age distribution.

- Population density is usually measured by determining the number of individuals of a species per unit of area (or volume), though for plants and some animals, total mass is a more useful measure than the number of individuals.

- To study the dynamics (changes over time and space) of a population, demographers measure its rates of birth, death, immigration, and emigration.

- A life table for a population is constructed by determining for a cohort (a group of individuals born at the same time) the number still alive at specific times and the number of offspring they produced during each time interval. Life tables are used to predict future trends in populations.

- Survivorship curves show the number of individuals in a cohort still alive at different times over the life span. Graphs of survivorship in real populations often resemble one of three types.

  1. Most individuals survive for the entire potential life span and die at about the same age.

  2. Survivorship is the same throughout the life span.

  3. Survivorship of the young is very low, but is high for most of the remainder of the life span.

- The age distribution of a population describes the proportions of individuals in all age categories. It reveals much about the recent history of births and deaths in the population. In populations of long-lived species such as humans, the timing of births and deaths may influence age distributions for many years.

Populations interact with one another in a variety of ways.

- Interactions of populations of different species fall into five general categories:

  1. Mutually beneficial (mutualism)

  2. Beneficial to one participant while leaving the other unaffected (commensalism)

  3. Harmful to one participant while leaving the other unaffected (amensalism)

  4. Beneficial to one participant but harmful to the other (predator–prey, parasite–host interactions)

  5. Mutually harmful (competition)

- These five types of species interactions, as well as the effect of the physical environment, determine the range of environmental conditions under which a species can persist.

Four factors influence the density of a population.

- Species that use abundant resources frequently reach higher population densities than species that use scarce resources.

- Species with small individuals typically reach higher population densities than species with large individuals.

- Recently introduced species often reach high population densities.

- Complex social organization may favor high densities.

All populations have the potential to grow exponentially, but because of environmental limitations, no population can maintain exponential growth for very long.

- In exponential growth, the per capita rate of increase remains constant per unit of time while the number of individuals added to the population accelerates. This occurs because, as the population grows, more and more individuals are alive to reproduce.

- Exponential growth can be expressed mathematically as

$$\Delta N / \Delta t = (b - d)\, N$$

where $\Delta N / \Delta t$ is the rate of change in the population size (ΔN = change in number of individuals; Δt = change in time), b and d are the per capita birth and death rates, respectively, and N is the population size.

- The difference between the per capita birth rate (b) and the per capita death rate (d) is the net reproductive rate (r). Where there are no limits on population growth, r has its highest value, designated r_{max}, the intrinsic rate of increase. Therefore, the rate of growth of a population under optimal conditions is

$$\Delta N / \Delta t = r_{max}\, N$$

- The environmental carrying capacity (K) is the maximum population size that the environment can support. It is determined by resource availability, as well as by disease, predators, and, in some cases, social interactions. Because growth tends to slow as the carrying capacity is approached, an S-shaped growth curve is characteristic of many populations growing in environments with limited resources. This type of growth can be expressed mathematically by the logistic growth equation:

$$\Delta N / \Delta t = r\,((K - N)/K)\,N$$

Population growth stops when $N = K$ because then $(K - N)/K = 0$.

- Density-dependent environmental factors—such as resource depletion, predation, and disease—cause per capita birth or death rates of a population to change in response to changes in population density. As a result, population growth slows down as density increases. A population whose dynamics are influenced primarily by density-dependent factors is said to be regulated.

- Factors (such as adverse weather) that change per capita birth or death rates of a population irrespective of its density are said to be density-independent. A population whose dynamics are controlled by density-independent factors is unregulated.

- The density of a population is determined by the combined effects of all density-dependent and density-independent factors affecting it.

Populations do not fluctuate as much as theoretically possible, but some fluctuate much more than others.

- Populations of species with short-lived individuals that have high reproductive rates typically fluctuate more than those of species with long-lived individuals that have low reproductive rates.

- Population fluctuations may be strongly influenced by years of good reproduction.

- Populations of species that depend on just a few resources are more prone to fluctuations than those of species that use a greater variety of resources.

- Predator–prey interactions often lead to cyclic fluctuations in both predator and prey populations.

- In species in which populations are fragmented into separated subpopulations, the individual fragments are prone to fluctuation leading to extinction. In the rescue effect, immigrants from other subpopulations prevent declining fragments from becoming extinct.

Some species have very small geographical ranges, whereas others are very widespread.

- Many recently-formed species and species declining toward extinction have very small ranges.

- The range of a species is often limited by its dispersal ability.

- Predators may eliminate a prey species in some areas and thus restrict its range.

- Competition between species may restrict the range of one or both of the interacting species.

Humans use the principles of population dynamics to manage and control populations.

- To maximize the number of individuals that can be harvested from a population, the population should be held far enough below carrying capacity to have high birth and growth rates.

- Demographic traits, particularly reproductive capacity, determine how heavily a population can be exploited.

- The populations of undesirable species are most effectively controlled by lowering the carrying capacity of the environment for those species. For non-native pest species, the introduction of predators and parasites from the pest's original habitat is sometimes an effective control measure.

- To preserve a rare species, the most important step usually is to provide it with a suitable habitat.

- Over the past 10,000 years, Earth's carrying capacity for humans has been increased several times by technological advances: the agricultural revolution, the tapping of fossil fuels, and the development of modern medicine. Earth's human population grew explosively in the twentieth century and now exceeds 6 billion people.

Local population densities may be influenced by remote events.

- Both local conditions and distant events, perhaps even occurring on a different continent, may affect the dynamics of a population.

The Big Picture

- As you learned in previous chapters, some mutualistic relationships have extraordinary evolutionary or ecological significance. Examples are the role of endosymbiosis in the evolution of eukaryotic cells and the association of nitrogen-fixing bacteria with their host plants.

- An understanding of concepts of population ecology such as exponential growth and carrying capacity is fundamental to your ability to think intelligently about issues such as human population growth, the preservation of biodiversity, and the control of undesirable species.

- Darwin's realization that all populations have the inherent capacity for exponential growth was crucial to the development of his theory of natural selection.

Common Problem Areas

- Students frequently have problems interpreting graphs depicting concepts of population ecology. Look carefully at how the axes of a graph are labeled. Ask yourself questions such as: Are both scales arithmetic? Is the fate of a cohort of the population being traced, or is the growth pattern of the entire population being described?

Study Strategies

- Redrawing graphs from the textbook is a good way to reinforce your understanding of the concepts being presented.

- Take advantage of laboratory activities involving computer simulations of population growth or other aspects of population ecology.

- Review the following tutorials and activities on the Student Website/CD:
  Tutorial 54.1 Exponential Growth Simulation
  Tutorial 54.2 Logistic Growth Simulation
  Tutorial 54.3 Predator–Prey Simulation
  Tutorial 54.4 Habitat Fragmentation
  Activity 54.1 Ecological Interactions
  Activity 54.2 Logistic Growth

Test Yourself

Knowledge and Synthesis Questions

1. The aspects of population structure studied by ecologists include the
   a. distribution of genotypes within a population.
   b. age distribution of a population.
   c. spacing of population members.
   d. Both b and c

2. Based on the following life table, during what time interval is survivorship greatest?

Age (years)	Number Alive
0	800
1	770
2	550
3	125
4	75
5	0

   a. 0–1 years
   b. 1–2 years
   c. 2–3 years
   d. 3–4 years
   e. 4–5 years

3. For a population that is stable in size, the following age distribution indicates that the population's

   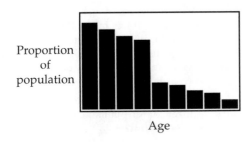

   a. birth and death rates are both high.
   b. birth and death rates are both low.
   c. birth rate is low but its death rate is high.
   d. birth rate is high but its death rate is low.

4. A bird eats the fruit of a plant species. The seeds are not digested and germinate in the bird's excrement at some distance from the parent plant. This is an example of
   a. predation.
   b. competition.
   c. commensalism.
   d. mutualism.

5. Certain birds follow swarms of foraging army ants and prey upon the insects that the ants flush. The relationship between these birds and the ants is an example of
   a. competition.
   b. commensalism.
   c. mutualism.
   d. Both a and b

6. Which of the following would *not* tend to be associated with the ability of a species to attain high population densities?
   a. Large body size
   b. Utilization of abundant resources
   c. Complex social organization
   d. Recent introduction into a new region

7. In the following graph, which of the expressions (a through d) of the exponential growth equation should be increased in order for curve 1 to become more like curve 2?

   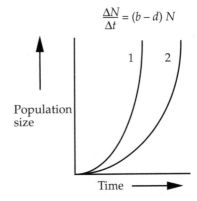

   a. $\Delta N/\Delta t$
   b. d
   c. b
   d. $(b-d)$

8. Species A and B have intrinsic rates of increase (r_{max}) of A = 0.25 and B = 0.50. In reference to the graph shown in Question 7, the population growth curve for A should be more like curve _____.

9. In the logistic population growth curve shown below, the rate of growth is greatest at which point?

   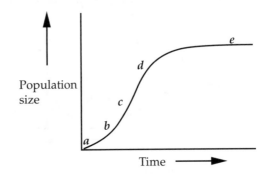

10. At which point in the graph shown above would there be zero population growth ($\Delta N/\Delta t = 0$)?

11. Based on the graph shown below, which of the following events is *least* likely to be true of this population?

    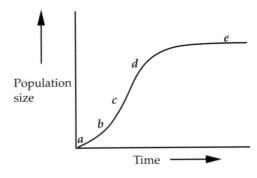

    a. Parasites spread between population members.
    b. Increased competition for food causes some individuals to delay reproduction.

c. The number of predators in the area varies with population density.

d. Territorial disputes led to injury and death of some males.

12. In the graph shown in Question 11, the population size at which the curves for the birth and death rates intersect is
a. an estimate of the environmental carrying capacity.
b. an estimate of the intrinsic rate of increase.
c. the point at which density-dependent regulation begins.
d. the point at which density-independent regulation begins.

13. Which of the following would probably *not* be true of a population whose dynamics are primarily influenced by density-independent factors?
a. The population's growth pattern is similar to the logistic growth curve.
b. The birth rate of the population is dependent on the nutritional status of its adult females.
c. The most important source of mortality in the population is unfavorable weather conditions.
d. Both a and b

14. A species will tend to have high population fluctuations if its members
a. have long life spans.
b. have a high reproductive rate.
c. eat a varied diet.
d. form a large, unfragmented population.

15. Which vertical line in the following graph represents the time at which the predator population (gray) is increasing and the prey population (black) is decreasing?

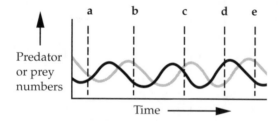

16. If a species of plant that is trampled by an animal eventually evolves sharp spines that prevent trampling, we can say that its association with the animal has changed from
a. amensalism to competition.
b. amensalism to commensalism.
c. commensalism to competition.
d. commensalism to mutualism.

17. The best way to decrease the population size of a pest species is to
a. poison it.
b. introduce additional predators.
c. decrease the carrying capacity of the habitat for it.
d. add competitors.

Application Questions

1. Draw and explain survivorship curves characteristic of humans in a wealthy country, wild birds, and annual plants.

2. How does the division of some populations into discrete subpopulations relate to the process known as the rescue effect?

3. The zebra mussel was not found in North America before 1985, yet it is now an abundant pest species in the Great Lakes region and Mississippi River drainage. What principles regarding species abundance and range size apply to the explosive increase of the zebra mussel?

4. Describe how you would assess the degree of interspecific competition that exists between two rodent species living in the same habitat. How you would interpret the results?

5. Describe the history of human population growth over the past 12,000 years. What factors have raised Earth's carrying capacity for our species?

Answers

Knowledge and Synthesis Answers

1. **d.** The aspects of population structure studied by ecologists include the spacing and age distribution of the members of the population but not the distribution of genotypes.

2. **a.** The decrease in consecutive age classes is least between 0 and 1 year, so the survivorship is greatest during that interval.

3. **a.** When a population's birth rate and death rate are both high, the age distribution is dominated by young individuals. A similar age distribution would occur in a population with a high birth rate and a low death rate, but such a population would not be stable in size.

4. **d.** Both species have benefited; the bird dispersed and provided fertilizer for the plant's seed, and the plant provided food for the bird.

5. **d.** Both ants and birds are competing for the same food supply; the birds also benefit from the ants' activity in flushing the insects, but the ants might not be affected.

6. **a.** Figure 54.4 shows the strong tendency for small species to reach greater population densities than species with larger body sizes.

7. **b.** Both curves show exponential growth, but the rate of exponential growth for curve 2 is less than the rate for curve 1. Because the rate of growth is determined by the expression $(b - d)$, increasing d, the death rate, would cause curve 1 to be more like curve 2.

8. **Curve 2.** Because $r_{max} = (b - d)$, you would expect the growth curve for species A to correspond to the curve with the smaller rate of exponential growth.

9. **c.** The curve showing the relationship between population size and time is steepest at point c, so the growth rate would be greatest at that point.

10. **e.** The growth rate at e would be zero. This is the point called the environmental carrying capacity, the point at which the birth and death rates are equal.

11. **b.** The graph indicates that the birth rate is independent of population size, so event b would not be true. The graph shows that the death rate is density-dependent

12. **a.** The environmental carrying capacity is the equilibrium population size when the birth rate equals the death rate. The birth rate equals the death rate at the intersection of the two curves.

13. **d.** The logistic growth curve describes density-dependent growth because the rate of increase in the size of the

population steadily decreases as the carrying capacity is approached. The nutritional status of the females in the population would depend on the availability of food and hence would be density-dependent. Unfavorable weather, on the other hand, generally occurs without any predictable relationship to the size of a population and thus is a density-independent factor.

14. **b.** Species with long-lived individuals that have low reproductive rates generally have more stable populations than species with short-lived individuals that have high reproductive rates. Species that depend on only a few kinds of foods and that have a fragmented population structure tend to be less stable than those that use a greater variety of resources and that have larger and less-fragmented populations.

15. **c.** The key to answering this question is realizing that prey cycles lead predator cycles, so the black line indicates the prey, and the gray line, the predator.

16. **a.** Initially the plant is harmed and the animal is unaffected (amensalism). After the evolution of sharp spines, the two species are competitors for space.

17. **c.** The best way to control a pest species is to reduce the carrying capacity of the habitat for the species by removing the resources it depends on.

Application Answers

1. See Figure 54.1 of the textbook. All of the graphs depict the survivorship of a group of individuals born at the same time (a cohort) from their time of birth until all members of the group have died. In the human population in the United States, death rates are low until old age. In many species of wild birds, the probability of an individual's dying at a particular age is almost the same over most of its lifetime. In annual plants (and in other organisms that produce many offspring), poor survivorship of young individuals is followed by much lower death rates in the middle part of the life span.

2. The rescue effect occurs if a subpopulation that has undergone a large decline is saved from extinction by the immigration of individuals from other subpopulations.

3. The failure of the zebra mussel to reach North America until 1985 was due to its inability to disperse across the Atlantic Ocean from Europe. Its failure demonstrates how dispersal abilities can restrict geographic ranges. Its rapid spread in North America since 1985 shows the ability of this species to disperse quickly in suitable freshwater habitat. The high population density of zebra mussels in this continent also illustrates that species introduced into a new region often reach levels of abundance above those found in their native ranges because their normal diseases and predators are absent.

4. A removal experiment could be done in which one or the other species is removed in two similar areas. One then looks for changes in the abundance of the remaining species. The species with the largest increase in population density was most affected by competition with the removed species.

5. See Figure 54.20 in the textbook.

55 Communities and Ecosystems

Important Concepts

An ecological community consists of all the organisms that live together in a particular area. The community and its physical environment constitute an ecosystem.

- The species that make up a community interact with one another via the processes of competition, predation, parasitism, commensalism, mutualism, and amensalism.

- In most cases, species enter and drop out of communities independently over environmental gradients. Where environmental conditions change abruptly, however, the ranges of many species may end at the same place.

- In some cases, interactions among only a few of the many species in a community may exert the strongest effects on community structure. An example is the interaction of mice, gypsy moths, and oak trees in forests in the eastern United States.

Nearly all ecological processes depend, directly or indirectly, on the amount and pattern of solar energy input and supply of water.

- With very few exceptions, energy flow in ecosystems originates with photosynthesis, which captures about 5 percent of the solar energy that arrives on Earth.

- Gross primary productivity is defined as the *rate* at which plants assimilate energy through photosynthesis; the accumulated energy is called primary production.

- Net primary production is the portion of assimilated energy left over after subtracting the energy plants use to power their own metabolism. All other organisms in an ecosystem directly or indirectly derive their energy from net primary production, which takes the form of new plant tissues.

- The major limits on primary production in terrestrial ecosystems are low temperatures and lack of moisture. In aquatic systems, production is limited by light, nutrients, and temperature.

- Species richness—the number of species living in a community—increases with primary productivity up to a point, after which it declines. One hypothesis to explain this decline postulates that interspecific competition becomes more intense when productivity is higher, resulting in competitive exclusion of some species.

- Energy flows through ecosystems when organisms eat one another. Organisms are grouped into trophic levels according to the number of steps through which energy passes to get to them. Energy passes, in sequence, from primary producers (photosynthesizers), to herbivores, to primary carnivores, to secondary carnivores, and so on. Detritivores (decomposers) feed on the dead bodies and waste products of other organisms. Omnivores obtain their food from more than one trophic level.

- A food chain is a sequence of feeding linkages beginning with a primary producer. Food chains in a community are usually interconnected in a complex food web.

- Most of the energy captured at any trophic level is used for respiration. Only the net production (growth plus reproduction) of organisms at a trophic level is available to organisms at the next level. Largely because of this loss of energy, most communities have only three to five trophic levels.

- A pyramid of energy shows how energy (production) decreases in moving from lower to higher trophic levels. A pyramid of biomass shows the mass of organisms existing at different trophic levels at a given moment of time. For the same ecosystem, these pyramids usually have similar shapes. Major exceptions include aquatic communities, in which a small mass of rapidly dividing unicellular photosynthesizers can support a much larger biomass of herbivores.

- Detritivores are essential for the continued productivity of ecosystems because they release nutrients from dead bodies that can be taken up again by plants.

- Communities with more species tend to be more productive and more stable than communities with fewer species.

- Keystone species have effects on ecological communities that are greater than would be expected on the basis of their abundance. Examples of keystone species are large grazing or browsing mammals that change vegetation structure and that are predators of dominant competitors.

All ecological communities are subjected to a variety of disturbances.

- A disturbance is an event that changes the survival rate of one or more species in a community. Typically, small disturbances are much more common than large ones.

- According to the intermediate disturbance hypothesis, communities subjected to very high or very low levels of disturbance have fewer species than those that experience intermediate levels.

- Ecological succession is a change in community composition following a disturbance. The patterns and causes of succession are varied, but species that first colonize a site often alter the conditions so that they become favorable for other species.

- Primary succession begins with the establishment of organisms on newly available sites that previously had no organisms.

- Secondary succession begins when organisms reestablish themselves on disturbed sites where some organisms survived the disturbance. This type of succession may also begin with the dead parts of organisms.

Large-scale, long-term ecological and evolutionary processes influence the composition of ecological communities.

- An example of such processes was the formation of the Central American land bridge connecting North and South America. After it formed, many species of mammals that had evolved on one continent colonized the other. This dispersal caused many species (especially South American forms) to become extinct and many new species to evolve.

The Big Picture

- The productivity of ecosystems, both terrestrial and aquatic, is relevant to the problem of providing food for the exploding human population (discussed in Chapter 54).

- In urging a cautious approach to human manipulation of the natural environment, ecologists like to say that "You can never do just one thing." The descriptions in this chapter of the effects of keystone species and of other complex community interactions illustrate the meaning of this saying.

- As mentioned in Chapter 54, if we wish to preserve a rare species, usually the most important step is to protect its habitat. This often requires an understanding of (and manipulation of) ecological succession in order to prevent changes in habitat unfavorable to the rare species.

Common Problem Areas

- Students sometimes have difficulty understanding the difference between pyramids of energy and pyramids of biomass. Pyramids of energy can *never* be inverted because there must always be a lower rate of energy flow through any given trophic level than through the trophic level below it. Pyramids of biomass *can* be inverted because they do not represent energy flow rates, but rather the amount of living matter present in the various trophic levels of the community at a given moment in time.

Study Strategies

- This chapter contains descriptions of a number of studies of community structure, including several that relate species richness to other community characteristics. As you study these descriptions, focus on what question the study was designed to answer and how the results support the conclusions drawn from the study.

- Review the following tutorial and activities on the Student Website/CD:
  Tutorial 55.1 Primary Succession on a Glacial Moraine

Activity 55.1 Ecosystem Energy Flow
Activity 55.2 The Major Trophic Levels

Test Yourself

Knowledge and Synthesis Questions

1. Which of the following could *not* be considered an ecosystem?
   a. A small pond
   b. All the fish in a coral reef
   c. Earth
   d. A pile of dung in a pasture

2. Which environmental factor frequently limits primary production in both terrestrial and aquatic ecosystems?
   a. Temperature
   b. Moisture
   c. Light
   d. Nutrient supply

3. A plant in the dark uses 0.02 ml of O_2 per minute. The same plant in sunlight releases 0.14 ml of O_2 per minute. A correct estimate of its rate of *gross* primary production is _____ ml of O_2 per minute.
   a. 0.02
   b. 0.12
   c. 0.14
   d. 0.16

4. Examine the following food web. Organism 9 is a(n)

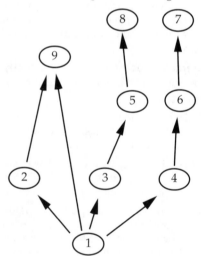

   a. herbivore.
   b. primary carnivore.
   c. secondary carnivore.
   d. omnivore.

5. The food web shown in Question 4 has _____ trophic levels.
   a. three
   b. two
   c. four
   d. five

6. In which of the following ecosystems would you expect to observe the lowest biomass of primary consumers relative to the biomass of the primary producers?
   a. Tropical rainforest
   b. Temperate grassland
   c. Freshwater lake
   d. Open ocean

7. Which of the following is most characteristic of a keystone species?
   a. Very abundant
   b. Removal having a great effect on community structure
   c. Herbivore
   d. Uniform distribution

8. Species richness is
   a. positively correlated with stability of annual net primary production.
   b. often greatest in communities with an intermediate level of primary production.
   c. typically greatest in communities experiencing the least amount of disturbance.
   d. Both a and b

9. If the following list of stages of ecological succession occurring on glacial moraines in Alaska were put in correct chronological sequence, which would be third?
   a. Arrival of willows and alders
   b. Arrival of lichens
   c. Arrival of conifers
   d. Increase in soil nitrogen content

10. Which of the following would represent primary succession?
   a. The development of an aquatic community in a newly excavated farm pond, eventually followed by the filling in of the pond with sediment and its conversion to a forest
   b. The gradual establishment of a forest following a fire that destroyed the previous community
   c. The invasion of weeds followed by shrubs and trees on an abandoned farm field
   d. Both a and b

Application Questions

1. Some ecologists have regarded a community (especially a plant community) as a tightly integrated "superorganism," whereas other ecologists have argued that each species in a community is individually distributed according to its particular interactions with the environment. How has this question been investigated? Which view of communities appears to be better supported by the evidence?

2. Describe how the interactions among gypsy moths, mice, and oak trees influence the frequency of gypsy moth outbreaks in the forests of eastern North America.

3. If the dry biomass of the primary producers of an ecosystem is 1000 g per square meter and the efficiency of transfer between trophic levels is 20 percent, what dry biomass weight would you expect at the secondary carnivore level?

4. Where do biomass pyramids like the one shown below occur and what are the conditions that create them?

Dry Weight (g/m²)

Consumers   4

Producers   21

5. About four million years ago, a land bridge formed between North and South America. Describe the effects of this event on the mammalian fauna that had evolved in South America prior to the formation of the bridge.

Answers

Knowledge and Synthesis Answers

1. **b.** An ecosystem includes the environment and the biological community living there. A coral reef ecosystem would include much more than just the fish.

2. **a.** Temperature, nutrient supply, and light can limit production in aquatic ecosystems; temperature and moisture most often limit production on land.

3. **d.** The amount of production used in maintenance and biosynthesis is added to net primary production to determine gross primary production. Photosynthetic release of O_2 in the light is an estimate of net production; O_2 use in the dark is an estimate of maintenance and biosynthesis costs. You would add 0.02 ml to 0.14 ml to get an estimate of gross primary production of 0.16 ml per minute.

4. **d.** Because organism 9 eats from both the primary producer level (1) and the herbivore level (2), it would be an omnivore.

5. **c.** Four trophic levels are depicted. The levels are primary producer (1), herbivore (2, 3, 4), primary carnivore (5, 6), and secondary carnivore (7, 8). The omnivore (9) occupies either the herbivore or primary carnivore level, depending on whether it is feeding on a primary producer or on an herbivore.

6. **a.** Compared to grasslands or aquatic ecosystems, forests typically have a lower ratio of primary consumer biomass to primary producer biomass because trees (the dominant primary producers) store energy for long periods of time in difficult-to-digest forms.

7. **b.** By definition, removal of a keystone species has a larger effect on the community than would be expected based on its abundance.

8. **d.** Of the first three choices, only c is incorrect. The highest levels of species richness are associated with intermediate levels of disturbance.

9. **d.** Of the stages listed, the arrival of lichens would be first, followed by the arrival of willows and alders, an increase in soil nitrogen content (the correct answer), and finally the arrival of conifers.

10. **a.** Succession that begins on a newly available site, like a freshly excavated pond, is primary. Succession that occurs after disturbances such as fire or the conversion of a natural community to farmland is considered secondary.

Application Answers

1. The question of whether plant communities are "superorganisms" or loose assemblages of species has been investigated by studying how the species composition of communities changes along environmental gradients, such as soil moisture. The results show that each species of plant has its own unique response to environmental factors. Thus the plant community found at a particular location is best regarded not as a "superorganism," but rather as a loose assemblage of species that happen to be adapted to the environmental conditions found there (see Figure 55.1).

2. See Figure 55.2 in the textbook.

3. The answer is 8 grams. The flow of energy is from primary producers, to herbivores, to primary carnivores, to secondary carnivores. If each of these three energy transfers is only 20 percent efficient, then the biomass of secondary carnivores should be:

$$1000 \text{ g} \times 0.2 = 200 \text{ g}$$
$$200 \text{ g} \times 0.2 = 40 \text{ g}$$
$$40 \text{ g} \times 0.2 = 8 \text{ g}$$

4. In most aquatic ecosystems, the dominant primary producers are unicellular algae. Their populations multiply so rapidly and are cropped so efficiently by slower-growing herbivores that they frequently support a larger herbivore mass than their own. Note that the diagram depicts an inverted pyramid of biomass. A pyramid of energy can never be inverted.

5. The land bridge permitted the mammalian faunas of both North America and South America to disperse to the other continent. Many colonizing North American species were able to compete successfully with South American forms such as marsupials and ground sloths and to drive them extinct. Subsequently, the northern invaders formed new species found today only in South America.

56 *Biogeography*

Important Concepts

The goals of biogeography are to document and explain patterns in the distribution of life on Earth. Biogeographers divide Earth into several biogeographic regions.

- If a species occupies an area, either it evolved there, or it evolved elsewhere and dispersed to that area.
- If a species is not found in an area, either it evolved elsewhere and never dispersed to that area, or it was once present in that area but no longer lives there.
- Biogeographers divide Earth into a number of biogeographic regions on the basis of the taxonomic similarities among the organisms living in them. The biotas of the biogeographic regions differ because barriers such as oceans restrict the dispersal of organisms.
- Species found in only one place are said to be endemic to that location. Remote islands contain many endemic species.
- Ecological biogeographers study how current ecological interactions influence where species are found today.

Past events influence today's patterns of distribution.

- Continental drift has influenced the evolution and mixing of species throughout the history of life on Earth.
- Biogeographers often analyze species distributions by converting a taxonomic phylogeny into an area phylogeny, a process that involves replacing the names of the taxa with the names of the places where those taxa live or lived.
- Historical biogeographers infer the times of separation of taxa within a lineage on the basis of evidence provided by molecular clocks, fossils, and the distributions of living species.
- A split distribution of a species can be accounted for in two ways.
  1. A barrier may appear that splits a species' distribution (a vicariant event).
  2. A species may cross an existing barrier and establish a new population (dispersal).
- The longer an area has been isolated, the more endemic taxa it is likely to have.
- To explain biogeographic patterns, scientists apply the parsimony principle. That is, they prefer the explanation that requires the smallest number of unobserved events to account for the pattern.

Climate exerts a powerful influence on the geographic distributions of organisms.

- Solar energy varies with latitude. Regions near the poles receive less energy per unit of ground area than regions near the equator because of the lower angle of the sun. At high latitudes there is also more variation over the course of a year in both day length and the angle of arriving solar energy than at latitudes closer to the equator.
- Rising air expands and cools, releasing moisture, whereas descending air warms and dries, taking up more moisture.
- Unequal heating of the atmosphere at low and high latitudes produces vertical and latitudinal movements of air masses. These movements cause very moist climates to occur at the equator and at 60° north and south latitudes (where air rises) and arid climates at about 30° north and south latitudes and near the poles (where air descends).
- In the Tropics, the intertropical convergence zone is caused by the replacement of warm, rising air by air that flows toward the equator from north and south. The position of this zone shifts latitudinally during the course of the year, with the result that the Tropics and Subtropics experience rainy and dry seasons.
- The spinning of Earth on its axis causes air masses moving latitudinally to be deflected to the right in the Northern Hemisphere and to the left in the Southern Hemisphere. Thus, winds blowing toward the equator at low latitudes veer to become the northeast and southeast trade winds, whereas winds blowing away from the equator at midlatitudes are deflected to become the prevailing westerlies.
- The movement of air over mountains often results in a rain shadow—a dry area on the leeward side of the range.
- Ocean currents are driven primarily by prevailing winds but are deflected by continents. The poleward movement of ocean water warmed in the Tropics is a major mechanism of heat transfer to high latitudes.

A terrestrial biome is a major ecosystem type that differs from other types in the structure of its dominant vegetation.

- The distribution of biomes on Earth is strongly influenced by annual patterns of temperature and precipitation.
- Tundra is found at high latitudes in the Arctic and in high mountains. It is a treeless biome dominated by short perennial plants. In the Arctic, permanently frozen soil called permafrost underlies tundra vegetation.

- Boreal forests are located equatorward from, or at lower elevations on mountains than, tundra. The short summers favor evergreen trees, which are the dominant vegetation.

- Temperate deciduous forests are dominated by deciduous trees, which produce leaves that photosynthesize rapidly during the warm, moist summers and lose their leaves during the cold winters.

- Temperate grasslands occur in areas that are relatively dry for much of the year. Grasses dominate this biome because they are well adapted to grazing and fire.

- Cold deserts are found in dry regions at middle to high latitudes. Seasonal changes in temperature are great. They are dominated by a few species of low-growing shrubs.

- Hot deserts, characterized by very warm and dry conditions year-round, are found in two belts around the 30°N and 30°S latitudes. Succulent plants that store large quantities of water in their stems are conspicuous.

- The chaparral biome, found on the west side of continents at midlatitudes, is dominated by evergreen shrubs and low trees that are adapted to survive periodic fires. Winters are cool and wet; summers are hot and dry.

- Thorn forests and savannas are found in semiarid climates on the equatorial side of hot deserts. Savanna, a grassland punctuated by scattered trees, is maintained by grazing, browsing, and burning. In their absence, it reverts to dense thorn forest dominated by small, spiny shrubs and trees.

- Tropical deciduous forests are dominated by trees that lose their leaves during the long, hot, dry season. Because their soils are less leached of nutrients than the soils of wetter areas, most of these forests have been cleared for grazing cattle and growing crops.

- Tropical evergreen forests, found in equatorial regions where annual rainfall exceeds 250 cm annually, have the greatest species richness of all biomes and the highest productivity of all ecological communities. Nevertheless, their soils are poor and usually cannot support long-term agriculture unless massively fertilized. Epiphytes—plants that grow on other plants, deriving their nutrients and water from the atmosphere—thrive in tropical mountain forests where clouds form regularly.

Aquatic ecosystems occupy three-fourths of Earth's surface.

- Earth's oceans form one interconnected water mass, with only partial barriers to dispersal. For freshwater organisms that cannot survive out of water, terrestrial habitats are barriers to dispersal.

- Although freshwater lakes, ponds, and streams contain only about 2.5 percent of Earth's water, they are the habitat for about 10 percent of all aquatic species.

- Though the oceans are connected, living successfully in a particular region requires that organisms have specific physiological tolerances and morphological characteristics. Thus, most marine organisms have restricted ranges.

- Boundaries between biogeographic regions in the oceans typically occur where the temperature and salinity of surface water changes abruptly as a result of ocean currents.

- Few species that live in shallow water are able to disperse across wide deep-water barriers.

Biogeographers have discovered several regional patterns of species richness.

- Species richness increases rapidly when the area being sampled crosses a biogeographic boundary.

- Species richness varies latitudinally, with more species living in tropical than in high-latitude regions.

- According to the MacArthur-Wilson model, equilibrium species richness (the number of species living in an area) on islands is determined by the rates of arrival of new species and extinction of species already present.

- The MacArthur-Wilson model predicts that the equilibrium number of species should increase with island size and decrease with distance from the mainland. Biogeographers have confirmed the predictions of the model.

The distributions of plants, animals, and continents have had a strong influence on human history.

- Eurasia happened to have a larger number of species of plants and animals suitable for domestication than Africa or the Americas.

- Eurasian peoples acquired immunity to many human diseases as a result of their association with domestic animals. When Europeans colonized the New World, they brought with them diseases that devastated indigenous peoples.

- Dispersal of people and their domesticated plants and animals was relatively easy in Eurasia because most mountain ranges are oriented in an east-west direction.

The Big Picture

- As you will learn in Chapter 57, the MacArthur-Wilson island biogeographic model has been applied to the problem of saving endangered species from extinction.

- Much evidence supporting the theory of continental drift is based on the distributions of organisms on Earth. By the same token, the acceptance by geologists and biologists of the reality of continental drift revolutionized the field of historical biogeography.

- The distributions of particular plant and animal species that have been domesticated in various regions of the world have had major consequences for human history and culture. So have the distributions of many species of microorganisms and animals that cause or carry disease.

Common Problem Areas

- The graphical representations of the concepts of the MacArthur-Wilson model are troublesome for some students. Try redrawing the graphs for yourself, paying particular attention to how the axes of the graphs are labeled.

Study Strategies

- In attempting to master the information concerning the many biomes described in this chapter, focus on the graphs that summarize the essential points about each one.

- You also should study Figure 56.9 to get a sense of the geographical location and extent of each biome. A useful exercise to help you learn the relative locations of biomes is to imagine a journey (e.g., from equatorial South America to Alaska, or from Maine to California), naming the biomes that you would pass through.

- Review the following tutorials and activity on the Student Website/CD:
  Tutorial 56.1 Rain Shadow
  Tutorial 56.2 Biomes
  Tutorial 56.3 Biogeography Simulation
  Activity 56.1 Major Biogeographic Regions

Test Yourself

Knowledge and Synthesis Questions

1. The following diagram shows the present locations of four continents (W, X, Y, Z) and the number of endemic beetle families found on each. Three were part of a supercontinent in the past; the other was always separate.

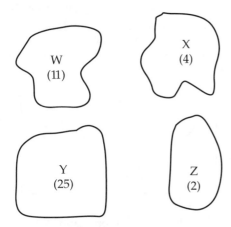

Based on this information, choose the correct conclusions.
a. W was always separate; Y separated from the supercontinent first.
b. X was always separate; Y separated from the supercontinent first.
c. Y was always separate; W separated from the supercontinent first.
d. Z was always separate; X separated from the supercontinent first.

2. The following diagram shows three islands. Islands A and B were connected in the past, C was always separate. A species of land snail is found on all three islands.

Which of the following does *not* correctly describe the distribution of this snail relative to the three islands?
a. Vicariant distribution relative to A and B
b. Dispersal distribution relative to A and C

c. Dispersal distribution relative to B and C
d. Vicariant distribution relative to A and C

3. If the species of snail mentioned in Question 2 is found *only* on the three islands shown in the preceding diagram, which statement would be correct concerning the endemism of this species?
a. Endemic relative to A and B
b. Endemic relative to C
c. Endemic relative to A, B, and C
d. All of the above

4. Which of the following biogeographical regions represents the largest area?
a. Nearctic
b. Palearctic
c. Neotropical
d. Oriental

5. In comparing an acre of land in Colombia to an acre of land in Michigan, which of the following would *not* differ?
a. The angle of the sun reaching the ground in the month of July
b. The solar energy flux in the month of July
c. The annual solar energy flux
d. The total hours of daylight per year

6. If Earth did not spin on its axis, from what direction would the northeast trade winds blow?
a. Northeast
b. South
c. North
d. East

7. Ocean circulation patterns are influenced by all of the following *except*
a. circulation of Earth's atmosphere.
b. deflection by land masses.
c. upwelling of deeper water.
d. rotation of Earth on its axis.

8. The following diagram shows a mountain with a sea breeze blowing as indicated by the arrow. Circle the letter for the area with air that would be *both* relatively warm and dry.

9. In what area of the diagram in Question 8 is a process occurring that is similar to the process that occurs in the intertropical convergence zone?

10. The biome that is maintained by browsers, grazers, or fire is the
a. tundra.
b. savanna.
c. cold desert.
d. tropical deciduous forest.

11. Match the letters of the following biomes with the descriptions that follow.
    a. Tundra
    b. Boreal forest
    c. Hot desert
    d. Chaparral
    e. Tropical deciduous forest

    _____ Mostly coniferous, wind-pollinated and wind-dispersed tree species
    _____ Leaves lost during dry season; agriculturally desirable land
    _____ Cool winters, hot dry summers; maritime climate
    _____ Succulent plants prominent; found at 30°N and 30°S latitudes
    _____ Distribution altitudinally or latitudinally determined; permafrost present

12. The following climograph shows yearly variation in rainfall and temperature for four biomes. Select the correct curve for each of the following biomes.

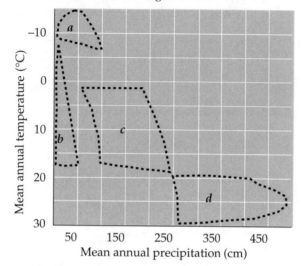

    _____ Tundra
    _____ Tropical evergreen forest
    _____ Temperate deciduous forest
    _____ Cold desert

13. The tropical evergreen forest biome has relatively constant _____, but _____ varies seasonally; the temperate deciduous forest biome has relatively constant _____, but _____ varies seasonally.

14. Which of the following comparisons of freshwater and marine habitats is *not* true?
    a. About 10 percent of all aquatic species live in freshwater habitats.
    b. There are many more insect species in freshwater habitats than in marine habitats.
    c. The global volume of marine habitats is much greater than that of freshwater habitats.
    d. Freshwater species richness is less than marine species richness in proportion to the relative extent of the two habitats.

15. Which of the following is most likely to be true of a family of freshwater fish with species distributed on several continents?
    a. They are a recently derived group.
    b. They have an ancient lineage that was widely distributed in Laurasia or Gondwana.

c. Their distribution was not affected by continental drift.
d. Their distribution can be explained by movements of glaciers.

16. Which of the following statements about the biogeography of marine biomes is *false*?
    a. Species adapted for life in shallow water are seldom able to disperse across wide deep-water barriers.
    b. In interpreting the biogeography of the marine biome, oceanic currents are more important than physical barriers.
    c. Organisms adapted for life in nutrient-poor waters usually flourish in nutrient-rich areas.
    d. Continental drift does not explain the distribution of many marine organisms.

17. Which of the following locations would you expect to have the greatest number of species?
    a. Mountainous forest area on a mainland in the temperate zone
    b. Mountainous forest area on a mainland in the tropics
    c. Lowland forest area on a mainland in the tropics
    d. Mountainous forest area on a peninsula in the tropics

18. According to the island biogeographic model, which of the statements about the following graph showing the effect of species number on arrival and extinction rates is *not* true?

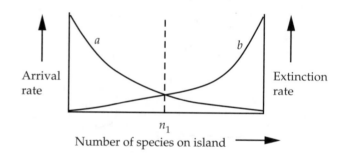

    a. Curve *a* is the arrival rate curve.
    b. Curve *b* is the extinction rate curve.
    c. The arrival rate equals the extinction rate at n_1.
    d. The extinction rate is zero at n_1.

19. The species numbers of many islands of different sizes are plotted against their distance from the mainland. Data points for islands of similar size were connected to form the four curves shown in this figure.

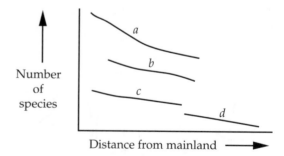

Circle the letter of the curve corresponding to the group of islands with the smallest size.

20. The island biogeographic model makes predictions about the effects of species number on the rate of extinction. In the following figure, the solid curve shows this relation-

ship for a large island. Which of the curves shows the expected relationship for a small island?

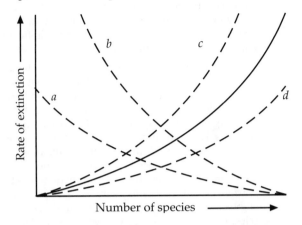

a. Curve *a*
b. Curve *b*
c. Curve *c*
d. Curve *d*

Application Questions

1. Based on an examination of the following phylogenetic tree and information on the distribution of the taxa members on three continents, discuss what can be inferred about the histories of the three continents.

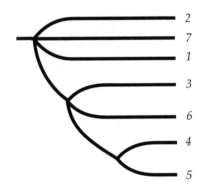

Species	Continent		
	A	*B*	*C*
1	✓	✓	✓
2	✓	✓	✓
3	✓	✓	
4		✓	
5		✓	
6	✓	✓	
7	✓	✓	✓

2. Many species of large mammals were domesticated in Eurasia, but none in Africa and only one in the Americas. Why? What were the consequences of this biogeographical pattern?

Answers

Knowledge and Synthesis Answers

1. **c.** Because the number of endemic taxa in an area is related to the period of time that the area has been separated from other areas, we would expect that continent Y

had always been separate and continent W had been separated first from the other two.

2. **d.** A vicariant distribution requires that the species once had a continuous distribution in what are now separate areas.

3. **c.** The snail is endemic to the group of islands, but not to any single one or pair of them because it is also found on the remaining islands or island.

4. **b.** See Figure 56.1 in the textbook.

5. **d.** All areas on Earth receive equal hours of daylight per year. The seasonal distribution of those hours, the solar energy input, and the sun angle do vary latitudinally.

6. **c.** If Earth did not spin on its axis, air flowing south toward the equator would not be deflected to the right. Therefore, the northeast trade winds would blow directly from the north instead of from the northeast.

7. **c.** The upwelling of deeper water, especially on the western side of continents, is an effect of the pattern of ocean current circulation and not a cause of the pattern.

8. **e.** The air would have lost most of its moisture while rising on the windward side of the mountain, and in descending to area *e* it would have warmed again.

9. **b.** In the intertropical convergence zone, warm air rises and loses much of its moisture. These same events occur when moist air rises over a mountain.

10. **b.** If savanna vegetation is not grazed, browsed, or burned, it typically reverts to dense thorn forest.

11. **b.** Mostly coniferous, wind-pollinated and wind-dispersed tree species
    e. Leaves lost during dry season; agriculturally desirable land
    d. Cool winters, hot dry summers; maritime climate
    c. Succulent plants prominent; found at 30°N and 30°S latitudes
    a. Distribution altitudinally or latitudinally determined; permafrost present

12. **a.** Tundra
    d. Tropical evergreen forest
    c. Temperate deciduous forest
    b. Cold desert

13. The tropical evergreen forest biome has relatively constant **temperature,** but **rainfall** varies seasonally; the temperate deciduous forest biome has relatively constant **rainfall** (or precipitation, because snow may occur in winter) but **temperature** varies seasonally.

14. **d.** Even though freshwater habitats occupy a very small global area, they have about 10 percent of all aquatic species, which, by proportion, gives them a very large species richness.

15. **b.** The most likely explanation for families of freshwater fish with representatives on several continents is that their lineages originated prior to the breakup of Laurasia or Gondwana.

16. **c.** Usually organisms adapted for one zone do not do well in a different zone.

17. **b.** Species richness increases toward the equator. It is also higher in mountainous regions than in lowland areas and is higher on mainland areas than on peninsulas.

18. **d.** The extinction rate is nonzero and equals the arrival rate at n_1, which is therefore the equilibrium species number on the island.

19. **d.** Each curve corresponds to a group of similar-sized islands whose species number is plotted against their distance from the mainland. For each curve, species number decreases with distance from the mainland, and the curve that is lowest relative to the vertical axis would be the group of smallest islands.

20. **c.** With less space available, populations of the different species would be smaller. This would subject them to higher extinction rates than would be expected on larger islands.

Application Answers

1. Species 2, 7, and 1 represent the lineage that first separated from the others. That each species in this lineage is found on all these continents suggests that the three continents were once joined. The next lineage includes species 3 and 6, which are found on continents A and B. This suggests that the original supercontinent split to form two continents. One became continent C. Species 3 and 6 evolved on the other continent, which then split to form continents A and B. Species 4 and 5, in the final lineage, evolved on continent B after the split.

2. Many large mammal species that happened to be found in Eurasia had several social characteristics that made them amenable to domestication: They lived in herds with male dominance hierarchies and were nonterritorial. All the large mammals of Africa lacked one or more of these traits. Domestication of these mammals provided several advantages to Eurasian cultures, the most obvious of which was increased efficiency of agricultural production. Others were the availability of horses for use in warfare and immunity to many diseases (such as smallpox and measles) that originated in domesticated species. When Europeans colonized the New World, these diseases devastated the indigenous peoples.

57 *Conservation Biology*

Important Concepts

Conservation biology is the scientific study of how to preserve the diversity of life.

- The science of conservation biology draws on concepts and knowledge from population genetics, evolution, ecology, biogeography, wildlife management, economics, and sociology. The development of this science has been spurred by the accelerating pace of human-caused extinctions of species.

There are many reasons to value biodiversity.

- Many individual species are important because they supply useful products such as medicinal drugs.
- Extinctions lessen the aesthetic pleasure many people derive from interacting with other organisms. Extinctions caused by human activities also raise ethical issues.
- Extinctions deprive us of opportunities to study the structure and functioning of communities.
- Ecosystem processes provide many services that can be replaced only by expensive and continuing human effort. Examples of such services are the maintenance of fertile soils, detoxification and recycling of waste products, control of agricultural pests, and pollination.

Conservation biologists use several methods to estimate the rates of extinction.

- Estimates of current rates of extinction worldwide are based primarily on species–area relationships and rates of tropical deforestation.
- To estimate the risk of extinction of a particular population, biologists develop models that incorporate information about its size, its genetic variation, and the morphology, physiology, and behavior of its members.
- Endangered species are those that are in imminent danger of extinction over all or much of their range. Threatened species are those that are likely to become endangered in the near future.

Conservation biologists determine which human activities are endangering species and use that information to devise actions to preserve the species.

- Habitat destruction is the most important cause of species extinction today. As habitats are fragmented into small patches, populations of species that require large areas cannot be maintained.

- The proportion of a habitat patch subject to detrimental edge effects increases as patch size decreases. For example, species from surrounding habitats may invade the edges of the patch to compete with or prey upon the patch inhabitants. The persistence of species in small patches may be improved if the patches are connected by corridors of suitable habitat through which individuals can disperse.
- Exotic pests, predators, and competitors introduced by humans have driven many species to extinction. Biologists use information on species that have become invasive to identify species likely to become invasive if introduced.
- Overhunting was, until recently, the most important cause of extinction. Overexploitation for commercial reasons continues to threaten some species, such as elephants, orchids, and parrots. Certification programs have been devised to ensure that forest and marine species are harvested in ways that both avoid overexploitation and protect biodiversity and ecosystem productivity.
- Truly endangered species cannot withstand any rate of harvest. The Convention on International Trade in Endangered Species (CITES) determines which species are banned in international trade.
- Humans often reduce the frequency and intensity of disturbances such as fires and floods and thereby endanger species dependent on these disturbances. Reestablishment of historic disturbance patterns, for example by controlled burning in forests, may help preserve such species.
- As rapid global warming occurs, species with poor dispersal abilities may become extinct because of their inability to keep pace with climate change by shifting their ranges. High sea surface temperatures attributed to global warming are already causing increased mortality of reef-forming corals.

The field of restoration ecology studies methods of restoring natural habitats.

- Restoration of degraded ecosystems is an important component of conservation biology but is often difficult to accomplish.
- For a small fraction of endangered species, such as the California condor, captive propagation can help to prevent extinction by maintaining the species during

critical periods, and by providing individuals for reintro-
duction into the wild.

**The new field of conservation medicine tries to identify
the causes of increases in wildlife diseases and to find
effective solutions.**

- Biodiversity is increasingly threatened by outbreaks of
disease among wild organisms. Some of these diseases,
such as the West Nile virus, can be transmitted to
humans.

- Biologists use molecular techniques to identify microbial
pathogens and study the life histories of disease vectors
to devise the best strategy to limit transmission of
pathogens.

**Society will determine what rate of species extinction is
acceptable.**

- Science can supply information about the accelerating
rate of species extinctions, but it is up to society as a
whole to determine what rate of species loss due to
human activities is acceptable.

The Big Picture

- Conservation biology provides an excellent example of
the importance of the synthetic approach to science. In
this case, knowledge and concepts derived from different
areas of biology, as well as the social sciences, are being
applied to the solution of the problem of species and
ecosystem preservation.

- Preserving biological diversity is considered by many
informed individuals to be one of the most important
challenges facing humanity in the twenty-first century.

Study Strategies

- This chapter refers to and relies on a number of concepts
introduced in earlier chapters on ecology, such as: invasive species, habitat fragmentation, and the management
of populations (Chapter 54); species richness and disturbance (Chapter 55); and species–area relationships and
island biogeography (Chapter 56). You may find it useful
to review those concepts as you study the material in this
chapter.

- Review the following activity on the Student
Website/CD:
Activity 57.1 Concept Matching

Test Yourself

Knowledge and Synthesis Questions

1. The concepts of conservation biology come mainly from
all of the following fields *except*
   a. ecology.
   b. evolutionary biology.
   c. population genetics.
   d. immunology.

2. The fynbos shrub community in South Africa
   a. is threatened by introduced species of taller, faster-
      growing plants.
   b. is a fire-adapted community.
   c. helps to maintain a regional supply of high-quality
      water.
   d. All of the above

3. Which of the following is *not* currently a major cause of
the global reduction in biodiversity?
   a. Overexploitation
   b. Global warming
   c. Habitat destruction
   d. Introduction of foreign predators and disease

4. Based on species–area relationships, ecologists predict
that
   a. about one million tropical evergreen forest species may
      become extinct in the next hundred years.
   b. a 90 percent loss of habitat will result in loss of about
      9 percent of the species living there.
   c. the area required by most species will need to be
      reduced.
   d. extinction is inevitable.

5. In the United States, the groups of organisms with the
highest proportion of endangered or extinct species live in
   a. grasslands.
   b. the deciduous forest biome.
   c. freshwater habitats.
   d. deserts.

6. In the following graph, select the curve (*a*, *b*, or *c*) that
correctly shows the expected relationship between habitat patch area and the proportion influenced by edge
effects.

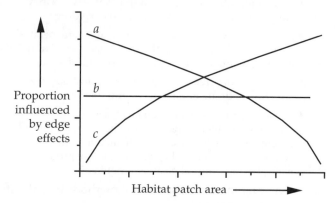

7. Which of the following is *not* likely to be a characteristic
of an invasive plant species?
   a. Small seeds
   b. Large range in its native continents
   c. Short generation time
   d. Specific root symbionts and seed dispersers

8. Why do conservation biologists believe that global warming may lead to extensive decimation of species?
   a. Because little change in plant community composition
      has occurred in the past, we cannot expect present
      communities to adapt to climate change.
   b. The magnitude of climate change will be much greater
      than past periods of climatic change.
   c. Many plant species may not be able to shift their
      ranges at the same pace as the northern movement of
      temperature zones.
   d. All of the above

9. "Bleaching" of coral
   a. occurs when dinoflagellate symbionts of the coral die.
   b. is most severe in areas with cloudy water.
   c. is associated with high surface water temperatures.
   d. All of the above

10. Experiments on wetland restoration have demonstrated that planting a richer mixture of species is associated with
    a. faster accumulation of belowground nitrogen.
    b. more complex vegetation structure.
    c. more rapid development of vegetation cover.
    d. All of the above

11. Which of the following about the California Condor is *false*?
    a. It is being introduced into regions that were not part of its historic geographical range as part of the effort to save the species.
    b. It became endangered partly because of high mortality that resulted from its eating carcasses containing lead shot.
    c. It has experienced an increase in population size in the last decade as a result of captive propagation.
    d. All of the above

Application Questions

1. Aside from the aesthetic benefits of biodiversity, list three economic reasons people should care about species extinctions.

2. Under the Endangered Species Act of 1973, some species are listed as endangered, whereas others are deemed threatened. What is the distinction between these two categories?

3. What are corridors, and how have experiments demonstrated that they help species persist in patchy environments?

4. Why are exotic species frequently a threat to the biological diversity of the region where they are introduced?

5. In recent years, severe forest fires in many parts of the western United States have raised controversy concerning the use of controlled burning as a forest management tool. What are the benefits of this practice?

6. How are the techniques and strategies of conservation medicine being used to combat West Nile virus, which has spread rapidly through the United States and Canada in recent years?

Answers

Knowledge and Synthesis Answers

1. **d.** Although immunology is sometimes a useful tool in assessing the amount of genetic variation that exists in a population, its importance is less than the other fields listed.

2. **d.** See Figure 57.1 in the textbook.

3. **b.** Global warming is predicted to be a major cause of extinction in the future, but is not yet having an impact on populations.

4. **a.** Based on estimates of current and future disappearance of tropical evergreen forests, about one million species that live in these communities could become extinct. This loss is not inevitable if we reduce the rate at which these forests are converted to pasture and cropland.

5. **c.** Habitat destruction and pollution have caused extinction or endangerment of a very high proportion of aquatic freshwater species.

6. **a.** Because the edge of a habitat patch equals the perimeter of the patch, it is proportionally greater for smaller habitat patches. Therefore, the relationship between edge effect and habitat patch is an inverse relationship, as shown by curve *a*.

7. **d.** A species that requires specific mutualists, such as root symbionts and seed dispersers, is not likely to be successful when introduced to a new geographical region because the species on which it is dependent are unlikely to occur in its new location.

8. **c.** Although the expected magnitude of the climate change due to global warming may be similar to past climate changes, the rate of warming will be greater. This may make it impossible for many species to extend their ranges at the same rate as the northward movement of the temperature zones.

9. **c.** High surface water temperatures cause bleaching of coral.

10. **d.** When a species-rich mixture is introduced, wetland restoration is more successful for all of the reasons listed.

11. **a.** The California condor's historic range included the regions of California and Arizona, where it is currently being reintroduced.

Application Answers

1. A list of economic reasons people should care about species extinctions would include the importance of natural products, ecosystems services, and nature tourism.

2. Endangered species are in imminent danger of extinction over all or a significant portion of their range. Threatened species are those likely to become endangered in the near future.

3. Corridors are relatively thin strips of habitat of a particular type that connect larger patches of the same type of habitat. Their importance in permitting individuals to disperse from one patch to another was demonstrated in the experiment described in Figure 54.14 in the textbook.

4. Native species that live in a particular community have evolved to cope successfully with the particular predators, competitors, and diseases that are part of its environment. Introduced species represent a change in the environment for which native species may not have been prepared by natural selection.

5. In ecosystems that naturally experience periodic fires, controlled burning prevents the excessive accumulation of fuel (dead branches, leaf litter, etc.) and thereby lessens the danger that catastrophic, tree-consuming canopy fires will occur. In addition, because many species in such ecosystems require periodic fires for successful establishment and reproduction, controlled burning may be necessary to maintain species richness.

6. Precise identification of the West Nile virus itself was accomplished using molecular techniques such as those discussed in Chapters 13, 16, and 17. Because the disease is mosquito-borne, biologists are now studying the life history of the mosquito vectors to discover how best to prevent transmission of the virus.

58 *Earth System Science*

Important Concepts

The Earth's system is composed of cycles of materials, inputs of solar energy, and interactions between living organisms and the physical environment.

- Earth is essentially a closed system with respect to atomic matter, but it is an open system with respect to energy.

- Energy from the sun, combined with the energy of radioactive decay in Earth's interior, drives the processes, including biogeochemical cycles, that move materials around the planet.

- Earth is an unusual planet in having life, oceans, a moderate surface temperature, continental drift, and a large moon. The effects of the moon include: stabilizing the tilt of Earth's axis and thereby influencing climate; helping to produce tides; and slowing Earth's rotation.

- The elements on which life depends cycle among the four compartments of Earth's physical environment: oceans, fresh waters, atmosphere, and land.

- Concentrations of mineral nutrients are very low in most ocean waters because most elements that enter the oceans gradually sink to the seafloor. Primary production in the oceans is highest in zones of upwelling adjacent to continents, where nutrient-rich cold bottom water rises to the surface.

- Most mineral nutrients that enter fresh water are released by the weathering of rocks and are carried to lakes and rivers via groundwater or by surface flow.

- In temperate lakes, turnovers that occur in the spring and fall when the water temperature is uniformly $4°C$ allow nutrients and oxygen to become evenly distributed in the water column. The bottom waters of permanently stratified tropical lakes lack oxygen.

- The lowest layer of the atmosphere is the troposphere, where almost all water vapor is located and where most global air circulation takes place. Above the troposphere is the stratosphere, where incoming ultraviolet radiation is absorbed by ozone (O_3).

- The most abundant gases in the atmosphere are nitrogen (about 78 percent) and oxygen (about 21 percent). Although carbon dioxide constitutes only 0.03 percent of the atmosphere, it is the source of the carbon used by terrestrial photosynthetic organisms and of the dissolved carbonate used by marine producers. The greenhouse gases (carbon dioxide, water vapor, and ozone being the most important) trap outgoing infrared (heat) radiation emitted by Earth and thus raise its surface temperature.

- About one-fourth of Earth's surface is covered by land. The type of soil that forms in an area and the elements it contains depend on the underlying rock, as well as on climate, topography, the organisms living there, and the length of time that soil-forming processes have been acting.

The movement of water and the occurrence of fires are important components of Earth's system.

- The pattern of movement of a chemical element through organisms and compartments of the physical environment is called its biogeochemical cycle.

- The hydrological cycle is driven by evaporation, most of it from ocean surfaces. Human activities, such as dam construction and pumping of groundwater for irrigation, have altered the flux of water from the land to the oceans. If current water consumption trends continue, about half of the world's population will have an inadequate supply of water by 2025.

- Fire is an important disturbance for ecosystems and releases chemical elements from vegetation. Biomass burning also contributes a significant quantity of carbon dioxide and other greenhouse gases to the atmosphere.

The carbon cycle

- Nearly all the carbon in organisms comes from carbon dioxide in the atmosphere or dissolved bicarbonate (HCO_3^-) in water. Carbon is incorporated into organic molecules by photosynthesis; respiration and combustion break down these organic molecules and return carbon to the atmosphere and water.

- Fossil fuels exist because in the remote past large quantities of organic carbon were removed from the carbon cycle by burial of organisms in sediments lacking oxygen. The ever-increasing rate of burning of fossil fuels over the past 150 years has elevated the carbon dioxide concentration of the atmosphere.

- Historical records and computer models indicate that Earth is warmer when atmospheric carbon dioxide levels are higher and cooler when they are lower. The global warming caused by a doubling of the atmospheric CO_2 concentration would shift climates toward the poles and disrupt current precipitation patterns. It might also melt the Greenland and Antarctic ice caps, thereby causing the flooding of coastal regions.

- Several physical and biological processes act to lower atmospheric CO_2 levels. These include: sedimentation of the calcium carbonate shells of marine organisms; the transfer of carbon to deep water via the sinking of dense, saline water in the North Atlantic Ocean; and increased carbon storage in terrestrial vegetation, principally in forests and savannas.

- Because the adverse effects of global warming are likely to be severe, humans should take steps to reduce the rate of release of CO_2 to the atmosphere.

The nitrogen cycle.

- Nitrogen gas (N_2) makes up 78 percent of Earth's atmosphere, but nitrogen can be converted into biologically useful forms by only a few species of microorganisms that can carry out nitrogen fixation, which converts elemental nitrogen into ammonia. Thus, nitrogen is often in limited supply in ecosystems. Microorganisms also perform denitrification, which removes nitrogen from the biosphere and returns it to the atmosphere.

- Total nitrogen fixation by humans as result of the use of fertilizers and burning of fossil fuels equals global natural nitrogen fixation. Human-caused perturbations of the nitrogen cycle have a variety of adverse effects, including groundwater contamination, increased air pollution, and eutrophication (defined below).

The sulfur cycle.

- Sulfur is released as the gasses sulfur dioxide and hydrogen sulfide by volcanoes and fumaroles. Sulfur is always abundant enough to meet the needs of organisms.

- The burning of fossil fuels produces emissions that form sulfuric and nitric acid in the atmosphere. The resulting acid precipitation affects all industrialized countries and has been shown to have harmful effects on lake ecosystems and on terrestrial plants.

The phosphorus cycle.

- The phosphorus cycle differs from the cycles of carbon, nitrogen, and sulfur in that it lacks a gaseous phase. On land, most phosphorus becomes available through the weathering of phosphate-containing rocks.

- Phosphorus is often a limiting nutrient in soils and lakes. However, adding phosphorus (from detergents and fertilizers) to lakes is a major cause of eutrophication (nutrient enrichment). Algal blooms and the depletion of oxygen in affected lakes are among the undesirable consequences of this process.

- Recovery and recycling of phosphorus from sewage and animal wastes could reduce the amount of phosphorus entering lakes and rivers while supplying much of the needs of the fertilizer and detergent industries.

Biogeochemical cycles interact with one another and are being altered by human activities.

- Human activities are increasing the cycling of other elements such as arsenic. Human-caused global warming is also leading to the spread of diseases affecting both humans and other organisms.

To live on Earth in a manner that maximizes human potential in a sustainable manner, we must make major changes in how we use environmental resources.

- Actions people undertake today could greatly influence the future quality of human life and the welfare of other species that share our small planet with us. A sign that such actions are possible is the voluntary decline in fertil-

ity rates in both rich and poor countries that has led to the current reduction in the global rate of human population growth.

The Big Picture

- Knowledge of how materials move through biogeochemical cycles is crucial for understanding and predicting the effects of human alterations of these cycles. Global warming and eutrophication of lakes are two important examples.

- The important roles of nitrogen-fixers and other microorganisms in the nitrogen cycle illustrate the sometimes overlooked ways in which rather obscure members of the biotic community may be essential for the healthy functioning of ecosystems.

Study Strategies

- In studying the material on biogeochemical cycles, focus on these basic questions: Where are the abiotic reserves of an element located? How does the element leave the reserve and enter living organisms? Why do living organisms need the element? How does the element return to its abiotic reserve?

- Review the following tutorials on the Student Website/CD:
  Tutorial 58.1 The Global Hydrological Cycle
  Tutorial 58.2 The Global Carbon Cycle
  Tutorial 58.3 The Ocean Conveyor Belt
  Tutorial 58.4 The Global Nitrogen Cycle

Test Yourself

Knowledge and Synthesis Questions

1. Zones of upwelling near the shores of land masses have high primary production because such zones
   a. bring nutrients from the seafloor up to the surface.
   b. are characterized by clear water that permits light to penetrate to an unusually great depth.
   c. bring warm water to the surface.
   d. trap nutrients washed into the ocean from nearby land masses.

2. Which of the following features would be characteristic of oceans but *not* of freshwater ecosystems?
   a. Receives material from land mostly via groundwater
   b. Seasonal mixing of materials
   c. Elements buried in bottom sediments for long periods of time
   d. Bottom waters frequently lacking oxygen

3. In the following diagram, the temperature of a freshwater lake is plotted against depth.

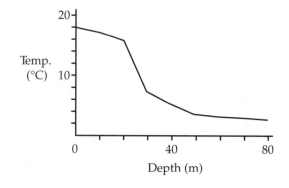

This temperature profile would be most characteristic of the lake during
a. winter.
b. spring.
c. summer.
d. fall.

4. Based on data in the graph from Question 3, the thermocline for the lake is located between
a. 0 and 20 meters.
b. 20 and 30 meters.
c. 30 and 50 meters.
d. 50 and 80 meters.

5. Which of the following features would be characteristic of the stratosphere but *not* of the troposphere?
a. Most ultraviolet radiation is absorbed here.
b. Most water vapor resides here.
c. Most of the mass of the atmosphere lies here.
d. Circulation of this layer influences ocean currents.

6. Which of the following compartments of the global ecosystem is characterized by very slow movement of materials within it?
a. Oceans
b. Fresh waters
c. Atmosphere
d. Land

7. In which of the following compartments of the global ecosystem would circulation of materials be affected by Earth's revolution around the sun?
a. Oceans
b. Fresh waters
c. Atmosphere
d. All of the above

8. Which of the following statements about biogeochemical cycles is *not* true?
a. Most elements remain longest in the living portion of their cycle.
b. Gaseous elements cycle more quickly than elements without a gaseous phase.
c. You may have some atoms in your body that were once part of a dinosaur.
d. Biogeochemical cycles all include both organismal and nonliving components.

9. Next to each of the following features, place a *c, n, p,* or *s* if it is characteristic of the biogeochemical cycles of carbon, nitrogen, phosphorus, or sulfur. (Note: Some features may apply to more than one cycle.)
_____ Major reservoir is atmospheric
_____ Major reservoir is in sedimentary rock
_____ Often in short supply in ecosystems
_____ Fossil fuel reserve is part of this cycle
_____ Involved in cloud formation
_____ Lacks a gaseous phase
_____ Includes a form that is a greenhouse gas
_____ Most fluxes involve organisms

10. Which of the following statements is *not* a major concern about our altering of the carbon cycle?
a. The Greenland and Antarctic ice caps may melt if global warming continues.
b. The burning of fossil fuels adds compounds of sulfur to the atmosphere.
c. The increase in atmospheric CO_2 exceeds the ability of the oceans to absorb the increase.
d. CO_2 is a gas that traps infrared radiation.

11. Which of the following statements concerning acid precipitation is *false*?
a. Acids that enter the atmosphere primarily affect ecosystems located less than 100 kilometers from the source of the pollution.
b. The effects of acidification of lakes appear to be quickly reversible if their pH can be returned to normal.
c. Regulation of emission sources has raised the pH of precipitation in many parts of the eastern United States in the last two decades.
d. Sulfuric acid and nitric acid from the burning of fossil fuels are the major causes of acid precipitation.

12. The process by which a lake ecosystem is altered by eutrophication involves several stages, each causing the next. Which of the following stages would occur *second* in the chain of causation?
a. Algal blooms occur.
b. Oxygen levels drop in deeper water.
c. Phosphorus input from sewage and agricultural runoff increases.
d. Respiratory demand from decomposers increases.

Application Questions

1. Diagram a generalized biogeochemical cycle, showing the compartments in which elements may reside and the processes by which they move from one compartment to another.

2. How may human-set fires be contributing to global warming? What steps can be taken to reduce the frequency and intensity of fires started by humans?

3. How do farming methods that require the large-scale use of irrigation alter the hydrological cycle?

4. How do farming methods that require the large-scale use of manufactured fertilizers contribute to eutrophication and other adverse environmental effects?

5. Why would you expect a small lake to be able to recover more quickly from eutrophication than a large lake such as Lake Erie?

6. Why may global warming lead to the spread of human disease?

Answers
Knowledge and Synthesis Answers

1. **a.** Low concentrations of mineral nutrients limit the primary production of most ocean waters. In zones of upwelling, nutrient-rich water is brought to the surface from the ocean bottom, thereby enhancing the growth of photosynthesizing plankton.

2. **c.** Most elements remain in bottom sediments of the ocean for millions of years, until they are elevated above sea level by movements of Earth's crust.

3. **c.** The steep thermocline evident in this temperature profile would typically be established only by mid- to late summer.

4. **b.** The range of depths over which the temperature changes most abruptly is 20 to 30 meters.

5. **a.** The ozone that shields the surface of Earth from most incoming ultraviolet radiation is located in the stratosphere.

6. **d.** Unlike their behavior in air and water, elements on land move slowly and usually only short distances.

7. **d.** The revolution of Earth around the sun (combined with the tilt of Earth's axis) is the cause of the seasons. Circulation patterns of materials in the oceans, fresh waters, and the atmosphere are all affected by seasonal changes in temperature, precipitation, and prevailing wind direction.

8. **a.** Most elements cycle through organisms more quickly than they cycle through the nonliving world.

9. **n.** Major reservoir is atmospheric
   c., p. Major reservoir is in sedimentary rock
   n., p. Often in short supply in ecosystems
   c. Fossil fuel reserve is part of this cycle
   s. Involved in cloud formation
   p. Lacks a gaseous phase
   n., c. Includes a form that is a greenhouse gas
   n. Most fluxes involve organisms

10. **b.** The addition of compounds of sulfur to the atmosphere has more to do with acid precipitation than it does with global warming.

11. **a.** Acid precipitation is a regional, not a local, environmental problem. Acids may travel hundreds of kilometers before they settle to Earth in precipitation or as dry particles.

12. **a.** The correct sequence is:
    1. Phosphorus input from sewage and agricultural runoff increases.
    2. Algal blooms occur.
    3. Respiratory demand from decomposers increases.
    4. Oxygen levels drop in deeper water.

Application Answers

1. See Figure 58.1 in the text.

2. Fires that burn biomass, most of which are deliberately started, contribute large fractions of the annual production of several greenhouse gases, particularly carbon dioxide, carbon monoxide, ammonia, and tropospheric ozone. Steps that we can take to reduce the frequency and severity of fires include controlled burning of fire-prone vegetation (to reduce the risk of high-intensity canopy fires), improved productivity of tropical agriculture (to reduce the rate at which forests are burned to create more agricultural land), and greater use of solar cookers (to reduce the demand for firewood).

3. The demand for water for irrigation often results in the building of dams, canals, and reservoirs. These structures cause more water to be evaporated from the land and less water to flow to the ocean. Dams also cause seasonal fluctuations of water flows to be lessened, with possible undesirable effects on seawater salinity at the mouth of the dammed river. Excessive pumping of groundwater for irrigation results in the depletion of this resource.

4. If farmers apply fertilizers to croplands in quantities in excess of the ability of the plants to absorb them, some of the excess nitrates and phosphates leave the soil and enter bodies of water. This process contributes to eutrophication of lakes and the creation of "dead zones" in the sea, such as the one in the Gulf of Mexico around the mouth of the Mississippi River. Excessive use of nitrogen-containing fertilizers has also caused contamination of groundwater and has been linked to outbreaks of toxic dinoflagellates in estuaries on the Atlantic coast.

5. The rate at which a lake can recover from eutrophication depends on the rate of replacement of its water. A small lake is likely to have a quicker replacement time of its water than a very large body of water like Lake Erie.

6. Cold temperatures in winter kill many pathogens. The milder winters associated with global warming will allow increasing numbers of pathogens to survive; hence the diseases they cause will become more common. An example of this process is the spread of dengue fever, a tropical human disease that is expanding its range to higher latitudes. Some diseases of plants and other organisms are also expected to spread as a consequence of global warming.